physica status solidi (b)

ISSN 0370-1972 · SUBJECT AND AUTHOR INDEX (b) VOLUMES 115 TO 120 · 1983

ISSN 0031-8965 · SUBJECT AND AUTHOR INDEX (a) VOLUMES 75 TO 80 · 1983

physica status solidi (a)

AKADEMIE-VERLAG · BERLIN

Classification Scheme

1. Structure of Crystalline Solids
   - 1.1 Perfectly Periodic Structure
   - 1.2 Solid-State Phase Transformations
   - 1.3 Alloys. Metallurgy
   - 1.4 Microstructure (Magnetic Domains See 18; Ferroelectric Domains See 14.4.1)
   - 1.5 Films
   - 1.6 Surfaces
2. Non-Crystalline State
3. Crystal Growth
4. Bonding Properties
5. Mössbauer Spectroscopy
6. Lattice Dynamics. Phonons
7. Acoustic Properties
8. Thermal Properties
9. Diffusion
10. Defect Properties (Irradiation Defects See 11)
    - 10.1 Metals
    - 10.2 Non-Metals
11. Irradiation Effects (X-Ray Diffraction Investigations See 1 and 10)
12. Mechanical Properties (Plastic Deformations See 10)
    - 12.1 Metals
    - 12.2 Non-Metals
13. Electron States
    - 13.1 Band Structure
    - 13.2 Fermi Surfaces
    - 13.3 Surface and Interface States
    - 13.4 Impurity and Defect States
    - 13.5 Elementary Excitations (Phonons See 6)
      - 13.5.1 Excitons
      - 13.5.2 Plasmons
      - 13.5.3 Polarons
      - 13.5.4 Magnons
14. Electrical Properties. Transport Phenomena
    - 14.1 Metals. Semi-Metals
    - 14.2 Superconductivity. Superconducting Materials and Devices
    - 14.3 Semiconductors
      - 14.3.1 Films
      - 14.3.2 Surfaces and Interfaces
      - 14.3.3 Devices. Junctions (Contact Problems See 14.3.4)
      - 14.3.4 High-Field Phenomena, Space-Charge Effects, Inhomogeneities, Injected Carriers (Electroluminescence See 20.3; Junctions See 14.3.3)
    - 14.4 Dielectrics
      - 14.4.1 Ferroelectrics
15. Thermoelectric and Thermomagnetic Properties
16. Photoconductivity. Photovoltaic Effects
17. Emission of Electrons and Ions
    - 17.1 Field Emission Microscope Investigations
18. Magnetic Properties
    - 18.1 Paramagnetic Properties
    - 18.2 Ferromagnetic Properties
      - 18.2.1 Ferromagnetic Films
    - 18.3 Ferrimagnetic Properties
    - 18.4 Antiferromagnetic Properties

(Continued on cover three)

physica status solidi (b)

basic research

physica status solidi (a)

applied research

SUBJECT AND AUTHOR INDEX

Volumes 115 to 120, 1983

Volumes 75 to 80, 1983

ISSN 0370-1972/0031-8965

Edited by

E. BORCHARDT

AKADEMIE-VERLAG · BERLIN

Subscriptions and orders for this Subject and Author Index or single copies of physica status solidi (b) and physica status solidi (a) should be directed

in the GDR:

to the Postzeitungsvertrieb or
to the Akademie-Verlag, DDR-1086 Berlin, Leipziger Str. 3 – 4;

in the other socialist countries:

to a book-shop for foreign language literature or
to the competent news-distributing agency;

in the FRG and BERLIN (WEST):

to a book-shop or
to the wholesale distributing agency Kunst und Wissen, Erich Bieber OHG, Wilhelmstr. 4 – 6, D-7000 Stuttgart 1;

in the other Western European countries:

to Kunst und Wissen Erich Bieber GmbH, Dufourstr. 51, CH-8008 Zürich;

in USA and CANADA:

to Verlag Chemie International Inc., 303 N.W. 12th Avenue, Deerfield Beach, FL 33441, USA;

in other countries:

to the international book- and journal-selling trade,
to Buchexport, Volkseigener Außenhandelsbetrieb der Deutschen Demokratischen Republik, DDR-7010 Leipzig, Postfach 160, or
to the Akademie-Verlag, DDR-1086 Berlin, Leipziger Straße 3 – 4.

Schriftleiter und verantwortlich für den Inhalt der Zeitschriften:
Professor Dr. Dr. h. c. P. Görlich, DDR-1086 Berlin, Leipziger Str. 3—4, Postfach 1233 bzw. DDR-6900 Jena, Schillbachstr. 24.

Herausgeber und verantwortlich für den Inhalt des Registers:
Dipl.-Phys. E. Borchardt, DDR-1086 Berlin, Leipziger Str. 3—4, Posthach 1233.

Verlag:
Akademie-Verlag, DDR-1086 Berlin, Leipziger Str. 3—4; Fernruf: 2236221 und 2236229; Telex-Nr.: 114420; Bank: Staatsbank der DDR, Berlin, Kto.-Nr.: 6836-26-20712.

Veröffentlicht unter der Lizenznummer 1310 und 1620 des Presseamtes beim Vorsitzenden des Ministerrates der Deutschen Demokratischen Republik.

Gesamtherstellung:
VEB Druckerei „Thomas Müntzer", DDR-5820 Bad Langensalza.

Bestellnummer des Registers 1068/85 R16.

Printed in the German Democratic Republic.

AN (EDV) 20831/20735

Preface

The sixteenth index volume contains volumes 115 to 120 of "physica status solidi (b) — basic research" as well as volumes 75 to 80 of "physica status solidi (a) — applied research", i.e. all volumes having appeared in 1983 containing 1481 papers.

The general scheme of last index volume was maintained. The basis for quoting is the subject classification given every papers in the respective issue. Nevertheless, before taking over these subject classifications into the index volume they were once more carefully revised and in some cases inproved and completed in order to get a most uniform and clear classification for all papers. The average number of quotations per paper in this index volume is nearly the same as in the last one, namely about 3.0.

Again I have to express my thanks to the readers for their interest, and I will always readily take up any critical remarks and proposals for improvements. I am especially obliged to Prof. Dr. P. Görlich, Prof. Dr. E. Gutsche, Dr. H.-J. Hänsch, Dr. H. Lange, Dr. S. Oberländer, and Dipl. Phys. K. Müller for giving me advice and unselfish help. Furthermore I have to thank Mrs. E. Schwan for her exact and reliable secretary's work at the manuscript.

Berlin, October 1984

Eva Borchardt

Classification Scheme

1. Structure of Crystalline Solids
   - 1.1 Perfectly Periodic Structure
   - 1.2 Solid-State Phase Transformations
   - 1.3 Alloys. Metallurgy
   - 1.4 Microstructure (Magnetic Domains See 18; Ferroelectric Domains See 14.4.1)
   - 1.5 Films
   - 1.6 Surfaces
2. Non-Crystalline State
3. Crystal Growth
4. Bonding Properties
5. Mössbauer Spectroscopy
6. Lattice Dynamics. Phonons
7. Acoustic Properties
8. Thermal Properties
9. Diffusion
10. Defect Properties (Irradiation Defects See 11)
    - 10.1 Metals
    - 10.2 Non-Metals
11. Irradiation Effects (X-Ray Diffraction Investigations See 1 and 10)
12. Mechanical Properties (Plastic Deformations See 10)
    - 12.1 Metals
    - 12.2 Non-Metals
13. Electron States
    - 13.1 Band Structure
    - 13.2 Fermi Surfaces
    - 13.3 Surface and Interface States
    - 13.4 Impurity and Defect States
    - 13.5 Elementary Excitations (Phonons See 6)
      - 13.5.1 Excitons
      - 13.5.2 Plasmons
      - 13.5.3 Polarons
      - 13.5.4 Magnons
14. Electrical Properties. Transport Phenomena
    - 14.1 Metals. Semi-Metals
    - 14.2 Superconductivity. Superconducting Materials and Devices
    - 14.3 Semiconductors
      - 14.3.1 Films
      - 14.3.2 Surfaces and Interfaces
      - 14.3.3 Devices Junctions (Contact Problems See 14.3.4)
      - 14.3.4 High-Field Phenomena, Space-Charge Effects, Inhomogeneities, Injected Carriers (Electroluminescence See 20.3; Junctions See 14.3.3)
    - 14.4 Dielectrics
      - 14.4.1 Ferroelectrics
15. Thermoelectric and Thermomagnetic Properties
16. Photoconductivity. Photovoltaic Effects
17. Emission of Electrons and Ions
    - 17.1 Field Emission Microscope Investigations

18. Magnetic Properties
    18.1 Paramagnetic Properties
    18.2 Ferromagnetic Properties
        18.2.1 Ferromagnetic Films
    18.3 Ferrimagnetic Properties
    18.4 Antiferromagnetic Properties

19. Magnetic Resonance

20. Optical Properties
    20.1 Spectra. Optical Constants (X-Ray Spectra See 20)
    20.2 Lasers
    20.3 Luminescence (X-Ray Spectra See 20)

21. Metals. Alloys
    21.1 Transition Metals (Group Sc to Zn) and Their Alloys
        21.1.1 Fe and Fe Alloys
    21.2 Alkali-Metals
    21.3 Alkali-Earth Metals
    21.4 Rare-Earth Metals
    21.5 Actinides
    21.6 Noble Metals
    21.7 Semi-Metals (Semiconducting Alloys See 22.7)

22. Semiconductors and Ionic Crystals
    22.1 Elements
        22.1.1 Germanium
        22.1.2 Silicon
        22.1.3 Group VI Elements
    22.2 III–V Compounds
        22.2.1 Arsenides
        22.2.2 Phosphides
        22.2.3 Antimonides
        22.2.4 Mixed Crystals
    22.3 IV–IV, II–V, and III–VI Compounds
        22.3.1 Mixed Crystals
    22.4 II–VI Compounds
        22.4.1 Sulphides
        22.4.2 Selenides
        22.4.3 Tellurides
        22.4.4 Mixed Crystals
    22.5 Halides
        22.5.1 Silver Halides
        22.5.2 Alkali-Halides
        22.5.3 Alkali-Earth Halides
        22.5.4 Mixed Crystals
    22.6 Simple Oxides (CdO, ZnO See 22.4)
        22.6.1 Mixed Crystals
    22.7 Semiconducting Intermetallic Compounds
        22.7.1 Mixed Crystals
    22.8 Tertiary and Higher Compounds
        22.8.1 Salts of Oxy-Acids (Carbonates, Phosphates, Silicates, Including Mica etc).
        22.8.2 Spinels, Garnets, Ferrites
    22.9 Organic Semiconductors

23. Solidified Gases

For comments on the use of the classification scheme see p. 6.

Some Comments on the Use of the Subject and Author Index

The classification system[1]) comprises 20 broad sections each devoted to a major field of solid state physics and 3 broad sections devoted to the main groups of substances. A section may contain a number of subsections dealing with aspects of this field or special substance, respectively, which tend to receive particular attention. A paper dealing with this field will either be classified under one or more of the subsections or, if none of these are relevant, it will just appear under the general section heading (i.e. sections and subsections are equivalent in this sense). For example, a paper dealing with problems of paramagnetism would appear under 18.1 but not 18, whereas a paper on diamagnetism would be quoted under 18; a paper dealing with GaP would appear under 22.2.2, not under 22 or 22.2, whereas a paper on AlN would be quoted under 22.2, and a paper on a II–IV or a IV–VI compound would be quoted under 22.

All papers will be quoted under each relevant section. There is made a distinction between one or two principal entries (refering to the main subject matter of the paper) and secondary entries (refering to subsidiary subject matter and substance, respectively). In the principal entries the title of the paper is given in its original language and in English. In addition, every principal entry is marked by * in front of the title of the paper. In the secondary entries the title is only given in English.

Each entry gives the volume, the number of the issue, the initial and final page numbers, and the year in which the paper appeared. A paper of physica status solidi (b) of physica status solidi (a) is marked by (b) or (a) in front of the line, respectively. The letter "K" before the page numbers indicates a short note published in the offset part of the journal. Each entry is followed by references of the other entries. These references give the subject classifications and the pages in the index where the other entries appear. The references to the principal entries are given in italics.

Review articles appear, at the head of the list of entries under a given subject and are given the subtitle "Review Articles". These articles are listed in the chronological order of their publication, irrespective of being a paper of physica status solidi (b) or physica status solidi (a). The subject classification for a review article consisting of several parts is based on the article as a whole rather than on the individual parts.

The entries for original papers, short notes, and errata are made together in a single list for each subject and this follows the entries for review articles. These are again ordered according to data of publication, irrespective of being a paper of physica status solidi (b) or physica status solidi (a). Original papers are sometimes published in several parts under the same principal title (in general with special subtitles). The various parts of papers of this type are treated separately and each part is classified according to its particular contents.

In quoting errata the following rule is obeyed. If the erratum refers to a paper quoted in the preceding index, it is listed in the chronological order of its publication otherwise the erratum is quoted together with the paper to which it refers.

---

[1]) See cover two and three, and pages 4 and 5. For the case of losing the cover by bookbinding we repeat the classification scheme on pages 4 and 5.

In using the subject index attention should be given to the following details:

subject	quoted in	remarks
representations of space groups	1 and 13.1	
Alfen waves, helicon waves	13 and 20	
hot electrons	13 and 14.3.4	
transport phenomena	14 or subsection of 14	if not clearly to be assigned to 8 or 15
galvanomagnetic effects	14 or subsection of 14	not in 18
electro-optical effects	14 and 20 (or subsection of 14 and 20)	
piezoelectricity, electrostriction	14 or subsection of 14	not in 12 or subsection of 12; different electromechanical effects are quoted in 12 or subsection of 12
ionic conduction	14.4	not in 14.3
magneto-optical effects	18 and 20 (or subsection of 18 and 20)	possibly in additional sections, e.g. Faraday effect in 13
ferromagnetic films	18.2.1	not additionally in 1.5
X-ray spectra	20	but not X-ray studies of structure
MIS and MOS structures	the section assigned to the semiconductor	not in the section assigned to the oxide or the matal
MIM structures	the section assigned to the insulator	generally not in the section assigned to the metal
epitaxial layers	the section assigned to the layer	not in the section assigned to the substrate

The author index gives the names of the authors in alphabetical order. Each name is followed by the bibliographic details of the paper and the numbers of the pages on which the paper appears in the subject index. A paper of physica status solidi (b) or physica status solidi (a) is again marked by (b) or (a) in front of the number of the volume, respectively. The principal entry is again given in italics. When several papers have been published by the same author they are arranged according to the dates of publication, irrespective of being a paper of physica status solidi (b) or physica status solidi (a).

Subject Index

1. Structure of Crystalline Solids

Review Articles

Original Papers and Short Notes

A. N. Anagnostopoulos, C. Manolikas, and D. Papadopoulos
*Comparison of the Structure and the Electric Properties of $ZnIn_2S_4$(III)- and $CdInGaS_4$-Layered Crystals
(a) Vol. **77**, No. 2, p. 595–601 (1983).
See also *14.3*, p. *221*; 22.8, p. 506.

D. M. Vardanyan and H. M. Manoukyan
*Extinction and Absorption of a Plane Monochromatic X-Ray Wave on Dynamic Reflection at a Laminar-Crystalline Medium in the Bragg Case
(a) Vol. **79**, No. 2, p. 617–622 (1983).

Yu. P. Khapachev
*The Theory of Dynamical X-Ray Diffraction on a Superlattice
(b) Vol. **120**, No, 1, p. 155–163 (1983).
See also *20*, p. *311*.

A. Brau, J. P. Farges, A. Filhol, and H. Grassi
*Order–Disorder Phase Transition in $TEA^+(TCNQ)_2^-$
(b) Vol. **120**, No. 2, p. 547–554 (1983).
See also 8, p. 92; *14.3*, p. *227*; 22.9, p. 532.

1.1 Perfectly Periodic Structure

Review Articles

B. PAŁÓSZ
*Reasons for Polytypism of Crystals of the Type MX_2 (I): Polytypism and Structure Behaviour of Inorganic Crystals: General Problems
(a) Vol. 77, No. 1, p. 11–34 (1983).
See also *1*, p. *9*; 10, p. 97.

B. PAŁÓSZ
*Reasons for Polytypism of Crystals of the Type MX_2 (II): Classification of Faults and Structural Series of Polytypes; Conditions of Polytypic Growth of CdI_2, PbI_2, $CdBr_2$, $SnSe_2$ and $Ti_{1.2}S_2$
(a) Vol. 80, No. 1, p. 11–41 (1983).
See also *1*, p. *9*; 10.2, p. 108; 22, p. 403; 22.5, p. 473.

Original Papers and Short Notes

G. J. ASHWELL, J. G. ALLEN, E. P. GOODINGS, D. A. KENNEDY, and I. W. NOWELL
*Crystal Structure of (DEPA) $(TCNQ)_4$ and Stoichiometric Variation of Some Related TCNQ Salts
(a) Vol. 75, No. 2, p. 663–667 (1983).
See also 14.3, p. 217; 22.9, p. 526.

N. E. BELOVA, F. EICHHORN, V. A. SOMENKOV, K. UTEMISOV, und S. SH. SHILSHTEIN
*Analyse der Neigungsmethode zur Untersuchung von Pendellösungsinterferenzen von Neutronen und Röntgenstrahlen
(Analysis of the Inclination Method for Investigating Pendellösung Interferences of Neutrons and X-Rays)
(a) Vol. 76, No. 1, p. 257–265 (1983).
See also 22.1.1, p. 419; 22.1.2, p. 427.

K. H. J. BUSCHOW and P. G. VAN ENGEN
*Note on the Magnetic and Magneto-Optical Properties of Ni_2In Type 3d Transition Metal Compounds
(a) Vol. 76, No. 2, p. 615–620 (1983).
See also *18.2*, p. *282*; *20*, p. *308*; 21, p. 353; 21.1, p. 367; 21.1.1, p. 381.

E. K. POLYCHRONIADIS and P. DELAVIGNETTE
*Accurate Characterization of Subgrain Boundaries in a TEM
(a) Vol. 77, No. 1, p. 291–308 (1983).
See also *1.4*, p. *31*.

A. I. PROKHVATILOV and A. P. ISAKINA
*Lattice Parameters, Thermal Expansion Coefficients, and Vacancy Density in Solid CD_4
(a) Vol. 78, No. 1, p. 147–155 (1983).
See also 8, p. 88; *10.2*, p. *114*; 12.2, p. 143; 22.9, p. 529.

K. Lal and P. Thoma
*Study of Microstructural Changes in Silicon Single Crystals Subject to High Electric Fields by Using a High Resolution X-Ray Diffraction Method
(a) Vol. 80, No. 2, p. 491–502 (1983).
See also *14.4*, p. *254*; 22.1.2, p. 439.

1.2 Solid-State Phase Transformations

Original Papers and Short Notes

Y. Y. Abolinsh, S. V. Karpov, and A. A. Shultin
*Raman Studies of Ammonium Ion Motion in NH_4NO_3–V Crystals
(b) Vol. **115**, No. 1, p. K1 – K5 (1983).
See also *6*, p. *70*; 20.1, p. 313; 22.8.1, p. 512.

W. Rehwald, A. Vonlanthen, and W. Meyer
*Single Crystal 2,4-Hexadiynylene-Bis (p-Toluenesulfonate): Elastic Properties and Phase Transitions in the Monomer and Polymer
(a) Vol. **75**, No. 1, p. 219 – 226 (1983).
See also 7, p. 82; *12.2*, p. *141*; 22.9, p. 525.

A. V. Salker, K. Seshan, and H. V. Keer
*Phase Transition Behaviour of VO_2
(a) Vol. **75**, No. 1, p. K37 – K40 (1983).
See also 8, p. 85; 22.6, p. 490.

I. P. Aleksandrova, S. Grande, Yu. N. Moskvich, A. I. Krieger, and V. A. Koptsik
*Second-Order Quadrupole Shift in the Incommensurate Structures of Rb_2ZnBr_4 and Rb_2ZnCl_4
(b) Vol. **115**, No. 2, p. 603 – 613 (1983).
See also 14.4.1, p. 256; *19*, p. *300*; 22.8, p. 503.

A. I. Lotkov, V. N. Grishkov, A. V. Kuznetsov, and S. N. Kulkov
*TiNi Aging and Its Effect on the Start Temperature of the Martensitic Transformation
(a) Vol. **75**, No. 2, p. 373 – 377 (1983).
See also 21.1, p. 364.

M. Ishigame and S. Yamashita
Raman Scattering in K_2SO_4 at High Temperatures
(b) Vol. **116**, No. 1, p. 49 – 56 (1983).
See also *6*, p. *71*; *20.1*, p. *315*; 22.8.1, p. 513.

S. K. Khanna
*On the Elastic Anisotropy of Ferroelectric Triglycine Selenate
(b) Vol. **116**, No. 1, p. 173 – 178 (1983).
See also 12.2, p. 142; *14.4.1*, p. *256*; 22.9, p. 526.

V. A. Klimenko
*Modification of the Pair Approximation of the Quasi-Chemical Ordering Theory of Binary Systems
(a) Vol. **76**, No. 1, p. 45 – 55 (1983).
See also *1.3*, p. *22*; 21, p. 352.

V. P. Barkhatov, V. F. Balakirev, Yu. V. Golikov, and E. G. Kostitsin
*X-Ray Diffraction Investigation of Cation Distribution in High-Temperature Cubic Spinels of Mg–Mn–O System
(a) Vol. **76**, No. 1, p. 57–63 (1983).
See also *1.3*, p. *22*; 22.8.2, p. 519.

S. V. Sudareva, V. A. Rassokhin, and A. F. Prekul
*The Structure of Cr–Al Alloys Exhibiting Anomalous Physical Properties
(a) Vol. **76**, No. 1, p. 101–106 (1983).
See also *1.4*, p. *29*; 21, p. 352; 21.1, p. 366.

A. Rajewska, B. Pura, J. Przedmojski, and R. Dąbrowski
*X-Ray Study of the Smectic A–Nematic Phase Transition in 4-n-Pentylphenyl-4-(Trans-4′-n-Pentylcyclohexyl) Benzoate
(a) Vol. **76**, No. 1, p. 127–130 (1983).
See also 2, p. 45; 22.9, p. 526.

V. B. Dudnikova and V. S. Urusov
*Solvus of KCl:Ca^{2+} Solid Solution by Flotation Measurements of Density of Single Crystals
(a) Vol. **76**, No. 1, p. 217–221 (1983).
See also *10.2*, p. *111*; 22.5.2, p. 478.

M. M. Hafiz, A. A. Ammar, A. I. Al-Adl, and Aboutaleb Mohamed
*Effect of Heat Treatment on the Conduction and Structure of Ge–Se–Te Amorphous Alloys
(a) Vol. **76**, No. 1, p. 319–325 (1983).
See also *2*, p. *46*; *14.3*, p. *217*; 22.8, p. 504.

J. W. Brightwell, C. N. Buckley, L. S. Miller, and B. Ray
*Electrical Transport and Phase Stability in Silver Iodide–Cadmium Iodide
(a) Vol. **76**, No. 1, p. 391–398 (1983).
See also *14.4*, p. *250*; 22.5.4, p. 488.

Y. Watanabe, J. Takeuchi, and M. Oka
*Microwave Transport Properties around the Structural Phase Transition of MnSe
(a) Vol. **76**, No. 1, p. K13–K15 (1983).
See also *14.3*, p. *218*; 22.4.2, p. 466.

N. Kamegashira and Y. Miyazaki
*Jahn-Teller Transition in $NdMnO_3$
(a) Vol. **76**, No. 1, p. K39–K42 (1983).
See also *14.3*, p. *218*; 22.8.1, p. 513.

P. Duhaj and F. Hanic
*Electron Microscopy and Electron Diffraction Study of Crystallization of Metastable Phases in Amorphous Fe–B Alloys
(a) Vol. **76**, No. 2, p. 467–477 (1983).
See also 1.3, p. 23; *2*, p. *47*; 21.1.1, p. 381.

O. Bostanjoglo
*Multiple Phase Transitions in GeTe Films. Dynamics and Film Structure
(a) Vol. **76**, No. 2, p. 525–531 (1983).
See also *1.5*, p. *36*; 2, p. 47; 3, p. 58; 22, p. 406.

T. K. Chattopadhyay, E. Gmelin, and H. G. von Schnering
*Heat Capacity Study of the Phase Transitions in As_4S_3 and As_4S_4
(a) Vol. **76**, No. 2, p. 543–551 (1983).
See also *8*, p. *87*; 22, p. 406.

K. Sumiyama, M. Shiga, and Y. Nakamura
*Characteristic Martensitic Transformation of the Fe–Pt Disordered Alloys around the γ–α Phase Boundary
(a) Vol. **76**, No. 2, p. 747–751 (1983).
See also 18.2, p. 282; 21.1.1, p. 381.

Z. Czapla, O. Czupiński, L. Sobczyk, J. Mróz, and H. Pykacz
*Dielectric Properties and Ferroelectric Phase Transitions in Mixed $Rb_x(NH_4)_{1-x}HSeO_4$ Crystals
(a) Vol. **76**, No. 2, p. K155–K157 (1983).
See also *14.4.1*, p. *257*; 22.8.1, p. 513.

L. G. Khvostantsev and N. A. Nikolaev
*Electroresistance of Cerium at Phase Transitions under Hydrostatic Pressure up to 8.0 GPa. *P–T* Diagram of the Cerium Phase Transitions
(a) Vol. **77**, No. 1, p. 161–166 (1983).
See also 12.1, p. 138; *14.1*, p. *207*; 21.4, p. 391.

U. Heinecke, C. Cruz, E. Wieser, and C. Bernhardt
*Structural Transformations and Connected Changes of Magnetic Properties by Mechanical Treatment of $SrFe_{12}O_{19}$
(a) Vol. **77**, No. 1, p. 225–232 (1983).
See also 5, p. 66; *18.3*, p. *292*; 22.8.2, p. 521.

L. A. Shebanov, V. Ya. Fritsberg, and A. P. Gaevskis
*Crystallographic Properties and Superconductivity of Solid Solutions of the $BaBi_xPb_{1-x}O_3$ System
(a) Vol. **77**, No. 1, p. 369–373 (1983).
See also *14.2*, p. *213*; 22.8.1, p. 514.

K. Biedrzycki, J. Chrzanowski, and B. Sujak
*Some Electrical and Optical Effects Observed in Polycrystalline Ice at 200 K
(a) Vol. **77**, No. 1, p. K95–K98 (1983).
See also 14, p. 204; 20, p. 308; 22.6, p. 493.

B. Vengalis and K. Valatska
*Phase Segregation in Tellurium-Saturated GeTe
(b) Vol. **117**, No. 2, p. 471–476 (1983).
See also 8, p. 88; 13.4, p. 179; 14.3, p. 220; 20.1, p. 319; 22.3, p. 457.

G. Theodorou
*Critical Behaviour at the Incommensurate–Commensurate Transition of Ho
(b) Vol. **117**, No. 2, p. K67 – K69 (1983).
See also 21.4, p. 391.

S. Trimper and A. Kühnel
*On the Phase Transition Temperature for a Heisenberg Ferromagnet
(b) Vol. **117**, No. 2, p. K75 – K79 (1983).
See also *18.2*, p. *284*.

B. A. Men, M. L. Levitan, and A. N. Men
*Influence of the Ground State Symmetry on Qualitative Form of Ordering Phase Diagram
(a) Vol. **77**, No. 2, p. 455 – 461 (1983).
See also 18, p. 274; 21, p. 355.

T. Huang and A. L. Ruoff
*Pressure-Induced Phase Transitions of HgTe
(a) Vol. **77**, No. 2, p. K193 – K197 (1983).
See also *12.2*, p. *143*; 22.4.3, p. 468.

S. Deonarine
*The Nature of Stage I–II Structural Phase Transition in the Layer Compound $RbVF_4$
(a) Vol. **77**, No. 2, p. K199 – K201 (1983).
See also 22.8, p. 506.

V. E. Antonov, T. E. Antonova, I. T. Belash, E. G. Ponyatovskii, and V. I. Rashupkin
*The Pd–Pt–H System: Phase Transformations at High Pressure and Superconductivity
(a) Vol. **78**, No. 1, p. 137 – 146 (1983).
See also *14.2*, p. *213*; 21, p. 356; 21.1, p. 371.

H. Hedler and G. Götz
*Incorporation of Sb in Si during CW-Laser Stimulated Solid Phase Epitaxy
(a) Vol. **78**, No. 1, p. 237 – 241 (1983).
See also 11, p. 129; 14.3, p. 221; 22.1.2, p. 432.

T. Kala
*Contribution to the Study of Tetragonal and Rhombohedral Phase Coexistence in the $PbZrO_3$–$PbTiO_3$ System
(a) Vol. **78**, No. 1, p. 277 – 282 (1983).
See also *14.4.1*, p. *258*; 22.8.1, p. 515.

A. Le Lann and A. Dubertret
*A Structural Description for the F.C.C. → B.C.C. Martensitic Transformation in Terms of $\{011\}\ \langle 011\rangle_{f.c.c.}$ Simultaneous Double Shear. Relation with the Atomic Configuration in $\{3, 15, 10\}_{f.c.c.}$ Habit Plane
(a) Vol. **78**, No. 1, p. 309 – 321 (1983).
See also 1.4, p. 32; 21, p. 356.

H. J. Fischbeck
*On the Translation Mode of the Kink Lattice of a Quasi-One-Dimensional Peierls-Fröhlich System with a Nearly Half-Filled Band
(b) Vol. **118**, No. 2, p. 595–603 (1983).
See also *6*, p. *76*; 7, p. 83; 13.1, p. 160.

Ch. Seidel and V. N. Prigodin
*The Impurity Effect on the Phase Transitions in Quasi-One-Dimensional Conductors
(b) Vol. **118**, No. 2, p. 713–724 (1983).
See also *10.2*, p. *115*; 14.2, p. 213; 22.9, p. 529.

H. Mashiyama
*Modified Phase Modulation-Only Approximation of Incommensurate Phases in K_2SeO_4-Type Crystals
(b) Vol. **118**, No. 2, p. 757–767 (1983).
See also 8, p. 89; 14.4, p. 252; 22.8.1, p. 515.

G. Wendrock
*Study of the Behaviour of Al–Zn Alloys by SAXS in the Temperature Range of the Instable Solid Solution
(a) Vol. **78**, No. 2, p. 497–506 (1983).
See also 1.3, p. 25; 21, p. 357; 21.1, p. 373.

M. Krupski
*Rigid Sphere Model Applied to the Phase Transitions of Complex Compounds of the Type K_2PtCl_6
(a) Vol. **78**, No. 2, p. 751–758 (1983).
See also 6, p. 77; 22.8, p. 508.

K. Osuch
*On the Physical Tensor Field Criterion in Birman's Theory of Symmetry Changes
(b) Vol. **119**, No. 1, p. 375–380 (1983).

P. Muralt, E. Voit, and R. Kind
*Cl-NQR Study of the Low Temperature Phase of Rb_2ZnCl_4
(b) Vol. **119**, No. 1, p. K65–K67 (1983).
See also *19*, p. *304*; 22.8, p. 508.

M. Hidaka, Y. Okamoto, and Y. Zikumaru
*Structural Phase Transition of $CsPbCl_3$ below Room Temperature
(a) Vol. **79**, No. 1, p. 263–269 (1983).
See also 6, p. 78; 22.8, p. 509.

M. T. Sebastian and P. Krishna
*Mechanism of Solid State Transformations in Single Crystals of $Zn_xCd_{1-x}S$
(a) Vol. **79**, No. 1, p. 271–280 (1983).
See also 10.2, p. 118; 22.4.4, p. 471.

J. W. Brightwell, L. S. Miller, A. Munday, and B. Ray
*The Silver Iodide–Thallium Iodide Pseudo-Binary System
(a) Vol. **79**, No. 1, p. 293 – 300 (1983).
See also 14.4, p. 253; 22.5, p. 475; 22.5.1, p. 476.

A. Boeuf, R. Caciuffo, J. M. Fournier, L. Manes, J. Rebizant, F. Rustichelli, J. C. Spirlet, and A. Wright
*Study of the 25 K NpO_2 Phase Transition by a Neutron Oscillation Camera
(a) Vol. **79**, No. 1, p. K1 – K3 (1983).
See also 22.6, p. 496.

S. K. Decheva-Plachkova
*On the Fundamental Energy Gap in the System $(GeTe)_{1-x}(AgSbTe_2)_x$
(b) Vol. **119**, No. 2, p. K97 – K100 (1983).
See also *13.1*, p. *164*; 22.8, p. 509.

M. T. Dove, I. U. Heilmann, J. K. Kjems, J. Kurittu, and G. S. Pawley
*A Neutron Scattering Study of Phonons in Per-Deuterated S-Triazine
(b) Vol. **120**, No. 1, p. 173 – 181 (1983).
See also *6*, p. *79*; 12.2, p. 144; 22.9, p. 531.

D. I. Uzunov
*Critical Region and Characteristic Lengths for Zero-Temperature Structural Phase Transitions
(b) Vol. **120**, No. 1, p. K39 – K43 (1983).
See also 8, p. 91.

Ch. Seidel and V. N. Prigodin
*Three-Dimensional CDW Ordering and Impurity Effect in Quasi-One-Dimensional Conductors
(b) Vol. **120**, No. 1, p. K73 – K77 (1983).
See also *10.1*, p. *106*; 14.2, p. 214.

J. Berdowski
*The Influence of Isotope Effects on Anomalous Changes of Dielectric and Electric Properties of DADA Crystals in the Range of the Phase Transition
(a) Vol. **80**, No. 1, p. 65 – 72 (1983).
See also *14.4.1*, p. *259*; 22.8.1, p. 517.

A. Aresti, F. Meloni, G. G. Pegna, and M. Serra
*Microwave Radiometric Study of Ferro–Paraelectric Phase Transitions in KNO_3
(a) Vol. **80**, No. 1, p. 119 – 126 (1983).
See also *14.4.1*, p. *259*; 22.8.1, p. 518.

R. Tognato
*Why Metals Do Not Show Overheating
(a) Vol. **80**, No. 1, p. 147 – 157 (1983).
See also *10.1*, p. *107*; 21, p. 360.

1.3 Alloys. Metallurgy

Review Articles

R. E. Hummel
*Differential Reflectometry and Its Application to the Study of Alloys, Ordering, Corrosion, and Surface Properties
(a) Vol. **76**, No. 1, p. 11–44 (1983).
See also 10.1, p. 99; 13.1, p. 155; 13.3, p. 167; *20.1*, p. *312*; 21, p. 349; 21.1, p. 363; 22.6, p. 490.

L. N. Aleksandrov and F. L. Edelman
*Shock Crystallization of Films
(a) Vol. **76**, No. 2, p. 409–427 (1983).
See also *1.5*, p. *35*; *3*, p. *57*; 22.1.1, p. 418; 22.1.2, p. 423; 22.1.3, p. 440; 22.6, p. 490.

Original Papers and Short Notes

V. A. Maslov, I. I. Naumov, and V. E. Panin
*On the Problem of Electron Phase Stability
(b) Vol. **115**, No. 1, p. 317–324 (1983).
See also *13.2*, p. *166*; 21.1, p. 363.

A. S. Lileev and V. P. Menushenkov
*The Crystal Structure and Coercive Force of $SmCo_5$ Sintered Permanent Magnets
(a) Vol. **75**, No. 1, p. 341–346 (1983).
See also *18.2*, p. *280*; 21.1, p. 364; 21.4, p. 390.

F. Abd El-Salam, A. I. Eatah, and A. Tawfik
*Effect of Li Loss on Some Physical Properties of Al–Li Alloy
(a) Vol. **75**, No. 2, p. 379–384 (1983).
See also 10.1, p. 100; 12.1, p. 137; 14.1, p. 206; 21, p. 351; 21.2, p. 387.

J. M. Lang, U. Dahmen, and K. H. Westmacott
*The Origin of Mo_2C Precipitate Morphology in Molybdenum
(a) Vol. **75**, No. 2, p. 409–420 (1983).
See also 1.4, p. 29; 21, p. 351.

A. Baldan
*Dislocation Structures in Aluminium-Rich Ni_3(Al, Ta) and Ni_3(Al, Hf)
(a) Vol. **75**, No. 2, p. 441–449 (1983).
See also *10.1*, p. *100*; 21, p. 351; 21.1, p. 364.

W. Thulke, R. Frahm, R. Haensel, and P. Rabe
*X-Ray Study of Niobium Disulfide Intercalated with Copper
(a) Vol. **75**, No. 2, p. 501–509 (1983).
See also *4*, p. *60*; 20, p. 307; 22.8, p. 503.

G. Weber, B. Gillot, and P. Barret
*Interface Structure in Relation with the Mechanisms in the Reaction Copper–Silicon
(a) Vol. **75**, No. 2, p. 567–576 (1983).
See also *1.6*, p. *39*; 22.1.2, p. 426.

D. Schryvers, G. van Tendeloo, J. van Landuyt, and S. Amelinckx
*High-Resolution Study of the One- and Two-Dimensional Long-Period Antiphase Boundary Structures in Au_4Zn
(a) Vol. **75**, No. 2, p. 607–616 (1983).
See also *1.4*, p. *29*; 21.1, p. 365; 21.6, p. 396.

K. H. J. Buschow, J. H. N. van Vucht, P. G. van Engen, D. B. de Mooij, and A. M. van der Kraan
*Structure, Magnetic and Magneto-Optical Properties, and ^{57}Fe Mössbauer Effect in a New Variety of Heusler Alloy: PtFeSb
(a) Vol. **75**, No. 2, p. 617–623 (1983).
See also 5, p. 65; *18.2*, p. *281*; 20.1, p. 314; 21.1.1, p. 379.

V. Navrátil, M. Hamerský, P. Lukáč, V. P. Soldatov, and V. I. Startsev
*Solid Solution Hardening of Cd–Zn Single Crystals at 1.5 K
(a) Vol. **75**, No. 2, p. K133–K135 (1983).
See also *10.1*, p. *100*; 21, p. 351; 21.1, p. 365.

J. Imhof
*An Iteration Procedure in the Texture Analysis
(a) Vol. **75**, No. 2, p. K187–K189 (1983).
See also *1.4*, p. *29*.

V. A. Klimenko
*Modification of the Pair Approximation of the Quasi-Chemical Ordering Theory of Binary Systems
(a) Vol. **76**, No. 1, p. 45–55 (1983).
See also 1.2, p. 14; 21, p. 352.

V. P. Barkhatov, V. F. Balakirev, Yu. V. Golikov, and E. G. Kostitsin
*X-Ray Diffraction Investigation of Cation Distribution in High-Temperature Cubic Spinels of Mg–Mn–O System
(a) Vol. **76**, No. 1, p. 57–63 (1983).
See also *1.2*, p. *15*; 22.8.2, p. 519.

P. Lošťák, J. Horák, and L. Koudelka
*Some Physical Properties and Point Defects in $Bi_{2-x}In_xTe_3$ Mixed Crystals
(a) Vol. **76**, No. 1, p. K71–K75 (1983).
See also 14.3, p. 218; *20.1*, p. *316*; 22.7.1, p. 501.

K.-H. Steinmetz, H.-P. Fischer, M. Martin, and E. Nembach
*Positron-Annihilation Measurements of the Fermi Surface of Copper-Base Solid Solutions with Gallium, Germanium, and Nickel
(b) Vol. **116**, No. 2, p. 575–583 (1983).
See also *13.2*, p. *166*; 21.1, p. 367.

J. Richter and G. Schubert
*The Electrical Resistivity of Short-Range Ordered Alloys and Its Temperature Dependence
(b) Vol. **116**, No. 2, p. 597–605 (1983).
See also *14.1*, p. *207*; 21, p. 352.

P. Duhaj and F. Hanic
*Electron Microscopy and Electron Diffraction Study of Crystallization of Metastable Phases in Amorphous Fe–B Alloys
(a) Vol. **76**, No. 2, p. 467–477 (1983).
See also *1.2*, p. *15*; *2*, p. *47*; 21.1.1, p. 381.

D. Schryvers, J. van Landuyt, G. van Tendeloo, and S. Amelinckx
*Electron Microscopy Study of Two New Long Period Structures in Pt_xTi Alloys ($3 \leqq x \leqq 8$)
(a) Vol. **76**, No. 2, p. 575–593 (1983).
Erratum
(a) Vol. **78**, No. 2, p. 761 (1983).
See also *1.4*, p. *30*; 21, p. 353; 21.6, p. 397.

J. M. López and J. A. Alonso
*Semiempirical Theory of Solid Solubility in Metallic Alloys
(a) Vol. **76**, No. 2, p. 675–682 (1983).
See also 8, p. 88; 21, p. 353; 21.1, p. 368.

K. Kambas
*Preparation and Optical Investigation of the $CdInGaS_4$ Compound
(a) Vol. **76**, No. 2, p. 735–741 (1983).
See also *6*, p. *73*; *20.1*, p. *317*; 22.8, p. 505.

W. Bażela-Wróbel and A. Szytuła
*Neutron Diffraction Study of the $Ni_2Mn_{1-x}Cr_xSn$ System
(a) Vol. **76**, No. 2, p. K117–K120 (1983).
See also 18.2, p. 283; 21.1, p. 368.

V. E. Antonov, T. E. Antonova, I. T. Belash, E. G. Ponyatovskii, V. I. Rashupkin, and V. G. Thiessen
*The Pd–Ni–H System at High Pressure
(a) Vol. **77**, No. 1, p. 71–79 (1983).
See also 14.2, p. 212; 18.2, p. 283; 21, p. 354; 21.1, p. 369.

G. I. Gudzenko, V. B. Molodkin, and V. G. Novitskii
*Dynamical Diffraction in Alloys. Influence of Long-Range Order
(a) Vol. **77**, No. 1, p. 115–120 (1983).
See also *10.1*, p. *102*; 21, p. 354; 21.1, p. 369.

R. Deblieck, G. A. Wiegers, K. D. Bronsema, D. van Dyck, G. van Tendeloo, J. van Landuyt, and S. Amelinckx
*Clustering and "Phase Ordering" Transitions in Molybdenum (III) Sulphide, $Mo_{2.065}S_3$
(a) Vol. **77**, No. 1, p. 249–261 (1983).
See also 22, p. 407.

M. Otha, T. Kanadani, A. Sakakibara, H. Yamada, and M. Yamada
*Effects of Surface and Thickness of Specimens on the Aging in Al–Zn Alloys
(a) Vol. **78**, No. 1, p. K23 – K26 (1983).
See also 10.1, p. 104; 12.1, p. 138; 21, p. 357; 21.1, p. 372.

G. Wendrock
Study of the Behaviour of Al – Zn Alloys by SAXS in the Temperature Range of the Instable Solid Solution
(a) Vol. **78**, No. 2, p. 497 – 506 (1983).
See also *1.2*, p. *18*; 21, p. 357; 21.1, p. 373.

W. Thulke, R. Haensel, and P. Rabe
*Anisotropic Absorption of Unpolarized X-Radiation in Copper-Intercalated Niobium Disulfide
(a) Vol. **78**, No. 2, p. 539 – 545 (1983).
See also *4*, p. *63*; 20, p. 310; 22.8, p. 508.

Shu Zhen and G. J. Davies
Calculation of the Lennard-Jones *n–m* Potential Energy Parameters for Metals
(a) Vol. **78**, No. 2, p. 595 – 605 (1983).
See also *4*, p. *63*; 12.1, p. 139; 21, p. 358.

A. A. Mohamed, S. A. Saleh, E. A. Abou-Saif, M. Abd-Rabo, and G. A. El-Ganainy
*On the Identification of New and Normal Phases in the System Cadmium–Antimony Alloys
(a) Vol. **78**, No. 2, p. 705 – 716 (1983).
See also 21, p. 358; 21.7, p. 401.

S. F. Kolodeznaya and E. V. Kozlov
*Calculation of the Properties and the Structure of Ordering Alloys of the Cu–Au System
(b) Vol. **119**, No. 1, p. 131 – 139 (1983).
See also 13.1, p. 162; 21.1, p. 374; 21.6, p. 398.

Yu. S. Stark, A. S. Steinberg, V. M. Vasilev, and V. I. Dimitrov
*Monte Carlo Determination of Long-Range Pair Interaction Energies from Diffuse Scattering of X-Ray by Alloys
(b) Vol. **119**, No. 1, p. 147 – 151 (1983).
See also *4*, p. *63*; 21.1, p. 374; 21.6, p. 398.

N. Shohoji
*Statistical Thermodynamics of Hydrogen Solution in Some Lanthanides
(b) Vol. **119**, No. 1, p. K87 – K90 (1983).
See also 4, p. 63; 21.4, p. 392.

V. V. Kelarev, V. V. Chuev, A. N. Pirogov, and S. K. Sidorov
Anisotropy and Exchange Effects in Heavy Rare-Earth-Cobalt Compounds of the RCo_5 Type
(a) Vol. **79**, No. 1, p. 57 – 66 (1983).
See also *18.3*, p. *293*; 21.1, p. 374; 21.4, p. 392.

A. Maciel, S. G. Fries, J. A. H. da Jornada, R. P. Livi, and E. R. Fraga
*Study of the Nuclear Quadrupole Interaction in Tl_5Te_3
(a) Vol. **80**, No. 1, p. 245 – 248 (1983).
See also *13.4*, p. *189*; 22.7, p. 499.

S. K. Pabi
*On the Variation of Interface Composition around a Dissolving Precipitate
(a) Vol. **80**, No. 1, p. 305 – 316 (1983).
See also 9, p. 95; 21, p. 361; 21.6, p. 399.

V. B. Dudnikova, V. S. Urusov, and L. N. Bannykh
*Solvus of KCl:Ba^{2+} Solid Solution from Flotation Measurements of Single Crystal Density
(a) Vol. **80**, No. 1, p. 399 – 402 (1983).
See also *10.2*, p. *120*; 22.5.2, p. 484.

P. Meisterle and W. Pfeiler
*Short-Range Ordering in the Matrix of a Two-Phase AgAl Alloy
(a) Vol. **80**, No. 1, p. K37 – K40 (1983).
See also *14.1*, p. *210*; 21, p. 361.

B. Rauschenbach, A. Kolitsch, and K. Hohmuth
*Iron Nitride Phases Formed by Nitrogen Ion Implantation and Thermal Treatment
(a) Vol. **80**, No. 2, p. 471 – 482 (1983).
See also 1.5, p. 38; *11*, p. *135*; 21, p. 362; 21.1.1, p. 385.

C. Manolikas and A. N. Anagnostopoulos
*Electrical Effects Associated with the Ordering Process in $CdInGaS_4$ Crystals (I): Electron Microscopy and Electron Diffraction Study
(a) Vol. **80**, No. 2, p. 503 – 511 (1983).
See also 14.3, p. 228; 22.8, p. 510.

J. Teuho
*A Study of Non-Periodic Antiphase Boundaries in Long-Period Ordered $Au_3Zn(H)$ (II): Dissociated Antiphase Boundaries
(a) Vol. **80**, No. 2, p. 589 – 600 (1983).
See also *1.4*, p. *34*; 21.1, p. 378; 21.6, p. 400.

H. Kido, T. Hoshikawa, M. Shimada, and M. Koizumi
*Preparation and Magnetic and Electrical Properties of RZnSi
(a) Vol. **80**, No. 2, p. 601 – 605 (1983).
See also 14.1, p. 211; *18.1*, p. *279*; 21.1, p. 378; 21.4, p. 394.

J. Tanaka, K. Takahashi, K. Yukino and S. Horiuchi
*Electrical Conduction of $(La_{0.8}Ca_{0.2})MnO_3$ with Homogeneous Ionic Distribution
(a) Vol. **80**, No. 2, p. 621 – 630 (1983).
See also *14.3*, p. *228*; 22.8.1, p. 518.

F. Vasiliu, P. Gr. Lucuta, and F. Constantinescu
*Structure – Sintering Temperature Relationship in PZT Solid Solutions at the Morphotropic Phase Boundary
(a) Vol. **80**, No. 2, p. **637 – 642** (1983).
See also *14.4.1*, p. *260*; 22.8, p. 511.

M. Jurczyk
*Magnetic Properties of Substituted $Pr_2(Co_{1-x}M_x)_{17}$ Compounds (M = Fe, Mn, and Cr)
(a) Vol. **80**, No. 2, p. **657 – 662** (1983).
See also *18.2*, p. *289*; 21.1, p. 378; 21.4, p. 394.

L. Amaral, F. P. Livi, and A. A. Gomes
*Remarks on Alloying-Induced Lattice Parameter Changes in Intermetallic Compounds
(a) Vol. **80**, No. 2, p. **669 – 677** (1983).
See also 21.1, p. 378; 21.1.1, p. 385; 21.4, p. 394.

1.4 Microstructure (Magnetic Domains See 18; Ferroelectric Domains See 14.4.1)

Review Article

A. E. ROMANOV and V. I. VLADIMIROV
Disclinations in Solids
(a) Vol. **78**, No. 1, p. 11 – 34 (1983).
See also *10*, p. *97*; 12, p. 136.

Original Papers and Short Cotes

SH. SH. IBRAGIMOV, O. P. MAKSIMKIN, and SH. B. SHIGANAKOV
*Grain-Boundary Deformation in Nickel Irradiated by α-Particles
(a) Vol. **75**, No. 1, p. 55 – 58 (1983).
See also *11*, p. *122*; 21.1, p. 363.

J. M. LANG, U. DAHMEN, and K. H. WESTMACOTT
The Origin of Mo_2C Precipitate Morphology in Molybdenum
(a) Vol. **75**, No. 2, p. 409 – 420 (1983).
See also *1.3*, p. *21*; 21, p. 351.

K.-H. BÄTHER, W. BRÜCKNER, W. MOLDENHAUER, and H.-P. BRÜCKNER
*Microscopic Degradation Model for ZnO Varistors
(a) Vol. **75**, No. 2, p. 465 – 472 (1983).
See also *14.3.3*, p. *237*; 22.4, p. 460.

H. O. MARTIKAINEN, M. A. KORHONEN, and V. K. LINDROOS
*Determination of Grain Boundary Rotation Parameters from Kikuchi Patterns (I): A Simplified Method for Determining the Accurate Orientation Relationship of Two Grains
(a) Vol. **75**, No. 2, p. 559 – 566 (1983).
See also 21.1.1, p. 379.

D. SCHRYVERS, G. VAN TENDELOO, J. VAN LANDUYT, and S. AMELINCKX
*High-Resolution Study of the One- and Two-Dimensional Long-Period Antiphase Boundary Structures in Au_4Zn
(a) Vol. **75**, No. 2, p. 607 – 616 (1983).
See also *1.3*, p. *22*; 21.1, p. 365; 21.6, p. 396.

J. IMHOF
*An Iteration Procedure in the Texture Analysis
(a) Vol. **75**, No. 2, p. K187 – K189 (1983).
See also *1.3*, p. *22*.

S. V. SUDAREVA, V. A. RASSOKHIN, and A. F. PREKUL
*The Structure of Cr–Al Alloys Exhibiting Anomalous Physical Properties
(a) Vol. **76**, No. 1, p. 101 – 106 (1983).
See also *1.2*, p. *15*; 21, p. 352; 21.1, p. 366.

R .S. BHATTACHARYA, A. K. RAI, S. C. LING, and P. P. PRONKO
*Epitaxial Regrowth of Si Implanted (100) and (211) GaAs
(a) Vol. **76**, No. 1, p. 131–136 (1983).
See also *11*, p. *125*; 22.2.1, p. 443.

S. B. NEWCOMB, W. M. STOBBS, and J. A. LITTLE
*The Transmission Electron Microscopy of 1 to 2 nm Rhodium Particles Dispersed on γ-Al_2O_3
(a) Vol. **76**, No. 1, p. 191–196 (1983).
See also 22.6, p. 492.

U. MESSERSCHMIDT, Y. NISHINO, T. IMURA, and H. SAKA
X-Ray Topographic In-Situ Observation of Slip Band Propagation in MgO Single Crystals
(a) Vol. **76**, No. 1, p. 277–284 (1983).
Erratum
(a) Vol. **80**, No. 1, p. 411 (1983).
See also *10.2*, p. *111*; 22.6, p. 492.

NGUYEN AN, R. HERRMANN, and G. WORM
*Investigation of Asymmetrical Grain Boundaries and Their Coincidence Site Lattices
(b) Vol. **116**, No. 2, p. 501–510 (1983).
See also 10, p. 97.

V. V. ARISTOV, M. N. ZOLOTUKHIN, V. V. KVEDER, YU. A. OSIPYAN, I. I. SNIGHIREVA, and I. I. KHODOS
*The Effect of Annealing on the Dislocation Dissociation in Plastically Deformed Silicon
(a) Vol. **76**, No. 2, p. 485–491 (1983).
See also *10.2*, p. *112*; 22.1.2, p. 428.

M. MÜHLBERG and D. HESSE
*TEM Precipitation Studies in Te-Rich As-Grown PbTe Single Crystals
(a) Vol. **76**, No. 2, p. 513–524 (1983).
See also 10.2, p. 112; 22, p. 406.

D. SCHRYVERS, J. VAN LANDUYT, G. VAN TENDELOO, and S. AMELINCKX
*Electron Microscopy Study of Two New Long Period Structures in Pt_xTi Alloys ($3 \leq$ $\leq x \leq 8$)
(a) Vol. **76**, No. 2, p. 575–593 (1983).
Erratum
(a) Vol. **78**, No. 2, p. 761 (1983).
See also *1.3*, p. *23*; 21, p. 353; 21.6, p. 397.

A. R. LANG
*The Correct Rules for Determining the Sign of Fault Vectors in X-Ray Topography
(a) Vol. **76**, No. 2, p. 595–599 (1983).

H. O. MARTIKAINEN, M. A. KORHONEN, and V. K. LINDROOS
*Determination of Grain Boundary Rotation Parameters from Kikuchi Patterns (II): Experimental Optimization of the Analysis of Grain Boundary Orientation Relationships and Verification of the Accuracy of the Proposed Method
(a) Vol. **76**, No. 2, p. 709–719 (1983).
See also 21.1.1, p. 381.

D. SUNDARARAMAN, A. L. E. TERRANCE, G. VAN TENDELOO, and J. VAN LANDUYT
*High-Resolution Electron Microscopy Study of a Ti–1.6 at% N Alloy
(a) Vol. 76, No. 2, p. K109–K111 (1983).
See also 21, p. 353.

A. D'AMICO, G. PETROCCO, G. STOCHINO, A. SALSANO, and G. LUZZI
*Grain Boundaries as Possible Origin of Flicker Noise in GaP Polycrystalline Material
(a) Vol. 76, No. 2, p. K125–K129 (1983).
See also *14.3.2*, p. *233*; 22.2.2, p. 449.

R. Z. VALIEV, V. YU. GERTSMAN, O. A. KAIBYSHEV, and SH. KH. KHANNANOV
*Non-Equilibrium State and Recovery of Grain Boundary Structure (I): General Analysis, Crystallogeometrical Aspects
(a) Vol. 77, No. 1, p. 97–105 (1983).
See also 10, p. 98.

T. GESSNER, M. PASEMANN, and B. SCHMIDT
Calculation of Neutron-Induced Defect Clusters in Silicon and Comparison with TEM Investigations
(a) Vol. 77, No. 1, p. 133–138 (1983).
See also *11*, p. *127*; 22.1.2, p. 429.

J. TEUHO and K. MÄKINEN
*A Study of Non-Periodic Antiphase Boundaries in Long-Period Ordered $Au_3Zn(H)$ (I): Antiphase Boundary Planes
(a) Vol. 77, No. 1, p. 263–268 (1983).
See also *1.3*, p. *24*; 21.1, p. 369; 21.6, p. 397.

B. COCHRANE and P. J. GOODHEW
TEM Images of Faceted Bubbles and Voids
(a) Vol. 77, No. 1, p. 269–282 (1983).
See also 10.1, p. 102; *11*, p. *128*; 21.1, p. 369.

E. K. POLYCHRONIADIS and P. DELAVIGNETTE
*Accurate Characterization of Subgrain Boundaries in a TEM
(a) Vol. 77, No. 1, p. 291–308 (1983).
See also *1.1*, p. *11*.

C. LACOUR, A. DUBON et P. M. GRANGEON
*Aspect aléatoire de la formation des composés par frittage réactif
(Random Aspects of the Formation of Compounds by Reactive Sintering)
(a) Vol. 77, No. 1, p. 309–322 (1983).
See also 22.8, p. 505.

G. VAN TENDELOO and S. AMELINCKX
*Real Space Evidence for the Occurrence of Microdomains in „Disordered" Cu_3Pd
(a) Vol. 77, No. 1, p. K9–K11 (1983).
See also 21.1, p. 369.

M. ČERŇANSKÝ
*Erratum to: Broadening and Shift of Diffraction Lines Due to Partial Coherence of X-Rays
(b) Vol. **117**, No. 2, p. 755 (1983).
See also *10.1*, p. *103*; 21.1, p. 370.

W. DRABIK and J. TORUŃ
*On the Analytical Methods of Line Defect and Beam Direction Determinations
(a) Vol. **77**, No. 2, p. 513–519 (1983).
See also 22.2.4, p. 454.

R. Z. VALIEV, V. YU. GERTSMAN, and O. A. KAIBYSHEV
*Non-Equilibrium State and Recovery of Grain Boundary Structure (II): Energetic Analysis
(a) Vol. **78**, No. 1, p. 177–186 (1983).
See also 10, p. 98.

R. S. BHATTACHARYA, A. K. RAI, and P. P. PRONKO
*Oxygen Implantation in GaAs: Damage Studies
(a) Vol. **78**, No. 1, p. 283–286 (1983).
See also *11*, p. *129*; 22.2.1, p. 445.

E. K. POLYCHRONIADIS and P. DELAVIGNETTE
*Dislocation Structures and Subgrain Boundaries in Al_3Zr
(a) Vol. **78**, No. 1, p. 287–301 (1983).
See also *10.1*, p. *104*; 21, p. 356.

C. BOULESTEIX and B. YANGUI
*Star Patterns and Dissociation of Interfaces without Spontaneous Strain Compatibility between Orientation Domains in Ferroelastic Crystals. Case of a Monoclinic Rare Earth Sesquioxide
(a) Vol. **78**, No. 1, p. 303–308 (1983).
See also 22.6, p. 495.

A. LE LANN and A. DUBERTRET
A Structural Description for the F.C.C. → B.C.C. Martensitic Transformation in Terms of $\{011\}$ $\{011\}_{f.c.c.}$ Simultaneous Double Shear. Relation with the Atomic Configuration in $\{3, 15, 10\}_{f.c.c.}$ Habit Plane
(a) Vol. **78**, No. 1, p. 309–321 (1983).
See also *1.2*, p. *17*; 21, p. 356.

G. D. SUKHOMLIN and A. V. ANDREEVA
*Particular Properties of $\Sigma = 3^n$ Boundaries in F.C.C. Polycrystals (I): Crystallographical Parameters and Grain Boundary Faceting during Annealing
(a) Vol. **78**, No. 1, p. 333–341 (1983).
See also 21.1, p. 372; 21.1.1, p. 383.

L. MALICSKÓ and R. SCHOLZ
*Microtwins in $LiNbO_3$ Crystals Observed by High-Voltage and High-Resolution Electron Microscopy
(a) Vol. **78**, No. 1, p. K1–K4 (1983).
See also 10.2, p. 115; 22.8.1, p. 515.

A. M. ARUSTAMYAN and F. N. CHUKHOVSKII
*Diffraction Contrast from Partial Dislocations in the Two-Beam Dynamical Theory
(a) Vol. **78**, No. 1, p. K35 – K38 (1983).
See also *10*, p. *98*.

NGUYEN AN and R. HERRMANN
*Distribution of the Atoms on Grain Boundaries and in the Bravais Cell of the Coincidence Site Lattice
(b) Vol. **118**, No. 2, p. 691 – 702 (1983).
See also *1.1*, p. *12*.

P. H. BEZIRGANYAN and V. G. ASLANYAN
*Formation and Observation of Moiré Distribution of X-Ray Waves in Bicrystal Interferometers
(a) Vol. **78**, No. 2, p. 415 – 421 (1983).

R. Z. VALIEV, O. A. KAIBYSHEV, V. V. ASTANIN, and A. K. EMALETDINOV
*The Nature of Grain Boundary Sliding and the Superplastic Flow
(a) Vol. **78**, No. 2, p. 439 – 448 (1983).
See also *10.1*, p. *104*; 21.1, p. 373.

R. KÖHLER and W. MÖHLING
*Sensitivity of Plane Wave Topography to Microdefects
(a) Vol. **78**, No. 2, p. 489 – 496 (1983).
See also 10.2, p. 116; 22.1.2, p. 433.

T. NATTERMANN
*Impurity Pinning, Interface Roughening, and the Incommensurate–Commensurate Transition
(b) Vol. **119**, No. 1, p. 209 – 213 (1983).

C. F. NILSEN and K. N. SUBRAMANIAN
Deformation of α–β Brass with Oriented and Equiaxed Duplex Boundaries
(a) Vol. **79**, No. 1, p. 281 – 292 (1983).
See also *12.1*, p. *140*; 21.1, p. 375.

S. K. MAKSIMOV and N. I. BORGARDT
*Changes in the Electron Wave Phases at the Exit Surface of a Scattering Crystal Located near the Reflection Position
(b) Vol. **119** No. 2, p. 685 – 692 (1983).

J. IMHOF
*Texture Analysis by Iteration (I): General Solution of the Fundamental Problem
(b) Vol. **119**, No. 2, p. 693 – 701 (1983).
See also *1.3*, p. *26*.

S. S. SHEININ and R. PEREZ
*Dynamical Theory of Twin Boundary Contrast Including Effects of Common Reflections
(a) Vol. **79**, No. 2, p. 367 – 376 (1983).

J. Šternberk, E. Kratochvílová, J. Hřebík, and A. Gemperle
*Coercivity and Microstructure of Low-Alloy Cr–Mo Steel
(a) Vol. **79**, No. 2, p. 523–529 (1983).
See also *18.2*, p. *287*; 21.1.1, p. 384.

J. Imhof
*Texture Analysis by Iteration (II): Special Cases of the General Solution
(b) Vol. **120**, No. 1, p. 321–328 (1983).
See also *1.3*, p. *26*.

M. Pasemann, D. Hoehl, A. L. Aseev, and O. P. Pchelyakov
*Analysis of Rod-Like Defects in Silicon and Germanium by Means of High-Resolution Electron Microscopy
(a) Vol. **80**, No. 1, p. 135–139 (1983).
See also *11*, p. *133*; 22.1.1, p. 422; 22.1.2, p. 438.

A. Charai and C. Boulesteix
*Visualization and Study of the Shape and Size of Very Small Coherent Precipitates by a Phase Contrast Method
(a) Vol. **80**, No. 1, p. 333–341 (1983).
See also *10.1*, p. *107*; 21, p. 361.

A. K. Rai, R. S. Bhattacharya, and P. P. Pronko
*Annealing of Amorphous Layer in Si Implanted Ga As
(a) Vol. **80**, No. 1, p. 367–372 (1983).
See also 10.2, p. 120; 11, p. 134; 22.1.2, p. 438.

A. E. Nosenko, A. I. Otko, V. V. Kravchishyn, and L. V. Kostyk
Growth Macrodefects in $Ca_3Ga_2Ge_3O_{12}$ Single Crystals
(a) Vol. **80**, No. 1, p. 373–386 (1983).
See also *10.2*, p. *120*; 22.8.2, p. 523.

E. I. Goldman, Yu. V. Markin, P. S. Sulzhenko, and A. G. Zhdan
Grain Boundary Tunneling Conductivity of Semiconductor Poly- and Bicrystals
(a) Vol. **80**, No. 2, p. 423–429 (1983).
See also *14.3.2*, p. *235*; 22, p. 415.

J. Teuho
*A Study of Non-Periodic Antiphase Boundaries in Long-Period Ordered Au_3Zn(H) (II): Dissociated Antiphase Boundaries
(a) Vol. **80**, No. 2, p. 589–600 (1983).
See also *1.3*, p. *27*; 21.1, p. 378; 21.6, p. 400.

C. Hidalgo and N. de Diego
*Positron Trapping in the Alloy Fe–1.25 wt% C
(a) Vol. **80**, No. 2, p. K145–K148 (1983).
See also *10.1*, p. *107*; 21.1.1, p. 386.

1.5 Films

Review Article

L. N. ALEKSANDROV and F. L. EDELMAN
*Shock Crystallization of Films
(a) Vol. **76**, No. 2, p. 409–427 (1983).
See also 1.3, p. 21; *3*, p. *57*; 22.1.1, p. 418; 22.1.2, p. 423; 22.1.3, p. 440; 22.6, p. 490.

Original Papers and Short Notes

R. MATTHEIS
The Influence of Substrate Temperature on the Oriented Growth of Ni and Ni_2Si on Si (111)
(a) Vol. **75**, No. 1, p. 195–205 (1983).
See also *3*, p. *57*; 22.1.2, p. 424.

W. KLEINN
*Erratum zu: Restwiderstandsbestimmung aus Strom–Spannungs-Kurven kurzer freitragender Elektortransportproben epitaktischer Goldschichten
(Erratum to: Determination of the Residual Resistivity from Current–Voltage Characteristics of Short Support-Free Samples of Epitaxial Gold Films)
(a) Vol. **75**, No. 1, p. 347 (1983).
See also *14.1*, p. *206*; 21.6, p. 396.

H. ATMANI, P. MICHON et C. VAUTIER
*Détermination des paramètres de la cristallisation de couches minces de selénium amorphe par microscopie optique
(Determination of the Crystallization Parameters of Amorphous Selenium Thin Films by Optical Microscopy)
(a) Vol. **75**, No. 1, p. K5–K8 (1983).
See also 22.3.1, p. 440.

BAI-XIN LIU
*A Metastable Electron Compound Formed by Ion Irradiation
(a) Vol. **75**, No. 1, p. K77–K81 (1983).
See also *11*, p. *123*; 21.1, p. 364; 21.6, p. 396.

B. GŁOWACKY
*X-Ray Diffraction Pattern and Morphology of Superconducting Nb_3Sn Layer Grown on a Polycrystalline Nb Substrate of (111) Main Texture
(a) Vol. **75**, No. 1, p. K103–K105 (1983).
See also *14.2*, p. *212*.

O. G. ALAVERDOVA, M. YA. FUKS, L. S. KHAZAN, L. P. KOVAL, L. A. MATVEEVA, I. F. MIKHAILOV, N. N. SOLDATENKO, and YU. A. TKHÓRIK
*Coherency of the Interphase Boundary and Elastic Strain in the Epitaxial System Ge/GaAs
(a) Vol. **75**, No. 2, p. 367–371 (1983).
See also *10.2*, p. *109*; 12.2, p. 141; 22.1.1, p. 419; 22.2.1, p. 443.

H. Gotoh, K. Sasamoto, S. Kuroda, and M. Kimata
*Molecular Beam Epitaxy of AlSb on GaAs and GaSb on AlSb Films
(a) Vol. **75**, No. 2, p. **641**–**645** (1983).
See also *3*, p. *57*; 14.3.1, p. 230; 22.2.1, p. 443; 22.2.3, p. 452.

V. I. Ivanov-Omskii, V. K. Ogorodnikov, and V. D. Rozumnyi
*Surface Defects of Closed-System-Grown $Cd_xHg_{1-x}Te$ Epitaxial Layers
(a) Vol. **76**, No. 1, p. 71–74 (1983).
See also 20.1, p. 316; 22.4.4, p. 470.

Yu. B. Bolkhovityanov
*The Lattice-Pulling Effect Induced by the Film Stresses. Conditions of Existence and Discovery
(a) Vol. **76**, No. 1, p. 85–93 (1983).
See also *3*, p. *57*; 22.2.1, p. 443; 22.2.2, p. 448.

L. N. Aleksandrov
*Formation of Semiconductor Epitaxial Films by Pulse Heating Crystallization or Regrowth
(a) Vol. **76**, No. 1, p. 179–190 (1983),
See also *3*, p. *57*; 8, p. 86; 22.1.1, p. 419; 22.1.2, p. 427.

S. Grosswig, J. Härtwig, U. Schellenberger, and W.-D. Zimmer
*Theoretical and Experimental Study of Growth Sector Boundaries in Deuterated Potassium Dihydrogen Phosphate Single Crystals
(a) Vol. **76**, No. 1, p. 241–247 (1983).
See also *3*, p. *57*; 10.2, p. 111; 22.8.1, p. 513.

E. M. Trukhanov
*Accommodation of Misfit in Heterostructures with Continuous and Island Films
(a) Vol. **76**, No. 2, p. 437–445 (1983).
See also *10.2*, p. *111*; 22.1.1, p. 420.

O. Bostanjoglo
*Multiple Phase Transitions in GeTe Films. Dynamics and Film Structure
(a) Vol. **76**, No. 2, p. 525–531 (1983).
See also *1.2*, p. *16*; 2, p. 47; 3, p. 58; 22, p. 406.

A. H. King
*Evidence of the Formation of Twins by Deformation and "Growth-Accidents" in Evaporated Thin Films of Gold
(a) Vol. **76**, No. 2, p. 629–636 (1983).
See also 10.1, p. 101; 21.6, p. 397.

M. Kittler and K.-W. Schröder
Determination of Semiconductor Parameters and of the Vertical Structure of Devices by Numerical Analysis of Energy-Dependent EBIC Measurements
(a) Vol. **77**, No. 1, p. 139–151 (1983).
See also 14.3.2, p. 233; *14.3.3*, p. *238*; 22.1.2, p. 430.

U. Rensch and G. Eichhorn
*Investigations on the Structure of MOCVD AIN Layers on Silicon
(a) Vol. **77**, No. 1, p. 195–199 (1983).
See also *3*, p. *58*; 22.2, p. 442.

H. G. Schneider, H. Karge, and B. Weber
*Characterization of Heteroepitaxial Silicon Layers on Sapphire by Rutherford-Backscattering (RBS) Analyses
(a) Vol. **77**, No. 1, p. 201–205 (1983).
See also 22.1.2, p. 430.

K. T.-Y. Kung, B. X. Liu, and M.-A. Nicolet
*Study of Ni–Nb System by Ion Mixing
(a) Vol. **77**, No. 1, p. 355–359 (1983).
See also *2*, p. *48*; 14.1, p. 207; 21, p. 354; 21.1, p. 369.

V. M. Mukhortov, Yu. I. Golovko, V. A. Aleshin, E. V. Sviridov, Vl. M. Mukhortov, V. P. Dudkevich, and E. G. Fesenko
Ferroelectric Phase Transition in Heteroepitaxial Films of (Ba, Sr) TiO_3
(a) Vol. **77**, No. 1, p. K37–K40 (1983).
See also *14.4.1*, p. *257*; 22.8.1, p. 514.

W. Kleinn
*Determination of Residual Resistivity and Observation of Resistance Recovery from Current–Voltage Curves of Short Epitaxial Gold Thin Films
(a) Vol. **77**, No. 2, p. K149–K153 (1983).
See also *14.1*, p. *208*; 21.6, p. 398.

K. Hübner, A. V. Shendrik, and A. M. Praulinsh
*Decomposition of SiO_x Films Due to Internal Stress
(b) Vol. **118**, No. 1, p. 239–244 (1983).
See also *4*, p. *61*; 19, p. 302; 20.1, p. 321; 22.6, p. 494

K. E. Mironov, V. K. Ogorodnikov, V. D. Rozumnyi, and V. I. Ivanov-Omskii
*Closed System LPE Growth of $Cd_xHg_{1-x}Te$
(a) Vol. **78**, No. 1, p. 125–131 (1983).
See also *3*, p. *58*; 22.4.4, p. 471.

V. M. Mukhortov, Yu. I. Golovko, V. A. Aleshin, E. V. Sviridov, Vl. M. Mukhortov, V. P. Dudkevich, and E. G. Fesenko
(Ba, Sr)TiO_3 Heteroepitaxial Films
(a) Vol. **78**, No. 1, p. 253–257 (1983).
See also *14.4.1*, p. *258*; 22.8.1, p. 515.

J. Woltersdorf and E. Pippel
*Influence of Substrate Deformations on Interface Structures (I): Theory
(a) Vol. **78**, No. 2, p. 475–483 (1983).
See also *10.1*, p. *105*; 21.6, p. 398.

H. Ezbakhe et A. Donnadieu
*Rugosité de surface et sélectivité de couches minces dioxyde molybdène
(Surface Roughness and Selectivity of Molybdenum Dioxyde Thin Films)
(a) Vol. **78**, No. 2, p. 607–614 (1983).
See also *20.1*, p. *324*; 22.6, p. 495.

O. K. Gorodnichenko, V. A. Presnov, and N. D. Vasilenko
*Deformation of Epitaxial Structures on the Basis of GaAs in the Process of Their Production
(a) Vol. **78**, No. 2, p. K171–K173 (1983).
See also *10.2*, p. *117*; 22.2.1, p. 446.

E. Pippel and J. Woltersdorf
*Influence of Substrate Deformations on Interface Structures (II): Experimental Investigations for Epitaxial Ag/Pd and Au/Pd Bicrystals
(a) Vol. **79**, No. 1, p. 189–198 (1983).
See also *10.1*, p. *106*; 21.6, p. 398.

L. Nanu, Doina Bărbulescu, I. N. Mihăilescu, V. Teodorescu, and Leona C. Nistor
*Annealing of Implanted Silicon under the Action of Microsecond Pulsed TEA-CO_2 Laser Radiation
(a) Vol. **79**, No. 1, p. K5–K8 (1983).
See also *11*, p. *132*; 22.1.2, p. 435.

M. Gasgnier, L. Nevot, P. Baillif, and J. Bardolle
*Characterization and Crystalline Structures of Tungsten Thin Films
(a) Vol. **79**, No. 2, p. 531–542 (1983).
See also 1.3, p. 26; 21, p. 360.

M. Gasgnier
*Complex Crystallization Phenomena in Rare-Earth Binary Alloys and Compounds as Thin Films
(a) Vol. **80**, No. 1, p. K1–K6 (1983).
See also 2, p. 55; 21.4, p. 393.

I. E. Bolotov and V. L. Prilepo
*On the Vacancy Mechanism of the Occurrence of Compressive Stresses during the Crystallization of Amorphous Films
(a) Vol. **80**, No. 1, p. K67–K70 (1983).
See also 3, p. 59; 10.2, p. 120; 12.2, p. 144; 22.1.3, p. 441; 22.4.1, p. 465.

B. Rauschenbach, A. Kolitsch, and K. Hohmuth
Iron Nitride Phases Formed by Nitrogen Ion Implantation and Thermal Treatment
(a) Vol. **80**, No. 2, p. 471–482 (1983).
See also *1.3*, p. *27*; *11*, p. *135*; 21, p. 362; 21.1.1, p. 385.

J. H. Basson and H. Booyens
*The Introduction of Misfit Dislocations in HgCdTe Epitaxial Layers
(a) Vol. **80**, No. 2, p. 663–668 (1983).
See also *10.2*, p. *121*; 22.4.4, p. 472.

1.6 Surfaces

Original Papers and Short Notes

V. Atzrodt, W. Titel, Th. Wirth, and H. Lange
*AES Studies on Thin Film $MoSi_2$
(a) Vol. **75**, No. 1, p. K15 – K20 (1983).
See also *11*, p. *123*; 21, p. 350.

G. Weber, B. Gillot, and P. Barret
*Interface Structure in Relation with the Mechanisms in the Reaction Copper–Silicon
(a) Vol. **75**, No. 2, p. 567 – 576 (1983).
See also *1.3*, p. *22*; 22.1.2, p. 426.

P. Streitenberger
*Generalized Pseudoatom Model of Inhomogeneous Systems and Its Application to a Simple-Metal Surface
(b) Vol. **116**, No. 1, p. 179 – 190 (1983).
See also *13*, p. *147*; 21, p. 352.

G. Dharmadurai
Phonon Transmission Theory of Heat Transfer through Solid–Gas Interfaces
(b) Vol. **116**, No. 1, p. 307 – 311 (1983).
See also *6*, p. *72*; *8*, p. *86*.

L. Komitov and A. G. Petrov
*Optical Method for Determination of Anchoring Energy of Tilted Nematic Layers
(a) Vol. **76**, No. 1, p. 137 – 143 (1983).
See also *2*, p. *45*; 22.6, p. 492.

T. Nanba and T. P. Martin
Raman Scattering from Metal Smokes
(a) Vol. **76**, No. 1, p. 235 – 239 (1983).
See also *6*, p. *72*; *20.1*, p. *316*; 21.1, p. 366; 21.6, p. 397.

A. I. Pekarev, G. F. Krasnova, and G. Z. Nemtsev
*Orientation Dependence of Oxidation Stacking Fault Density in Silicon
(a) Vol. **76**, No. 1, p. 327 – 330 (1983).
See also *10.2*, p. *111*; 22.1.2, p. 427.

D. Schulze, J. Finster, E. Hensel, W. Skorupa, and U. Kreissig
*XPS Studies of SiO_2 Surface Layers Formed by Oxygen Ion Implantation into Silicon
(a) Vol. **76**, No. 1, p. K21 – K24 (1983).
See also *17*, p. *269*; 22.6, p. 492.

Yu. N. Belyaev and A. V. Kolpakov
On the Theory of X-Ray Diffraction in a Perfect Crystal with Distorted Surface Layer
(a) Vol. **76**, No. 2, p. 641 – 646 (1983).
See also *1*, p. *9*.

K. Abdelmoula, J. J. Métois, and G. Nihoul
*Reconstructed Surface Structure of (111) Gold Platelets, Observed by High Resolution Techniques
(a) Vol. **77**, No. 1, p. K5–K7 (1983).
See also 21.6, p. 397.

A. L. Golovin and R. M. Imanov
*Investigation of the X-Ray Scattering Intensity for the Laue-Case Diffraction under Total-External-Reflection Conditions
(a) Vol. **77**, No. 1, p. K91–K94 (1983).
See also 22.1.2, p. 430.

C. M. van Baal
*The Influence of Texture on the X-Ray Determination of Residual Strains in Ground or Worn Surfaces
(a) Vol. **77**, No. 2, p. 521–526 (1983).
See also 10.1, p. 103; 12.1, p. 138; 21, p. 355.

P. Zaumseil and U. Winter
*Detection of Surface Imperfections at Polished Silicon Wafers by TCD Measurements
(a) Vol. **77**, No. 2, p. K139–K141 (1983).
See also *10.2*, p. *113*; 22.1.2, p. 431.

J. Vetter and H. Betghe
*Thermal Self-Decoration of GeS Crystals
(a) Vol. **78**, No. 1, p. 103–109 (1983).
See also 1.3, p. 24; 22, p. 409.

L. Z. Mezey, O. Nishikawa, and J. Giber
*Monotonic and Alternating Surface Segregation Profiles in Binary Alloys
(a) Vol. **78**, No. 1, p. 323–331 (1983).
See also *1.3*, p. *24*; 21.1, p. 372; 21.6, p. 398.

Th. Klupsch
*Towards a Molecular Theory of the Solid–Liquid Phase Boundary. Structural Theory and Effective Lattice Models
(b) Vol. **118**, No. 2, p. 681–690 (1983).
See also *2*, p. *50*; 3, p. 59.

P. J. Ratcliffe and R. A. Collins
*The Influence of Ion Implantation on the Thermal Oxidation of Copper
(a) Vol. **78**, No. 2, p. 547–553 (1983).
See also *11*, p. *130*; 21.1, p. 373.

Th. Klupsch
*Towards a Molecular Theory of the Solid–Liquid Phase Boundary. Elementary Kinetic Processes
(b) Vol. **119**, No. 1, p. 279–289 (1983).
See also 2, p. 51.

D. F. Lynch and A. E. Smith
*A Scattering Matrix Calculation of Very Low Energy Electron Reflection Intensities for Barium and Magnesium Oxides
(b) Vol. **119**, No. 1, p. 355–361 (1983).
See also 22.6, p. 496.

H. Tsuya and F. Shimura
*Transient Behaviour of Intrinsic Gettering in CZ Silicon Wafers
(a) Vol. **79**, No. 1, p. 199–206 (1983).
See also *10.2*, p. *118*; 14.3.3, p. 240; 22.1.2, p. 435.

A. Milchev
*Heat Capacity of Diatomic Molecules Adsorbed on a Periodic Substrate and the Localized-to-Mobile Transition
(a) Vol. **79**, No. 1, p. 301–308 (1983).
See also *8*, p. *90*.

A. Wolkenberg
*A Mechanism for the Effect of Doping on the Silicon Native Oxide Thickness
(a) Vol. **79**, No. 1, p. 313–322 (1983).
See also 10.2, p. 118; 22.1.2, p. 435; 22.6, p. 496.

G. S. Popkirov
*An Electrochemial Evidence of the O_2 Photodesorption from n-TiO_2 Surface
(a) Vol. **79**, No. 1, p. 323–326 (1983).
See also *14.3.3*, p. *240*; 22.6, p. 496.

G. K. Baranova
*On the Role of Surface Oxides in the Processes of Metal Etching
(a) Vol. **79**, No. 1, p. K9–K12 (1983).
See also 10.1, p. 106; 21, p. 359.

W. Beck and H. Kronmüller
The Influence of Surface Roughness on the Coercive Field of the Nearly Non-Magnetostrictive Amorphous Ferromagnetic Alloy $Co_{58}Ni_{10}Fe_5Si_{11}B_{16}$
(a) Vol. **79**, No. 2, p. 389–394 (1983).
See also 2, p. 53; *18.2*, p. *287*; 21.1, p. 376; 21.1.1, p. 384.

B. Rauschenbach and K. Hohmuth
Analysis of Heavy-Ion Sputtering Based on the Shock Wave Model
(a) Vol. **79**, No. 2, p. 395–401 (1983).
See also *11*, p. *132*; 21, p. 359.

H. Pfeiffer
*Analytical Theory of the Growth of Crystal Faces
(a) Vol. **79**, No. 2, p. 445–454 (1983).
See also *3*, p. *59*.

V. Atzrodt and H. Lange
*Principal Component Analysis as a Method for Silicide Investigation with Auger Electron Spectroscopy
(a) Vol. **79**, No. 2, p. 489–496 (1983).
See also 17, p. 270; 21, p. 360; 22.1.2, p. 436.

Min-shengLei and Zhi-cheng Wang
Space Distribution of the Chemical Composition of the Thermo-Grown SiO_2–SiO_x–Si System
(b) Vol. **120**, No. 1, p. 189–195 (1983).
See also *1.3*, p.*26*;*4*, p. *64*; 22.6, p. 497.

K. S. Bhadauria and S. B. Shrivastava
*Positron Annihilation Rate at Metal Surfaces
(b) Vol. **120**, No. 1, p. K27–K31 (1983).
See also 21.6, p. 399.

S. Surma, M. Błaszkiewicz, R. Męclewski, and W. Gubernator
*A Rearranged Structure on a Rhenium Surface
(a) Vol. **80**, No. 1, p. 269–276 (1983).
See also 17.1, p. 272; 21.4, p. 393.

A. L. Golovin and R. M. Imanov
*Obtaining Quantitative Information on Amorphous Layer Tickness on Crystal Surface Using X-Ray Diffraction under Specular Reflection Conditions
(a) Vol. **80**, No. 1, p. K63–K65 (1983).
See also 22.1.2, p. 438.

2. Non-Crystalline State

Review Articles

N. C. HALDER and K. C. PHILLIPS
*Electronic Density of States of Liquid Metals Studied by Perturbation and Green Function Methods
(b) Vol. **115**, No. 1, p. 9–35 (1983).
See also *13*, p. *146*; 21, p. 349.

K. H. FISCHER
Spin Glasses (I)
(b) Vol. **116**, No. 2, p. 357–414 (1983).
See also 8, p. 85; 13, p. 146; 14.1, p. 206; *18.2*, p. *280*; *18.4*, p. *295*; 21.1, p. 363; 21.1.1, p. 379.

Original Papers and Short Notes

M. STUTZMANN, J. STUKE, and H. DERSCH
Electron Spin Resonance of Doped Glow-Discharge Amorphous Germanium
(b) Vol. **115**, No. 1, p. 141–151 (1983).
See also *13.4*, p. *172*; 14.3, p. 215; *19*, p. *299*; 22.1.1, p. 418; 22.1.2, p. 423.

H. KAHNT and F. SCHIRRMEISTER
*Electric Field Dependent Conductivity in Amorphous Semiconductors
(b) Vol. **115**, No. 1, p. 171–179 (1983).
See also 13.4, p. 172; *14.3.4*, p. *243*; 22, p. 403.

A. BOBÁK
*Curie Temperature of Amorphous Ferromagnets
(b) Vol. **115**, No. 1, p. K21–K24 (1983).
See also *18.2*, p. *280*.

H. KERKOW, G. KREYSCH, B. LUKASCH, and G. LÜCK
Calorimetric Determination of the Formation Energy for an Amorphous Layer on a Crystalline Silicon Substrate
(a) Vol. **75**, No. 1, p. 251–254 (1983).
See also *11*, p. *122*; 22.1.2, p. 424.

F. T. PARKER and H. OESTERREICHER
Temperature Dependence of Intrinsic Magnetic Hardness in Some Rare-Earth Pseudobinary Compounds and Amorphous Materials
(a) Vol. **75**, No. 1, p. 273–282 (1983).
See also *18.2*, p. *280*; 21.1, p. 364; 21.4, p. 390.

S. TYAGI
*Ferromagnetic Resonance in Single Particles of Amorphous $Fe_{81.5}B_{14.5}Si_4$
(a) Vol. **75**, No. 1, p. K29–K32 (1983).
See also *18.2*, p. *281*; 19, p. 299; 21.1.1, p. 379.

G. Cocco, L. Schiffini, A. Lucci, L. Battezzati, and M. Sampoli
*Structural Features of $Pd_{76}B_{24}$ Amorphous Alloy
(a) Vol. **75**, No. 1, p. K53 – K56 (1983).
See also 21.6, p. 396.

C. Gunkel, D. Schalch, and A. Scharmann
Electrical Conduction in Silicon Nitride Thin Films under 15 keV He^+ Irradiation
(a) Vol. **75**, No. 1, p. K57 – K62 (1983).
See also *11*, p. *123*; *14.3.1*, p. *229*; 22, p. 403.

C. S. Sunandana and R. Singh
*An Infrared Study of the Glass System $CaO \cdot B_2O_3 \cdot Al_2O_3 \cdot xFe_2O_3$ ($0 \leqq x \leqq 1$)
(a) Vol. **75**, No. 1, p. K91 – K94 (1983).
See also *20.1*, p. *314*; 22.6.1, p. 498.

N. M. Ravindra, C. Ance, J. P. Ferraton, A. Donnadieu, and S. Robin
*Penn-Like Model Calculations. Application to Hydrogenated Amorphous Silicon Films Prepared by Glow Discharge
(b) Vol. **115**, No. 2, p. 347 – 352 (1983).
See also 13.1, p. 156; 14.4, p. 250; 22.1.2, p. 425.

K. H. J. Buschow, H.-J. Eifert, and B. Elschner
*Electronic Structure and Stability of Amorphous Pd–Si Alloys
(b) Vol. **115**, No. 2, p. 455 – 462 (1983).
See also *13*, p. *147*; 19, p. 299; 21, p. 350; 22.1.2, p. 425.

A. M. Korostil
*A New Approach to the Theory of Tunnelling Relaxation
(b) Vol. **115**, No. 2, p. K117 – K119 (1983).
See also 6, p. 71; *13*, p. *147*.

L. Tichý, H. Tichá, M. Frumar, and A. Tříska
*Compositional Dependence of the Average Gap in the Ge-Rich Region of the $Ge_{1-x}S_x$ Non-Crystalline System
(b) Vol. **115**, No. 2, p. K127 – K131 (1983).
See also *4*, p. *60*; 22, p. 404.

A. Apostolov, H. Hristov, M. Mikhov, and V. Skumriev
*Magnetic Properties of Bulk Amorphous Alloys Gd_4Co_3, Er_4Co_3, and Sm_4Co_3
(a) Vol. **75**, No. 2, p. 401 – 404 (1983).
See also *18.1*, p. *278*; 18.2, p. 281; 21.4, p. 390.

H. Pfleiderer and B. Rauscher
Capacitance of Amorphous Silicon pin Solar Cells
(a) Vol. **75**, No. 2, p. 537 – 545 (1983).
See also 13.4, p. 175; *14.3.1*, p. *230*; *14.3.3*, p. *237*; 16, p. 264; 22.1.2, p. 426.

F. Schauer, O. Zmeškal, and S. Nešpůrek
A New Approach to the Analysis of Steady-State Space-Charge-Limited Currents Using Their Activation Energies. An Application to a-As_2Se_3
(a) Vol. **75**, No. 2, p. 591–594 (1983).
See also *13.4*, p. *175*; *14.3.4*, p. *244*; 22, p. 404.

M. I. Mitkowa and W. Fallmann
*Some Properties of Silver Photodiffused Chalcogenide Glasses
(a) Vol. **75**, No. 2, p. K145–K148 (1983).
See also *14.3.3*, p. *237*; 20.1, p. 315; 22.8, p. 503.

G. A. Petrakovskii, K. A. Sablina, and V. P. Ikonnikov
*Amorphization of $RbFeS_2$
(a) Vol. **75**, No. 2, p. K165–K167 (1983).
See also *5*, p. *65*; 22.8, p. 504.

N. Ishii, M. Kumeda, and T. Shimizu
*Defect States in Amorphous Silicon
(b) Vol. **116**, No. 1, p. 91–100 (1983).
See also 10.2, p. 110; *13.4*, p. *176*; 22.1.2, p. 426.

L. Wojtczak and T. Balcerzak
*Reaction Field Approximation for Inhomogeneous Ferromagnets
(b) Vol. **116**, No. 1, p. 217–225 (1983).
See also *18.2*, p. *281*.

R. N. Joarder and R. V. Gopala Rao
*Temperature Dependence of the Structure Factor of Liquid Metals
(b) Vol. **116**, No. 1, p. 299–306 (1983).
See also 4, p. 61; 9, p. 93; 21.1, p. 366; 21.2, p. 387; 21.6, p. 396.

J. Katriel and G. F. Kventzel
The Constant-Coupling Approximation: Some Exact Results
(b) Vol. **116**, No. 1, p. K51–K56 (1983).
See also *13*, p. *148*; *18.2*, p. *281*.

A. Rajewska, B. Pura, J. Przedmojski, and R. Dąbrowski
X-Ray Study of the Smectic A–Nematic Phase Transition in 4-n-Pentylphenyl-4-(Trans-4′-n-Pentylcyclohexyl) Benzoate
(a) Vol. **76**, No. 1, p. 127–130 (1983).
See also *1.2*, p. *15*; 22.9, p. 526.

L. Komitov and A. G. Petrov
*Optical Method for Determination of Anchoring Energy of Tilted Nematic Layers
(a) Vol. **76**, No. 1, p. 137–143 (1983).
See also *1.6*, p. *39*; 22.6, p. 492.

V. M. Pan, I. Ya. Dekhtyar, M. E. Osinovskii, S. E. Litvin, B. G. Nikitin, M. M. Nishenko, and M. P. Voronko
*Thermodynamical and Structural Properties of Amorphous $Fe_{80}B_{20}$ and $Fe_{80}P_{13}C_7$ Alloys
(a) Vol. **76**, No. 1, p. 197–206 (1983).
See also 5, p. 65; 8, p. 87; 14.1, p. 206; *18.2.1*, p. *290*; 21.1.1, p. 380.

S. Neov, I. Gerassimova, and B. Sydzhimov
*Neutron Diffraction Study of 2 $TeO_2 \cdot V_2O_5$ Glass
(a) Vol. **76**, No. 1, p. 297–301 (1983).
See also **4**, p. 61; 22.6.1, p. 498.

M. M. Hafiz, A. A. Ammar, A. I. Al-Adl, and Aboutaleb Mohamed
*Effect of Heat Treatment on the Conduction and Structure of Ge–Se–Te Amorphous Alloys
(a) Vol. **76**, No. 1, p. 319–325 (1983).
See also 1.2, p. 15; *14.3*, p. *217*; 22.8, p. 504.

A. G. Kazanskii, V. F. Kiselev, E. A. Silaev, and V. S. Vavilov
On the Role of Surface Phenomena in the Staebler-Wronski Effect
(a) Vol. **76**, No. 1, p. 337–343 (1983).
See also *13.3*, p. *168*; *14.3.2*, p. *233*; 16, p. 265; 22.1.2, p. 427.

W. Matz, B. Kunsch, and E. Wieser
*The Structure of Liquid $Fe_{75}B_{25}$ by Neutron Scattering
(a) Vol. **76**, No. 1, p. K17–K20 (1983).
See also 21.1.1, p. 380.

H. J. Goldsmid, M. M. Kaila, and G. L. Paul
*Thermal Conductivity of Amorphous Silicon
(a) Vol. **76**, No. 1, p. K31–K33 (1983).
See also *8*, p. *87*; 22.1.2, p. 427.

J. Weis and I. Červeň
*Structural Changes in Amorphous $Fe_{83}B_{17}$ by Neutron Irradiation
(a) Vol. **76**, No. 1, p. K49–K51 (1983).
See also *11*, p. *126*; 21.1.1, p. 380.

P. Czarnecki and G. Fritzkowski
*Transverse Mangetoresistance of Amorphous $Fe_{40-x}Ni_{40}Cr_xB_{20}$ Alloys in Magnetic Fields up to 20 T
(a) Vol. **76**, No. 1, p. K93–K96 (1983).
See also *14.1*, p. *207*; 18.2, p. 282; 21.1, p. 366; 21.1.1, p. 380.

K. Handrich
Phonon Lifetime and Thermal Conductivity in Amorphous Solids
(b) Vol. **116**, No. 2, p. 533–538 (1983).
See also *6*, p. *72*; *8*, p. *87*; 22.9, p. 527.

S. N. Kaul
*Two-Current Conduction Model and Anisotropic Magnetoresistance in Transition Metal-Metalloid Glasses
(b) Vol. **116**, No. 2, p. K99–K104 (1983).
See also *14.1*, p. *207*; 21.1, p. 367; 21.1.1, p. 381.

B. JHA, B. BHATTACHARJEE, S. PAUL, and R. PAUL
*X-Ray Diffraction Studies on the Mesophases of n–p-Cyanobenzylidene-p-Octyloxy Aniline in Magnetic Field
(a) Vol. **76**, No. 2, p. 461–466 (1983).
See also 22.9, p. 527.

P. DUHAJ and F. HANIC
*Electron Microscopy and Electron Diffraction Study of Crystallization of Metastable Phases in Amorphous Fe–B Alloys
(a) Vol. **76**, No. 2, p. 467–477 (1983).
See also *1.2*, p. *15*; 1.3, p. 23; 21.1.1, p. 381.

H. BÖRNER and R. HÖHNE
*Study of Magnetic After-Effects in the Amorphous Alloy $Fe_{40}Ni_{40}P_{14}B_6$
(a) Vol. **76**, No. 2, p. 499–503 (1983).
See also *18.2*, p. *282*; 21.1, p. 367; 21.1.1, p. 381.

O. BOSTANJOGLO
Multiple Phase Transitions in GeTe Films. Dynamics and Film Structure
(a) Vol. **76**, No. 2, p. 525–531 (1983).
See also *1.2*, p. *16*; *1.5*, p. *36*; 3, p. 58; 22, p. 406.

C. SANCHEZ, R. MORINEAU, and J. LIVAGE
Electrical Conductivity of Amorphous V_2O_5
(a) Vol. **76**, No. 2, p. 661–666 (1983).
See also *14.3*, p. *218*; 22.6, p. 493.

G. COCCO, L. SCHIFFINI, M. SAMPOLI, A. LUCCI, and R. RIONTINO
*Compositional Effects in Pd–B Glassy Alloys from X-Ray Determined Radial Distribution Functions
(a) Vol. **76**, No. 2, p. 753–762 (1983).
See also 21, p. 353.

J. SCHNEIDER, J. KAMARAD, Z. ARNOLD, and M. BITTERLICH
*The Pressure Dependence of the Curie Temperature of Amorphous Fe–P–Ga and Fe–P–Ge Alloys
(a) Vol. **76**, No. 2, p. K183–K188 (1983).
See also *18.2*, p. *283*; 21.1.1, p. 382.

J. KOČKA, O. ŠTIKA, and J. STUCHLÍK
Comments on the Interpretation of DLTS Spectra of a-Si:H
(a) Vol. **76**, No. 2, p. K217–K220 (1983).
See also *13.4*, p. *178*; *14.3.2*, p. *233*; 22.1.2, p. 429.

H. HERMANN
*Influence of Directed Chemical Bonding on the Structure of Metallic Glasses. Two-Dimensional Model
(b) Vol. **117**, No. 1, p. 185–195 (1983).
See also **4**, p. 61; 21, p. 353.

H. BERGNER, V. BRÜCKNER, D. DIETRICH, F. KERSTAN, and W. NOWICK
Picosecond Reflectivity Measurements on Glow Discharge a-Si:H
(b) Vol. **117**, No. 1, p. 197–202 (1983).
See also *20.1*, p. *318*; 22.1.2, p. 429.

A. L. AGEEV, YU. A. BABANOV, V. V. VASIN, N. V. ERSHOV, and A. V. SERIKOV
*Amorphous Problem in EXAFS Data Analysis
(b) Vol. **117**, No. 1, p. 345–350 (1983).
See also *20*, p. *308*; 21.1.1, p. 382.

C. ANCE and N. M. RAVINDRA
*Departure of Hydrogen from a-Si:H. A Spectroscopic Study
(a) Vol. **77**, No. 1, p. 241–248 (1983).
See also *10.2*, p. *112*; 22.1.2, p. 430.

K. T.-Y. KUNG, B. X. LIU, and M-A. NICOLET
*Study of Ni–Nb System by Ion Mixing
(a) Vol. **77**, No. 1, p. 355–359 (1983).
See also 1.5, p. 37; 14.1, p. 207; 21, p. 354; 21.1, p. 369.

G. JUŠKA and K. ARLAUSKAS
*Features of Hot Carriers in Amorphous Selenium
(a) Vol. **77**, No. 1, p. 387–391 (1983).
See also *14.3.4*, p. *245*; 22.1.3, p. 440.

S. L. DRECHSLER and G. M. VUJIČIĆ
On the Influence of Tunneling States on the Transition Temperature of Metallic Glassy Superconductors
(b) Vol. **117**, No. 2, p. 569–574 (1983).
See also *14.2*, p. *213*; 21.1, p. 370.

H. BERGNER, V. BRÜCKNER, F. KERSTAN, and W. NOWICK
*Comparative Studies of Fast Recombination Processes in Amorphous Silicon Films
(b) Vol. **117**, No. 2, p. 603–610 (1983).
See also *20.3*, p. *339*; 22.1.2, p. 430.

H. GRIMM and H. KRONMÜLLER
Investigation of Structural Defects in the Amorphous Ferromagnetic Alloy $Fe_{40}Ni_{40}P_{14}B_6$
(b) Vol. **117**, No. 2, p. 663–674 (1983).
See also *10.1*, p. *103*; *18.2*, p. *284*; 21.1, p. 370; 21.1.1, p. 382.

A. LEHMANN, L. SCHUMANN, and K. HÜBNER
Optical Phonons in Amorphous Silicon Oxides (I): Calculation of the Density of States and Interpretation of LO–TO Splittings of Amorphous SiO_2
(b) Vol. **117**, No. 2, p. 689–698 (1983).
See also *6*, p. *74*; *20.1*, p. *320*; 22.6, p. 494.

A. J. COHEN and G. G. JANEZIC
*Relationships among Trapped Hole and Trapped Electron Centers in Oxidized Soda–Silica Glasses of High Purity
(a) Vol. **77**, No. 2, p. 619–624 (1983).
See also *13.4*, p. *180*; *20.1*, p. *321*; 22.6.1, p. 498.

B. BRIDGE, N. D. PATEL, and D. N. WATERS
*On the Elastic Constants and Structure of the Pure Inorganic Oxide Glasses
(a) Vol. 77, No. 2, p. 655–668 (1983).
See also *12.2*, p. *142*; 22.6, p. 494.

H. RATAJCZAK, P. SLANČO, T. TIMA, A. ZENTKO, E. DEMKOVICSOVÁ, and M. KONČ
*Electrical Resistivity and Hall Effect in Thin Amorphous UFe Films
(a) Vol. 77, No. 2, p. 785–791 (1983).
See also *14.1*, p. *208*; 18.2, p. 284; 21.1.1, p. 382.

R. N. MUKHERJEE, D. N. S. SRIVASTWA, and S. C. SEN
Ultraviolet Absorption of Tl^+ Ion in K_2O–KBr–B_2O_3 and Na_2O–NaBr–B_2O_3 Glasses
(a) Vol. 77, No. 2, p. 799–802 (1983).
See also *20.1*, p. *321*; 22.8.1, p. 514.

N. P. KALMYKOVA, B. T. KOLOMIETS, V. KH. SHPUNT, and E. A. SMORGONSKAYA
*On the Inflence of the Non-Homogeneous Structure on the Current–Voltage Behaviour of Chalcogenide Vitreous Semiconductors
(a) Vol. 77, No. 2, p. 803–808 (1983).
See also *14.3.4*, p. *246*; 22, p. 408.

C. JÄGER and M. MÜLLER
*Crystallization in a Fluorozirconate Glass: Determination of the ^{19}F Chemical Shift in β-$BaZrF_6$
(a) Vol. 77, No. 2, p. K167–K170 (1983).
See also *19*, p. *302*; 22.8, p. 506.

K. SCHIELE and S. TRIMPER
*On the Elastic Constants of a Nematic Liquid Crystal
(b) Vol. 118, No. 1, p. 267–274 (1983).
See also *12*, p. *136*.

A. BOBÁK
*The Behaviour of an Amorphous Ferromagnet with Dilution
(b) Vol. 118, No. 1, p. K61–K63 (1983).
See also *18.2*, p. *284*.

N. ISHII, M. KUMEDA, and T. SHIMIZU
Defect States in Amorphous Germanium
(b) Vol. 118, No. 1, p. K65–K68 (1983).
See also *13.4*, p. *182*; 19, p. 302; 22.1.1, p. 421.

M. SUHRKE
*On the Time Dependence of the Luminescence Spectrum in Chalcogenide Glasses
(b) Vol. 118, No. 1, p. K79–K83 (1983).
See also *20.3*, p. *341*; 22, p. 409.

C. M. MÔ and P. MOSER
*Hydrogen in $Fe_{40}Ni_{40}P_{14}B_6$ Amorphous Alloy Studied by Magnetic After Effect, Internal Friction, and Positron Annihilation
(a) Vol. 78, No. 1, p. 201–210 (1983).
See also 10.1, p. 103; *18.2*, p. *285*; 21.1, p. 371; 21.1.1, p. 383.

R. Höhne, H. Börner, and M. Krötzsch
*On the Annealing Behaviour and Crystallization of the Amorphous Alloy $Fe_{40}Ni_{40}P_{14}B_6$
(a) Vol. 78, No. 1, p. 211–216 (1983).
See also *18.2*, p. *285*; 21.1, p. 371; 21.1.1, p. 383.

W. Burkhardt and A. Feltz
*The Influence of the Structure of Different Arsenic Chalcogenide Glasses on Refractive Index and Dispersion
(b) Vol. 118, No. 2, p. 653–660 (1983).
See also *20.1*, p. *323*; 22, p. 410.

H. Kronmüller
*The Rôle of Two-Level Systems in Amorphous Metallic Alloys (I): Induced Anisotropy in Amorphous Ferromagnets
(b) Vol. 118, No. 2, p. 661–673 (1983).
See also *18.2*, p. *285*; 21.1.1, p. 383.

Th. Klupsch
*Towards a Molecular Theory of the Solid–Liquid Phase Boundary. Structural Theory and Effective Lattice Models
(b) Vol. 118, No. 2, p. 681–690 (1983).
See also *1.6*, p. *40*; 3, p. 59.

T. Kaneyoshi
*On an Anomalous Behaviour of Spin-Wave Stiffness Constant in Amorphous Ferromagnets
(b) Vol. 118, No. 2, p. 751–755 (1983).
See also *18.2*, p. *285*.

K. Hricovíni, P. Mikušík, and J. Krempaský
*Photoemission Studies of $Pd_{80-x}Si_{20}Co_x$ Amorphous Alloys
(b) Vol. 118, No. 2, p. 823–827 (1983).
See also 13.1, p. 161; *17*, p. *270*; 21.1, p. 372.

B. V. Kobrin, R. M. Kupriyanova, V. S. Minaev, E. P. Prokopiev, and V. P. Shantarovich
*Microstructure and Imperfections of Glassy and Glass-Ceramic Telluride Semiconductors on the Basis of Positron Annihilation
(a) Vol. 78, No. 2, p. 385–389 (1983).
See also *10.2*, p. *116*; 22, p. 410.

E. Bonetti, E. Evangelista, A. Cavallini, and C. Patuelli
*Internal Friction Peaks and Structural Relaxation on Some Electroless Deposited Amorphous Alloys
(a) Vol. 78, No. 2, p. 615–621 (1983).
See also *12.1*, p. *139*; 21.1, p. 373.

B. Rauscher, H. Pfleiderer, and B. Bullemer
*Tail–Tail Luminescence in Amorphous Silicon
(a) Vol. 78, No. 2, p. 623–632 (1983).
See also *20.3*, p. *343*; 22.1.2, p. 433.

K. G. Chakraborty and J. F. Baret
*Spin-One Lattice Gas Model for Isotherm-Reversal in Spin-Labelled Lipid Monolayers
(b) Vol. **119**, No. 1, p. 61–71 (1983).
See also *13*, p. *150*; 22.9, p. 529.

S. Katsura and A. Matsuno
*Phase Diagram of a Spin Glass of $Eu_pSr_{1-p}S$
(b) Vol. **119**, No. 1, p. 73–77 (1983).
See also *18*, p. *275*; 18.1, p. 279; 18.2, p. 286; 22, p. 410.

A. S. T. Pires and F. C. Sá Barreto
*Solitons in Ferroelectric Liquid Crystal
(b) Vol. **119**, No. 1, p. 121–130 (1983).
See also *14.4.1*, p. *258*; 22.9, p. 530.

Th. Klupsch
Towards a Molecular Theory of the Solid–Liquid Phase Boundary. Elementary Kinetic Processes
(b) Vol. **119**, No. 1, p. 279–289 (1983).
See also *1.6*, p. *40*.

E. Canessa and J. Vignolo
Electrical Resistivities of the Liquid Alkali Metals
(b) Vol. **119**, No. 1, p. 429–436 (1983).
See also *14.1*, p. *209*; 21.2, p. 388.

E. Canessa and J. Vignolo
*The Resistivity of Liquid Mercury at 150 °C
(b) Vol. **119**, No. 1, p. K83–K85 (1983).
See also *14.1*, p. *209*; 21, p. 358.

A. V. Dvurechenskii, I. A. Ryazantsev, L. S. Smirnov, H. Klose, and M. Rieth
Amorphous Silicon–Phosphorus Layers
(a) Vol. **79**, No. 1, p. 83–86 (1983).
See also 13.4, p. 186; *14.3.1*, p. *231*; 16, p. 267; 19, p. 304; 22.1.2, p. 434.

W. Beck and H. Kronmüller
On the Coercive Field of the Nearly Non-Magnetostrictive Amorphous Ferromagnetic Alloy $Co_{58}Ni_{10}Fe_5Si_{11}B_{16}$
(a) Vol. **79**, No. 1, p. 109–114 (1983).
See also 10.1, p. 105; *18.2*, p. *286*; 21.1, p. 375; 21.1.1, p. 383.

B. Rauscher and B. Bullemer
*Photoluminescence Decay in Amorphous Silicon Alloys
(a) Vol. **79**, No. 1, p. 161–166 (1983).
See also *20.3*, p. *344*; 22.1.2, p. 434.

S. Różański
Determination of Pitch and Chiral Smectic C DOBAMBC
(a) Vol. **79**, No. 1, p. 309–312 (1983).
See also *14.4.1*, p. *258*; 22.9, p. 530.

M. MERTIG, G. POMPE, and E. HEGENBARTH
*Low-Temperature Specific Heat of Vapour-Deposited Ge_xSe_{1-x} Glass ($x = 0.3$)
(a) Vol. **79**, No. 1, p. K49–K52 (1983).
See also *8*, p. *90*; 22, p. 412.

B. T. KOLOMIETS, T. N. MAMONTOVA, A. V. CHERNYSHOV, G. Z. VINOGRADOVA, and N. V. DEMOKRITOVA
*The Effect of Low Oxygen Concentrations on the Photoluminescence of Vitreous Selenium
(a) Vol. **79**, No. 1, p. K89–K92 (1983).
See also *10.2*, p. *118*; *20.3*, p. *345*; 22.3, p. 458.

P. CHANDRASEKHARAM, D. RAJA REDDY, and B. K. REDDY
*Electrical Properties of Vapour Phase Grown $(ZnSe)_x(CdTe)_{1-x}$ Single Crystals
(a) Vol. **79**, No. 1, p. K103–K107 (1983).
See also *14.3*, p. *224*; 22.4.4, p. 471.

V. D. ISKRA
*On the Problem of an Intermediate Order in Amorphous Semiconductors
(b) Vol. **119**, No. 2, p. 515–524 (1983).
See also 20.1, p. 327; 22, p. 412.

G. POMPE, M. GAAFAR, P. BÜTTNER, and TH. FRANCKE
*Low Temperature Thermal Conductivity of Amorphous (Fe, Ni, Co) (P, B, Si) Alloys and Their Change by Heat Treatment
(b) Vol. **119**, No. 2, p. 579–587 (1983).
See also *8*, p. *90*; 21.1, p. 375; 21.1.1, p. 384.

B. A. VAID, K. C. SHARMA, and V. K. SYAL
*A Disorder Model of Melting
(b) Vol. **119**, No. 2, p. 703–708 (1983).

A. MORIMOTO, S. OOZORA, M. KUMEDA, and T. SHIMIZU
*Annealing Behaviour of Hydrogenated Amorphous Silicon–Nitrogen Alloy Films Prepared by Sputtering
(b) Vol. **119**, No. 2, p. 715–720 (1983).
See also *10.2*, p. *119*; 19, p. 304; 20.1, p. 327; 22, p. 412.

I. MUŠEVIČ, B. ŽEKŠ, R. BLINC, TH. RASING, and P. WYDER
Dielectric Study of the Modulated Smectic C*–Uniform Smectic C Transition in a Magnetic Field
(b) Vol. **119**, No. 2, p. 727–733 (1983).
See also *14.4.1*, p. *259*; 22.9, p. 530.

H. TICHÁ, L. TICHÝ, and E. ČERNOŠKOVÁ
*Temperature Dependence of the Optical Gap in Glassy GeS_2
(b) Vol. **119**, No. 2, p. K135–K137 (1983).
See also **6**, p. 79; *20.1*, p. *327*; 22, p. 412.

J. RICHTER
*Disorder and Competing Interactions in Spin Systems: Finite-Cluster Calculations
(b) Vol. **119**, No. 2, p. K159–K163 (1983).
See also *18.4*, p. *297*; 21, p. 359.

Yu. G. Chukalkin, V. R. Shtirts, and B. N. Goshchitskii
Amorphization of Neutron-Irradiated Yttrium Iron Garnet
(a) Vol. **79**, No. 2, p. 361–366 (1983).
See also *11*, p. *132*; 18.3, p. 293; 22.8.2, p. 523.

W. Beck and H. Kronmüller
The Influence of Surface Roughness on the Coercive Field of the Nearly Non-Magnetostrictive Amorphous Ferromagnetic Alloy $Co_{58}Ni_{10}Fe_5Si_{11}B_{16}$
(a) Vol. **79**, No. 2, p. 389–394 (1983).
See also 1.6, p. 41; *18.2*, p. *287*; 21.1, p. 376; 21.1.1, p. 384.

W. Gräpe
*A Method for the Determination of a Static Fatigue Limit of Glasses
(a) Vol. **79**, No. 2, p. 403–410 (1983).
See also *10.2*, p. *119*; 22.6, p. 496.

J. Filipecki, B. Całusiński, and Z. Mandecki
*Investigation of Electron Phenomena in the Metallic Glasses $Fe_xNi_{78-x}B_{10}Si_{12}$ with the Use of Positron Annihilation Methods before and after Their Crystallization
(a) Vol. **79**, No. 2, p. 585–591 (1983).
See also 13, p. 152; 21.1, p. 376; 21.1.1, p. 384.

L. Komitov and A. G. Petrov
*Quasihomeotropic, Reversely Pretilted Nematic Liquid Crystal Layers. Electrooptical Behaviour
(a) Vol. **79**, No. 2, p. 623–629 (1983).
See also 14.4, p. 253; 20.1, p. 328; 22.9, p. 531.

J. Weis, D. Barančok, and I. Červeň
*RDF Analysis of Structural Evolution during the Primary Crystallization of the $Fe_{83}B_{17}$ Glassy Alloy
(a) Vol. **79**, No. 2, p. K117–K122 (1983).
See also 21.1.1, p. 384.

A. N. Rodionov, R. I. Kalendarev, A. V. Shendrik, and Yu. R. Zakis
*Photoinduced Paramagnetic States of Yellow Arsenic
(a) Vol. **79**, No. 2, p. K151–K154 (1983).
See also *19*, p. *305*; 22.1, p. 417.

A. Milchev and I. Avramov
*On the Influence of Amorphization on Atomic Diffusion in Condensed Systems
(b) Vol. **120**, No. 1, p. 123–130 (1983).
See also *9*, p. *95*.

W. Fuhs, H. M. Welsch, and D. C. Booth
Optical Quenching of Photoconductivity in a-Si: H Films
(b) Vol. **120**, No. 1, p. 197–205 (1983).
See also 14.3.1, p. 232; *16*, p. *267*; 22.1.2, p. 437.

M. Stutzmann and J. Stuke
Temperature Dependence of ESR Spectra of Doped Amorphous Germanium
(b) Vol. **120**, No. 1, p. 225–234 (1983).
See also *13.4*, p. *189*; *19*, p. *305*; 22.1.1, p. 421.

J. Šlechta
*Comparison between the Self-Consistent Continued-Fraction and Other Continued-Fraction Methods
(b) Vol. **120**, No. 1, p. 329–339 (1983).
See also *13*, p. *152*.

A. Alegria, J. M. Barandiarán, and J. Colmenero
*Dielectric Relaxation at the Glass Transition as a Free Volume Process. A Single Relaxation Time Approach
(b) Vol. **120**, No. 1, p. 349–360 (1983).
See also 14.4, p. 254; 22.9, p. 531.

C. Regnaut, E. Fusco, and J. P. Badiali
*Structure and Thermodynamics of the Liquid Noble Metals from the Generalized OPW and Resonant Modell Approaches
(b) Vol. **120**, No. 1, p. 373–382 (1983).
See also 8, p. 91; 13.1, p. 164; 21.1, p. 376; 21.6, p. 339.

H.-J. Unger
*Localization of Electronic States in Disordered Tight-Binding Systems (I): Derivation and Study of the Model
(b) Vol. **120**, No. 1, p. 401–413 (1983).
See also *13.4*, p. *189*; 22, p. 413.

M. Chaturvedi and V. Srivastava
*Studies on Quantum Percolation
(b) Vol. **120**, No. 1, p. 421–433 (1983).
See also *13*, p. *152*.

D. Della Sala, F. Evangelisti, P. Fiorini, and C. Giovannella
Influence of Oxygen on the Photoconductivity of Hydrogenated Amorphous Silicon
(b) Vol. **120**, No. 1, p. K59–K62 (1983).
See also *16*, p. *267*; 22.1.2, p. 437.

G. L. Whittle and A. M. Stewart
Mössbauer Effect Study of $Fe_{85-x}Ni_xB_{15}$ Metallic Glasses
(a) Vol. **80**, No. 1, p. 101–107 (1983).
See also *5*, p. *68*; 18.2, p. 288; 21.1, p. 377; 21.1.1, p. 385.

M. Vázquez, W. Fernengel, and H. Kronmüller
The Effect of Tensile Stresses on the Magnetic Properties of $Co_{58}Fe_5Ni_{10}Si_{11}B_{16}$ Amorphous Alloys
(a) Vol. **80**, No. 1, p. 195–204 (1983).
See also 12.1, p. 140; *18.2*, p. *288*; 21.1, p. 377; 21.1.1, p. 385.

P. Duhaj and P. Švec
*Transmission Electron Microscopy Study of Crystallization in Amorphous $Fe_{40}Ni_{40}B_{20}$ Alloy
(a) Vol. **80**, No. 1, p. 231–236 (1983).
See also *3*, p. *59*; 21.1.1, p. 385.

M. Gasgnier
Complex Crystallization Phenomena in Rare-Earth Binary Alloys and Compounds as Thin Films
(a) Vol. **80**, No. 1, p. K1–K6 (1983).
See also *1.5*, p. *38*; 21.4, p. 393.

F. D. Doenitz
*The Tetradecahedron as a Really Existing Structural Unit in Metallic Glasses of the Interstitial Type
(a) Vol. **80**, No. 1, p. K53–K56 (1983).

H. Bergner, V. Brückner, D. Dietrich, and W. Nowick
Picosecond Recombination of Photoexcited Carriers in Glow Discharge a-Si:H Films of Different Deposition Temperature
(b) Vol. **120**, No. 2, p. 655–658 (1983).
See also *20.1*, p. *330*; 22.1.2, p. 439.

O. Uemura, Y. Wada, T. Takahara, and T. Satow
*The Semiconduction in the Molten Tl–$CuTlTe_2$ System
(b) Vol. **120**, No. 2, p. 685–693 (1983).
See also *14.3*, p. *227*; 15, p. 263; 18, p. 276; 22.8, p. 510.

V. I. Motorin
*Vitrification Kinetics of Pure Metals
(a) Vol. **80**, No. 2, p. 447–455 (1983).
See also 21, p. 362; 21.1, p. 377; 21.6, p. 399.

W. Burckhardt and A. Feltz
*The Influence of the Structure of Different Germanium Chalcogenide Glasses on the Molar Refraction and Its Wavelength Dependence
(a) Vol. **80**, No. 2, p. 463–470 (1983).
See also *20.1*, p. *330*; 22, p. 415.

M. Vázquez, W. Fernengel and H. Kronmüller
Influence of Applied Stresses on the Magnetic Properties of Annealed $Co_{58}Fe_5Ni_{10}Si_{11}B_{16}$ Amorphous Alloy
(a) Vol. **80**, No. 2, p. 513–518 (1983).
See also 12.1, p. 140; *18.2*, p. *289*; 21.1, p. 378; 21.1.1, p. 385.

M. Ellner, M. El-Boragy und B. Predel
*Zur Bildung palladiumreicher metallischer Gläser in binären und ternären Systemen mit Kupfer, Silber und Gold als einer sowie mit Phosphor oder Arsen als weiterer Komponente
(On the Formation of Palladium-Rich Metallic Glasses in Binary and Ternary Systems with Copper, Silver, and Gold as One Component and Phosphorus and Arsenic as Further Component)
(a) Vol. **80**, No. 2, p. 519–533 (1983).
See also 13.1, p. 165; 21.1, p. 378; 21.6, p. 399.

L. ČERVINKA, V. CHÁB, and S. KOC
*XPS Investigation of the Amorphous $TlSe_x$ System
(a) Vol. **80**, No. 2, p. 685–689 (1983).
See also *4*, p. *64*; 17, p. 271; 22.3, p. 458.

P. KOŠTIAL, L. MALÍK, and L. BOTH
*New Aspect of Ultrasonic Detection of Some Significant Structural Parameters of Amorphous Se–Te Alloys
(a) Vol. **80**, No. 2, p. K167–K170 (1983).
See also *7*, p. *84*; 22.1.3, p. 441.

P. N. DASHUK, A. V. KOLOBOV, B. T. KOLOMIETS, and V. M. LYUBIN
*Plasma-Induced Processes in Chalcogenide Vitreous Semiconductors
(a) Vol. **80**, No. 2, p. K187–K192 (1983).
See also *11*, p. *135*; 22, p. 416.

3. Crystal Growth

Review Article

L. N. Aleksandrov and F. L. Edelman
*Shock Crystallization of Films
(a) Vol. **76**, No. 2, p. 409–427 (1983).
See also 1.3, p. 21; *1.5*, p. *35*; 22.1.1, p. 418; 22.1.2, p. 423; 22.1.3, p. 440; 22.6, p. 490.

Original Papers and Short Notes

R. Mattheis
*The Influence of Substrate Temperature on the Oriented Growth of Ni and Ni_2Si on Si (111)
(a) Vol. **75**, No. 1, p. 195–205 (1983).
See also 1.5, p. 35; 22.1.2, p. 424.

H. Gotoh, K. Sasamoto, S. Kuroda, and M. Kimata
*Molecular Beam Epitaxy of AlSb on GaAs and GaSb on AlSb Films
(a) Vol. **75**, No. 2, p. 641–645 (1983).
See also *1.5*, p. *36*; 14.3.1, p. 230; 22.2.1, p. 443; 22.2.3, p. 452.

Y. Iwama and Y. Takeno
*Formation Process of MnBi Thin Films by Williams' Method
(a) Vol. **76**, No. 1, p. 75–83 (1983).
See also *18.2.1*, p. *290*; 22.7, p. 499.

Yu. B. Bolkhovityanov
*The Lattice-Pulling Effect Induced by the Film Stresses. Conditions of Existence and Discovery
(a) Vol. **76**, No. 1, p. 85–93 (1983).
See also *1.5*, p. *36*; 22.2.1, p. 443; 22.2.2, p. 448.

M. E. Fleet
*Preferred Crystallographic Orientation for Crystallisation under Nonhydrostatic Stress
(a) Vol. **76**, No. 1, p. 151–156 (1983).

L. N. Aleksandrov
*Formation of Semiconductor Epitaxial Films by Pulse Heating Crystallization or Regrowth
(a) Vol. **76**, No. 1, p. 179–190 (1983).
See also *1.5*, p. *36*; 8, p. 86; 22.1.1, p. 419; 22.1.2, p. 427.

S. Grosswig, J. Härtwig, U. Schellenberger, and W.-D. Zimmer
*Theoretical and Experimental Study of Growth Sector Boundaries in Deuterated Potassium Dihydrogen Phosphate Single Crystals
(a) Vol. **76**, No. 1, p. 241–247 (1983).
See also *1.5*, p. *36*; 10.2, p. 111; 22.8.1, p. 513.

O. Bostanjoglo
Multiple Phase Transitions in GeTe Films. Dynamics and Film Structure
(a) Vol. **76**, No. 2, p. 525–531 (1983).
See also *1.2*, p. *16*; *1.5*, p. *36*; 2, p. 47; 22, p. 406.

M. Heming, G. Lehmann, K. Recker, and F. Wallrafen
EPR and Luminescence of Mn^{2+} in Distorted Tetrahedral Symmetry in $CaZnF_4$ Crystals
(b) Vol. **117**, No. 1, p. 271–276 (1983).
See also *19*, p. *301*; *20.3*, p. *338*; 22.5, p. 474.

T. A. Cherepanova and A. S. Koziejowska
*Sublattice Models of the Binary Crystal Growth (I): The Growth Kinetics of Ordering Binary Crystals from Solution and Melt
(a) Vol. **77**, No. 1, p. 35–44 (1983).

U. Rensch and G. Eichhorn
*Investigations on the Struçture of MOCVD AIN Layers on Silicon
(a) Vol. **77**, No. 1, p. 195–199 (1983).
See also *1.5*, p. *37*; 22.2, p. 442.

Z. Kroczak and M. Subotowicz
*Structural and Galvanomagnetic Properties of $Pb_{1-x}Mn_xTe$ Single Crystals Grown by the Bridgman-Stockbarger Method
(a) Vol. **77**, No. 2, p. 497–503 (1983).
See also *1.3*, p. *24*; *14.3*, p. *220*; 22, p. 408.

T. A. Cherepanova and A. S. Koziejowska
*Sublattice Models of the Binary Crystal Growth (II):: Influence of Diffusion in the Melt on the Crystallization Kinetics of Binary Alloys
(a) Vol. **77**, No. 2, p. 561–570 (1983).
See also 9, p. 94.

T. Taguchi
*Crystal Growth and Neutral-Acceptor Bound-Exciton Emission of ZnCdTe by THM with Te Solvent
(a) Vol. **77**, No. 2, p. K115–K119 (1983).
See also 13.4, p. 181; 13.5.1, p. 196; *20.3*, p. *341*; 22.4.4, p. 471.

K. E. Mironov, V. K. Ogorodnikov, V. D. Rozumnyi, and V. I. Ivanov-Omskii
*Closed System LPE Growth of $Cd_xHg_{1-x}Te$
(a) Vol. **78**, No. 1, p. 125–131 (1983).
See also *1.5*, p. *37*; 22.4.4, p. 471.

Yu. B. Bolkhovityanov, B. A. Bobylev, V. I. Korzhov, V. I. Yudaev, and E. H. Hairi
*Growth Peculiarities of Graded Band Gap GaAsSb/GaAs Heterostructures and Effective Light Modulation in the 1.06 μm Wavelength Region
(a) Vol. **78**, No. 1, p. 343–347 (1983).
See also 20.1, p. 323; 22.2.4, p. 455.

TH. KLUPSCH
Towards a Molecular Theory of the Solid–Liquid Phase Boundary. Structural Theory and Effective Lattice Models
(b) Vol. **118**, No. 2, p. 681 – 690 (1983).
See also *1.6*, p. *40*; *2*, p. *50*.

A. A. EL-SHARKAWY, M. S. ZAGHLOUL, and A. M. SANAD
*Study of Thermal Properties, Electrical Conductivity, and the Crystallization Process of the Glassy Alloy $SeGe_{20}$
(a) Vol. **78**, No. 2, p. 457 – 462 (1983).
See also 8, p. 89; 14.3, p. 222; 22.1.1, p. 421; 22.1.3, p. 440.

H. PFEIFFER
*Analytical Theory of the Growth of Crystal Faces
(a) Vol. **79**, No. 2, p. 445 – 454 (1983).
See also *1.6*, p. *41*.

P. DUHAJ and P. ŠVEC
*Transmission Electron Microscopy Study of Crystallization in Amorphous $Fe_{40}Ni_{40}B_{20}$ Alloy
(a) Vol. **80**, No. 1, p. 231 – 236 (1983).
See also *2*, p. *55*; 21.1.1, p. 385.

U. DAHMEN and K. H. WESTMACOTT
*Ledge Structure and the Mechanism of θ′ Precipitate Growth in Al–Cu
(a) Vol. **80**, No. 1, p. 249 – 262 (1983).
See also 10.1, p. 107; 21, p. 361; 21.1, p. 377.

A. A. KAMINSKII, B. V. MILL, G. G. KHODZHABAGYAN, A. F. KONSTANTINOVA, A. I. OKOROCHKOV, and I. M. SILVESTROVA
*Investigation of Trigonal $(La_{1-x}Nd_x)_3Ga_5SiO_{14}$ Crystals (I): Growth and Optical Properties
(a) Vol. **80**, No. 1, p. 387 – 398 (1983).
See also *20.1*, p. *329*; 22.8.1, p. 518.

I. E. BOLOTOV and V. L. PRILEPO
On the Vacancy Mechanism of the Occurrence of Compressive Stresses during the Crystallization of Amorphous Films
(a) Vol. **80**, No. 1, p. K67 – K70 (1983).
See also *1.5*, p. *38*; 10.2, p. 120; 12.2, p. 144; 22.1.3, p. 441; 22.4.1, p. 465.

R. B. LAL, M. D. AGGARWAL, R. L. KROES, and W. R. WILCOX
*A New Technique of Solution Crystal Growth
(a) Vol. **80**, No. 2, p. 547 – 552 (1983).
See also 14.4.1, p. 259; 22.9, p. 532.

J. K. JESZKA, J. ULAŃSKI, and M. KRYSZEWSKI
*TTT · TCNQ CT Complex Crystallized in Polymer Matrix
(a) Vol. **80**, No. 2, p. K141 – K143 (1983).
See also *14.3*, p. *228*; 22.9, p. 532.

4. Bonding Properties

Original Papers and Short Notes

J. Shanker and R. Sundaraj
Dielectric and Anharmonic Behaviour of Silver Halides, Thallous Halides, Alkaline Earth Fluorides, and Lead Fluoride Crystals
(b) Vol. **115**, No. 1, p. 67–74 (1983).
See also 13, p. 146; *14.4*, p. *250*; 22.5, p. 473; 22.5.1, p. 476; 22.5.3, p. 486.

C. Engler and W. Lorenz
Erratum to: Calculation of Local Densities of States, Local Charges, and Adsorption Energy on Large Semiconductor Substrate with Single Adsorbate
(b) Vol. **115**, No. 1, p. 325 (1983).
See also *13.1*, p. *155*; 13.3, p. 167; 22.2.2, p. 448.

J. Shanker and K. Singh
*Analysis of the Equation of State of Lithium and Sodium Halides
(b) Vol. **115**, No. 2, p. 381–387 (1983).
See also 22.5.2, p. 477.

R. K. Singh and N. K. Pandey
*Study of Thermophysical and Anharmonic Properties of Fluorite Compounds
(b) Vol. **115**, No. 2, p. 555–562 (1983).
See also 6, p. 70; 12.2, p. 141; 22.5.3, p. 486.

V. Ya. Antonchenko, A. S. Davydov, and A. V. Zolotariuk
*Solitons and Proton Motion in Ice-Like Structures
(b) Vol. **115**, No. 2, p. 631–640 (1983).
See also *6*, p. *71*; 13.5.4, p. 203; 22.6, p. 491.

L. Tichý, H. Tichá, M. Frumar, and A. Tříska
*Compositional Dependence of the Average Gap in the Ge-Rich Region of the $Ge_{1-x}S_x$ Non-Crystalline System
(b) Vol. **115**, No. 2, p. K127–K131 (1983).
See also *2*, p. *44*; 22, p. 404.

W. Thulke, R. Frahm, R. Haensel, and P. Rabe
*X-Ray Study of Niobium Disulfide Intercalated with Copper
(a) Vol. **75**, No. 2, p. 501–509 (1983).
See also *1.3*, p. *21*; 20, p. 307; 22.8, p. 503.

R. K. Singh, D. K. Neb, and S. P. Sanyal
*Effect of Three-Body Forces on Thermophysical and Anharmonic Properties of Rare Gas Solid Mixtures
(b) Vol. **116**, No. 1, p. 289–297 (1983).
See also 6, p. 72; 8, p. 86; 12.2, p. 142; 23, p. 533.

R. N. JOARDER and R. V. GOPALA RAO
Temperature Dependence of the Structure Factor of Liquid Metals
(b) Vol. **116**, No. 1, p. 299–306 (1983).
See also *2*, p. *45*; 9, p. 93; 21.1, p. 366; 21.2, p. 387; 21.6, p. 396.

S. NEOV, I. GERASSIMOVA, and B. SYDZHIMOV
Neutron Diffraction Study of 2 $TeO_2 \cdot V_2O_5$ Glass
(a) Vol. **76**, No. 1, p. 297–301 (1983).
See also *2*, p. *46*; 22.6.1, p. 498.

CH. OPITZ and F. DÜBLER
Modelling and Illustration of a Penetration Mechanism for Hydrogen into Metal
(a) Vol. **76**, No. 2, p. 505–512 (1983).
See also *9*, p. *94*; 13, p. 148; 22.5, p. 473.

H. HERMANN
Influence of Directed Chemical Bonding on the Structure of Metallic Glasses. Two-Dimensional Model
(b) Vol. **117**, No. 1, p. 185–195 (1983).
See also *2*, p. *47*; 21, p. 353.

G. BABONAS, N. S. KORETS, and S. MARCINKEVIČIUS
*Dispersion of Optical Constants in CdP_2
(b) Vol. **117**, No. 2, p. 477–481 (1983).
See also *20.1*, p. *319*; 22.3, p. 457.

T. SOMA, Y. KONNO, and H.-M. KAGAYA
*Local Heine-Abarenkov Potential with Volume-Dependent Parameters of Al
(b) Vol. **117**, No. 2, p. 743–748 (1983).
See also 12.1, p. 138; 21, p. 355.

J. CLAVERIE, G. CAMPET, D. CONTE, G. LE FLEM, and P. HAGENMULLER
*Influence of Covalent Bonding on the Photoelectrochemical Properties of Some Perovskite-Type Related Compounds
(a) Vol. **77**, No. 2, p. 603–609 (1983).
See also 13.4, p. 180; *16*, p. *266*; 22.6.1, p. 498.

K. HÜBNER, A. V. SHENDRIK, and A. M. PRAULINSH
*Decomposition of SiO_x Films Due to Internal Stress
(b) Vol. **118**, No. 1, p. 239–244 (1983).
See also *1.5*, p. *37*; 19, p. 302; 20.1, p. 321; 22.6, p. 494.

T. SOMA and H.-MATSUO KAGAYA
*The Metallic and Ionic Contributions to Lattive Vibrations of III–V Covalent Crystals
(b) Vol. **118**, No. 1, p. 245–254 (1983).
See also 6, p. 75; 22.2.1, p. 445; 22.2.3, p. 452.

A. V. MALAKHOVSKII and G. G. VASILEV
Covalency as Source of Intensity for d–d Transitions in Octahedral Complexes
(b) Vol. **118**, No. 1, p. 337–344 (1983).
See also *20.1*, p. *322*; 22.5, p. 474.

A. K. Ghosh, S. Chakraborty, and A. Manna
Density Functional Approach for Metals. The Matheiss Prescription
(b) Vol. **118**, No. 1, p. 373–379 (1983).
See also *13*, p. *150*; 21.1, p. 371; 21.1.1, p. 382; 21.6, p. 398.

K. C. Gupta, R. M. Agrawal, and R. P. S. Rathore
*Incomplete Gamma Functions and Lattice Sums for B.C.C. Structures
(b) Vol. **118**, No. 1, p. 399–404 (1983).
See also *6*, p. *75*; 22.5.2, p. 480.

B. R. K. Gupta and R. P. Goyal
*A Theoretical Analysis of Interionic Potentials in Rutile-Type Crystals
(b) Vol. **118**, No. 1, p. K7–K12 (1983).
See also *12.2*, p. *143*; 22.6, p. 494.

A. P. Deshpande, V. B. Sapre, and C. Mande
*A Study of the Chemical Shifts in X-Ray Absorption Spectra and the Sublattice Distortion in Some Ternary Chalcopyrite Compounds of the Type $CuMS_2$
(b) Vol. **118**, No. 1, p. K39–K42 (1983).
See also *20*, p. *309*; 22.8, p. 507.

N. Shohoji
*Statistical Thermodynamic Approach to Some Hyperstoichiometric Dihydrides of Lanthanides (La, Ce, Pr, Nd, Sm) and Actinides (Np, Pu, Am)
(b) Vol. **118**, No. 2, p. 811–822 (1983).
See also *8*, p. *89*; 10.1, p. 104; 21.4, p. 392; 21.5, p. 395.

Ch. Opitz, H. Müller, and H. Miessner
*EHT Calculations for Interstital Hydrogen Bonding in Nickel
(b) Vol. **118**, No. 2, p. K85–K87 (1983).
See also *10.1*, p. *104*; 13.4, p. 184; 21.1, p. 372.

V. Heera, G. Seifert, and P. Ziesche
Semirelativistic SCF-SW-Xα Calculations of Uranium Compounds. Electronic Structure and Chemical Bond
(b) Vol. **118**, No. 2, p. K107–K112 (1983).
Erratum
(b) Vol. **120**, No. 2, p. K227 (1983).
See also *13.1*, p. *161*; 21.5, p. 395.

N. Agrawal, R. P. Goyal, B. R. K. Gupta, and R. K. Singh
*Evaluation of Cohesive Energies for NaCl–NaBr Mixed Crystals
(b) Vol. **118**, No. 2, p. K141–K146 (1983).
See also 22.5.4, p. 488.

T. Soma, Y. Konno, and H.-M. Kagaya
*Compression Effect on the Bulk Modulus of Al.
(b) Vol. **118**, No. 2, p. K155–K157 (1983).
See also *12.1*, p. *139*; 21, p. 357.

W. Thulke, R. Haensel, and P. Rabe
*Anisotropic Absorption of Unpolarized X-Radiation in Copper-Intercalated Niobium Disulfide
(a) Vol. **78**, No. 2, p. 539–545 (1983).
See also *1.3*, p. *25*; 20, p. 310; 22.8, p. 508.

Shu Zhen and G. J. Davies
*Calculation of the Lennard-Jones *n–m* Potential Energy Parameters for Metals
(a) Vol. **78**, No. 2, p. 595–605 (1983).
See also 1.3, p. 25; 12.1, p. 139; 21, p. 358.

G. D. Jain and J. Shanker
*Use of Harrison's Potential in Mixed Alkali Halide Crystals
(b) Vol. **119**, No. 1, p. 79–82 (1983).
See also 12.2, p. 143; 22.5.4, p. 488.

Yu. S. Stark, A. S. Steinberg, V. M. Vasilev, and V. I. Dimitrov
*Monte Carlo Determination of Long-Range Pair Interaction Energies from Diffuse Scattering of X-Rays by Alloys
(b) Vol. **119**, No. 1, p. 147–151 (1983).
See also *1.3*, p. *25*; 21.1, p. 374; 21.6, p. 398.

S. P. Singh and R. P. S. Rathore
*Van der Waals' Forces in Face Centred Cubic Cobalt
(b) Vol. **119**, No. 1, p. 343–348 (1983).
See also 6, p. 77; 12.1, p. 139; 21.1, p. 374.

T. Soma and H.-Maksuo Kagaya
*Binding Force Including Ionic Interactions and Bulk Properties for Tetrahedral Compounds
(b) Vol. **119**, No. 1, p. K41–K45 (1983).
See also 12.2, p. 143; 22.2, p. 442; 22.4, p. 461.

N. Shohoji
Statistical Thermodynamics of Hydrogen Solution in Some Lanthanides
(b) Vol. **119**, No. 1, p. K87–K90 (1983).
See also *1.3*, p. *25*; 21.4, p. 392.

T. Soma and H.-Matsuo Kagaya
*Local Heine-Abarenkov Model Potential for III–V and II–VI Covalent Compounds
(b) Vol. **119**, No. 2, p. 547–553 (1983).
See also 13.1, p. 163; 22.2, p. 442; 22.4, p. 461.

P. Gluck
Polarization Forces between Macroscopic Bodies
(b) Vol. **119**, No. 2, p. K131–K133 (1983).
See also *13.3*, p. *170*.

U. Pietsch
*Hydrostatic Pressure and Uniaxal Stress Dependence of the Silicon (222) X-Ray Reflection Power
(b) Vol. **120**, No. 1, p. 183–188 (1983).
See also *13*, p. *152*; 22.1.2, p. 437.

MIN-SHENG LEI and ZHI-CHENG WANG
Space Distribution of the Chemical Composition of the Thermo-Grown SiO_2–SiO_x–Si System
(b) Vol. **120**, No. 1, p. 189 – 195 (1983).
See also *1.3*, p. *26*; 1.6, p. 42; 22.6, p. 497.

U. PIETSCH and K. UNGER
The Influence of Free Carriers on the Equilibrium Lattice Parameter of Semiconductor Materials
(a) Vol. **80**, No. 1, p. 165 – 172 (1983).
See also *1.1*, p. *12*; 22.1.1, p. 422; 22.1.2, p. 438; 22.2.1, p. 447; 22.2.2, p. 451; 22.2.4, p. 455.

I. SAVATINOVA and E. ANACHKOVA
Correlations between OH(OD) Stretching Vibrations and Structural Characteristics of Hydrogen Bonds in $K_4Fe(CN)_6 \cdot 3\,H_2O$ Crystal
(b) Vol. **120**, No. 2, p. 539 – 545 (1983).
See also *6*, p. *80*; *20.1*, p. *329*; 22.8, p. 510.

P. ENDERS
*Structure and Bonding in Cubic IV–VI Crystals (I): Evidence for Covalent Bonding from LCAO Parameters
(b) Vol. **120**, No. 2, p. 735 – 744 (1983).
See also *13*, p. *153*; 22, p. 414.

L. ČERVINKA, V. CHÁB, and S. KOC
*XPS Investigation of the Amorphous $TlSe_x$ System
(a) Vol. **80**, No. 2, p. 685 – 689 (1983).
See also *2*, p. *56*; 17, p. 271; 22.3, p. 458.

5. Mössbauer Spectroscopy

Review Article

V. S. Shpinel
Exchange Interaction and Hyperfine Fields at Nonmagnetic Impurities in Rare-Earth Metals
(b) Vol. **118**, No. 1, p. 11–24 (1983).
See also *18.2*, p. *280*; *18.4*, p. *295*; 21.4, p. 390.

Original Papers and Short Notes

M. Ya. Flyagin and A. N. Men
*Calculation of Magnetic Hyperfine Interactions in the Rare-Earth Orthoferrites with Model Covalency Parameter
(b) Vol. **115**, No. 1, p. 277–284 (1983).
See also *18.4*, p. *295*; 22.8.2, p. 519.

W. Bergholz, S. Damgaard, J. W. Petersen, and G. Weyer
*A Mössbauer-Spectroscopy Study of the Annealing of Supersaturated Solutions of ^{57}Co in Silicon
(a) Vol. **75**, No. 1, p. 289–300 (1983).
See also 22.1.2, p. 424.

A. V. Mitin and N. V. Polyakov
*γ-Magnetic Resonance in the Presence of Electronic Spin Relaxation
(b) Vol. **115**, No. 2, p. 477–484 (1983).
See also 19, p. 300.

K. H. J. Buschow, J. H. N. van Vucht, P. G. van Engen, D. B. de Mooij, and A. M. van der Kraan
Structure, Magnetic and Magneto-Optical Properties, and ^{57}Fe Mössbauer Effect in a New Variety of Heusler Alloy: PtFeSb
(a) Vol. **75**, No. 2, p. 617–623 (1983).
See also *1.3*, p. *22*; *18.2*, p. *281*; 20.1, p. 314; 21.1.1, p. 379.

R. Nagarajan, E. V. Sampathkumaran, R. Vijayaraghavan, and Bhaktdarshan
*^{151}Eu Mössbauer Effect Measurements in Eu_4As_3
(a) Vol. **75**, No. 2, p. K149–K151 (1983).
See also 21.4, p. 390.

G. A. Petrakovskii, K. A. Sablina, and V. P. Ikonnikov
*Amorphization of $RbFeS_2$
(a) Vol. **75**, No. 2, p. K165–K167 (1983).
See also *2*, p. *45*; 22.8, p. 504.

V. M. Pan, I. Ya. Dekhtyar, M. E. Osinovskii, S. E. Litvin, B. G. Nikitin, M. M. Nishenko, and M. P. Voronko
Thermodynamical and Structural Properties of Amorphous $Fe_{80}B_{20}$ and $Fe_{80}P_{13}C_7$ Alloys
(a) Vol. **76**, No. 1, p. 197–206 (1983).
See also *2*, p. *45*; 8, p. 87; 14.1, p. 206; *18.2.1*, p. *290*; 21.1.1, p. 380.

G. L. Whittle
*Magnetic Order in Isoelectronic Co(GaVFe) Alloys
(a) Vol. **76**, No. 1, p. 351–356 (1983).
See also *18*, p. *274*; 21.1, p. 366.

P. C. M. Gubbens, W. Ras, A. M. van der Kraan, and H. K. J. Buschow
*Magnetization and ^{161}Dy Mössbauer Effect Study of $DyMn_2$ and Dy_6Mn_{23} and Their Ternary Hydrides
(b) Vol. **117**, No. 1, p. 277–282 (1983).
See also *18.3*, p. *292*; 21.4, p. 391.

N. N. Delyagin, V. I. Nesterov, and S. I. Semenov
*The Oscillations of the Mössbauer Line Shift near T_c in the Ferromagnetic Ni–Rh Alloy
(b) Vol. **117**, No. 1, p. K27–K29 (1983).
See also *18.2*, p. *283*; 21.1, p. 368.

A. K. Zhetbaev and A. N. Ozernoi
*Mössbauer Study of Radiation Effect of the ^{181}Ta(p, n)^{181}W Reaction
(a) Vol. **77**, No. 1, p. 63–66 (1983).
See also 11, p. 127; 21, p. 354.

Ramani, M. P. Sathyavathiamma, N. G. Puttaswamy, and R. M. Mallya
*Mössbauer Effect Study of the Thermal Decomposition in Hydrogen of Homogeneously Mixed Ferrous Nickel Oxalates with Different Iron to Nickel Ratios
(a) Vol. **77**, No. 1, p. 87–96 (1983).
See also 22.8, p. 505.

U. Heinecke, C. Cruz, E. Wieser, and C. Bernhardt
*Structural Transformations and Connected Changes of Magnetic Properties by Mechanical Treatment of $SrFe_{12}O_{19}$
(a) Vol. **77**, No. 1, p. 225–232 (1983).
See also *1.2*, p. *16*; *18.3*, p. *292*; 22.8.2, p. 521.

S. Ligenza, M. Łukasiak, Z. Kucharski, and J. Suwalski
*Study of the $3d^5$ Shell Deformation of Fe^{3+} Ions in $ZnFe_2O_4$ by High-Field Mössbauer Spectroscopy
(b) Vol. **117**, No. 2, p. 465–470 (1983).
See also 18.3, p. 292; 22.8.2, p. 521.

A. Dragūnas, K. Makariūnas, and M. Balčiūniene
*Correlations of the ^{119}Sn Mössbauer Isomer Shift with Interatomic Distances for Complex Tetrahedral Semiconductors
(a) Vol. **77**, No. 2, p. 463–466 (1983).
See also 22.7.1, p. 501.

J. L. Dormann, A. Tomas, and M. Nogues
*Cation Ordering in $LiFe_5O_8$ Studied by Mössbauer Spectroscopy and X-Ray Crystallography
(a) Vol. **77**, No. 2, p. 611–618 (1983).
See also *1.3*, p. *24*; 22.8.2, p. 521.

A. A. Bahgat and M. K. Fayek
*Erratum to: Absolute Evaluation of Ferrous and Ferric Concentration in Ca Amphibole
(a) Vol. **77**, No. 2, p. 809 (1983).
See also 22.8, p. 506.

V. S. Patil, R. G. Pillay, P. N. Tandon, and H. G. Devare
*Hyperfine Field Investigations in Some Ru and Rh Based Heusler Alloys
(b) Vol. **118**, No. 1, p. 57–61 (1983).
See also 18.2, p. 284; 21, p. 355; 21.1.1, p. 382.

A. R. Arakelyan, R. G. Gabrielyan, A. R. Mkrtchyan, H. A. Aslanyan, and Kh. V. Kotandyan
*On a Certain Inverse Problem in γ-Acoustic Spectroscopy
(b) Vol. **118**, No. 1, p. K69–K74 (1983).
See also *7*, p. *83*; 22.9, p. 529.

G. Dehe and U. Behn
*Contribution of the Direct B–B Interaction to the Supertransferred Hyperfine Field at Diamagnetic Antimony in Oxidic Spinels Determined by ^{121}Sb Mössbauer Spectroscopy
(a) Vol. **78**, No. 2, p. 485–488 (1983).
See also *18.3*, p. *293*; 22.8.2, p. 522.

D. Fruchart, B. Malaman, G. Le Caer, and B. Roques
Magnetic Studies of Monoclinic FeGe
(a) Vol. **78**, No. 2, p. 555–569 (1983).
See also *18.4*, p. *296*; 21.1.1, p. 383.

J. Pebler
*Paramagnetic Hyperfine Structure of Iron (57) in $V_{1-x}Fe_xO_2$
(a) Vol. **78**, No. 2, p. 589–594 (1983).
See also *18.1*, p. *279*; 22.6.1, p. 498.

Y. Muraoka, M. Shiga, and Y. Nakamura
*Magnetic Properties and Thermal Expansion of $(Zr_{1-x}Ta_x)$ Fe_2
(a) Vol. **78**, No. 2, p. 717–722 (1983).
See also 8, p. 89; *18.2*, p. *285*; 18.4, p. 297; 21.1.1, p. 383.

O. Schneeweiss, T. Zemčík, T. Žák, and S. Mager
*Atomic Structure and Magnetic Properties of the Pseudobinary Alloys Fe_3(Al, Si)
(a) Vol. **79**, No. 1, p. 125–129 (1983).
See also *1.3*, p. *26*; 14.1, p. 209; 18.2, p. 286; 21.1.1, p. 384.

D. K. Chakrabarty, A. Bandyopadhyay, S. B. Patil, and S. N. Shringi
*Magnetic and Mössbauer Studies of Some Mixed Cobaltites of the Lanthanides and Alkali Metals
(a) Vol. **79**, No. 1, p. 213–222 (1983).
See also *18.2*, p. *286*; 22.8, p. 508.

G. Dehe and J. Suwalski
*^{121}Sb and ^{57}Fe Mössbauer Effect in Sb Substituted Lithium Ferrite in the Presence of a Magnetic Field
(b) Vol. **119**, No. 2, p. K155–K158 (1983).
See also *18.3*, p. *293*; 22.8.2, p. 523.

R. Olivella, E. Molins, and J. Tejada
*Mössbauer Second-Order Doppler Shift in MgO:^{57}Co
(b) Vol. **119**, No. 2, p. K165–K168 (1983).
See also 22.6, p. 496.

E. Fritzsch and C. Pietzsch
*Lamb Mössbauer Fraction and Mean Square Displacement of Iron in Pyrite (FeS_2)
(a) Vol. **79**, No. 2, p. K113–K115 (1983).
See also 6, p. 79; 22, p. 413.

A. Antonini, V. Fano, and I. Ortalli
*Mössbauer Study of the $Pb_xSn_{1-x}{}^{119}Te$ System
(b) Vol. **120**, No. 1, p. K15–K18 (1983).
See also 22, p. 413.

G. L. Whittle and A. M. Stewart
*Mössbauer Effect Study of $Fe_{85-x}Ni_xB_{15}$ Metallic Glasses
(a) Vol. **80**, No. 1, p. 101–107 (1983).
See also 2, p. 54; 18.2, p. 288; 21.1, p. 377; 21.1.1, p. 385.

E. A. Vasilev and V. A. Virchenko
*Hyperfine Interactions of ^{57}Fe and ^{119}Sn in Iron Antimonide Base Alloys
(a) Vol. **80**, No. 1, p. K41–K44 (1983).
See also 21.1.1, p. 385.

B. Głowacki and J. Chojcan
*Studies on the Condition of Obtaining a Single Phase. Superconducting Nb_3Sn Layer by Mössbauer Spectroscopy
(a) Vol. **80**, No. 1, p. K93–K96 (1983).
See also 9, p. 96; 14.2, p. 214; 21, p. 361.

R. G. Gabrielyan, L. A. Kocharyan, A. R. Mkrtchyan, and H. A. Unanyan
*Time Dependence of Mössbauer γ-Radiation Modulated by Acoustic Excitations
(b) Vol. **120**, No. 2, p. K123–K128 (1983).

J. Kansy, T. J. Panek, and M. Szuszkiewicz
*Influence of Non-Homogeneity of Sample Thickness on the Determination of Absolute Mössbauer Fraction
(a) Vol. **80**, No. 2, p. 643–655 (1983).
See also 22.8, p. 511.

6. Lattice Dynamics. Phonons

Review Articles

V. Kh. Brikenshtein, V. A. Benderskii, and P. G. Filippov
Fluorescence of Molecular Crystals at High Levels of Optical Pumping
(b) Vol. **117**, No. 1, p. 9–39 (1983).
See also *13.5.1*, p. *194*; 20.2, p. 332; *20.3*, p. *334*; 22.9, p. 525.

K. N. Shrivastava
*Theory of Spin-Lattice Relaxation
(b) Vol. **117**, No. 2, p. 437–458 (1983).
See also *13*, p. *146*; 20.1, p. 312; 22.5, p. 473; 22.8.1, p. 512.

M. Cardona
*Vibrational Spectra of Hydrogen in Silicon and Germanium
(b) Vol. **118**, No. 2, p. 463–481 (1983).
See also *20.1*, p. *312*; 22.1.1, p. 418; 22.1.2, p. 423.

Original Papers and Short Notes

A. S. Davydov and A. V. Zolotariuk
*Autolocalized Collective Excitations in Molecular Chains with Cubic Anharmonicity
(b) Vol. **115**, No. 1, p. 115–125 (1983).
See also *13.5.1*, p. *194*.

O. Ziep
*Estimation of Second-Order Auger Recombination in Lead Chalcogenides
(b) Vol. **115**, No. 1, p. 161–170 (1983).
See also *13.1*, p. *155*; 13.5.2, p. 200; 22.8, p. 502.

E. Gurauskas, J. Kavaliauskas, G. Krivaitė, and A. Šileika
*Infrared Absorption Spectra of $Cd_{0.2}Hg_{0.8}Te$ Crystals
(b) Vol. **115**, No. 1, p. 195–202 (1983).
See also *20.1*, p. *312*; 22.4.4, p. 470.

V. Schulz and L. Merten
*Dispersion Curves of Stimulated Raman Effect in Piezoelectric Cubic Crystals
(b) Vol. **115**, No. 1, p. 225–233 (1983).
See also 13.5, p. 192; *20*, p. *307*; 22.2.2, p. 448.

T. Soma, H.-Matsuo Kagaya, and M. Nishigaki
*Mean-Square Displacement and Lindeman's Criterion for Melting of Alkali Metals
(b) Vol. **115**, No. 1, p. 273–276 (1983).
See also 8, p. 85; 21.2, p. 387.

Y. Y. Abolinsh, S. V. Karpov, and A. A. Shultin
*Raman Studies of Ammonium Ion Motion in NH_4NO_3–V Crystals
(b) Vol. **115**, No. 1, p. K1 – K5 (1983).
See also *1.2*, p. *14*; 20.1, p. 313; 22.8.1, p. 512.

V. Kalesinskas, J. Grigas, R. Jankevičius, and A. Audzionis
Soft Mode in the Microwave Dielectric Spectra of the SbSI–BiSI System
(b) Vol. **115**, No. 1, p. K11 – K13 (1983).
See also *14.4.1*, p. *256*; 22.8, p. 502.

H. Sobotta, H. Neumann, N. N. Syrbu, and V. Riede
*Infrared Lattice Vibration Spectra of Monoclinic ZnP_2
(b) Vol. **115**, No. 1, p. K55 – K58 (1983).
See also *20.1*, p. *313*; 22.3, p. 456.

H.-J. Fitting and J. Boyde
*Monte-Carlo Calculation of Electron Attenuation in SiO_2
(a) Vol. **75**, No. 1, p. 137 – 142 (1983).
See also *14.3.4*, p. *243*; 22.6, p. 490.

G. D. Khattak
*Specific Heat of Uranium Dioxide (UO_2) between 0.3 and 50 K
(a) Vol. **75**, No. 1, p. 317 – 321 (1983).
See also *8*, p. *85*; 18.4, p. 295; 22.6, p. 490.

C. Malinowska-Adamska, M. Owsiejczyk, and L. Wojtczak
*Thermodynamics of Strongly Anharmonic Crystals
(b) Vol. **115**, No. 2, p. 335 – 345 (1983).
See also 8, p. 85; 21, p. 350; 21.2, p. 387.

H. Köster, Jr., I. N. Iassievich, and K. Hübner
*Thermally Stimulated Excitation of Electrons from Deep Traps of Insulators in Electric Fields
(b) Vol. **115**, No. 2, p. 409 – 414 (1983).
See also *13.4*, p. *174*; 14.3.4, p. 243; 22.6, p. 491.

H. Rauh, R. Geick, N. Lehner, J. Bouillot, and G. Eckold
*Phonon Dispersion in the Perovskite-Type Layer Material K_2ZnF_4
(b) Vol. **115**, No. 2, p. 463 – 470 (1983).
See also 22.8, p. 503.

A. N. Nikitina, I. S. Osadko, I. A. Vasilieva, and L. A. Yanovskaya
*Vibronic and Electron–Phonon Coupling in Spectra of Substituted Polyenes in Rigid Matrices
(b) Vol. **115**, No. 2, p. 493 – 498 (1983).
See also *13.4*, p. *174*; 20.1, p. 314; 20.3, p. 335; 22.9, p. 525.

R. K. Singh and N. K. Pandey
Study of Thermophysical and Anharmonic Properties of Fluorite Compounds
(b) Vol. **115**, No. 2, p. 555 – 562 (1983).
See also *4*, p. *60*; 12.2, p. 141; 22.5.3, p. 486.

M. Bancewicz and A. Stasch
*Theoretical Investigation of the Resonant Magnon–Phonon Interaction in FeF_2 in High Magnetic Field
(b) Vol. **115**, No. 2, p. 563–574 (1983).
See also *13.5.4*, p. *203*; 18, p. 273; 22, p. 404.

D. Sen, S. K. Sarkar, S. Sengupta, and D. Roy
*Consistency Condition in the Energy Expression of the Pseudopotential Theory of Metals and Unified Study of Alkali Metals
(b) Vol. **115**, No. 2, p. 593–602 (1983).
See also *12.1*, p. *137*; 21.2, p. 387.

V. Ya. Antonchenko, A. S. Davydov, and A. V. Zolotariuk
*Solitons and Proton Motion in Ice-Like Structures
(b) Vol. **115**, No. 2, p. 631–640 (1983).
See also *4*, p. *60*; 13.5.4, p. 203; 22.6, p. 491.

M. A. Aldzhanov, A. M. Aliev, R. K. Veliev, K. K. Mamedov, M. I. Mekhtiev, and V. Ya. Steinshreiber
*Phonon Density of States and Heat Capacity of $CdIn_2S_4$
(b) Vol. **115**, No. 2, p. K75–K78 (1983).
See also *8*, p. *86*; 22.8, p. 503.

V. K. Arora
*Nonohmic Conductivity for Acoustic-Phonon Scattering
(b) Vol. **115**, No. 2, p. K107–K112 (1983).
See also *14.3*, p. *216*; 22, p. 404.

A. M. Korostil
A New Approach to the Theory of Tunnelling Relaxation
(b) Vol. **115**, No. 2, p. K117–K119 (1983).
See also *2*, p. *44*; *13*, p. *147*.

M. Ishigame and S. Yamashita
*Raman Scattering in K_2SO_4 at High Temperatures
(b) Vol. **116**, No. 1, p. 49–56 (1983).
See also 1.2, p. 14; *20.1*, p. *315*; 22.8.1, p. 513.

T. Soma, H.-M. Kagaya, and Y. Kimura
*Thermal Expansion Coefficient of Alkali Metals
(b) Vol. **116**, No. 1, p. 57–61 (1983).
See also 8, p. 86; 21.2, p. 387.

J. M. Wesselinowa
*Spin–Phonon Interactions in the Ising Model with a Transverse Field
(b) Vol. **116**, No. 1, p. 71–82 (1983).
See also *14.4.1*, p. *256*; 22.8.1, p. 513.

M. A. Ansari, Anil Kumar, and B. K. Srivastava
Thermal Conductivity of Germanium Containing Dislocations at Low Temperatures
(b) Vol. **116**, No. 1, p. 163–167 (1983).
See also *8*, p. *86*; 22.1.1, p. 419.

I. T. Godmanis, A. N. Trukhin, and K. Hübner
*Exicton–Phonon Interaction in Crystalline and Vitreous SiO_2
(b) Vol. **116**, No. 1, p. 279–287 (1983).
See also *13.5.1*, p. *195*; 20.1, p. 315; 22.1.1, p. 419; 22.6, p. 492.

R. K. Singh, D. K. Neb, and S. P. Sanyal
Effect of Three-Body Forces on Thermophysical and Anharmonic Properties of Rare Gas Solid Mixtures
(b) Vol. **116**, No. 1, p. 289–297 (1983).
See also *4*, p. *60*; 8, p. 86; 12.2, p. 142; 23, p. 533.

G. Dharmadurai
*Phonon Transmission Theory of Heat Transfer through Solid–Gas Interfaces
(b) Vol. **116**, No. 1, p. 307–311 (1983).
See also 1.6, p. 39; *8*, p. *86*.

D. Bose, A. Ghosh, and A. N. Basu
*Perturbation-Theoretical Model Calculation of the Lattice Mechanics of Low Polarisability Ionic Crystals
(b) Vol. **116**, No. 1, p. 313–322 (1983).
See also 12.2, p. 142; 22.5.2, p. 478.

T. Nanba amd T. P. Martin
*Raman Scattering from Metal Smokes
(a) Vol. **76**, No. 1, p. 235–239 (1983).
See also 1.6, p. 39; *20.1*, p. *316*; 21.1, p. 366; 21.6, p. 397.

P. Čapková, R. Kužel, and J. Šedivý
*Thermal Vibrations and Static Displacements of Atoms in Nonstoichiometric Titanium Carbide
(a) Vol. **76**, No. 1, p. 383–390 (1983).
See also 8, p. 87; 21.1, p. 366.

N. M. Gasanly, A. F. Goncharov, N. N. Melnik, A. S. Ragimov, and V. I. Tagirov
*Optical Phonons and Structure of $TlGaS_2$, $TlGaSe_2$, and $TlInS_2$ Layer Single Crystals
(b) Vol. **116**, No. 2, p. 427–443 (1983).
See also *20.1*, p. *316*; 22.8, p. 504.

F. Comas and C. Trallero Giner
Low-Frequency Secondary Radiation in Polar Semiconductors at Finite Temperatures
(b) Vol. **116**, No. 2, p. 445–453 (1983).
See also *20*, p. *308*; 22, p. 405.

K. Handrich
*Phonon Lifetime and Thermal Conductivity in Amorphous Solids
(b) Vol. **116**, No. 2, p. 533–538 (1983).
See also 2, p. 46; *8*, p. *87*; 22.9, p. 527.

D. Ihle and B. Lorenz
*Theory of the Verwey Transition in the Mixed Valence Compound Eu_3S_4
(b) Vol. **116**, No. 2, p. 539–546 (1983).
Se also *13.5.3*, p. *201*; 18.2, p. 282; 22, p. 405.

V. VORLÍČEK, I. GREGORA, and D. CHVOSTOVÁ
*Raman Scattering in GeS Single Crystals at Low Temperatures
(b) Vol. **116**, No. 2, p. 639–643 (1983).
See also *20.1*, p. *316*; 22, p. 405.

S. PAČESOVÁ, B. BŘEZINA, and L. JASTRABÍK
*Optical Properties of K_2SeO_4 in the Vicinity of Phase Transitions
(b) Vol. **116**, No. 2, p. 645–652 (1983).
See also 14.4.1, p. 257; *20.1*, p. *316*; 22.8.1, p. 513.

T. SOMA, H.-MATSUO KAGAYA, and M. NISHIGAKI
*Volume Dependence of the Melting Temperature for Alkali Metals with Debye's Model
(b) Vol. **116**, No. 2, p. 673–676 (1983).
See also 8, p. 87; 21.2, p. 387.

M. YA. VALAKH, A. P. LITVINCHUK, and G. G. TARASOV
*Effect of the Pumping Intensity on the Resonant Secondary Emission Spectra of Optically Aligned Hot Excitons in ZnTe Crystals
(b) Vol. **116**, No. 2, p. K61–K65 (1983).
Erratum
(b) Vol. **118**, No. 2, p. K167 (1983).
See also 13.5.1, p. 196; *20.3*, p. *337*; 22.4.3, p. 468.

K. TAKARABE and K. WAKAMURA
*Polarized Raman Spectra of InS Single Crystal
(b) Vol. **116**, No. 2, p. K79–K83 (1983).
See also *20.1*, p. *317*; 22.3, p. 456.

N. GOPI KRISHNA and D. B. SIRDESHMUKH
*X-Ray Determination of the Mean-Square Atomic Displacement and Associated Debye Temperature of Erbium
(b) Vol. **116**, No. 2, p. K105–K108 (1983).
See also *8*, p. *87*; 21.4, p. 390.

K. KAMBAS
*Preparation and Optical Investigation of the $CdInGaS_4$ Compound
(a) Vol. **76**, No. 2, p. 735–741 (1983).
See also 1.3, p. 23; *20.1*, p. *317*; 22.8, p. 505.

L. WENDLER
*Effects of a Transition Layer on Slab Polaritons
(b) Vol. **117**, No. 1, p. 241–246 (1983).
See also 13.5, p. 192; *20*, p. *308*.

A. M. BEREZHKOVSKII and A. A. OVCHINNIKOV
*Temperature Dependence of the Wannier-Stark Level Width in a Semiconductor in a Strong Electric Field
(b) Vol. **117**, No. 1, p. 289–299 (1983).
See also *13*, p. *148*; 14.3.4, p. 244; 22.2.3, p. 452.

V. N. KASHCHEEV
*Phonon Relaxation Due to Anharmonism and Central Peak in Structural Phase Transitions (I): Soft Optic Modes and Acoustic Phonons in Cubic Crystals
(b) Vol. **117**, No. 2, p. 483–492 (1983).
See also 14.4.1, p. 257.

H. ZIMMERMANN and R. BOYN
On the Configuration Coordinate Diagram of ZnS:Fe
(b) Vol. **117**, No. 2, p. 611–618 (1983).
See also *13.4*, p. *179*; 20.1, p. 320; 20.3, p. 339; 22.4.1, p. 463.

S. MRÓZ, A. MRÓZ, A. GRUDNIEWSKI, and E. BĄK
*Effective Debye Temperature of the (110) Ni Face for Different Azimuths
(b) Vol. **117**, No. 2, p. 683–688 (1983).
See also 21.1, p. 370.

A. LEHMANN, L. SCHUMANN, and K. HÜBNER
*Optical Phonons in Amorphous Silicon Oxides (I): Calculation of the Density of States and Interpretation of LO–TO Splittings of Amorphous SiO_2
(b) Vol. **117**, No. 2, p. 689–698 (1983).
See also 2, p. 48; *20.1*, p. *320*; 22.6, p. 494.

P. KLEINERT
*Cluster–Bethe–Lattice Treatment of the Silicon Pair Defect in Gallium Arsenide
(b) Vol. **117**, No. 2, p. K91–K94 (1983).
See also *20.1*, p. *320*; 22.2.1, p. 444.

D. W. FIELD
*A Reexamination of Diffraction Data on Zinc to Determine Anharmonic Components in the One-Particle-Potential Model
(a) Vol. **77**, No. 2, p. 733–740 (1983).
See also 21.1, p. 370.

P. K. ACHARYYA, A. K. TEWARI, S. K. ROY, and D. L. BHATTACHARYA
*Anharmonic Study on the Debye-Waller Factor for a Vacancy-Trapped Mössbauer Impurity in Solid Solution of Aluminium
(b) Vol. **118**, No. 1, p. 41–50 (1983).
See also 10.1, p. 103; 21, p. 355.

M. D. KAPLAN
*The Dynamical Electron–Phonon Mode Coupling in Jahn-Teller Crystals with Antiferroelastic Ordering
(b) Vol. **118**, No. 1, p. 81–87 (1983).
See also 7, p. 83; 14.4.1, p. 257.

J. C. UPADHYAYA and M. L. SHARMA
*Model Potential Calculation of Phonon Dispersion in Tb and Ho
(b) Vol. **118**, No. 1, p. 121–128 (1983).
See also 21.4, p. 392.

N. M. Plakida, A. V. Belushkin, I. Natkaniec, and T. Wasiutyński
*The Internal Soft-Mode Phase Transition in Solid Biphenyl
(b) Vol. **118**, No. 1, p. 129–133 (1983).
See also 22.9, p. 528.

K. Regiński
*Some Extensions of the Shell Model of Ionic Crystals
(b) Vol. **118**, No. 1, p. 147–157 (1983).
See also 22, p. 409.

G. Jeandel, A. Warin et G. Morlot
*Analyse du spectre d'absorption infrarouge de $BaF_2:Y^{3+}$
(Analysis of the Infrared Absorption Spectrum of $BaF_2:Y^{3+}$)
(b) Vol. **118**, No. 1, p. 167–172 (1983).
See also *20.1*, p. *321*; 22.5.3, p. 486.

E. L. Albuquerque
*Surface-Ripple Brillouin Scattering from a Glass–Metallic Liquid Interface
(b) Vol. **118**, No. 1, p. 223–227 (1983).
See also 7, p. 83.

T. Soma and H.-Matsuo Kagaya
The Metallic and Ionic Contributions to Lattice Vibrations of III–V Covalent Crystals
(b) Vol. **118**, No. 1, p. 245–254 (1983).
See also *4*, p. *61*; 22.2.1, p. 445; 22.2.3, p. 452.

P. Kleinert
*Vibrational Behaviour of Mixed Diatomic Crystals $AB_{1-c}C_c$ within the Bethe Lattice
(b) Vol. **118**, No. 1, p. 283–291 (1983).
See also *20.1*, p. *322*; 22.2.4, p. 455; 22.4.4, p. 471.

H. Köster, Jr. and K. Hübner
*Statistics and Transport Behaviour of Electrons in SiO_2
(b) Vol. **118**, No. 1, p. 293–301 (1983).
See also *13*, p. *150*; 14.4, p. 252; 22.6, p. 494.

A. P. Fedotov and V. F. Shabanov
*Deuteron Dynamics in Crystals of Sodium Trideuterium Selenite
(b) Vol. **118**, No. 1, p. 345–351 (1983).
See also 20.1, p. 322; 22.8.1, p. 515.

K. C. Gupta, R. M. Agrawal, and R. P. S. Rathore
*Incomplete Gamma Functions and Lattice Sums for B.C.C. Structures
(b) Vol. **118**, No. 1, p. 399–404 (1983).
See also *4*, p. *62*; 22.5.2, p. 480.

O. A. C. Nunes
On the Multiphoton Laser Absorption in Free-Carrier Semiconductors
(b) Vol. **118**, No. 1, p. K25–K29 (1983).
See also *13*, p. *150*; *20.1*, p. *322*; 22, p. 409.

M. El Sherif, F. Brehat, and B. Wyncke
*A Simple Best Fit Method for the Far Infrared Reflectivity Spectra by Using the Four-Parameter Semi-Quantum Model: Application on the Sodium Nitrite Spectrum
(b) Vol. **118**, No. 1, p. K47 – K50 (1983).
See also *20.1*, p. *322*; 22.8.1, p. 515.

H. Neumann, R. D. Tomlinson, W. Kissinger, and N. Avgerinos
*Polarization-Dependent Infrared Reflectivity Spectra of $CuInSe_2$
(b) Vol. **118**, No. 1, p. K51 – K56 (1983).
See also *20.1*, p. *322*; 22.8, p. 507.

A. Kumar, A. K. Sinha, and M. A. Ansari
Thermal Conductivity of LiF Containing Dislocations at Low Temperatures
(b) Vol. **118**, No. 2, p. 483 – 491 (1983).
See also *8*, p. *88*; 22.5.2, p. 481.

T. D. Sokolovskii
*Phonon Spectrum of Boron Nitride with Wurtzite Structure
(b) Vol. **118**, No. 2, p. 493 – 498 (1983).
See also 8, p. 88; 22, p. 409.

C. Malinowska-Adamska
*Dynamics and Thermodynamics of Anharmonic Crystals near the Instability Point
(b) Vol. **118**, No. 2, p. 535 – 539 (1983).
See also 8, p. 89; 12.1, p. 139; 21.2, p. 387.

E. P. Pokatilov and S. I. Beril
*Electron–Phonon Interaction in Periodic Two-Layer Structures
(b) Vol. **118**, No. 2, p. 567 – 573 (1983).
See also 13.3, p. 169; 22.2, p. 442.

K. Regiński
*Phonon Polaritons in the Presence of a Magnetic Field
(b) Vol. **118**, No. 2, p. 587 – 593 (1983).
See also *13.5*, p. *193*; 18, p. 275; 20, p. 309; 22.5.2, p. 481.

H. J. Fischbeck
*On the Translation Mode of the Kink Lattice of a Quasi-One-Dimensional Peierls-Fröhlich System with a Nearly Half-Filled Band
(b) Vol. **118**, No. 2, p. 595 – 603 (1983).
See also *1.2*, p. *18*; 7, p. 83; 13.1, p. 160.

C. Razzetti, P. P. Lottici, L. Zanotti, and M. Curti
*A Raman Study of Compositional Disorder in Defect Chalcopyrites
(b) Vol. **118**, No. 2, p. 743 – 749 (1983).
See also *20.1*, p. *324*; 22.8, p. 508.

J. Zieliński and J. Kowalska
*Role of Phonon-Induced Hybridization in the Spinless Anderson Model
(b) Vol. **118**, No. 2, p. 805 – 810 (1983).
See also *13*, p. *150*.

M. Krupski
Rigid Sphere Model Applied to the Phase Transitions of Complex Compounds of the Type K_2PtCl_6
(a) Vol. **78**, No. 2, p. 751–758 (1983).
See also *1.2*, p. *18*; 22.8, p. 508.

K. S. Dubey
Effect of Empty Spaces on the Lattice Thermal Conductivity of Non-Crystalline Material at Low Temperatures. Application to Borosilicate Glass
(a) Vol. **78**, No. 2, p. K195–K200 (1983).
See also *8*, p. *89*; 22.8.1, p. 516.

M. A. Ansari, A. Kumar, and B. K. Srivastava
Lattice Thermal Conductivity of Crystals with Dislocations
(b) Vol. **119**, No. 1, p. 113–119 (1983).
See also *8*, p. *89*; *10*, p. *98*.

B. H. Bairamov, A. Heinrich, G. Irmer, V. V. Toporov, and E. Ziegler
*Raman Study of the Phonon Halfwidths and the Phonon–Plasmon Coupling in ZnO
(b) Vol. **119**, No. 1, p. 227–234 (1983).
See also 13.5.3, p. 202; *20.1*, p. *325*; 22.4, p. 460.

S. P. Singh and R. P. S. Rathore
Van der Waals' Forces in Face Centred Cubic Cobalt
(b) Vol. **119**, No. 1, p. 343–348 (1983).
See also *4*, p. *63*; 12.1, p. 139; 21.1, p. 374.

Wan-Sun Tse and Chia-Nan Chang
*Lattice Vibrations of Solid Hydrofluoric Acid
(b) Vol. **119**, No. 1, p. 349–354 (1983).
See also 22, p. 411.

T. M. Gasymov, A. A. Katanov, and M. M. Babaev
Thermopower and Thermomagnetic Effects in Kane Semiconductors under Conditions of Electron and Phonon Heating and Their Mutual Drag
(b) Vol. **119**, No. 1, p. 391–399 (1983).
See also *14.3.4*, p. *247*; *15*, p. *262*; 22, p. 411.

I. J. Lalov, K. T. Stoychev, and S. P. Penchev
Non-Linear Optical Susceptibilities of Molecular Crystals in the Region of Vibronic Excitations (I): The Three-Dimensional Case
(b) Vol. **119**, No. 1, p. 419–427 (1983).
See also *13.5.1*, p. *198*; 20.1, p. 326; 22, p. 411.

P. Kleinert
*Cluster Bethe Lattice Treatment for the Localized Vibrational Mode of Carbon in GaAs
(b) Vol. **119**, No. 1, p. K37–K40 (1983).
See also 22.2.1, p. 446.

V. A. Korshunov, M. A. Aldzhanov, K. K. Mamedov, and T. G. Kerimov
*Phonon Densities of States of GaS and GaSe Layer Crystals
(b) Vol. **119**, No. 1, p. K47 – K50 (1983).
See also 22.3, p. 458.

V. V. Mazurenko, V. V. Petrov, and V. D. Tkachev
*Investigation of the Single-Phonon Absorption Band at 230 cm^{-1} in Silicon Irradiated by Neutrons
(b) Vol. **119**, No. 1, p. K69 – K71 (1983).
See also *11*, p. *131*; 20.1, p. 326; 22.1.2, p. 434.

D. Chattopadhyay
Transient Behaviour of Hot Electrons in Quantum-Well Structures of Polar Semiconductors
(b) Vol. **119**, No. 1, p. K77 – K81 (1983).
See also *13*, p. *151*; *14.3*, p. *223*; 22, p. 411.

M. Hidaka, Y. Okamoto, and Y. Zikumaru
Structural Phase Transition of $CsPbCl_3$ below Room Temperature
(a) Vol. **79**, No. 1, p. 263 – 269 (1983).
See also *1.2*, p. *18*; 22.8, p. 509.

V. M. Fomin and E. P. Pokatilov
Nonlinear Kinetics and Optical Properties of Band Charge Carriers
(b) Vol. **119**, No. 2, p. 483 – 492 (1983).
See also 13, p. 151; 14.3.4, p. 247; *20.1*, p. *327*; 22, p. 412.

V. G. Plekhanov, V. V. Shepelev, and A. U. Grinfelds
Spectroscopic Manifestation of Self-Trapping Excitons in Ionic Insulators
(b) Vol. **119**, No. 2, p. 493 – 502 (1983).
See also *13.5.1*, p. *198*; 20.3, p. 345; 22.5.2, p. 483.

V. N. Kashcheev
*Phonon Relaxation Due to Anharmonism and Central Peak in Structural Phase Transitions (II): Soft Optic Modes in Uniaxial Ferroelectrics and Soft Acoustic Modes in Cubic Crystals
(b) Vol. **119**, No. 2, p. 507 – 514 (1983).
See also 14.4.1, p. 259.

A. A. Kiselev
*Circular Phonons and Optical Magnetization of Crystals
(b) Vol. **119**, No. 2, p. 525 – 530 (1983).
See also *18*, p. *274*; 20, p. 310; 22.5.2, p. 484.

G. Mützenich, R. Faymonville, and P. Grosse
*One-Phonon and Multiphonon Absorption in EuSe and $Eu_{1-x}Sr_xS$ $(0 < x < 1)$
(b) Vol. **119**, No. 2, p. 531 – 539 (1983).
See also *20.1*, p. *327*; 22, p. 412.

N. S. ORLOVA
*Variation of Phonon Dispersion Curves with Temperature in Indium Arsenide Measured by X-Ray Thermal Diffuse Scattering
(b) Vol. **119**, No. 2, p. 541–546 (1983).
See also 20, p. 310; 22.2.1, p. 446.

G. IRMER, V. V. TOPOROV, B. H. BAIRAMOV, and J. MONECKE
Determination of the Charge Carrier Concentration and Mobility in n-GaP by Raman Spectroscopy
(b) Vol. **119**, No. 2, p. 595–603 (1983).
See also 13.5.2, p. 200; *14.3*, p. *224*; *20.1*, p. *327*; 22.2.2, p. 450.

K. H. GOETZ and W. RICHTER
*Interaction of Free Hole Intraband Excitations with the A_1-Phonon in Highly Doped (p-Type) Tellurium
(b) Vol. **119**, No. 2, p. 631–638 (1983).
See also *13.1*, p. *163*; 20.1, p, 327; 22.1.3, p. 440.

M. C. OLIVEROS and D. R. TILLEY
*Quantization Procedure for Phonons in an Isotropic Elastic Half-Space
(b) Vol. **119**, No. 2, p. 675–683 (1983).

H. TICHÁ, L. TICHÝ, and E. ČERNOŠKOVÁ
Temperature Dependence of the Optical Gap in Glassy GeS_2
(b) Vol. **119**, No. 2, p. K135–K137 (1983).
See also *2*, p. *52*; *20.1*, p. *327*; 22, p. 412.

E. FRITZSCH and C. PIETZSCH
Lamb Mössbauer Fraction and Mean Square Displacemnet of Iron in Pyrite (FeS_2)
(a) Vol. **79**, No. 2, p. K113–K115 (1983).
See also *5*, p. *68*; 22, p. 413.

J. KOWALSKA and J. ZIELIŃSKI
*Spinless Anderson Model with a Local Phonon Mode
(b) Vol. **120**, No. 1, p. 99–104 (1983).
See also *13.4*, p. *188*.

V. S. GORELIK, B. S. UMAROV, and M. UMAROV
*On the Connection between Isofrequency Temperature Dependences of Inelastic Light Scattering and Dielectric Anomalies in Lithium Niobate Crystals
(b) Vol. **120**, No. 1, p. 131–136 (1983).
See also 14.4.1, p. 259; *20.1*, p. *328*; 22.8.1, p. 517.

N. M. GASANLY, A. F. GONCHAROV, N. N. MELNIK, and A. S. RAGIMOV
*Optical Phonons in $GaS_{1-x}Se_x$ Layer Mixed Crystals
(b) Vol. **120**, No. 1, p. 137–147 (1983).
See also *20.1*, p. *328*; 22.3.1, p. 459.

M. T. DOVE, I. U. HEILMANN, J. K. KJEMS, J. KURITTU, and G. S. PAWLEY
*A Neutron Scattering Study of Phonons in Per-Deuterated S-Triazine
(b) Vol. **120**, No. 1, p. 173–181 (1983).
See also *1.2*, p. *19*; 12.2, p. 144; 22.9, p. 531.

B. Pödör
*Zone Edge Phonons in Gallium Phosphide
(b) Vol. **120**, No. 1, p. 207–213 (1983).
See also 22.2.2, p. 451.

L. Wendler and E. Jäger
*Phonon-Polaritons in Bilayer Systems. Surface Phonon-Polaritons
(b) Vol. **120**, No. 1, p. 235–247 (1983).
See also 13.5, p. 193; *20.1*, p. *328*; 22.2.1, p. 446; 22.2.2, p. 451; 22.5.3, p. 487.

M. N. Sapozhnikov and V. I. Alekseev
*Study of Optical Relaxation and Inhomogeneous Distribution Function of Impurity Organic Molecules by Site-Selective Laser Spectrosocpy
(b) Vol. **120**, No. 1, p. 435–444 (1983).
See also 13.4, p. 189; 20.1, p. 329; *20.3*, p. *346*; 22.9, p. 531.

V. Riede, H. Neumann, F. Lévy, and H. Sobotta
*Two-Phonon Absorption Spectra and Lattice Vibration Anisotropy in HfS_2
(b) Vol. **120**, No. 1, p. K9–K14 (1983).
See also *20.1*, p. *329*; 22, p. 413.

A. A. Abdumalikov, F. Kh. Abdullaev, and S. Stamenković
Influence of the Thermostat on Kink Dynamics in the Φ^4 Model
(b) Vol. **120**, No. 1, p. K33–K37 (1983).
See also *9*, p. *95*.

L. Merten and R. Kotzott
*Excitation of Longitudinal Optical Phonons by Stimulated Raman Effect in Piezoelectric Cubic Crystals
(b) Vol. **120**, No. 2, p. 481–490 (1983).
See also 13.5, p. 193; *20*, p. *311*; 22, p. 414.

A. N. Botvich, V. G. Podoprigora, V. F. Shabanov, N. P. Shestakov, and A. N. Vtyurin
*Optical Properties of Molecular Crystals. Phenomenological Approach
(b) Vol. **120**, No. 2, p. 491–501 (1983).
See also *20.1*, p. *329*; 22.9, p. 532.

P. B. Allen
*A Tetrahedron Method for Doubly Constrained Brillouin Zone Integrals. Application to Silicon Optic Phonon Decay
(b) Vol. **120**, No. 2, p. 529–538 (1983).
See also *13.1*, p. *165*; 22.1.2, p. 439.

I. Savatinova and E. Anachkova
*Correlations between OH(OD) Stretching Vibrations and Structural Characteristics of Hydrogen Bonds in $K_4Fe(CN)_6 \cdot 3\,H_2O$ Crystal
(b) Vol. **120**, No. 2, p. 539–545 (1983).
See also 4, p. 64; *20.1*, p. *329*; 22.8, p. 510.

V. A. Kochelap and V. N. Sokolov
*Deformational Phase Transitions and Periodic Structures in a Magnetic Field
(b) Vol. **120**, No. 2, p. 565–575 (1983).
See also *13*, p. *153*; 22, p. 414.

C. Malinowska-Adamska
*Self-Consistent Calculations of the Lattice Dynamics and Thermodynamics of Anharmonic Noble Metals
(b) Vol. **120**, No. 2, p. 601–610 (1983).
See also 8, p. 92; 12.1, p. 140; 21.6, p. 399.

D. Suisky
Multiphonon Structure of Optical Transitions to Impurity Resonance States. The Influence of Anharmonicity
(b) Vol. **120**, No. 2, p. 645–654 (1983).
See also *13.4*, p. *190*; *20.1*, p. *330*; 22.4, p. 461.

I. J. Lalov, K. T. Stoychev, and S. P. Penchev
Non-Linear Optical Susceptibilities of Molecular Crystals in the Region of Vibronic Excitations (II): Totally Symmetric Vibrations
(b) Vol. **120**, No. 2, p. 671–678 (1983).
See also *13.5.1*, p. *199*; 20.1, p. 330; 22, p. 414.

A. Kumar and P. Padhi
*Phonon Conductivity of Ice Single Crystals
(b) Vol. **120**, No. 2, p. 679–683 (1983).
See also *14*, p. *205*; 22.6, p. 497.

S. Ichinose and I. Tamura
*Acoustic Phonon Anomaly in Mixed-Valent TmSe
(b) Vol. **120**, No. 2, p. 703–710 (1983).
See also 22, p. 414.

7. Acoustic Properties

Original Papers and Short Notes

W. Rehwald, A. Vonlanthen, and W. Meyer
Single Crystal 2,4-Hexadiynylene-Bis (p-Toluenesulfonate): Elastic Properties and Phase Transitions in the Monomer and Polymer
(a) Vol. **75**, No. 1, p. 219–226 (1983).
See also *1.2*, p. *14*; *12.2*, p. *141*; 22.9, p. 525.

W. Westera
*Electro-Acoustic Effects in CdSe
(a) Vol. **75**, No. 2, p. 405–408 (1983).
See also *14.3.4*, p. *243*; 22.4.2, p. 466.

V. V. Zoninashvili, I. A. Naskidashvili, U. Messerschmidt, V. A. Melik-Shakhnazarov, N. M. Yastrebova, and L. M. Kolesnikova
*The Dependence of the Elastic Relaxation on a Bias Stress in Pure Nb in the Temperature Region of the α-Maximum
(a) Vol. **75**, No. 2, p. K205–K208 (1983).
See also *10.1*, p. *100*; 12.1, p. 137; 21, p. 351.

Hoang Ngoc Cam, Nguyen Van Hieu, and Ha Vinh Tan
*On the Theory of the Nonlinear Acousto-Optical Effect in Semiconductors
(b) Vol. **116**, No. 1, p. 25–29 (1983).
See also 13.5.1, p. 195; *20*, p. *307*; 22.2, p. 442; 22.4, p. 460.

K. P. Maheṣhwari and Ganesh Tarey
*Excitation of Magnetoacoustic Waves by Two Microwave Beams in InSb
(b) Vol. **116**, No. 1, p. 339–346 (1983).
See also 13, p. 147; 18, p. 274; *20*, p. *307*; 22, p. 405.

Y. Kawai and T. Ogawa
ΔE Effect of a Mn Ferrite Single Crystal at Temperatures between 150 and 300 K
(a) Vol. **76**, No. 1, p. 375–381 (1983).
See also *12.2*, p. *142*; *18.3*, p. *291*; 22.8.2, p. 520.

D. Adlienė, L. Pranevičius, and G. Zubauskas
*Acoustic Emission Initiated by Ion Implantation in Bubble Garnet Films
(a) Vol. **76**, No. 1, p. K35–K37 (1983).
See also *11*, p. *125*; 18.3, p. 291; 22.8.2, p. 520.

A. L. Tronconi, K. Skeff-Neto, and M. A. Amato
*Piezoelectric Detection of the Photoacoustic Signal
(a) Vol. **76**, No. 2, p. K209–K211 (1983).
See also 22.8.1, p. 514.

Y. Itoh and C. Hamaguchi
*Resonant Brillouin Scattering in HgI_2 Investigated by Acoustic Domain Injection Method
(b) Vol. **117**, No. 2, p. 725–732 (1983).
See also *20.1*, p. *320*; 22.5, p. 474.

M. D. Kaplan
The Dynamical Electron–Phonon Mode Coupling in Jahn-Teller Crystals with Antiferroelastic Ordering
(b) Vol. **118**, No. 1, p. 81–87 (1983).
See also *6*, p. *74*; 14.4.1, p. 257.

E. L. Albuquerque
Surface-Ripple Brillouin Scattering from a Glass–Metallic Liquid Interface
(b) Vol. **118**, No. 1, p. 223–227 (1983).
See also *6*, p. *75*.

J. Ulner
*Theorie des durch das äußere Magnetfeld induzierten Jahn-Teller-Effektes in orthorhombischen $J = 1$-Paramagneten
(Theory of the Jahn-Teller Effect Induced by an External Magnetic Field in Orthorhombic $J = 1$ Paramagnets)
(b) Vol. **118**, No. 1, p. 389–398 (1983).
See also *18.1*, p. *278*.

A. R. Arakelyan, R. G. Gabrielyan, A. R. Mkrtchyan, H. A. Aslanyan, and Kh. V. Kotandyan
*On a Certain Inverse Problem in γ-Acoustic Spectroscopy
(b) Vol. **118**, No. 1, p. K69–K74 (1983).
See also *5*, p. *67*; 22.9, p. 529.

H. J. Fischbeck
On the Translation Mode of the Kink Lattice of a Quasi-One-Dimensional Peierls-Fröhlich System with a Nearly Half-Filled Band
(b) Vol. **118**, No. 2, p. 595–603 (1983).
See also *1.2*, p. *18*; *6*, p. *76*; 13.1, p. 160.

E. M. Gololobov, E. L. Mager, Z. V. Mezhevich, and L. K. Pan
*Elastic Properties and Root-Mean-Square Dynamic Displacement of Niobium Atoms at High Pressures
(b) Vol. **119**, No. 2, p. K139–K142 (1983).
See also *12.1*, p. *140*; 21.1, p. 376.

N. A. Tyapunina and Yu. A. Ivashkin
*Excess Concentration of Point Defects in Alkali Halide Crystals Exposed to Ultrasonic Waves
(a) Vol. **79**, No. 2, p. 351–359 (1983).
See also *10.2*, p. *119*; 22.5.2, p. 484.

A. V. Zamkov, I. T. Kokov, and A. T. Anistratov
*The Acousto-Optical Properties and Photoelasticity of $PbBr_2$ Single Crystals
(a) Vol. **79**, No. 2, p. K177–K180 (1983).
See also 12.2, p. 144; *20*, p. *311*; 22, p. 413.

B.-P. Kietis, R. Reksnys, and K. Tumkevičius
*Hall Effect und Acoustoelectric Phenomena in PbTe Film
(a) Vol. **80**, No. 1, p. K57 – K58 (1983).
See also *14.3.1*, p. *232*; 22, p. 414.

V. S. Arakelyan and A. A. Avakyan
*Diffusion of Interstitial Admixture in Solids Due to Standing Ultrasound Waves
(a) Vol. **80**, No. 1, p. K71 – K74 (1983).
See also *9*, p. *96*; 22.1.2, p. 438; 22.2.1, p. 447.

Ya. M. Olikh
*Characteristics of Sound Waves Propagating in the Symmetry Plane of Pyrargyrite and Proustite Crystals
(a) Vol. **80**, No. 1, p. K81 – K85 (1983).
See also 22.8, p. 510.

P. Koštial, L. Malík, and L. Both
*New Aspect of Ultrasonic Detection of Some Significant Structural Parameters of Amorphous Se–Te Alloys
(a) Vol. **80**, No. 2, p. K167 – K170 (1983).
See also *2*, p. *56*; 22.1.3, p. 441.

8. Thermal Properties

Review Article

K. H. FISCHER
Spin Glasses (I)
(b) Vol. **116**, No. 2, p. 357–414 (1983).
See also 2, p. 43; 13, p. 146; 14.1, p. 206; *18.2*, p. *280*; *18.4*, p. *295*; 21.1, p. 363; 21.1.1, p. 379.

Original Papers and Short Notes

O. N. DUBEY and R. S. TRIPATHI
*Susceptibility and Specific Heat of Spin Glasses in a Magnetic Field
(b) Vol. **115**, No. 1, p. 63–66 (1983).
See also *18*, p. *273*.

T. SOMA, H.-MATSUO KAGAYA, and M. NISHIGAKI
*Mean-Square Displacement and Lindeman's Criterion for Melting of Alkali Metals
(b) Vol. **115**, No. 1, p. 273–276 (1983).
See also 8, p. 85; 21.2, p. 387.

A. FRREIRA DA SILVA, M. FABBRI, and I. C. DA CUNHA LIMA
*Electron Hopping Energy Influence on the Specific Heat of Phosphorus-Doped Silicon
(b) Vol. **115**, No. 1, p. 311–316 (1983).
See also 13.4, p. 172; 22.1.2, p. 423.

G. D. KHATTAK
*Specific Heat of Uranium Dioxide (UO_2) between 0.3 and 50 K
(a) Vol. **75**, No. 1, p. 317–321 (1983).
See also *6*, p. *70*; 18.4, p. 295; 22.6, p. 490.

A. V. SALKER, K. SESHAN, and H. V. KEER
Phase Transition Behaviour of VO_2
(a) Vol. **75**, No. 1, p. K37–K40 (1983).
See also *1.2*, p. *14*; 22.6, p. 490.

C. MALINOWSKA-ADAMSKA, M. OWSIEJCZYK, and L. WOJTCZAK
Thermodynamics of Strongly Anharmonic Crystals
(b) Vol. **115**, No. 2, p. 335–345 (1983).
See also *6*, p. *70*; 21, p. 350; 21.2, p. 387.

H. MÜLLER, E. HEGENBARTH, and S. MAHMOUD
*Thermal Conductivity of the Intermetallic Rare Earth Compound $PrAl_2$ in the Temperature Range between 4.2 and 100 K
(b) Vol. **115**, No. 2, p. K69–K73 (1983).
See also 21, p. 350; 21.4, p. 390.

M. A. Aldzhanov, A. M. Aliev, R. K. Veliev, K. K. Mamedov, M. I. Mekhtiev, and V. Ya. Steinshreiber
*Phonon Density of States and Heat Capacity of $CdIn_2S_4$
(b) Vol. **115**, No. 2, p. K75–K78 (1983).
See also *6*, p. *71*; 22.8, p. 503.

H. Gruber and E. Krautz
Magnetoresistance and Conductivity in the Binary System Titanium–Oxygen (II): Semiconductive Titanium Oxides
(a) Vol. **75**, No. 2, p. 511–518 (1983).
See also *14.1*, p. *206*; *14.3*, p. *216*; 15, p. 261; 18, p. 273; 22.6, p. 491.

P. Türkes
*An Ion-Implanted Resistor as Thermal Transient Sensor for the Determination of the Thermal Diffusivity in Silicon
(a) Vol. **75**, No. 2, p. 519–523 (1983).
See also 22.1.2, p. 426.

T. Soma, H.-M. Kagaya, and Y. Kimura
Thermal Expansion Coefficient of Alkali Metals
(b) Vol. **116**, No. 1, p. 57–61 (1983).
See also *6*, p. *71*; 21.2, p. 387.

M. A. Ansari, Anil Kumar, and B. K. Srivastava
*Thermal Conductivity of Germanium Containing Dislocations at Low Temperatures
(b) Vol. **116**, No. 1, p. 163–167 (1983).
See also 6, p. 71; 22.1.1, p. 419.

R. K. Singh, D. K. Neb, and S. P. Sanyal
Effect of Three-Body Forces on Thermophysical and Anharmonic Properties of Rare Gas Solid Mixtures
(b) Vol. **116**, No. 1, p. 289–297 (1983).
See also *4*, p. *60*; 6, p. 72; 12.2, p. 142; 23, p. 533.

G. Dharmadurai
*Phonon Transmission Theory of Heat Transfer through Solid–Gas Interfaces
(b) Vol. **116**, No. 1, p. 307–311 (1983).
See also 1.6, p. 39; *6*, p. *72*.

L. Wojtczak, A. Urbaniak-Kucharczyk, J. Mielnicki, and B. Mrygon
*The Magnetic Contribution to Specific Heat in Thin Ferromagnetic Films
(a) Vol. **76**, No. 1, p. 113–120 (1983).
See also 18.2.1, p. 290.

L. N. Aleksandrov
Formation of Semiconductor Epitaxial Films by Pulse Heating Crystallization or Regrowth
(a) Vol. **76**, No. 1, p. 179–190 (1983).
See also *1.5*, p. *36*; *3*, p. *57*; 22.1.1, p. 419; 22.1.2, p. 427.

V. M. Pan, I. Ya. Dekhtyar, M. E. Osinovskii, S. E. Litvin, B. G. Nikitin, M. M. Nishenko, and M. P. Voronko
Thermodynamical and Structural Properties of Amorphous $Fe_{80}B_{20}$ and $Fe_{80}P_{13}C_7$ Alloys
(a) Vol. **76**, No. 1, p. 197–206 (1983).
See also *2*, p. *45*; 5, p. 65; 14.1, p. 206; *18.2.1*, p. *290*; 21.1.1, p. 380.

P. Čapková, R. Kužel, and J. Šedivý
Thermal Vibrations and Static Displacements of Atoms in Monstoichiometric Titanium Carbide
(a) Vol. **76**, No. 1, p. 383–390 (1983).
See also *6*, p. *72*; 21.1, p. 366.

H. Misiorek, T. Zakrzewski, and J. Rafałowicz
*The Lorenz Function for Aluminium of Different Chemical Purity, Neutron Irradiated or Plastically Deformed
(a) Vol. **76**, No. 1, p. K25–K26 (1983).
See also *11*, p. *125*; 12.1, p. 138; 21, p. 352.

H. J. Goldsmid, M. M. Kaila, and G. L. Paul
*Thermal Conductivity of Amorphous Silicon
(a) Vol. **76**, No. 1, p. K31–K33 (1983).
See also *2*, p. *46*; 22.1.2, p. 427.

L. M. Avkhutskii, R. L. Davidovich, L. A. Zemnukhova, P. S. Gordienko, V. Urbonavičius, and J. Grigas
Peculiarities of Phase Transitions and Physical Properties in $(NH_4)_2SbF_5$
(b) Vol. **116**, No. 2, p. 483–488 (1983).
See also *14.4.1*, p. *256*; 19, p. 301; 22.8, p. 505.

K. Handrich
*Phonon Lifetime and Thermal Conductivity in Amorphous Solids
(b) Vol. **116**, No. 2, p. 533–538 (1983).
See also 2, p. 46; *6*, p. *72*; 22.9, p. 527.

T. Soma, H.-Matsuo Kagaya, and M. Nishigaki
Volume Dependence of the Melting Temperature for Alkali Metals with Debye's Model
(b) Vol. **116**, No. 2, p. 673–676 (1983).
See also *6*, p. *73*; 21.2, p. 387.

N. Gopi Krishna and D. B. Sirdeshmuhk
*X-Ray Determination of the Mean-Square Atomic Displacement and Associated Debye Temperature of Erbium
(b) Vol. **116**, No. 2, p. K105–K108 (1983).
See also *6*, p. *73*; 21.4, p. 390.

T. K. Chattopadhyay, E. Gmelin, and H. G. von Schnering
*Heat Capacity Study of the Phase Transitions in As_4S_3 and As_4S_4
(a) Vol. **76**, No. 2, p. 543–551 (1983).
See also 1.2, p. 16; 22, p. 406.

J. M. López and J. A. Alonso
Semiempirical Theory of Solid Solubility in Metallic Alloys
(a) Vol. **76**, No. 2, p. 675–682 (1983).
See also *1.3*, p. *23*; 21, p. 353; 21.1, p. 368.

S. M. Wasim, B. Fernández, and R. Aldana
*Thermal Conductivity of HgSe
(a) Vol. **76**, No. 2, p. 743–746 (1983).
See also 22.4.2, p. 466.

B. P. Bespalov and A. M. Vainrub
*Thermal Conductivity of Pressed Pellets of TCNQ Complexes. Manifestation of Amorphous Structure
(a) Vol. **77**, No. 1, p. K13–K15 (1983).
See also 22.9, p. 528.

B. Vengalis and K. Valatska
Phase Segregation in Tellurium-Saturated GeTe
(b) Vol. **117**, No. 2, p. 471–476 (1983).
See also *1.2*, p. *16*; 13.4, p. 179; 14.3, p. 220; 20.1, p. 319; 22.3, p. 457.

H. K. Phan and T. H. Nguyen
*Measurement of the Electrothermal Nonlinearity and Thermal Relaxation Times of Thin Metal Films
(a) Vol. **77**, No. 2, p. K135–K138 (1983).
See also *14.1*, p. *208*; 21.1, p. 371.

A. I. Prokhvatilov and A. P. Isakina
Lattice Parameters, Thermal Expansion Coefficients, and Vacancy Density in Solid CD_4
(a) Vol. **78**, No. 1, p. 147–155 (1983).
See also 1.1, p. 11; *10.2*, p. *114*; 12.2, p. 143; 22.9, p. 529.

H.-D. Geiler, K. Hehl, and D. Stock
*Deposition of Laser Energy into Inhomogeneous Layer Systems
(a) Vol. **78**, No. 1, p. 193–199 (1983).
See also *11*, p. *129*; 22.1.2, p. 432.

I. V. Bodnar and N. S. Orlova
*X-Ray Study of Thermal Expansion in $CuAlS_2$, $CuGaS_2$, and $CuInS_2$ Compounds over the Temperature Range from 90 to 650 K
(a) Vol. **78**, No. 1, p. K59–K63 (1983).
See also 22.8, p. 508.

A. Kumar, A. K. Sinha, and M. A. Ansari
*Thermal Conductivity of LiF Containing Dislocations at Low Temperatures
(b) Vol. **118**, No. 2, p. 483–491 (1983).
See also 6, p. 76; 22.5.2, p. 481.

T. D. Sokolovskii
Phonon Spectrum of Boron Nitride with Wurtzite Structure
(b) Vol. **118**, No. 2, p. 493–498 (1983).
See also *6*, p. *76*; 22, p. 409.

H. V. SHARMA and I. SINGH
Valence Fluctuations and Magnetic–Nonmagnetic Transitions in a Model for Intermediate-Valence Systems
(b) Vol. **118**, No. 2, p. 511–520 (1983).
See also *13*, p. *150*; 18.1, p. 278; *18.2*, p. *285*; 21.4, p. 392.

C. MALINOWSKA-ADAMSKA
Dynamics and Thermodynamics of Anharmonic Crystals near the Instability Point
(b) Vol. **118**, No. 2, p. 535–539 (1983).
See also *6*, p. *76*; 12.1, p. 139; 21.2, p. 387.

H. MASHIYAMA
Modified Phase Modulation-Only Approximation of Incommensurate Phases in K_2SeO_4-Type Crystals
(b) Vol. **118**, No. 2, p. 757–767 (1983).
See also *1.2*, p. *18*; 14.4, p. 252; 22.8.1, p. 515.

N. SHOHOJI
*Statistical Thermodynamic Approach to Some Hyperstoichiometric Dihydrides of Lanthanides (La, Ce, Pr, Nd, Sm) and Actinides (Np, Pu, Am)
(b) Vol. **118**, No. 2, p. 811–822 (1983).
See also *4*, p. *62*; 10.1, p. 104; 21.4, p. 392; 21.5, p. 395.

A. A. EL-SHARKAWY, M. S. ZAGHLOUL, and A. M. SANAD
Study of Thermal Properties, Electrical Conductivity, and the Crystallization Process of the Glassy Alloy $SeGe_{20}$
(a) Vol. **78**, No. 2, p. 457–462 (1983).
See also *3*, p. *59*; 14.3, p. 222; 22.1.1, p. 421; 22.1.3, p. 440.

Y. MURAOKA, M. SHIGA, and Y. NAKAMURA
Magnetic Properties and Thermal Expansion of $(Zr_{1-x}Ta_x)Fe_2$
(a) Vol. **78**, No. 2, p. 717–722 (1983).
See also *5*, p. *67*; *18.2*, p. *285*; 18.4, p. 297; 21.1.1, p. 383.

I. HENNING, E. HEGENBARTH, and B. BREZINA
*Specific Heat of a K_2SeO_4 Single Crystal in Temperature Range 0.6 to 7 K
(a) Vol. **78**, No. 2, p. K167–K170 (1983).
See also *14.4.1*, p. *258*; 22.8.1, p. 516.

K. S. DUBEY
*Effect of Empty Spaces on the Lattice Thermal Conductivity of Non-Crystalline Material at Low Temperatures. Application to Borosilicate Glass
(a) Vol. **78**, No. 2, p. K195–K200 (1983).
See also 6, p. 77; 22.8.1, p. 516.

M. A. ANSARI, A. KUMAR, and B. K. SRIVASTAVA
*Lattice Thermal Conductivity of Crystals with Dislocations
(b) Vol. **119**, No. 1, p. 113–119 (1983).
See also 6, p. 77; *10*, p. *98*.

R. SCHUMANN and K. ELK
*On Thermopower and Thermal Conductivity of the Periodic Anderson Model
(b) Vol. **119**, No. 1, p. 221 – 225 (1983).
See also *15*, p. *262*; 21, p. 358.

N. DASS and B. M. MUSTAFA
*A Study of the Anderson-Grüneisen Parameter in Sodium Chloride
(b) Vol. **119**, No. 1, p. K27 – K30 (1983).
See also 12.2, p. 143; 22.5.2, p. 483.

A. MILCHEV
Heat Capacity of Diatomic Molecules Adsorbed on a Periodic Substrate and the Localized-to-Mobile Transition
(a) Vol. **79**, No. 1, p. 301 – 308 (1983).
See also *1.6*, p. *41*.

M. MERTIG, G. POMPE, and E. HEGENBARTH
Low-Temperature Specific Heat of Vapour-Deposited Ge_xSe_{1-x} Glass ($x = 0.3$)
(a) Vol. **79**, No. 1, p. K49 – K52 (1983).
See also *2*, p. *52*; 22, p. 412.

B. LORENZ
*Thermodynamic Properties of Uniform and Dimerized Hubbard Chains
(b) Vol. **119**, No. 2, p. 555 – 563 (1983).
See also 18, p. 275; 22.9, p. 530.

G. POMPE, M. GAAFAR, P. BÜTTNER, and TH. FRANCKE
*Low Temperature Thermal Conductivity of Amorphous (Fe, Ni, Co) (P, B, Si) Alloys and Their Change by Heat Treatment
(b) Vol. **119**, No. 2, p. 579 – 587 (1983).
See also *2*, p. *52*; 21.1, p, 375; 21.1.1, p. 384.

P. DEUS and H. A. SCHNEIDER
*Thermal Expansion of the Diamond-Like Semiconducting Compound $ZnSiP_2$ within the Temperature Range 20 to 300 K
(a) Vol. **79**, No. 2, p. 411 – 415 (1983).
See also 22.8, p. 509.

E. GMELIN, N. LENGE, and H. KRONMÜLLER
*Vacany Effect on the Transition Temperature of Magnetite
(a) Vol. **79**, No. 2, p. 465 – 475 (1983).
See also 18.3, p. 293; 22.6, p. 496.

E. E. VELMRE and B. P. FREIDIN
Numerical Modeling of the Electrothermal Transient Process in Diode Structures Based on Direct-Gap Semiconductors
(a) Vol. **79**, No. 2, p. K131 – K134 (1983).
See also *14.3.3*, p. *241*; 22.2.1, p. 446.

P. G. RUSTAMOV, CH. I. ABILOV, and S. G. AGDAMSKAYA
Thermoelectric Properties of $(PbTe)_{1-x}[(Cr, Fe, Ni)_2Te_3]_x$
(a) Vol. **79**, No. 2, p. K135 – K137 (1983).
See also 14.3, p. 226; *15*, p. *263*; 22, p. 413.

C. REGNAUT, E. FUSCO, and J. P. BADIALI
Structure and Thermodynamics of the Liquid Noble Metals from the Generalized OPW and Resonant Model Approaches
(b) Vol. **120**, No. 1, p. 373 – 382 (1983).
See also *2*, p. *54*; 13.1, p. 164; 21.1, p. 376; 21.6, p. 399.

D. I. UZUNOV
Critical Region and Characteristic Lengths for Zero-Temperature Structural Phase Transitions
(b) Vol. **120**, No. 1, p. K39 – K43 (1983).
See also *1.2*, p. *19*.

M. MONDAL and K. P. GHATAK
*Effect of Size Quantization on the Electronic Heat Capacity in Ultrathin Films of Semiconductors
(b) Vol. **120**, No. 1, p. K63 – K68 (1983).
See also *13.3*, p. *170*; 22.2.1, p. 447.

P. DEUS, H. NEUMANN, G. KÜHN, and B. HINZE
*Low-Temperature Thermal Expansion in $CuInSe_2$
(a) Vol. **80**, No. 1, p. 205 – 209 (1983).
See also 22.8, p. 509.

L. N. VASILEV, V. M. GRABOV, A. V. GOLUBKOV, A. G. GOROBETS, V. S. OSKOTSKII, I. A. SMIRNOV, and V. V. TIKHONOV
*Physical Properties and Phase Transitions of the Rare Earth Monosulphides in the Homogeneity Range
(a) Vol. **80**, No. 1, p. 237 – 244 (1983).
See also *1.3*, p. *26*; 13, p. 152; 14.3, p. 226; 18.4, p. 297; 22, p. 413.

P. DEUS, U. VOLAND, and H. A. SCHNEIDER
*Thermal Expansion of GaP within 20 to 300 K
(a) Vol. **80**, No. 1, p. K29 – K32 (1983).
See also 22.2.2, p. 451.

C. COLINET and A. PASTUREL
*A Thermodynamic Study of Cerium Behaviour in Hexagonal $CeNi_5$ Compound
(a) Vol. **80**, No. 1, p. K75 – K79 (1983).
See also 21.1, p. 377; 21.4, p. 393.

S. K. PLACHKOVA
Thermoelectric Power in the System $(GeTe)_{1-x}(AgSbTe_2)_x$
(a) Vol. **80**, No. 1, p. K97 – K100 (1983).
See also *14.3*, p. *227*; *15*, p. *263*; 22.8, p. 510.

V. A. Sablikov and V. B. Sandomirskii
The Photoacoustic Effect in Semiconductors
(b) Vol. **120**, No. 2, p. 471–480 (1983).
See also *13.3*, p. *171*; *14.3*, p. *227*; 22, p. 414.

A. Brau, J. P. Farges, A. Filhol, and H. Grassi
Order–Disorder Phase Transition in $TEA^+(TCNQ)_2^-$
(b) Vol. **120**, No. 2, p. 547–554 (1983).
See also *1*, p. *10*; *14.3*, p. *227*; 22.9, p. 532.

C. Malinowska-Adamska
Self-Consistent Calculations of the Lattice Dynamics and Thermodynamics of Anharmonic Noble Metals
(b) Vol. **120**, No. 2, p. 601–610 (1983).
See also *6*, p. *81*; 12.1, p. 140; 21.6, p. 399.

W. Schumacher
*On the Influence of the Magnetic Ordering on Thermodynamic and Electrical Properties in the Hubbard Model
(b) Vol. **120**, No. 2, p. 621–629 (1983).
See also 14.1, p. 210; 18.4, p. 298.

R. Schumann
Temperature Dependence of the Thermopower of the Periodic Anderson Model
(b) Vol. **120**, No. 2, p. K183–K188 (1983).
See also 14.1, p. 210; *15*, p. *263*; 21, p. 361.

L. A. Zemnukhova, R. L. Davidovich, P. S. Gordienko, J. Grigas, A. N. Kovrianov, S. I. Kuznetsov, T. A. Kaidalova, and V. Urbonavičius
Investigation of Phase Transitions in K_2SbF_5
(a) Vol. **80**, No. 2, p. 553–558 (1983).
See also *1.2*, p. *20*; *14.4.1*, p. *259*; 19, p. 306; 22.8, p. 511.

9. Diffusion

Review Article

T. D. Dzhafarov
*Photostimulated Diffusion in Semiconductors
(a) Vol. **79**, No. 1, p. 11–45 (1983).
See also 11, p. 122; 14.3.3, p. 236; 14.3.4, p. 243; 16, p. 264; 22, p. 403.

Original Papers and Short Notes

R. H. Rautioaho
A Pair Potential Calculation of Vacancy Formation and Migration Energies for Aluminium and Copper
(b) Vol. **115**, No. 1, p. 95–103 (1983).
See also *10.1*, p. *99*; 21, p. 349; 21.1, p. 363.

P. Guerin et A. Laforgue
*Recherches théoriques sur l'autodiffusion dans les cristaux ioniques
(Theoretical Studies on the Selfdiffusion in Ionic Crystals)
(b) Vol. **115**, No. 1, p. 235–244 (1983).
See also 22.5.2, p. 477.

K. K. Misra and A. N. Nigam
*Development of Stress in Nickel during Hydrogen Diffusion
(a) Vol. **75**, No. 2, p. 635–639 (1983).
See also *12.1*, p. *137*; 21.1, p. 365.

R. N. Joarder and R. V. Gopala Rao
Temperature Dependence of the Structure Factor of Liquid Metals
(b) Vol. **116**, No. 1, p. 299–306 (1983).
See also *2*, p. *45*; 4, p. 61; 21.1, p. 366; 21.2, p. 387; 21.6, p. 396.

V. N. Pavlovich
*Photostimulated Diffusion of Light Substitutional Impurities in Semiconductors and Insulators
(b) Vol. **116**, No. 1, p. K9–K14 (1983).
See also 22, p. 405; 22.1.2, p. 426.

M. Halbwachs, P. Mazot, and J. Woirgard
Anelastic Relaxation Phenomena in Plasma-Sprayed $(Al_2O_3)_3MgO$ Spinel with Reference to Diffusion Process
(a) Vol. **76**, No. 1, p. 157–164 (1983).
See also *10.2*, p. *110*; *12.2*, p. *142*; 22.8.2, p. 520.

Dao Khac An, Le Hoang Mai, and Pham Hoi
*Concentration Dependence of the Boron Diffusion Coefficient in Silicon
(a) Vol. **76**, No. 1, p. K85–K88 (1983).
See also 22.1.2, p. 428.

M. E. Gurevich, L. N. Larikov, V. G. Novitskii, A. E. Pogorelov, and A. F. Zhuravlev
*Laser-Stimulated Mass Transfer in Metals
(a) Vol. **76**, No. 2, p. 479–484 (1983).
See also *11*, p. *126*; 21.1, p. 367; 21.1.1, p. 381.

Ch. Opitz and F. Dübler
*Modelling and Illustration of a Penetration Mechanism for Hydrogen into Metal
(a) Vol. **76**, No. 2, p. 505–512 (1983).
See also 4, p. 61; 13, p. 148; 22.5, p. 473.

E. Balanzat and M. Halbwachs
A Comparative Analysis of the Ordering Kinetic Laws for the Derivation of the Vacancy Mobility Parameters in Concentrated Alloys
(a) Vol. **76**, No. 2, p. 667–673 (1983).
See also *10.1*, p. *102*; 14.1, p. 207; 21.1, p. 367; 21.6, p. 397.

I. V. Antonova, K. B. Kadyrakunov, E. V. Nidaev, and L. S. Smirnov
*Diffusion of Iron and Gold in Silicon Annealed with Millisecond Pulses
(a) Vol. **76**, No. 2, p. K213–K215 (1983).
See also 22.1.2, p. 429.

F. Găiseanu
*On the Diffusion Profile of Boron in Silicon at High Concentrations
(a) Vol. **77**, No. 1, p. K59–K61 (1983).
See also 22.1.2, p. 430.

T. A. Cherepanova and A. S. Koziejowska
Sublattice Models of the Binary Crystal Growth (II): Influence of Diffusion in the Melt on the Crystallization Kinetics of Binary Alloys
(a) Vol. **77**, No. 2, p. 561–570 (1983).
See also *3*, p. *58*.

R. Chaoui and B. Tuck
*Very Short Diffusions of Silver in the III–V Semiconductors
(a) Vol. **77**, No. 2, p. K189–K192 (1983).
See also 22.2.1, p. 445; 22.2.2, p. 450.

Th. Jung and E. Nebauer
*Interdiffusion Profiles of AuGe/n-GaAs Ohmic Contacts Studied by AES
(a) Vol. **77**, No. 2, p. K203–K206 (1983).
See also 14.3.4, p. 246; 22.2.1, p. 445.

H. J. Fuchs, W. Karthe, and S. Börner
On the Kinetics of a Paramagnetic Oxygen Centre in CaF_2
(b) Vol. **118**, No. 1, p. 211–215 (1983).
See also *10.2*, p. *113*; 19, p. 302; 22.5.3, p. 487.

L. Bottyán, D. L. Beke, and K. Tompa
*Diffusion-Induced Quadrupole Relaxation of ^{27}Al Nuclei in Dilute Al–Ti, Al–Cr, Al–Mn, and Al–Cu Alloys at High Temperatures
(b) Vol. **118**, No. 2, p. 835 – 842 (1983).
See also 10.1, p. 104; *19*, p. *303*; 21, p. 357; 21.1, p. 372.

H.-J. Rockosch and Ch. Herzig
*Isotope Effect of Manganese and Simultaneous Impurity Diffusion of Chromium and Zinc in Copper
(b) Vol. **119**, No. 1, p. 199 – 207 (1983).
See also 10.1, p. 105; 21.1, p. 374.

L. B. Harris and S. L. Bewlay
*Diffusion of Water in NaCl Bicrystals
(a) Vol. **79**, No. 1, p. 243 – 250 (1983).
See also 22.5.2, p. 483.

V. G. Chudinov and V. I. Protasov
*Radiation Diffusion–Dislocation Creep in the Presence of Weertman Mechanism of Dislocation Multiplication
(a) Vol. **79**, No. 2, p. 433 – 443 (1983).
See also 10.1, p. 106; 21, p. 360.

H. Hahn, G. Frohberg, and H. Wever
*Self-Diffusion in the Intermetallic B2 Electron Compound PdIn
(a) Vol. **79**, No. 2, p. 559 – 565 (1983).
See also 21, p. 360.

A. Milchev and I. Avramov
*On the Influence of Amorphization on Atomic Diffusion in Condensed Systems
(b) Vol. **120**, No. 1, p. 123 – 130 (1983).
See also *2*, p. *53*.

A. A. Abdumalikov, F. Kh. Abdullaev, and S. Stamenković
Influence of the Thermostat on Kink Dynamics in the Φ^4 Model
(b) Vol. **120**, No. 1, p. K33 – K37 (1983).
See also *6*, p. *80*.

S. K. Pabi
On the Variation of Interface Composition around a Dissolving Precipitate
(a) Vol. **80**, No. 1, p. 305 – 316 (1983).
See also *1.3*, p. *27*; 21, p. 361; 21.6, p. 399.

A. Priemuth and H. Süssmann
*On the Relation between the Carrier Concentration and the Effective Ion Charge of Interstitial Cu at the Electromigration in V_2VI_3 Compounds
(a) Vol. **80**, No. 1, p. K7 – K9 (1983).
See also *14.3*, p. *227*; 22.7, p. 500.

H. Jain

*Haven Ratio for Sodium in Synthetic α-Quartz
(a) Vol. **80**, No. 1, p. K15 – K20 (1983).
See also 10.2, p. 120; 14.4, p. 254; 22.6, p. 497.

C. H. Ling

*Resistivity Change in Ti/Au Bimetal Films Annealed in the Temperature Range 300 to 400 °C
(a) Vol. **80**, No. 1, p. K59 – K62 (1983).
See also *14.3.2*, p. *235*; 21.1, p. 377; 21.6, p. 399.

V. S. Arakelyan and A. A. Avakyan

*Diffusion of Interstitial Admixture in Solids Due to Standing Ultrasound Waves
(a) Vol. **80**, No. 1, p. K71 – K74 (1983).
See also *7*, p. *84*; 22.1.2, p. 438; 22.2.1, p. 447.

B. Głowacki and J. Chojcan

Studies on the Condition of Obtaining a Single Phase Superconducting Nb_3Sn Layer by Mössbauer Spectroscopy
(a) Vol. **80**, No. 1, p. K93 – K96 (1983).
See also *5*, p. *68*; 14.2, p. 214; 21, p. 361.

K. Budde and H. Richtering

*Einbau und Diffusion von Cu(I)-Jonen in CsI anhand von NMR-Messungen
(Incorporation and Diffusion of Cu(I)-Ions in CsI by NMR Measurements)
(b) Vol. **120**, No. 2, p. 595 – 599 (1983).
See also 10.2, p. 120; *19*, p. *305*; 22.5.2, p. 485.

10. Defect Properties (Irradiation Defects See 11)

Review Articles

B. Pałósz
Reasons for Polytypism of Crystals of the Type MX_2 (I): Polytypism and Structure Behaviour of Inorganic Crystals: General Problems
(a) Vol. **77**, No. 1, p. 11 – 34 (1983).
See also *1*, p. *9*; *1.1*, p. *11*.

A. E. Romanov and V. I. Vladimirov
*Disclinations in Solids
(a) Vol. **78**, No. 1, p. 11 – 34 (1983).
See also *1.4*, p. *29*; 12, p. 136.

Original Papers and Short Notes

J. Rasek
*Application of Hyperbolic Temperature Variation to the Analysis of Continuous and Stepwise Isochronal Curves and to the Problem of Quenching in Vacancies
(a) Vol. **75**, No. 1, p. 49 – 54 (1983).

C. Pisani, R. Dovesi, and P. Ugliengo
*Comparison of Different Approaches to the Study of Local Defects in Crystals (I): Theoretical Considerations and Computational Schemes
(b) Vol. **116**, No. 1, p. 249 – 259 (1983).

A. I. Landau
*Analytical Calculation of Parameters of Thermally Activated Dislocation Motion through a Random Array of Point Obstacles
(a) Vol. **76**, No. 1, p. 207 – 216 (1983).

Nguyen An, R. Herrmann, and G. Worm
Investigation of Asymmetrical Grain Boundaries and Their Coincidence Site Lattices
(b) Vol. **116**, No. 2, p. 501 – 510 (1983).
See also *1.4*, p. *30*.

C. Pisani, R. Dovesi, and P. Ugliengo
*Comparison of Different Approaches to the Study of Local Defects in Crystals (II): Substitutional Impurities in the Tight-Binding Approximation
(b) Vol. **116**, No. 2, p. 547 – 556 (1983).

A. I. Osetskii
*Low-Temperature Fluctuational Motion of Dislocations in Crystals (I): Quantum and Dissipative Effects
(b) Vol. **117**, No. 1, p. 355 – 365 (1983).

R. Z. Valiev, V. Yu. Gertsman, O. A. Kaibyshev, and Sh. Kh. Khannanov
Non-Equilibrium State and Recovery of Grain Boundary Structure (I): General Analysis, Crystallogeometrical Aspects
(a) Vol. **77**, No. 1, p. 97–105 (1983).
See also *1.4*, p. *31*.

R. Z. Valiev, V. Yu. Gertsman, and O. A. Kaibyshev
Non-Equilibrium State and Recovery of Grain Boundary Structure (II): Energetic Analysis
(a) Vol. **78**, No. 1, p. 177–186 (1983).
See also *1.4*, p. *32*.

V. N. Lukerchenko
*On the Relation between the Plastic Strain Rate of a Crystal and Microscopic Characteristics of Dislocation Motion
(a) Vol. **78**, No. 1, p. K9–K12 (1983).
See also *12*, p. *136*.

A. M. Arustamyan and F. N. Chukhovskii
*Diffraction Contrast from Partial Dislocations in the two-Beam Dynamical Theory
(a) Vol. **78**, No. 1, p. K35–K38 (1983).
See also *1.4*, p. *33*.

M. J. Marcinkowski
*The Surface Dislocation and Its Relationship to the Boundary-Integral Equation
(a) Vol. **78**, No. 2, p. 401–414 (1983).

M. A. Ansari, A. Kumar, and B. K. Srivastava
Lattice Thermal Conductivity of Crystals with Dislocations
(b) Vol. **119**, No. 1, p. 113–119 (1983).
See also **6**, p. 77; *8*, p. *89*.

A. Y. Belov, V. A. Chamrov, V. L. Indenbom, and J. Lothe
*Elastic Fields of Dislocations Piercing the Interface of an Anisotropic Bicrystal
(b) Vol. **119**, No. 2, p. 565–578 (1983).
See also *12*, p. *136*.

J. P. Michel
Estimation of the Critical Resolved Shear Stress for Dislocation Glide through a Random Array of Distinct Obstacles
(b) Vol. **119**, No. 2, p. K149–K153 (1983).

10.1 Metals

Review Article

E. R. Hummel
Differential Reflectometry and Its Application to the Study of Alloys, Ordering, Corrosion, and Surface Properties
(a) Vol. **76**, No. 1, p. 11–44 (1983).
See also *1.3*, p. *21*; 13.1, p. 155; 13.3, p. 167; *20.1*, p. *312*; 21, p. 349; 21.1, p. 363; 22.6, p. 490.

Original Papers and Short Notes

R. H. Rautioaho
*A Pair Potential Calculation of Vacancy Formation and Migration Energies for Aluminium and Copper
(b) Vol. **115**, No. 1, p. 95–103 (1983).
See also 9, p. 93; 21, p. 349; 21.1, p. 363.

R. Khanna
*Huang Diffuse Scattering from Self-Interstitials in a Trigonal Lattice
(b) Vol. **115**, No. 1, p. 305–310 (1983).
See also 21, p. 349.

R. Gerlach and H. Löffler
*On the Structure of the Precipitation-Free Zone in Two Al–Zn–Mg Alloys after Various Heat Treatments
(a) Vol. **75**, No. 1, p. 169–178 (1983).
See also 21, p. 350; 21.1, p. 364; 21.3, p. 389.

F. Walz, H. J. Blythe, and F. Dworschak
*Annealing Behaviour of Dilute FeCu, FeMn, and FeTi Alloys in the Temperature Range up to Stage III, Following Low-Temperature Electron Irradiation
(a) Vol. **75**, No. 1, p. 235–242 (1983).
See also 11, p. 122; *18.2*, p. *280*; 21.1, p. 364; 21.1.1, p. 379.

R. H. J. Fastenau and M. I. Baskes
*A Combined Atomistic and Monte Carlo Simulation of Point Defect–Dislocation Interactions
(a) Vol. **75**, No. 1, p. 323–334 (1983).
See also 21, p. 350.

P. Feltham
*Low-Temperature Anomalies in the Plasticity of Crystalline Materials
(a) Vol. **75**, No. 1, p. K95–K98 (1983).
See also 21, p. 350.

L. Kornblit
*Divacancy Binding Energy in F.C.C. Metals
(b) Vol. **115**, No. 2, p. 485–491 (1983).
See also 21.1, p. 364; 21.6, p. 396.

R. H. Rautioaho
*Energies of Self-Interstitials for Aluminium and Copper
(b) Vol. **115**, No. 2, p. K121 – K125 (1983).
See also 21, p. 351; 21.1, p. 364.

F. Abd El-Salam, A. I. Eatah, and A. Tawfik
Effect of Li Loss on Some Physical Properties of Al–Li Alloy
(a) Vol. **75**, No. 2, p. 379 – 382 (1983).
See also *1.3*, p. *21*; 12.1, p. 137; 14.1, p. 206; 21, p. 351; 21.2, p. 387.

A. Baldan
*Dislocation Structures in Aluminium-Rich Ni_3(Al, Ta) and Ni_3(Al, Hf)
(a) Vol. **75**, No. 2, p. 441 – 449 (1983).
See also *1.3*, p. *21*; 21, p. 351; 21.1, p. 364.

K. K. Ray and A. K. Mallik
*An Investigation on the Activation Volume for Plastic Flow in Nickel Determined by Micro and Macro Tests
(a) Vol. **75**, No. 2, p. 451 – 458 (1983).
See also 12.1, p. 137; 21.1, p. 365.

A. Korner and H. P. Karnthaler
*Glide Dislocations on Cube Planes in a Low Stacking-Fault Energy Alloy
(a) Vol. **75**, No. 2, p. 525 – 535 (1983).
See also 12.1, p. 137; 21, p. 351; 21.1, p. 365.

P. Delobelle and C. Oytana
*Creep at Low and Average Stress of Several Highly Concentrated Solid Solutions of the Cubic Centered Type
(a) Vol. **75**, No. 2, p. 625 – 634 (1983).
See also 12.1, p. 137; 21.1, p. 365; 21.1.1, p. 380.

V. Navrátil, M. Hamerský, P. Lukáč, V. P. Soldatov, and V. I. Startsev
*Solid Solution Hardening of Cd–Zn Single Crystals at 1.5 K
(a) Vol. **75**, No. 2, p. K133 – K135 (1983).
See also *1.3*, p. *22*; 21, p. 351; 21.1, p. 365.

I. Kanazawa, H. Murakami, T. Kurihara, Y. Sakurai, and M. Doyama
*Positron Annihilation Study of Quenched-in Defects in Dilute Al–Ge and Al–Si Alloys
(a) Vol. **75**, No. 2, p. K175 – K177 (1983).
See also 21, p. 351.

V. V. Zoninashvili, I. A. Naskidashvili, U. Messerschmidt, V. A. Melik-Shakhnazarov, N. M. Yastrebova, and L. M. Kolesnikova
*The Dependence of the Elastic Relaxation on a Bias Stress in Pure Nb in the Temperature Region of the α-Maximum
(a) Vol. **75**, No. 2, p. K205 – K208 (1983).
See also 7, p. *82*; 12.1, p. 137; 21, p. 351.

I. Kanazawa, H. Murakami, and M. Tanaka
*Internal Friction by a Second-Phase Particle in Al–Ge Alloy
(a) Vol. **75**, No. 2, p. K211–K213 (1983).
See also *12.1*, p. *137*; 21, p. 352.

K. Masuda
*On the Interaction between a Vacancy and an Interstitial Impurity Atom in B.C.C. Transition Metals
(b) Vol. **116**, No. 1, p. 9–16 (1983).
See also 21.1, p. 365; 21.1.1, p. 380.

H. Mizubayashi, T. Arai, and S. Okuda
*Low-Temperature Dislocation Pinning and Breakaway Observed in Ta Containing H
(a) Vol. **76**, No. 1, p. 165–177 (1983).
See also *12.1*, p. *138*; 21, p. 352.

F. Walz, H. J. Blythe, and F. Dworschak
*Annealing Behaviour of Dilute FeTi, FeCu, and FeMn Alloys in the Temperature Range above Stage III Following Low-Temperature Electron Irradiation
(a) Vol. **76**, No. 1, p. 267–276 (1983).
See also 11, p. 125; *18.2*, p. *281*; 21.1.1, p. 380.

W. Zag and K. Urban
*Temperature Dependence of the Threshold Energy for Atom Displacement in Irradiated Molybdenum
(a) Vol. **76**, No. 1, p. 285–295 (1983).
See also 11, p. 125; 21, p. 352.

S. Takahashi and K. Ikeda
Magnetic Interactions in Plastically Deformed Ni_3Mn Alloy
(a) Vol. **76**, No. 1, p. K97–K100 (1983).
See also *18.2*, p. *282*; *18.4*, p. *295*; 21.1, p. 366.

B. I. Craig
*Structure of Muons Trapped in Aluminium
(b) Vol. **116**, No. 2, p. K91–K94 (1983).
See also *13.4*, p. *177*; 21, p. 353.

G. H. van Haastert and A. van den Beukel
*Static and Dynamic Strain Aging in a Ni(C) Solid Solution
(a) Vol. **76**, No. 2, p. 609–614 (1983).
See also 21.1, p. 367.

A. H. King
Evidence of the Formation of Twins by Deformation and "Growth-Accidents" in Evaporated Thin Films of Gold
(a) Vol. **76**, No. 2, p. 629–636 (1983).
See also *1.5*, p. *36*; 21.6, p. 397.

E. Balanzat and M. Halbwachs
*A Comparative Analysis of the Ordering Kinetic Laws for the Derivation of the Vacancy Mobility Parameters in Concentrated Alloys
(a) Vol. **76**, No. 2, p. 667–673 (1983).
See also **9**, p. 94; **14.1**, p. 207; **21.1**, p. 367; **21.6**, p. 397.

G. Knuyt, L. M. Stals, L. de Schepper, and H. van Swygenhoven
*The Inhomogeneity Effect in the Vacancy–Dislocation Interaction for α-Iron
(a) Vol. **76**, No. 2, p. K175–K178 (1983).
See also **21.1.1**, p. 382.

C. W. Lung and L. Y. Xiong
*The Dislocation Distribution Function in the Plastic Zone at a Crack Tip
(a) Vol. **77**, No. 1, p. 81–86 (1983).
See also **12.1**, p. 138; **21**, p. 354.

G. I. Gudzenko, V. B. Molodkin, and V. G. Novitskii
*Dynamical Diffraction in Alloys. Influence of Long-Range Order
(a) Vol. **77**, No. 1, p. 115–120 (1983).
See also *1.3*, p. *23*; **21**, p. 354; **21.1**, p. 369.

H.-H. Potthoff
*Velocities of Dislocation Groups in Very Thin Neutron-Irradiated Copper Single Crystals Measured by Slip Line Cinematography
(a) Vol. **77**, No. 1, p. 215–224 (1983).
See also **11**, p. 128; **21.1**, p. 369.

B. Cochrane and P. J. Goodhew
TEM Images of Faceted Bubbles and Voids
(a) Vol. **77**, No. 1, p. 269–282 (1983).
See also **1.4**, p. 31; *11*, p. *128*; **21.1**, p. 369.

M. Anglada and F. Guiu
*Temperature and Strain Rate Dependence of the Cyclic Flow Stress of Molybdenum Single Crystals
(a) Vol. **77**, No. 1, p. 283–289 (1983).
See also *12.1*, p. *138*; **21**, p. 354.

M. Ohta, T. Kanadani, A. Sakakibara, H. Yamada, and M. Yamada
Effects of Grain Boundary on Age Hardening and Reversion in Al–Zn Alloys
(a) Vol. **77**, No. 1, p. K49–K51 (1983).
See also *1.3*, p. *24*; **12.1**, p. 138; **21**, p. 354; **21.1**, p. 369.

P. Beauchamp and J. Prevost
*Influence of Interaction Range on Computed Peierls Stress and Dislocation Displacement in B.C.C. Lithium
(b) Vol. **117**, No. 2, p. 499–506 (1983).
See also **21.2**, p. 387.

A. I. Osetskii
*Low-Temperature Fluctuational Motion of Dislocations in Crystals (II): Manifestations of Dissipative and Quantum Effects in the Low-Temperature Plasticity
(b) Vol. **117**, No. 2, p. 589–602 (1983).
See also 21, p. 354; 21.1, p. 370.

H. Grimm and H. Kronmüller
*Investigation of Structural Defects in the Amorphous Ferromagnetic Alloy $Fe_{40}Ni_{40}P_{14}B_6$
(b) Vol. **117**, No. 2, p. 663–674 (1983).
See also 2, p. 48; *18.2*, p. *284*; 21.1, p. 370; 21.1.1, p. 382.

M. Čerňanský
*Erratum to: Broadening and Shift of Diffraction Lines Due to Partial Coherence of X-Rays
(b) Vol. **117**, No. 2, p. 755 (1983).
See also *1.4*, p. *32*; 21.1, p. 370.

C. M. van Baal
The Influence of Texture on the X-Ray Determination of Residual Strains in Ground or Worn Surfaces
(a) Vol. **77**, No. 2, p. 521–526 (1983).
See also *1.6*, p. *40*; 12.1, p. 138; 21, p. 355.

N. Kuramochi, H. Mizubayashi, and S. Okuda
*Dislocation Breakaway at Low Temperatures in Nb
(a) Vol. **77**, No. 2, p. 633–642 (1983).
See also *12.1*, p. *138*; 21, p. 355.

Z. Gierak, J. W. Moroń, and A. Lehr
*Recovery of Structural Defects in Molybdenum
(a) Vol. **77**, No. 2, p. 775–783 (1983).
See also 14.1, p. 208; 21, p. 355.

P. K. Acharyya, A. K. Tewari, S. K. Roy, and D. L. Bhattacharya
Anharmonic Study on the Debye-Waller Factor for a Vacancy-Trapped Mössbauer Impurity in Solid Solution of Aluminium
(b) Vol. **118**, No. 1, p. 41–50 (1983).
See also *6*, p. *74*; 21, p. 355.

U. Messerschmidt, F. Appel, M. Bartsch, and R. Gerlach
*Dislocation Motion in a Dispersion Hardened Al–Zn–Mg Alloy and Its Plastic Properties
(a) Vol. **78**, No. 1, p. 93–102 (1983).
See also 21, p. 356; 21.1, p. 371; 21.3, p. 389.

C. M. Mô and P. Moser
Hydrogen in $Fe_{40}Ni_{40}P_{14}B_6$ Amorphous Alloy Studied by Magnetic After Effect, Internal Friction, and Positron Annihilation
(a) Vol. **78**, No. 1, p. 201–210 (1983).
See also *2*, p. *49*; *18.2*, p. *285*; 21.1, p. 371; 21.1.1, p. 383.

G. Dlubek, O. Brümmer, R. Krause, A. Baranowski, and B. Rozenfeld
*A Positron Study of Guinier-Preston Zones and Precipitates in Al–Zn Alloys Having Mg or Ge Additions
(a) Vol. **78**, No. 1, p. 217–228 (1983).
See also 21, p. 356; 21.1, p. 371.

E. K. Polychroniadis and P. Delavignette
*Dislocation Structures and Subgrain Boundaries in Al_3Zr
(a) Vol. **78**, No. 1, p. 287–301 (1983).
See also *1.4*, p. *32*; 21, p. 356.

M. Ohta, T. Kanadani, A. Sakakibara, H. Yamada, and M. Yamada
Effects of Surface and Thickness of Specimens on the Aging in Al–Zn Alloys
(a) Vol. **78**, No. 1, p. K23–K26 (1983).
See also *1.3*, p. *25*; 12.1, p. 138; 21, p. 357; 21.1, p. 372.

W. Puff, P. Mascher, P. Kindl, and H. Sormann
*Precise Positron Lifetime Measurements in Indium
(b) Vol. **118**, No. 2, p. 799–803 (1983).
See also 21, p. 357.

N. Shohoji
Statistical Thermodynamic Approach to Some Hyperstoichiometrie Dihydrides of Lanthanides (La, Ce, Pr, Nd, Sm) and Actinides (Np, Pu, Am)
(b) Vol. **118**, No. 2, p. 811–822 (1983).
See also *4*, p. *62*; *8*, p. *89*; 21.4, p. 392; 21.5, p. 395.

L. Bottyán, D. L. Beke, and K. Tompa
Diffusion-Induced Quadrupole Relaxation of ^{27}Al Nuclei in Dilute Al–Ti, Al–Cr, Al–Mn, and Al–Cu Alloys at High Temperatures
(b) Vol. **118**, No. 2, p. 835–842 (1983).
See also *9*, p. *95*; *19*, p. *303*; 21, p. 357; 21.1, p. 372.

Ch. Opitz, H. Müller, and H. Miessner
*EHT Calculations for Interstitial Hydrogen Bonding in Nickel
(b) Vol. **118**, No. 2, p. K85–K87 (1983).
See also *4*, p. *62*; 13.4, p. 184; 21.1, p. 372.

O. V. Presnyakova, V. I. Zaitsev, and V. A. Fomchenko
*Creep of Aluminium Single Crystals under High Pressure
(a) Vol. **78**, No. 2, p. 377–383 (1983).
See also 12.1, p. 139; 21, p. 357.

R. Z. Valiev, O. A. Kaibyshev, V. V. Astanin, and A. K. Emaletdinov
*The Nature of Grain Boundary Sliding and the Superplastic Flow
(a) Vol. **78**, No. 2, p. 439–448 (1983).
See also *1.4*, p. *33*; 21.1, p. 373.

A. R. Ali and G. Said
*On the Recovery and Annealing Kinetics of Cold-Worked Ni–4.5 at % Cu Alloy
(a) Vol. **78**, No. 2, p. 463–467 (1983).
See also 14.1, p. 209; 21.1, p. 373.

J. Woltersdorf and E. Pippel
*Influence of Substrate Deformations on Interface Structures (I): Theory
(a) Vol. **78**, No. 2, p. 475–483 (1983).
See also *1.5*, p. *37*; 21.6, p. 398.

E. Nembach
*Precipitation Hardening Caused by a Difference in Shear Modulus between Particle and Matrix
(a) Vol. **78**, No. 2, p. 571–581 (1983).
See also 12.1, p. 139; 21.1, p. 373.

A. R. Yavari
*Hardening by Thermal Vacancies and Quenching Strains in Singly-Oriented Pb Single Crystals
(a) Vol. **78**, No. 2, p. 583–588 (1983).
See also 12.1, p. 139; 21, p. 357.

V. Navrátil, M. Hamerský, P. Lukáč, V. P. Soldatov, and V. I. Startsev
*Solid Solution Hardening of Cd–Zn Single Crystals at 1.5, 2.8, and 4.2 K
(a) Vol. **78**, No. 2, p. 633–638 (1983).
See also *12.1*, p. *139*; 21, p. 358; 21.1, p. 373.

H.-J. Rockosch and Ch. Herzig
Isotope Effect of Manganese and Simultaneous Impurity Diffusion of Chromium and Zinc in Copper
(b) Vol. **119**, No. 1, p. 199–207 (1983).
See also *9*, p. *95*; 21.1, p. 374.

N. V. Moseev, V. G. Chudinov, B. N. Goshchitskii, and V. I. Protasov
*Atomic Configurations of Interstitials in Intermetallic Compound Nb_3Sn
(a) Vol. **79**, No. 1, p. 67–70 (1983).
See also 21, p. 358; 21.7, p. 401.

W. Beck and H. Kronmüller
On the Coercive Field of the Nearly Non-Magnetostrictive Amorphous Ferromagnetic Alloy $Co_{58}Ni_{10}Fe_5Si_{11}B_{16}$
(a) Vol. **79**, No. 1, p. 109–114 (1983).
See also 2, p. 51; *18.2*, p. *286*; 21.1, p. 375; 21.1.1, p. 383.

K. Iwasaki
*High-Temperature Internal Friction Peaks of Pure Aluminium
(a) Vol. **79**, No. 1, p. 115–124 (1983).
See also *12.1*, p. *139*; 21, p. 359.

R. Ramlau and H. Löffler
*On the Size–Shape Relation of Guinier-Preston Zones and α'_R-Precipitates in an Al–6.8 at% Zn Alloy
(a) Vol. **79**, No. 1, p. 141–148 (1983).
See also *1.3*, p. *26*; 21, p. 359; 21.1, p. 375.

E. PIPPEL and J. WOLTERSDORF
Influence of Substrate Deformations on Interface Structures (II): Experimental Investigations for Epitaxial Ag/Pd and Au/Pd Bicrystals
(a) Vol. **79**, No. 1, p. 189 – 198 (1983).
See also *1.5*, p. *38*; 21.6, p. 398.

G. K. BARANOVA
On the Role of Surface Oxides in the Processes of Metal Etching
(a) Vol. **79**, No. 1, p. K9 – K12 (1983).
See also *1.6*, p. *41*; 21, p. 359.

V. G. CHUDINOV and V. I. PROTAȘOV
Radiation Diffusion–Dislocation Creep in the Presence of Weertman Mechanism of Dislocation Multiplication
(a) Vol. **79**, No. 2, p. 433 – 443 (1983).
See also *9*, p. *95*; 21, p. 360.

M. KATO
*Activation Energy for Dislocation Motion in a One-Dimensional Periodic Field
(a) Vol. **79**, No. 2, p. 517 – 522 (1983).
See also 21.1.1, p. 384; 21.2, p. 388.

V. I. STASHENKO, O. A. TROITSKII, and V. I. SPITSYN
*Action of Current Pulses on Zinc Single Crystals during Creep
(a) Vol. **79**, No. 2, p. 549 – 557 (1983).
See also 21.1, p. 376.

N. G. ALEXANDROPOULOS, G. BOULAKIS, D. PAPADIMITRIOU, and I. THEODORIDOU
*A Note on X-Ray Raman Scattering from Atomic Hydrogen in $TiH_{1.9}$
(b) Vol. **120**, No. 1, p. K19 – K22 (1983).
See also *20*, p. *311*; 21.1, p. 376.

CH. SEIDEL and V. N. PRIGODIN
*Three-Dimensional CDW Ordering and Impurity Effect in Quasi-One-Dimensional Conductors
(b) Vol. **120**, No. 1, p. K73 – K77 (1983).
See also *1.2*, p. *19*; 14.2, p. 214.

C. B. CARTER
*Multiple Ribbons in Face-Centered Cubic Materials (IV): The Interaction between Double Ribbons and Inclined Dislocations
(a) Vol. **80**, No. 1, p. 51 – 64 (1983).
See also 21, p. 360.

K. H. GEORGY and R. KAMEL
*The Transient Creep of Cobalt near the Temperature of Transformation
(a) Vol. **80**, No. 1, p. 95 – 100 (1983).
See also 21.1, p. 376.

R. Tognato
*Why Metals Do Not Show Overheating
(a) Vol. **80**, No. 1, p. 147–157 (1983).
See also 1.2, p. 19; 21, p. 360.

U. Dahmen and K. H. Westmacott
Ledge Structure and the Mechanism of θ′ Precipitate Growth in Al–Cu
(a) Vol. **80**, No. 1, p. 249–262 (1983).
See also *3*, p. *59*; 21, p. 361; 21.1, p. 377.

A. Charai and C. Boulesteix
*Visualization and Study of the Shape and Size of Very Small Coherent Precipitates by a Phase Contrast Method
(a) Vol. **80**, No. 1, p. 333–341 (1983).
See also *1.4*, p. *34*; 21, p. 361.

V. G. Chudinov, V. P. Gogolin, B. N. Goshchitskii, and V. I. Protasov
*Evolution of the Cascade Region in Nb_3Sn during the Thermal Spike under Fast-Neutron Irradiation
(a) Vol. **80**, No. 1, p. 349–352 (1983).
See also 11, p. 134; 21, p. 361; 21.7, p. 402.

J. M. Khalifeh
*Asymptotic Elastic Energy in Simple Metals
(b) Vol. **120**, No. 2, p. K161–K166 (1983).
See also 13.4, p. 190; 21, p. 361.

B. I. Craig
*Low-Temperature Trapping of Positive Muons in Metals
(b) Vol. **120**, No. 2, p. K211–K214 (1983).
See also *13.4*, p. *191*; 21, p. 362.

S. V. Tsivinskii
The Theory of Electric Resistance of Pure Metals and Metals with a Given Concentration of Impurities, Vacancies, and Dislocations
(a) Vol. **80**, No. 2, p. 439–446 (1983).
See also *14.1*, p. *210*; 21, p. 362; 21.1, p. 377; 21.1.1, p. 385; 21.2, p. 388; 21.6, p. 399.

M. Z. Butt, P. Feltham, and I. M. Ghauri
*Stress-Equivalence of Solid Solution Hardening in Concentrated Ag–Al and Ag–Zn Alloys
(a) Vol. **80**, No. 2, p. K125–K128 (1983).
See also 21, p. 362; 21.1, p. 378; 21.6, p. 400.

C. Hidalgo and N. de Diego
*Positron Trapping in the Alloy Fe–1.25 wt% C
(a) Vol. **80**, No. 2, p. K145–K148 (1983).
See also *1.4*, p. *34*; 21.1.1, p. 386.

10.2 Non-Metals

Review Articles

E. G. Sieverts
*Classification of Defects in Silicon after Their g-Values
(b) Vol. **120**, No. 1, p. 11–29 (1983).
See also 13.4, p. 172; 19, p. 299; 22.1.2, p. 423.

B. Pałósz
Reasons for Polytypism of Crystals of the Type MX_2 (II): Classification of Faults and Structural Series of Polytypes; Conditions of Polytypic Growth of CdI_2, PbI_2, $SnSe_2$ and $Ti_{1.2}S_2$
(a) Vol. **80**, No. 1, p. 11–41 (1983).
See also *1*, p. *9*; *1.1*, p. *11*; 22, p. 403; 22.5, p. 473.

Original Papers and Short Notes

J. Prakash
*Effect of Ultraviolet Irradiation on Se^{2-}–Anion Vacancy Dipoles in Potassium Iodide
(b) Vol. **115**, No. 1, p. 131–139 (1983).
See also *20.1*, p. *312*; 22.5.2, p. 477.

Fat Duen Ho
*Approximate Wave Functions for the F^+-Centers in $LiNbO_3$ and $LiTaO_3$
(b) Vol. **115**, No. 1, p. 299–303 (1983).
See also 13.4, p. 172; *20.1*, p. *313*; 22.8.1, p. 512.

G. Hoebeeck, J. Provoost, and G. Jacobs
*The F' Centre in RbBr
(b) Vol. **115**, No. 1, p. K33–K36 (1983).
See also *20.1*, p. *313*; 22.5.2, p. 477.

P. Ketolainen
*On Mollwo-Ivey Relations of Mixed Alkali Halides
(b) Vol. **115**, No. 1, p. K41–K44 (1983).
See also *20.1*, p. *313*; 22.5.4, p. 488.

K. Takiyama, T. Fujita, H. Nojima, and M. Nishi
Vacuum Ultraviolet Absorption in γ-Irradiated LiCl
(b) Vol. **115**, No. 1, p. K59–K61 (1983).
See also *20.1*, p. *313*; 22.5.2, p. 477.

P. L. Kukk and A. J. Erm
*Luminescent Centre Formation Processes in Activated CdS Phosphors
(a) Vol. **75**, No. 1, p. 91–100 (1983).
See also 13.4, p. 173; *20.3*, p. *334*; 22.4.1, p. 462.

V. V. Aristov, A. V. Zaretskii, Yu. A. Osipyan, V. F. Petrenko, G. K. Strukova, and I. I. Khodos
*Dislocations in Deformed ZnSe Single Crystals as Studied by Weak-Beam Electron Microscopy
(a) Vol. **75**, No. 1, p. 101 – 106 (1983).
See also 22.4.2, p. 466.

M. Sato
*Development of Faulted Dislocation Dipoles in Silicon during High Temperature Deformation
(a) Vol. **75**, No. 1, p. 107 – 116 (1983).
See also 22.1.2, p. 423.

A. R. Lakshmanan, Bhuwan Chandra, and R. C. Bhatt
*Thermoluminescence of Z Centres in LiF (TLD-100) at Elevated Irradiation Temperatures
(a) Vol. **75**, No. 1, p. 263 – 272 (1983).
See also 11, p. 123; *20.3*, p. *335*; 22.5.2, p. 477.

H. Koizumi and T. Suzuki
*Low-Temperature Plasticity of CsBr and Inertial Effect of Dislocation Motion
(a) Vol. **75**, No. 1, p. 301 – 309 (1983).
See also 12.2, p. 141; 22.5.2, p. 477.

J. L. Demenet, P. Grosbras et J. C. Desoyer
*Fluage du germanium sous différentes orientations
(Creept of Germanium under Different Orientations)
(a) Vol. **75**, No. 1, p. K33 – K36 (1983).
See also 22.1.1, p. 418.

O. G. Alaverdova, M. Ya. Fuks, L. S. Khazan, L. P. Koval, L. A. Matveeva, I. F. Mikhailov, N. N. Soldatenko, and Yu. A. Tkhorik
*Coherency of the Interphase Boundary and Elastic Strain in the Epitaxial System Ge/GaAs
(a) Vol. **75**, No. 2, p. 367 – 371 (1983).
See also *1.5*, p. *35*; 12.2, p. 141; 22.1.1, p. 419; 22.2.1, p. 443.

P. L. Kukk and A. J. Erm
*Isochronal Annealing of Quenched CdS Phosphors
(a) Vol. **75**, No. 2, p. 393 – 399 (1983).
See also 13.4, p. 175; *20.3*, p. *336*; 22.4.1, p. 462.

J. Reichel
*The Distinction of Different Kinds of Thermally Induced Defects in Silicon
(a) Vol. **75**, No. 2, p. 459 – 464 (1983).
See also 14.3, p. 216; 22.1.2, p. 425.

A. Mazel, B. Legros-De Mauduit, and F. Reynaud
*Electron Microscopic Investigation of the Contrast on Dislocation Loops Seen Edge-on in InP (B3 Structure)
(a) Vol. **75**, No. 2, p. K121 – K123 (1983).
See also 22.2.2, p. 448.

G. V. Berenshtein, A. M. Dyachenko, and I. E. Maronchuk
*The Degradation of Semiconductor Mechanical Properties Induced by Microinclusions
(a) Vol. **75**, No. 2, p. K129 – K132 (1983).
See also *12.2*, p. *141*; 22.2.1, p. 443.

O. Boehnke, E. Nowak, P. Paufler, and H. Neumann
*On a Slip-Induced Anisotropy of the Electrical Resistance of GaP
(a) Vol. **75**, No. 2, p. K137 – K139 (1983).
See also *14.3*, p. *217*; 22.2.2, p. 448.

Z. Błaszczak, M. Ludwiczak, and F. Kaczmarek
*Bleaching of F_2 Centers in LiF:F_2 Color Center Lasers
(a) Vol. **75**, No. 2, p. K195 – K198 (1983).
See also *20.1*, p. *315*; 20.2, p. 332; 22.5.2, p. 478.

N. Ishii, M. Kumeda, and T. Shimizu
Defect States in Amorphous Silicon
(b) Vol. **116**, No. 1, p. 91 – 100 (1983).
See also *2*, p. *45*; *13.4*, p. *176*; 22.1.2, p. 426.

E. L. Andronikashvili and M. V. Galustashvili
*Influence of the Orientation of Anisotropic Impurity–Vacancy Complexes upon Dislocation Charging in NaCl Crystals
(b) Vol. **116**, No. 1, p. 335 – 338 (1983).
See also 22.5.2, p. 478.

C. K. Ong and J. Adetunji
*The Optical Absorption of F and F_H Centres in Cesium Halides
(b) Vol. **116**, No. 1, p. K19 – K21 (1983).
See also *20.1*, p. *315*; 22.5.2, p. 478.

R. B. Pode, D. G. Wakde, and B. T. Deshmukh
*Effect of Z_1 Centres on the Isothermal Stability of Colour Centres in Microcrystalline Powder of KCl
(b) Vol. **116**, No. 1, p. K39 – K42 (1983).
See also *20.1*, p. *316*; 22.5.2, p. 478.

V. V. Kalinin and N. N. Gerasimenko
*The Interaction between the Point Defects Introduced by Implantation and Dislocations in Silicon
(a) Vol. **76**, No. 1, p. 65 – 70 (1983).
See also *11*, p. *124*; 22.1.2, p. 427.

M. Halbwachs, P. Mazot, and J. Woirgard
*Anelastic Relaxation Phenomena in Plasma-Sprayed $(Al_2O_3)_3MgO$ Spinel with Reference to Diffusion Process
(a) Vol. **76**, No. 1, p. 157 – 164 (1983).
See also 9, p. 93; *12.2*, p. *142*; 22.8.2, p. 520.

V. B. Dudnikova and V. S. Urusov
*Solvus of KCl: Ca^{2+} Solid Solution by Flotation Measurements of Density of Single Crystals
(a) Vol. **76**, No. 1, p. 217–221 (1983).
See also *1.2*, p. *15*; 22.5.2, p. 478.

S. Grosswig, J. Härtwig, U. Schellenberger, and W.-D. Zimmer
Theoretical and Experimental Study of Growth Sector Boundaries in Deuterated Potassium Dihydrogen Phosphate Single Crystals
(a) Vol. **76**, No. 1, p. 241–247 (1983).
See also *1.5*, p. *36*; *3*, p. *57*; 22.8.1, p. 513.

U. Messerschmidt, Y. Nishino, T. Imura, and H. Saka
*X-Ray Topographic In-Situ Observation of Slip Band Propagation in MgO Single Crystals
(a) Vol. **76**, No. 1, p. 277–284 (1983).
Erratum
(a) Vol. **80**, No. 1, p. 411 (1983).
See also 1.4, p. 30; 22.6, p. 492.

A. I. Pekarev, G. F. Krasnova, and G. Z. Nemtsev
*Orientation Dependence of Oxidation Stacking Fault Density in Silicon
(a) Vol. **76**, No. 1, p. 327–330 (1983).
See also *1.6*, p. *39*; 22.1.2, p. 427.

A. F. Rakov
*Neutron Induced Colour and Luminescence Centres in Yttrium–Aluminium Garnet Single Crystals
(a) Vol. **76**, No. 1, p. K57–K60 (1983).
See also *11*, p. *126*; 20.1, p. 316; 20.3, p. 337; 22.8.2, p. 520.

M. Nagabhooshanam and V. Hari Babu
*Conductivity and Hall Coefficient Studies in Plastically Deformed CdTe Single Crystals
(a) Vol. **76**, No. 1, p. K89–K92 (1983).
See also *14.3*, p. *218*; 22.4.3, p. 468.

E. M. Trukhanov
*Accommodation of Misfit in Heterostructures with Continuous and Island Films
(a) Vol. **76**, No. 2, p. 437–445 (1983).
See also *1.5*, p. *36*; 22.1.1, p. 420.

N. L. Sizova, V. D. Natsik, I. I. Gorina, and V. R. Regel
*The Investigation of Heat Release in Slip Bands in Alkali Halide Crystals
(a) Vol. **76**, No. 2, p. 447–451 (1983).
See also 22.5.2, p. 479.

G. N. Gaidukov, S. K. Maksimov, and A. A. Podrezov
*The Effect of the Crystal Anisotropy on the Energy of the Dislocation–Impurity Atmosphere Elastic Interaction in Silicon
(a) Vol. **76**, No. 2, p. 453–459 (1983).
See also *12.2*, p. *142*; 22.1.2, p. 428.

V. V. Aristov, M. N. Zolotukhin, V. V. Kveder, Yu. A. Osipyan, I. I. Snighireva, and I. I. Khodos
*The Effect of Annealing on the Dislocation Dissociation in Plastically Deformed Silicon
(a) Vol. **76**, No. 2, p. 485 – 491 (1983).
See also *1.4*, p. *30*; 22.1.2, p. 428.

M. Mühlberg and D. Hesse
TEM Precipitation Studies in Te-Rich As-Grown PbTe Single Crystals
(a) Vol. **76**, No. 2, p. 513 – 524 (1983).
See also *1.4*, p. *30*; 22, p. 406.

K. N. Reddy, MD. Iqbal Ahmed, and V. H. Babu
*On Production of Z_1 Centres in NaCl:Sm^{2+} Crystals
(a) Vol. **76**, No. 2, p. K167 – K170 (1983).
See also 20.1, p. 317; 20.3, p. 338; 22.5.2, p. 479.

K. Somaiah and V. Hari Babu
*Ionic Conductivity of BaFCl
(b) Vol. **117**, No. 1, p. 75 – 79 (1983).
See also *14.4*, p. *251*; 22.5, p. 474.

J. Prakash
*Photo-Chemical Reaction and Photo-Induced Polarization Studies in Sulphur-Doped Potassium Bromide
(b) Vol. **117**, No. 1, p. 161 – 169 (1983).
See also *20.1*, p. *317*; 22.5.2, p. 479.

S. B. S. Sastry and S. Nagarajan
Thermoluminescence and Optical Absorption Studies on Rubidium Bromide Doped with Barium
(b) Vol. **117**, No. 1, p. 171 – 175 (1983).
See also *20.1*, p. *318*; *20.3*, p. *338*; 22.5.2, p. 479.

C. Ance and N. M. Ravindra
*Departure of Hydrogen from a-Si:H. A Spectroscopic Study
(a) Vol. **77**, No. 1, p. 241 – 248 (1983).
See also *2*, p. *48*; 22.1.2, p. 430.

K. N. Subramanian
*Deformation Structure in LiF Single Crystals
(a) Vol. **77**, No. 1, p. 361 – 367 (1983).
Erratum
(a) Vol. **78**, No. 2, p. 760 (1983).
See also 22.5.2, p. 479.

H. B. Lal and O. P. Srivastava
Electrical Transport Studies in Copper Mercuric Iodide
(a) Vol. **77**, No. 1, p. 405 – 412 (1983).
See also *14.4*, p. *251*; *15*, p. *261*; 22.5.4, p. 488.

M. Casalboni, R. Francini, U. M. Grassano, and R. Pizzoferrato
*Nonlinear Excitation of New Emission Bands in KCl Crystals Containing Color Centers
(b) Vol. **117**, No. 2, p. 493–498 (1983).
See also *20.3*, p. *339*; 22.5.2, p. 479.

P. A. Frugoli and C. A. Pimentel
*Point Defect Aggregates in γ-Irradiated LiF Single Crystals
(b) Vol. **117**, No. 2, p. 549–556 (1983).
See also 22.5.2, p. 479.

K. N. Reddy, Md. Iqbal Ahmed, U. V. Subba Rao, and V. Hari Babu
Thermoluminescence and Optical Absorption Studies of Z_1-Centres in NaCl Crystals Doped with Terbium
(a) Vol. **77**, No. 2, p. 643–648 (1983).
See also *20.1*, p. *321*; *20.3*, p. *340*; 22.5.2, p. 480.

A. Szekeres and S. Alexandrova
*RF Annealing of Defects Induced in SiO_2 by Oxygen Plasma
(a) Vol. **77**, No. 2, p. 721–724 (1983).
See also 22.1.2, p. 431.

G. A. Andreev and V. A. Klimov
*Preferential Precipitation on Extended Edge Dislocations in NaCl Crystals
(a) Vol. **77**, No. 2, p. 793–798 (1983).
See also 22.5.2, p. 480.

O. Breitenstein and T. Wosinski
*Scanning-DLTS Investigation of the EL 2 Level in Plastically Deformed GaAs
(a) Vol. **77**, No. 2, p. K107–K110 (1983).
See also *13.4*, p. *181*; 14.3, p. 221; 22.2.1, p. 444.

P. Zaumseil and U. Winter
*Detection of Surface Imperfections at Polished Silicon Wafers by TCD Measurements
(a) Vol. **77**, No. 2, p. K139–K141 (1983).
See also *1.6*, p. *40*; 22.1.2, p. 431.

H. J. Fuchs, W. Karthe, and S. Börner
*On the Kinetics of a Paramagnetic Oxygen Centre in CaF_2
(b) Vol. **118**, No. 1, p. 211–215 (1983).
See also 9, p. 94; 19, p. 302; 22.5.3, p. 487.

U. Pilatus and F. Fischer
*Thermally Stimulated Depolarization of AsO_2^- Centres in Alkali Halide Crystals
(b) Vol. **118**, No. 1, p. 217–221 (1983).
See also *14.4*, p. *252*; 22.5.2, p. 480.

J. Prakash and Rahul
A Method for the Evaluation of Dielectric Relaxation Parameters in Electronic Excited State
(b) Vol. **118**, No. 1, p. 229–237 (1983).
See also *14.4*, p. *252*; 22.5.2, p. 480.

B. RAMBABU, C. RAMASASTRY, and B. V. R. CHOWDARI
*Radiation Damage Studies of Co^{2+} and Ni^{2+} Doped NH_4Cl Crystals
(b) Vol. 118, No. 1, p. 381–387 (1983).
See also 19, p. 302; *20.1*, p. *322*; 22.5, p. 474.

F. CALLENS, P. MATTHYS, and E. BOESMAN
*ESR of a New Sulphur Centre in KCl
(b) Vol. 118, No. 1, p. K35–K37 (1983).
See also *19*, p. *302*; 22.5.2, p. 480.

R. CAPELLETTI, O. CARRANZA TORRES, J. GARCÍA SOLÉ, and M. MANFREDI
*On the 430 nm Emission Band in the NaCl:Pb^{++} System
(a) Vol. 78, No. 1, p. 35–39 (1983).
See also 13.4, p. 182; *20.3*, p. *342*; 22.5.2, p. 481.

B. A. DEVLIN and J. CORISH
*Impurity–Vacancy Dipoles in Cadmium-Doped Silver Bromide Crystals
(a) Vol. 78, No. 1, p. 53–60 (1983).
See also 22.5.1, p. 476.

HJ. BERNHARDT
Vacuum-UV Absorption and Thermoluminescence of X-Rayed Quartz Crystals
(a) Vol. 78, No. 1, p. 61–67 (1983).
See also *20.1*, p. *323*; *20.3*, p. *342*; 22.6, p. 494.

J. POŹNIAK and G. BERG
*Theoretical Analysis of the Dipole Polarization and Depolarization in Me^{++}-Doped NaCl-Type Alkali Halide Crystals (II): Next-Nearest Neighbour Approach
(a) Vol. 78, No. 1, p. 69–76 (1983).
See also 14.4, p. 252; 22.5.2, p. 481.

G. BERG, A. PIPPEL, and F. FRÖHLICH
*Influence of the Variation of the Association Energy on the Equilibrium Concentration of Dipole Aggregates in Ionic Crystals
(a) Vol. 78, No. 1, p. 77–83 (1983).
See also 22.5.2, p. 481.

J. GARCÍA SOLÉ, C. ZALDO, and F. AGULLÓ-LÓPEZ
*Optical Spectra of Pb^{2+} to Monitor the Precipitation Behaviour of NaCl:Cd
(a) Vol. 78, No. 1, p. 85–91 (1983).
See also *20.1*, p. *323*; 20.3, p. 342; 22.5.2, p. 481.

A. I. PROKHVATILOV and A. P. ISAKINA
*Lattice Parameters, Thermal Expansion Coefficients, and Vacancy Density in Solid CD_4
(a) Vol. 78, No. 1, p. 147–155 (1983).
See also 1.1, p. 11; 8, p. 88; 12.2, p. 143; 22.9, p. 529.

T. KACZOR, S. NOWAK, and J. ŻMIJA
*Description of Coloration and Discoloration Processes in Inorganic Photochromic Materials
(a) Vol. 78, No. 1, p. 187–192 (1983).
See also *20*, p. *309*.

·L. Malicskó and R. Scholz
Microtwins in $LiNbO_3$ Crystals Observed by High-Voltage and High-Resolution Electron Microscopy
(a) Vol. 78, No. 1, p. K1–K4 (1983).
See also *1.4*, p. *32*; 22.8.1, p. 515.

S. J. Pearton and J. M. Kahn
*Dislocations in Germanium: Effects of Plasma Hydrogenation
(a) Vol. 78, No. 1, p. K65–K69 (1983).
See also *13.4*, p. *183*; 22.1.1, p. 421.

B. S. Acharya and A. K. Panda
*Optical Absorption Spectra of Mercury Centres in KCl and NaCl
(a) Vol. 78, No. 1, p. K81–K85 (1983).
See also *20.1*, p. *323*; 22.5.2, p. 481.

M. Wurlitzer
Richter Type Magnetic After-Effect in YIG at Helium Temperature
(a) Vol. 78, No. 1, p. K87–K91 (1983).
See also *18.3*, p. *293*; 22.8.2, p. 522.

K. O. Kechechyan and A. A. Kirakosyan
*Magneto-Resistance of a Semiconductor with the Dislocation Wall in the Quantum Limit
(b) Vol. 118, No. 2, p. 521–524 (1983).
See also *14.3*, p. *222*; 22, p. 409.

B. Rambabu, C. Ramasastry, and B. V. R. Chowdari
*Cobalt Centres in Ammonium Chloride: Optical Absorption Studies
(b) Vol. 118, No. 2, p. 541–549 (1983).
See also 13.4, p. 183; *20.1*, p. *323*; 22.5, p. 475.

N. R. Nurtdinov, R. Stegmann, and M. Griepentrog
The Influence of the Variation of the Chemical Composition and of the Nitrogen Doping Level on the Energy Spectra of the Bound Excitons in $GaAs_{1-x}P_x$:N
(b) Vol. 118, No. 2, p. 613–621 (1983).
See also *13.5.1*, p. *197*; *20.3*, p. *342*; 22.2.4, p. 455.

Debendranath Sahoo
*Heterodyne Correlation of Scattered Light from a Sr^{++}-Doped NaCl Crystal. A Feasibility Study
(b) Vol. 118, No. 2, p. 623–630 (1983).
See also *20*, p. *309*; 22.5.2, p. 482.

Ch. Seidel and V. N. Prigodin
*The Impurity Effect on the Phase Transitions in Quasi-One-Dimensional Conductors
(b) Vol. 118, No. 2, p. 713–724 (1983).
See also *1.2*, p. *18*; 14.2, p. 213; 22.9, p. 529.

S. Zazubovich, A. Usarov, and Zh. Egemberdiev
*Photothermal Creation of $Cu^{\circ}v_a^+$ Centres and Mobility of Anion Vacancies in X-Irradiated KCl:Cu Crystals
(b) Vol. **118**, No. 2, p. 789–798 (1983).
See also 20.1, p. 324; *20.3*, p. *342*; 22.5.2, p. 482.

H. Opyrchał, K. D. Nierzewski, and H. Drulis
Effect of γ-Irradiation on EPR Spectra of Eu^{2+} Doped KCl and NaCl Crystals
(b) Vol. **118**, No. 2, p. K125–K127 (1983).
See also *11*, p. *130*; *19*, p. *303*; 22.5.2, p. 482.

D. N. Popov, N. M. Nancheva, B. N. Zlatov, and G. P. Georgiev
*Positron Angular Correlation in Uncoloured and Neutron Irradiated LiF and LiF:OH
(b) Vol. **118**, No. 2, p. K129–K132 (1983).
See also *11*, p. *130*; 22.5.2, p. 482.

B. V. Kobrin, R. M. Kupriyanova, V. S. Minaev, E. P. Prokopiev, and V. P. Shantarovich
*Microstructure and Imperfections of Glassy and Glass-Ceramic Telluride Semiconductors on the Basis of Positron Annihilation
(a) Vol. **78**, No. 2, p. 385–389 (1983).
See also *2*, p. *50*; 22, p. 410.

R. Köhler and W. Möhling
Sensitivity of Plane Wave Topography to Microdefects
(a) Vol. **78**, No. 2, p. 489–496 (1983).
See also *1.4*, p. *33*; 22.1.2, p. 433.

R. Gleichmann, H. Blumtritt, and J. Heydenreich
*New Morphological Types of CuSi Precipitates in Silicon and Their Electrical Effects
(a) Vol. **78**, No. 2, p. 527–538 (1983).
See also 14.3.3, p. 240; 22.1.2, p. 433.

M. Suezawa and K. Sumino
*The Nature of Photoluminescence from Plastically Deformed Silicon
(a) Vol. **78**, No. 2, p. 639–645 (1983).
See also *20.3*, p. *343*; 22.1.2, p. 433.

Y. Nishino and T. Imura
*Dislocation Configurations Induced by Cyclic Deformation in Silicon Crystals
(a) Vol. **78**, No. 2, p. 655–663 (1983)
See also 22.1.2, p. 433.

J. Manam, R. K. Gartia, and V. V. Ratnam
*Thermoluminescence of Z_3 Centres in X-Irradiated Ca-Doped KCl Crystals
(a) Vol. **78**, No. 2, p. 685–688 (1983).
See also *20.3*, p. *343*; 22.5.2, p. 482.

G. Kameswara Rao
*$a\langle 100\rangle$ Dislocation Loops in MgO. Unusual Configurations
(a) Vol. **78**, No. 2, p. 697–703 (1983).
See also 22.6, p. 495.

P. V. Petrashen and I. L. Shulpina
*X-Ray Determination of the Surface Compression Force in the Ion-Doped Layer in Silicon
(a) Vol. **78**, No. 2, p. K105–K108 (1983).
See also *11*, p. *131*; 22.1.2, p. 434.

N. I. Berezhnov, V. F. Stelmakh, and A. R. Chelyadinskii
*The Interstitial Type Defects in Ion Implanted Silicon
(a) Vol. **78**, No. 2, p. K121–K125 (1983).
See also *11*, p. *131*; 22.1.2, p. 434.

K. Schmalz and P. Gaworzewski
*On the Effect of Ambients on the Formation of Oxygen-Related Donors in Cs-Si
(a) Vol. **78**, No. 2, p. K141–K145 (1983).
See also 13.4, p. 184; 14.3, p. 223; 22.1.2, p. 434.

O. K. Gorodnichenko, V. A. Presnov, and N. D. Vasilenko
*Deformation of Epitaxial Structures on the Basis of GaAs in the Process of Their Production
(a) Vol. **78**, No. 2, p. K171–K173 (1983).
See also *1.5*, p. *38*; 22.2.1, p. 446.

L. I. Datsenko, E. N. Kislovskii, V. I. Khrupa, and K. M. Shevtsov-Kazmirchuk
*Effect of Twisting of the Diffraction Net Planes on the Integral Intensities at Laue Diffraction of X-Rays in Elastically Bent Silicon Crystals
(a) Vol. **78**, No. 2, p. K175–K178 (1983).
See also *1.1*, p. *12*; 22.1.2, p. 434.

V. Ziraps
*Ionic and Electronic Processes during Frenkel Defect Annealing in KBr Crystals above LHeT
(b) Vol. **119**, No. 1, p. 49–59 (1983).
See also 11, p. 131; 13.4, p. 185; 20.1, p. 325; 22.5.2, p. 482.

T. Ito
*A Pseudopotential Approach to the Lattice Parameters for GaAs Containing Impurity Elements
(b) Vol. **119**, No. 1, p. 307–314 (1983).
See also 12.2, p. 143; 22.2, p. 442; 22.2.1, p. 446.

J. Horák, P. Lošťák, and A. Montaner
*Point Defects in Ga-Doped Bi_2Se_3 Single Crystals
(b) Vol. **119**, No. 1, p. K17–K21 (1983).
See also 14.3, p. 223; 20.1, p. 326; 22.7, p. 499.

V. V. Aristov, Yu. A. Osipyan, R. Scholz, I. I. Snighireva, I. I. Khodos, and S. A. Shevchenko
*The Effect of Annealing on the Defect Structure and Electrical Properties of Deformed Single Crystals of Ge
(a) Vol. **79**, No. 1, p. 47–55 (1983).
See also *14.3*, p. *224*; 22.1.1, p. 421.

J. Manam, R. K. Gartia, and V. V. Ratnam
*Effect of Post-Irradiation Deformation on Thermoluminescence Spectra of KCl:Ca and NaCl:Ca Crystals
(a) Vol. **79**, No. 1, p. 167–171 (1983).
See also *20.3*, p. *344*; 22.5.2, p. 483.

H. Tsuya and F. Shimura
*Transient Behaviour of Intrinsic Gettering in CZ Silicon Wafers
(a) Vol. **79**, No. 1, p. 199–206 (1983).
See also *1.6*, p. *41*; 14.3.3, p. 240; 22.1.2, p. 435.

M. T. Sebastian and P. Krishna
*Mechanism of Solid State Transformations in Single Crystals of $Zn_xCd_{1-x}S$
(a) Vol. **79**, No. 1, p. 271–280 (1983).
See also *1.2*, p. *18*; 22.4.4, p. 471.

A. Wolkenberg
A Mechanism for the Effect of Doping on the Silicon Native Oxide Thickness
(a) Vol. **79**, No. 1, p. 313–322 (1983).
See also *1.6*, p. *41*; 22.1.2, p. 435; 22.6, p. 496.

O. K. Gorodnichenko, E. A. Glushkov, V. F. Kovalenko, and N. D. Vasilenko
*Strengthening of GaAs by Impurity Atom–Vacancy Complexes
(a) Vol. **79**, No. 1, p. 331–335 (1983).
See also 20.3, p. 344; 22.2.1, p. 446.

K. Biskupska, J. Tatarkiewicz, and K. Wieteska
Comparison of Fast Electron and Thermal Annealing of Proton Bombarded Silicon – X-Ray and IR Spectroscopy Study
(a) Vol. **79**, No. 1, p. K13–K15 (1983).
See also *11*, p. *132*; 20.1, p. 326; 22.1.2, p. 435.

V. I. Barbashov and Yu. B. Tkachenko
*Dislocation Mobility in Hydrostatically Compressed KBr Crystals
(a) Vol. **79**, No. 1, p. K17–K19 (1983).
See also *12.2*, p. *144*; 22.5.2, p. 483.

V. P. Dobrego and A. K. Fedotov
*Influence of Growth Dislocations on Low Temperature Conductivity in Irradiated p-Germanium
(a) Vol. **79**, No. 1, p. K45–K48 (1983).
See also 11, p. 132; *14.3*, p. *224*; 22.1.1, p. 421.

B. T. Kolomiets, T. N. Mamontova, A. V. Chernyshov, G. Z. Vinogradova, and N. V. Demokritova
*The Effect of Low Oxygen Concentrations on the Photoluminescence of Vitreous Selenium
(a) Vol. **79**, No. 1, p. K89–K92 (1983).
See also 2, p. 52; *20.3*, p. *345*; 22.3, p. 458.

Z. G. Akhvlediani, I. G. Akhvlediani, and T. L. Kalabegishvili
*Effect of Combined Mechanical Load and Irradiation on Hydrogen Stabilization in LiF:OH
(b) Vol. **119**, No. 2, p. 503–506 (1983).
See also 19, p. 304; 20.1, p. 327; 22.5.2, p. 483.

A. Morimoto, S. Oozora, M. Kumeda, and T. Shimizu
*Annealing Behaviour of Hydrogenated Amorphous Silicon–Nitrogen Alloy Films Prepared by Sputtering
(b) Vol. **119**, No. 2, p. 715–720 (1983).
See also *2*, p. *52*; 19, p. 304; 20.1, p. 327; 22, p. 412.

N. A. Tyapunina and Yu. A. Ivashkin
*Excess Concentration of Point Defects in Alkali Halide Crystals Exposed to Ultrasonic Waves
(a) Vol. **79**, No. 2, p. 351–359 (1983).
See also *7*, p. *83*; 22.5.2, p. 484.

W. Gräfe
*A Method for the Determination of a Static Fatigue Limit of Glasses
(a) Vol. **79**, No. 2, p. 403–410 (1983).
See also *2*, p. *53*; 22.6, p. 496.

G. Ya. Akimov and I. Yu. Prokhorov
*Fracture of Hydrostatically Compressed NaCl Single Crystals
(a) Vol. **79**, No. 2, p. 423–431 (1983).
See also *12.2*, p. *144*; 22.5.2, p. 484.

S. V. Moharil
Track Interaction Model for Supralinearity and Sensitization in LiF-TLD 100
(a) Vol. **79**, No. 2, p. 599–605 (1983).
See also 13.4, p. 187; *20.3*, p. *345*; 22.5.2, p. 484.

F. H. Baumann and W. Schröter
*Deformation-Induced Defects in p-Type Germanium
(a) Vol. **79**, No. 2, p. K123–K126 (1983).
See also 22.1.1, p. 421.

K. D. Glinchuk, N. M. Litovchenko, Z. A. Salnik, and S. I. Skryl
*Effect of Heat Treatment on the Minority Carrier Lifetime in Oxygen-Containing Silicon
(a) Vol. **79**, No. 2, p. K159–K163 (1983).
See also *13.4*, p. *188*; 22.1.2, p. 436.

L. Bosi, P. Gagliardelli, and M. Nimis
*More on the F-Center Decay in Additively Colored NaF:LiF
(b) Vol. **120**, No. 1, p. 41–48 (1983).
See also *20.3*, p. *346*; 22.5.2, p. 484.

S. Bhan
Erratum to: Thermoluminescence of NaF:Ca^{2+} Single Crystals
(b) Vol. **120**, No. 1, p. 451 (1983).
See also *20.3*, p. *346*; 22.5.2, p. 484.

H. Müller, P. Haasen, and W. Schröter
*Thermopower of Plastically Deformed Cadmium Telluride
(a) Vol. **80**, No. 1, p. 223 – 230 (1983).
See also 14.3.4, p. 248; *15*, p. *263*; 22.4.3, p. 469.

M. Servidori, C. Dal Monte, and Q. Zini
*Double Crystal X-Ray Analysis of Phosphorus Precipitation in Supersaturated Si–P Solid Solutions
(a) Vol. **80**, No. 1, p. 277 – 285 (1983).
See also 11, p. 133; 22.1.2, p. 438.

D. Lapraz and A. Baumer
*Thermoluminescent Properties of Synthetic and Natural Fluorapatite, $Ca_5(PO_4)_3F$
(a) Vol. **80**, No. 1, p. 353 – 366 (1983).
See also *20.3*, p. *347*; 22.8, p. 509.

A. K. Rai, R. S. Bhattacharya, and P. P. Pronko
Annealing of Amorphous Layer in Si Implanted GaAs
(a) Vol. **80**, No. 1, p. 367 – 372 (1983).
See also *1.4*, p. *34*; 11, p. 134; 22.1.2, p. 438.

A. E. Nosenko, A. I. Otko, V. V. Kravchishyn, and L. K. Kostyk
*Growth Macrodefects in $Ca_3Ga_2Ge_3O_{12}$ Single Crystals
(a) Vol. **80**, No. 1, p. 373 – 386 (1983).
See also 1.4, p. 34; 22.8.2, p. 523.

V. B. Dudnikova, V. S. Urusov, and L. N. Bannykh
*Solvus of $KCl:Ba^{2+}$ Solid Solution from Flotation Measurements of Single Crystal Density
(a) Vol. **80**, No. 1, p. 399 – 402 (1983).
See also *1.3*, p. *27*; 22.5.2, p. 484.

H. Jain
Haven Ratio for Sodium in Synthetic α-Quartz
(a) Vol. **80**, No. 1, p. K15 – K20 (1983).
See also *9*, p. *96*; 14.4, p. 254; 22.6, p. 497.

I. E. Bolotov and V. L. Prilepo
On the Vacancy Mechanism of the Occurrence of Compressive Stresses during the Crystallization of Amorphous Films
(a) Vol. **80**, No. 1, p. K67 – K70 (1983).
See also *1.5*, p. *38*; 3, p. 59; 12.2, p. 144; 22.1.3, p. 441; 22.4.1, p. 465.

K. Budde and H. Richtering
Incorporation and Diffusion of Cu(I)-Ions in CsI by NMR Measurements
(b) Vol. **120**, No. 2, p. 595 – 599 (1983).
See also *9*, p. *96*; *19*, p. *305*; 22.5.2, p. 485.

P. K. Chakraborty and K. V. Rao
*F-Band Absorption and Thermoluminescence of KBr Single Crystals X-Ray Irradiated under High AC Fields
(b) Vol. **120**, No. 2, p. K107 – K110 (1983).
See also 11, p. 134; *20.3*, p. *347*; 22.5.2, p. 485.

P. D. Alekseev, G. I. Baranov, E. P. Kurakina, and K. A. Maltsev
Formation of Hydrogen Bonding in Doped OH-Group Alkali Halide Crystals by the Action of γ-Radiation
(b) Vol. **120**, No. 2, p. K119–K121 (1983).
See also *11*, p. *134*; *20.1*, p. *330*; 22.5.2, p. 485.

K. Asami and M. Ishiguro
*Pressure Effects on the Absorption and Emission of F_A Centers in KCl
(b) Vol. **120**, No. 2, p. K205–K209 (1983).
See also *13.4*, p. *191*; 20.1, p. 330; 20.3, p. 347; 22.5.2, p. 485.

J. H. Basson and H. Booyens
*The Introduction of Misfit Dislocations in HgCdTe Epitaxial Layers
(a) Vol. **80**, No. 2, p. 663–668 (1983).
See also *1.5*, p. *38*; 22.4.4, p. 472.

I. Manika and J. Teteris
*Compositional Dependence of Photo-Induced Microhardness Changes in As–Se Films
(a) Vol. **80**, No. 2, p. K121–K123 (1983).
See also *12.2*, p. *145*; 22, p. 415.

G. A. Andreev
*Annealing of Deformation-Induced Vacancy Clusters in NaCl Crystals
(a) Vol. **80**, No. 2, p. K129–K132 (1983).
See also 22.5.2, p. 485.

11. Irradiation Effects (X-Ray Diffraction Investigations See 1 and 10)

Review Article

T. D. DZHAFAROV
Photostimulated Diffusion in Semiconductors
(a) Vol. **79**, No. 1, p. 11 – 45 (1983).
See also *9*, p. *93*; 14.3.3, p. 236; 14.3.4; p. 243; 16, p. 264; 22, p. 403.

Original Papers and Short Notes

A. M. TARATIN and S. A. VOROBEV
*Electronic Energy Loss of Channeled Negative Pions
(b) Vol. **115**, No. 1, p. 37 – 46 (1983).
See also 13, p. 146; 22.1.2, p. 423.

SH. SH. IBRAGIMOV, O. P. MAKSIMKIN, and SH. B. SHIGANAKOV
*Grain-Boundary Deformation in Nickel Irradiated by α-Particles
(a) Vol. **75**, No. 1, p. 55 – 58 (1983).
See also *1.4*, p. *29*; 21.1, p. 363.

S. UNTERRICKER and F. SCHNEIDER
*TDPAC Investigation of ^{111}Cd in Silicon after Recoil Implantation of the Parent Nuclei ^{111}In
(a) Vol. **75**, No. 1, p. 155 – 158 (1983).
See also 22.1.2, p. 424.

B. RAUSCHENBACH and K. HOHMUTH
*A Simple Approach to the Analysis of Ion Collision Cascade in Solids Based on the Shock Wave Model
(a) Vol. **75**, No. 1, p. 159 – 168 (1983).
See also 21, p. 350.

F. WALZ, H. J. BLYTHE, and F. DWORSCHAK
Annealing Behaviour of Dilute FeCu, FeMn, and FeTi Alloys in the Temperature Range up to Stage III, Following Low-Temperature Electron Irradiation
(a) Vol. **75**, No. 1, p. 235 – 242 (1983).
See also *10.1*, p. *99*; *18.2*, p. *280*; 21.1, p. 364; 21.1.1, p. 379.

T. M. ROBINSON
*Xenon Ion Irradiation of α-Fe
(a) Vol. **75**, No. 1, p. 243 – 249 (1983).
See also 21.1.1, p. 379.

H. KERKOW, G. KREYSCH, B. LUKASCH, and G. LÜCK
*Calorimetric Determination of the Formation Energy for an Amorphous Layer on a Crystalline Silicon Substrate
(a) Vol. **75**, No. 1, p. 251 – 254 (1983).
See also 2, p. 43; 22.1.2, p. 424.

A. R. LAKSHMANAN, BHUWAN CHANDRA, and R. C. BHATT
Thermoluminescence of Z Centres in LiF (TLD-100) at Elevated Irradiation Temperatures
(a) Vol. **75**, No. 1, p. 263 – 272 (1983).
See also *10.2*, p. *109*; *20.3*, p. *335*; 22.5.2, p. 477.

V. ATZRODT, W. TITEL, TH. WIRTH, and H. LANGE
*AES Studies on Thin Film $MoSi_2$
(a) Vol. **75**, No. 1, p. K15 – K20 (1983).
See also *1.6*, p. *39*; 21, p. 350.

C. GUNKEL, D. SCHALCH, and A. SCHARMANN
*Electrical Conduction in Silicon Nitride Thin Films under 15 keV He^+ Irradiation
(a) Vol. **75**, No. 1, p. K57 – K62 (1983).
See also 2, p. 44; *14.3.1*, p. *229*; 22, p. 403.

S. A. AZIMOV, D. A. ARONOV, V. M. RUBINOV, and M. TUICHIEV
*Quantum Yield of Current Carriers Excited by Electron Beams and Electron Penetration Depth in Thin GaP Films
(a) Vol. **75**, No. 1, p. K63 – K68 (1983).
See also *14.3.1*, p. *229*; 22.2.2, p. 448.

V. N. BRUDNYI, A. I. POTAPOV, and YU. V. RUD
*Electrical Properties of H^+-Irradiated p-$ZnGeAs_2$
(a) Vol. **75**, No. 1, p. K73 – K76 (1983).
See also *14.3*, p. *215*; 22.8, p. 502.

BAI-XIN LIU
*A Metastable Electron Compound Formed by Ion Irradiation
(a) Vol. **75**, No. 1, p. K77 – K81 (1983).
See also *1.5*, p. *35*; 21.1, p. 364; 21.6, p. 396.

K. B. KADYRAKUNOV, E. V. NIDAEV, I. E. TYSHCHENKO, and L. S. SMIRNOV
*Decomposition of Supersaturated Antimony Solutions in Silicon Created by Pulsed Annealing of Ion-Doped Layers
(a) Vol. **75**, No. 1, p. K107 – K109 (1983).
See also *14.3*, p. *215*; 22.1.2, p. 425.

A. P. TYUTNEV, G. S. MINGALEEV, A. V. VANNIKOV, A. F. DUNAEV, V. S. SAENKO, and V. S. LIKHOVIDOV
*Kinetics of Radiation-Induced Conductivity in Some Polymers
(a) Vol. **75**, No. 2, p. 385 – 392 (1983).
See also *14.3*, p. *216*; 22.9, p. 525.

K. B. KADYRAKUNOV, E. V. NIDAEV, A. E. PLOTNIKOV, L. S. SMIRNOV, I. G. MELNIK, and M. V. MAKEEV
Flash Lamp Annealing of Ion-Implanted Polycrystalline Silicon
(a) Vol. **75**, No. 2, p. 483 – 488 (1983).
See also *14.3.1*, p. *229*; 22.1.2, p. 426.

G. Oelgart and H. Scholz
*Determination of Electron Depth–Dose Function for Kilovolt Electrons on GaAsP
(a) Vol. **75**, No. 2, p. 547–553 (1983).
See also *14.3.3*, p. *237*; 22.2.4, p. 454.

V. D. Akhmetov, V. V. Bolotov, L. S. Smirnov, and V. A. Kharchenko
The Defect Accumulation and Irradiation-Induced Lattice Strain in Si at High-Dose Neutron Irradiation
(a) Vol. **75**, No. 2, p. 601–606 (1983).
See also *13.4*, p. *175*; *20.1*, p. *314*; 22.1.2, p. 426.

G. Neubert and S. Rogaschewski
*Determination of the Electron Backscattering Coefficient for Thin Films Using Double Layers
(a) Vol. **75**, No. 2, p. K191–K193 (1983).
See also *17*, p. *269*.

T. G. Abdel-Malik, A. A. Aly, A. M. Abdeen, and H. A. Motaweh
The Influence of the Surface Barrier on the Trajectory of Sputtered Particles
(b) Vol. **116**, No. 1, p. 169–172 (1983).
See also *13.3*, p. *168*; 21.1, p. 365.

N. Takeuchi, K. Inabe, and S. Nakamura
*Paramagnetic Species in NaCl:Ca, Cu Crystals Irradiated at RT
(b) Vol. **116**, No. 1, p. K5–K8 (1983).
See also *19*, p. *301*; 22.5.2, p. 478.

Yu. V. Bulgakov, K. Lenkeit, and R. Stolle
*Energy Distribution Measurements of 300 keV Transmitted Protons at the Axial-to-Planar Channeling Transition in Silicon
(b) Vol. **116**, No. 1, p. K23–K26 (1983).
See also 22.1.2, p. 427.

Yu. V. Bulgakov and K. Lenkeit
*The Energy Dependence of the Angles of Resonance Dechanneling of Ions in the ⟨110⟩ → (110) Axial-to-Planar Channeling Transition in Silicon
(b) Vol. **116**, No. 1, p. K27–K30 (1983).
See also 22.1.2, p. 427.

S. V. Yalyshko and R. Wedell
*Stopping Power for N^+ Ions with Energies from 160 to 440 keV in Aluminium
(b) Vol. **116**, No. 1, p. K47–K50 (1983).
See also 21, p. 352.

V. V. Kalinin and N. N. Gerasimenko
*The Interaction between the Point Defects Introduced by Implantation and Dislocations in Silicon
(a) Vol. **76**, No. 1, p. 65–70 (1983).
See also *10.2*, p. *110*; 22.1.2, p. 427.

E. V. Anufriev and V. A. Gurtov
Accumulation and Discharging of the Hole Charge in Oxide Layers of MIS Structures
(a) Vol. **76**, No. 1, p. 107–111 (1983).
See also *13.3*, p. *168*; *14.3.3*, p. *237*; 22.6, p. 492.

R. S. Bhattacharya, A. K. Rai, S. C. Ling, and P. P. Pronko
*Epitaxial Regrowth of Si Implanted (100) and (211) GaAs
(a) Vol. **76**, No. 1, p. 131–136 (1983).
See also *1.4*, p. *30*; 22.2.1, p. 443.

H. Fischer, G. Götz, and H. Karge
*Radiation Damage in Ion-Implanted Quartz Crystals (I): Nuclear and Electronic Energy Deposition
(a) Vol. **76**, No. 1, p. 249–256 (1983).
See also 22.6, p. 492.

F. Walz, H. J. Blythe, and F. Dworschak
Annealing Behaviour of Dilute FeTi, FeCu, and FeMn Alloys in the Temperature Range above Stage III Following Low-Temperature Electron Irradiation
(a) Vol. **76**, No. 1, p. 267–276 (1983).
See also *10.1*, p. *101*; *18.2*, p. *281*; 21.1.1, p. 380.

W. Zag and K. Urban
Temperature Dependence of the Threshold Energy for Atom Displacement in Irradiated Molybdenum
(a) Vol. **76**, No. 1, p. 285–295 (1983).
See also *10.1*, p. *101*; 21, p. 352.

D. V. Krishna Sastry and P. Jayarama Reddy
Influence of Nitrogen Ion Implantation on the Electrical Transport Properties of InTe and InSe Thin Films
(a) Vol. **76**, No. 1, p. 345–350 (1983).
See also *14.3.1*, p. *230*; 22.3, p. 456.

S. Nojima
*Isolation Characteristics in Selectively O^+ and Cr^+ Implanted GaAs
(a) Vol. **76**, No. 1, p. K9–K12 (1983).
See also *14.3.3*, p. *238*; 22.2.1, p. 443.

H. Misiorek, T. Zakrzewski, and J. Rafałowicz
*The Lorenz Function for Aluminium of Different Chemical Purity, Neutron Irradiated or Plastically Deformed
(a) Vol. **76**, No. 1, p. K25–K26 (1983).
See also *8*, p. *87*; 12.1, p. 138; 21, p. 352.

D. Adlienė, L. Pranevičius, and G. Zubauskas
*Acoustic Emission Initiated by Ion Implantation in Bubble Garnet Films
(a) Vol. **76**, No. 1, p. K35–K37 (1983).
See also *7*, p. *82*; 18.3, p. 291; 22.8.2, p. 520.

J. WEIS and I. ČERVEŇ
*Structural Changes in Amorphous $Fe_{83}B_{17}$ by Neutron Irradiation
(a) Vol. **76**, No. 1, p. K49 – K51 (1983).
See also *2*, p. *46*; 21.1.1, p. 380.

A. F. RAKOV
*Neutron Induced Colour and Luminescence Centres in Yttrium–Aluminium Garnet Single Crystals
(a) Vol. **76**, No. 1, p. K57 – K60 (1983).
See also *10.2*, p. *111*; 20.1, p. 316; 20.3, p. 337; 22.8.2, p. 520.

J. BELZ, K. F. HEIDEMANN, H. F. KAPPERT, and E. TE KAAT
*Anomalous Defect Interaction and Amorphization during Self Irradiation of Si Crystals at 450 K
(a) Vol. **76**, No. 1, p. K81 – K84 (1983).
See also 22.1.2, p. 428.

E. KÜHRT and R. WEDELL
*Energy Losses in Crystalline and Amorphous Solids in the Intermediate Energy Region
(b) Vol. **116**, No. 2, p. 585 – 595 (1983).
See also 13.5.2, p. 200; 22.1.2, p. 428.

V. A. KHODYREV and E. I. SIROTININ
*Charged-Particle Stopping in Crystals
(b) Vol. **116**, No. 2, p. 659 – 671 (1983).
See also 13, p. 148.

M. E. GUREVICH, L. N. LARIKOV, V. G. NOVITSKII, A. E. POGORELOV, and A. F. ZHURAVLEV
*Laser-Stimulated Mass Transfer in Metals
(a) Vol. **76**, No. 2, p. 479 – 484 (1983).
See also *9*, p. *94*; 21.1, p. 367; 21.1.1, p. 381.

H. FISCHER, G. GÖTZ, and H. KARGE
*Radiation Damage in Ion-Implanted Quartz Crystals (II): Annealing Behaviour
(a) Vol. **76**, No. 2, p. 493 – 498 (1983).
See also 22.6, p. 493.

A. GRAS-MARTI
*Recoil Implantation Yields and Depth Profiles. Analytical Expressions for Primary and Cascade Contributions
(a) Vol. **76**, No. 2, p. 621 – 627 (1983).

J. DZIESATY, F.-G. KIRSCHT, M. MAI, and K.-R. SPRUNG
*Influence of Backside Argon Implantation on the Si–SiO_2 Interface
(a) Vol. **76**, No. 2, p. K105 – K108 (1983).
See also *14.3.2*, p. *233*; 22.1.2, p. 428.

F. EKPO ENO, R. GRASSER, and A. SCHARMANN
*Radiation Effects in $CaWO_4$ Crystals
(a) Vol. **76**, No. 2, p. K113 – K116 (1983).
See also *20.3*, p. *338*; 22.8.1, p. 513.

G. P. Srivastava, K. N. Tripathi, and N. K. Sehgal
*Effect of Laser Irradiation on the Electrical Properties of Polycrystalline Thin Films of Tellurium
(a) Vol. **76**, No. 2, p. K137–K140 (1983).
See also *14.3.1*, p. *231*; 22.1.3, p. 440.

H.-J. Fitting and R. Technow
*Electron Backscattering at Various Angles of Incidence
(a) Vol. **76**, No. 2, p. K151–K154 (1983).
See also 17, p. 269; 21, p. 353; 22.1.2, p. 428.

H. Beez, D. Fasold, and H. Karge
*Refractive Index Changes of Ion-Implanted Quartz
(a) Vol. **76**, No. 2, p. K171–K174 (1983).
See also *20.1*, p. *317*; 22.6, p. 493.

E. Wilk, W. Wesch, and K. Hehl
*Defect Diffusion during Ion Implantation into GaAs
(a) Vol. **76**, No. 2, p. K197–K201 (1983).
See also *20.1*, p. *317*; 22.1.2, p. 429.

H. Kerkow, B. Lukasch, and G. Kreysch
*Kinetics of Furnace Annealing of As-Implanted Silicon
(a) Vol. **76**, No. 2, p. K203–K207 (1983).
See also 22.1.2, p. 429.

E. G. Vyatkin, Yu. M. Filimonov, A. M. Taratin, and S. A. Vorobiev
*Molecular-Type Channeling of Relativistic Electrons in Crystals
(b) Vol. **117**, No. 1, p. 41–50 (1983).
See also 13, p. 148; 22.1.1, p. 420.

G. V. Dedkov and A. M. Gorshokov
*Study of Various Approximations for the Continuum Planar Potentials at Channeling (I): Monoatomic Crystals
(b) Vol. **117**, No. 1, p. 57–66 (1983).
See also 21.7, p. 401; 22.1, p. 417; 22.1.2, p. 429.

A. K. Zhetbaev and A. N. Ozernoi
Mössbauer Study of Radiation Effect of the ^{181}Ta(p, n)^{181}W Reaction
(a) Vol. **77**, No. 1, p. 63–66 (1983).
See also *5*, p. *66*; 21, p. 354.

D. Ghose, D. Basu, and S. B. Karmohapatro
*Sputtering of Germanium by 15 to 35 keV Noble Gas Ions
(a) Vol. **77**, No. 1, p. 121–125 (1983).
See also 22.1.1, p. 420.

T. Gessner, M. Pasemann, and B. Schmidt
*Calculation of Neutron-Induced Defect Clusters in Silicon and Comparison with TEM Investigations
(a) Vol. **77**, No. 1, p. 133–138 (1983).
See also 1.4, p. 31; 22.1.2, p. 429.

S. JETSCHKE, H. KARGE, and K. HEHL
*Anisotropic Effects in N^+-Implanted $LiNbO_3$
(a) Vol. **77**, No. 1, p. 207–214 (1983).
See also 20.1, p. 319; 22.8.1, p. 514.

H.-H. POTTHOFF
Velocities of Dislocation Groups in Very Thin Neutron-Irradiated Copper Single Crystals Measured by Slip Line Cinematography
(a) Vol. **77**, No. 1, p. 215–224 (1983).
See also *10.1*, p. *102*; 21.1, p. 369.

H. KLOSE, A. MERTENS, R. REETZ, and NG. TANG
*Ion Beam Contacting of Silicon (I): Aluminium in p-Silicon
(a) Vol. **77**, No. 1, p. 233–239 (1983).
See also *14.3.4*, p. *244*; 22.1.2, p. 430.

B. COCHRANE and P. J. GOODHEW
*TEM Images of Faceted Bubbles and Voids
(a) Vol. **77**, No. 1, p. 269–282 (1983).
See also 1.4, p. 31; 10.1, p. 102; 21.1, p. 369.

A. REMÓN, J. PIQUERAS, J. LLOPIS, and C. BALLESTEROS
Cathodoluminescence and Optical Absorption in 20 kV Electron Irradiated MgO Single Crystals
(a) Vol. **77**, No. 1, p. K29–K32 (1983).
See also *20.1*, p. *319*; *20.3*, p. *339*; 22.6, p. 493.

V. N. BRUDNYI, A. I. POTAPOV, YU. V. RUD, and M. SERGINOV
*Electrical Properties of Electron-Irradiated GaAs and $ZnSiAs_2$ Solid Solutions
(a) Vol. **77**, No. 1, p. K33–K36 (1983).
See also 13.4, p. 179; *14.3*, p. *219*; 22.2.1, p. 444; 22.8, p. 505.

A. M. KANLOEV
*Kumakhov Radiation Energy Losses of Ultrarelativistic Electrons and Positrons
(b) Vol. **117**, No. 2, p. 507–511 (1983).
See also *20*, p. *308*; 22.1.2, p. 430.

V. V. CHERDYNTSEV, G. P. POKHIL, and V. A. RYABOV
*Scattering of Axially Channeled Particles by Atomic Strings in Crystals
(b) Vol. **117**, No. 2, p. 653–662 (1983).

M. M. JAKAS, G. H. LANTSCHNER, J. C. ECKARDT, and V. H. PONCE
*Angular Dependence of Stopping Power
(b) Vol. **117**, No. 2, p. K131–K134 (1983).
Erratum
(b) Vol. **120**, No. 2, p. K225 (1983).
See also 21.6, p. 398.

D. PANKNIN, E. WIESER, R. KLABES, and H. SYHRE
*Dose Dependence of the Flash Lamp Annealing of Arsenic-Implanted Silicon
(a) Vol. **77**, No. 2, p. 553–559 (1983).
See also 14.3, p. 220; 22.1.2, p. 431.

L. NANAI, I. HEVESI, L. MICHAILOVITS, I. APOSTOL, M. DINESCU, I. N. MIHAILESCU, and A. V. SIDORIN
*Damage of V_2O_5 Single Crystals by Powerful Pulsed CO_2 Laser Irradiation
(a) Vol. 77, No. 2, p. 679–684 (1983).
See also 22.6, p. 494.

O. YU. BORKOVSKAYA, N. L. DMITRUK, R. V. KONAKOVA, V. I. FINEBERG, and V. I. SHAKHOVTSOV
*The Radiation Effects on the Metal–GaP Interface
(a) Vol. 77, No. 2, p. K129–K133 (1983).
See also *14.3.4*, p. *246*; 22.2.2, p. 450.

L. PRANEVIČIUS, A. SAKALAS, and G. ZUBAUSKAS
*Ion Beam Induced Mixing in Bubble Garnet Films
(a) Vol. 77, No. 2, p. K171–K172 (1983).
See also *18.3*, p. *292*; 22.8.2, p. 522.

V. A. MURALEV
*Channeling of Ultrarelativistic Electrons in Crystals
(b) Vol. 118, No. 1, p. 363–372 (1983).
See also 21, p. 356; 22.1.2, p. 432.

K. KAMARÁS, R. ŚWIETLIK, and A. GRAJA
Polarized IR Reflectance Spectra of Pure and Neutron Irradiated $Qn(TCNQ)_2$ Mosaics
(b) Vol. 118, No. 1, p. K43–K46 (1983).
See also *20.1*, p. *322*; 22.9, p. 528.

H.-D. GEILER, K. HEHL, and D. STOCK
*Deposition of Laser Energy into Inhomogeneous Layer Systems
(a) Vol. 78, No. 1, p. 193–199 (1983).
See also *8*, p. *88*; 22.1.2, p. 432.

H. HEDLER and G. GÖTZ
Incorporation of Sb in Si during CW-Laser Stimulated Solid Phase Epitaxy
(a) Vol. 78, No. 1, p. 237–241 (1983).
See also *1.2*, p. *17*; 14.3, p. 221; 22.1.2, p. 432.

R. S. BHATTACHARYA, A. K. RAI, and P. P. PRONKO
*Oxygen Implantation in GaAs: Damage Studies
(a) Vol. 78, No. 1, p. 283–286 (1983).
See also *1.4*, p. *32*; 22.2.1, p. 445.

M. KATAYAMA, H. MORIMOTO, and K. SUGAWARA
Electron Irradiation Effects on Amorphous Silicon Solar Cells
(a) Vol. 78, No. 1, p. K5–K8 (1983).
See also *14.3.3*, p. *240*; *16*, p. *266*; 22.1.2, p. 433.

B. N. MUSKAHEV, ZH. A. KUSAINOV, K. KH. NUSUPOV, and S. ZH. TOKMOLDIN
*Electrical Properties of Shallow Implanted Layers in Silicon
(a) Vol. 78, No. 1, p. K19–K22 (1983).
See also *14.3*, p. *221*; 22.1.2, p. 433.

S. A. Azimov, D. A. Aronov, V. M. Rubinov, and M. Tuichiev
*The Influence of Heating on the Characteristics of the Electron-Excited Conduction in Gallium Phosphide Films
(a) Vol. **78**, No. 1, p. K53–K58 (1983).
See also *14.3.1*, p. *231*; 22.2.2, p. 450.

V. N. Baier, V. M. Katkov, and V. M. Strakhovenko
*Radiation at Planar Channeling of Relativistic Electrons in Thick Crystals
(b) Vol. **118**, No. 2, p. 499–509 (1983).
See also 13, p. 150.

H. Opyrchal, K. D. Nierzewski, and H. Drulis
*Effect of γ-Irradiation on EPR Spectra of Eu^{2+} Doped KCl and NaCl Crystals
(b) Vol. **118**, No. 2, p. K125–K127 (1983).
See also 10.2, p. 116; *19*, p. *303*; 22.5.2, p. 482.

D. N. Popov, N. M. Nancheva, B. N. Zlatov, and G. P. Georgiev
*Positron Angular Correlation in Uncoloured and Neutron Irradiated LiF and LiF:OH
(b) Vol. **118**, No. 2, p. K129–K132 (1983).
See also *10.2*, p. *116*; 22.5.2, p. 482.

K. Hübner, H. Köster, Jr., B. Derlich, and W. Ecke
*DLTS Investigations of Si–SiO_2 Interface States of Electron-Beam Irradiated MOS Structures
(b) Vol. **118**, No. 2, p. K133–K136 (1983).
See also *13.3*, p. *169*; 14.3.2, p. 234; 22.1.2, p. 433; 22.6, p. 495.

M. M. Hafiz, H. M. Shafey, A. M. Hafez, and M. Dongol
*Irradiation-Induced Photodarkening in As–Se–Cu Amorphous Films
(a) Vol. **78**, No. 2, p. 449–455 (1983).
See also *20.1*, p. *324*; 22, p. 410.

G. Kreysch, H. Kerkow, R. I. Bogdanov, V. P. Petukhov, and E. A. Romanovskii
*Proton-Induced L-Shell X-Ray Emission Cross-Sections of Elements with $24 \leqq Z_2 \leqq 50$ for Projectile Energies between 30 and 350 keV
(a) Vol. **78**, No. 2, p. 507–525 (1983).
See also 20, p. 310; 21, p. 357.

P. J. Ratcliffe and R. A. Collins
*The Influence of Ion Implantation on the Thermal Oxidation of Copper
(a) Vol. **78**, No. 2, p. 547–553 (1983).
See also *1.6*, p. *40*; 21.1, p. 373.

A. P. Tyutnev, V. S. Saenko, A. V. Vannikov, and G. S. Mingaleev
Radiation-Induced Conductivity in Polyethylene
(a) Vol. **78**, No. 2, p. 689–696 (1983).
See also *14.3*, p. *222*; 22.9, p. 529.

P. V. PETRASHEN and I. L. SHULPINA
*X-Ray Determination of the Surface Compression Force in the Ion-Doped Layer in Silicon
(a) Vol. **78**, No. 2, p. K105 – K108 (1983).
See also *10.2*, p. *117*; 22.1.2, p. 434.

E. YU. BRAILOVSKII, G. N. ERITSYAN, and N. E. GRIGORYAN
*Optical Absorption in InP Crystals with Point Radiation Defects
(a) Vol. **78**, No. 2, p. K113 – K115 (1983).
See also *20.1*, p. *325*; 22.2.2, p. 450.

N. I. BEREZHNOV, V. F. STELMAKH, and A. R. CHELYADINSKII
*The Interstitial Type Defects in Ion Implanted Silicon
(a) Vol. **78**, No. 2, p. K121 – K125 (1983).
See also *10.2*, p. *117*; 22.1.2, p. 434.

K. KAWABATA
Intense Exoelectron Emission from Fine Particles of Al_2O_3 and Its Dependence on the Conductivity of the Substrate Material
(a) Vol. **78**, No. 2, p. K127 – K130 (1983).
See also 13.3, p. 170; *17*, p. *270*; 22.6, p. 495.

A. F. AKKERMAN, A. L. GIBREKHTERMAN, and A. V. DVURECHENSKII
*Absorbed Energy Distribution in Heterogeneous Semiconductor Structures
(a) Vol. **78**, No. 2, p. K131 – K135 (1983).
See also 22.1.1, p. 421; 22.1.2, p. 434.

N. N. GERASIMENKO, A. E. GERSHINSKII, A. YU. SURTAEV, S. A. SOKOLOV, E. I. CHEREPOV, B. I. FOMIN, and G. V. TIMOFEEVA
Silicon–Nickel Silicide Schottky Barriers Formed by Ion Mixing
(a) Vol. **78**, No. 2, p. K151 – K154 (1983).
See also *14.3.3*, p. *240*; 22.1.2, p. 434.

V. ZIRAPS
Ionic and Electronic Processes during Frenkel Defect Annealing in KBr Crystals above LHeT
(b) Vol. **119**, No. 1, p. 49 – 59 (1983).
See also *10.2*, p. *117*; 13.4, p. 185; 20.1, p. 325; 22.5.2, p. 482.

V. V. MAZURENKO, V. V. PETROV, and V. D. TKACHEV
*Investigation of the Single-Phonon Absorption Band at 230 cm^{-1} in Silicon Irradiated by Neutrons
(b) Vol. **119**, No. 1, p. K69 – K71 (1983).
See also *6*, p. *78*; 20.1, p. 326; 22.1.2, p. 434.

S. ROGASCHEWSKI
*Energy Spectra of Backscattered Electrons for Solid Films and Double Layers by Use of Werner's Analytic Model
(a) Vol. **79**, No. 1, p. 149 – 159 (1983).

L. Nanu, Doina Bărbulescu, I. N. Mihăilescu, V. Teodorescu, and Leona C. Nistor
*Annealing of Implanted Silicon under the Action of Microsecond Pulsed TEA-CO_2 Laser Radiation
(a) Vol. **79**, No. 1, p. K5–K8 (1983).
See also *1.5*, p. *38*; 22.1.2, p. 435.

K. Biskupska, J. Tatarkiewicz, and K. Wieteska
*Comparison of Fast Electron and Thermal Annealing of Proton Bombarded Silicon – X-Ray and IR Spectroscopy Study
(a) Vol. **79**, No. 1, p. K13–K15 (1983).
See also 10.2, p. 118; 20.1, p. 326; 22.1.2, p. 435.

J. Grňo, Š. Bederka, and M. Veselý
*Pulse Radiation Annealing of Capless GaAs
(a) Vol. **79**, No. 1, p. K41–K44 (1983).
See also *14.3*, p. *224*; 22.2.1, p. 446.

V. P. Dobrego and A. K. Fedotov
Influence of Growth Dislocations on Low Temperature Conductivity in Irradiated p-Germanium
(a) Vol. **79**, No. 1, p. K45–K48 (1983).
See also *10.2*, p. *118*; *14.3*, p. *224*; 22.1.1, p. 421.

V. G. Chudinov, V. P. Gogolin, B. N. Goshchitskii, and V. I. Protasov
*Simulation of Collision Cascades in Intermetallic V_2Hf Compounds
(a) Vol. **79**, No. 1, p. K57–K60 (1983).
See also 14.2, p. 213; 21, p. 359; 21.1, p. 375.

Yu. G. Chukalkin, V. R. Shtirts, and B. N. Goshchitskii
*Amorphization of Neutron-Irradiated Yttrium Iron Garnet
(a) Vol. **79**, No. 2, p. 361–366 (1983).
See also 2, p. 53; 18.3, p. 293; 22.8.2, p. 523.

L. A. Kazakevich and P. F. Lugakov
*The Effect of Dislocations on the Interaction Processes of Point and Group Radiation Defects in Silicon
(a) Vol. **79**, No. 2, p. 377–380 (1983).
See also 14.3, p. 225; 22.1.2, p. 436.

V. I. Kuznetsov and P. F. Lugakov
*Parameters of Defect Clusters in Irradiated Silicon and a Method for Their Determining
(a) Vol. **79**, No. 2, p. 381–388 (1983).
See also 14.3, p. 225; 22.1.2, p. 436.

B. Rauschenbach and K. Hohmuth
*Analysis of Heavy-Ion Sputtering Based on the Shock Wave Model
(a) Vol. **79**, No. 2, p. 395–401 (1983).
See also 1.6, p. 41; 21, p. 359.

S. E. Donnelly, A. A. Lucas, Ph. Lambin, and J. P. Vigneron
*A Possible Mechanism for Electron-Bombardment-Induced Loop Punching in Helium-Implanted Materials
(a) Vol. **79**, No. 2, p. 543–548 (1983).
See also 21, p. 360.

B. Gross, H. von Seggern, and D. A. Berkley
*Long Term Behaviour of Radiation-Induced Currents in Fluorinated Ethylene Propylene Copolymer
(a) Vol. **79**, No. 2, p. 607–615 (1983).
See also *14.4*, p. *253*; 22.9, p. 530.

H. Bahadur and R. Parshad
*Some New Results on Irradiation Characteristics of Synthetic Quartz Crystals and Their Application to Radiation Hardening
(a) Vol. **79**, No. 2, p. 639–644 (1983).
See also *12.2*, p. *144*; 22.6, p. 496.

A. P. Tyutnev, G. S. Mingaleev, V. S. Saenko, and A. V. Vannikov
*Radiation Induced Conductivity of Polyethyleneterephthalate and Polystyrene
(a) Vol. **79**, No. 2, p. 651–659 (1983).
See also *14.3*, p. *225*; 22.9, p. 531.

V. X. Quang, T. Kim Anh, N. Q. Liem, K. Maass, and B. Selle
*A New Luminescence Band Related to Radiation Defects in Ion-Implanted Zinc Sulphide
(a) Vol. **79**, No. 2, p. K181–K185 (1983).
See also *20.3*, p. *345*; 22.4.1, p. 464.

W. Frentrup, M. Griepentrog, and U. Müller-Jahreis
The Influence of Cesium Implanted in Silicon on the Secondary Ion Emission
(a) Vol. **79**, No. 2, p. K193–K196 (1983).
See also *17*, p. *270*; 22.1.2, p. 437.

M. Pasemann, D. Hoehl, A. L. Aseev, and O. P. Pchelyakov
*Analysis of Rod-Like Defects in Silicon and Germanium by Means of High-Resolution Electron Microscopy
(a) Vol. **80**, No. 1, p. 135–139 (1983).
See also *1.4*, p. *34*; 22.1.1, p. 422; 22.1.2, p. 438.

B. Rauschenbach and A. Kolitsch
*Formation of Compounds by Nitrogen Ion Implantation in Iron
(a) Vol. **80**, No. 1, p. 211–222 (1983).
See also *1.3*, p. *26*; 21.1.1, p. 385.

M. Servidori, C. Dal Monte, and Q. Zini
Double Crystal X-Ray Analysis of Phosphorus Precipitation in Supersaturated Si–P Solid Solutions
(a) Vol. **80**, No. 1, p. 277–285 (1983).
See also *10.2*, p. *120*; 22.1.2, p. 438.

D. J. Farnum, O. T. Inal, and R. J. Walko
*A Field-Ion Microscopy Study of Ion Damage to Tungsten
(a) Vol. **80**, No. 1, p. 287 – 303 (1983).
See also 17.1, p. 272; 21, p. 361.

V. G. Chudinov, V. P. Gogolin, B. N. Goshchitskii, and V. I. Protasov
Evolution of the Cascade Region in Nb_3Sn during the Thermal Spike under Fast-Neutron Irradiation
(a) Vol. **80**, No. 1, p. 349 – 352 (1983).
See also *10.1*, p. *107*; 21, p. 361; 21.7, p. 402.

A. K. Rai, R. S. Bhattacharya, and P. P. Pronko
Annealing of Amorphous Layer in Si Implanted GaAs
(a) Vol. **80**, No. 1, p. 367 – 372 (1983).
See also *1.4*, p. *34*; 10.2, p. 120; 22.1.2, p. 438.

T. Jung, L. Roth, and H. Wittrich
*Auger Analysis of Plasma-Grown GaAs Oxide Films
(a) Vol. **80**, No. 1, p. K11 – K14 (1983).
See also 22.2.1, p. 447; 22.6, p. 497.

J. Tatarkiewicz
*Fast Electron Annealing of IR Absorption Bands of Proton Bombarded Silicon
(a) Vol. **80**, No. 1, p. K49 – K51 (1983).
See also *20.1*, p. *329*; 22.1.2, p. 438.

P. K. Chakraborty and K. V. Rao
F-Band Absorption and Thermoluminescence of KBr Single Crystals X-Ray Irradiated under High AC Fields
(b) Vol. **120**, No. 2, p. K107 – K110 (1983).
See also *10.2*, p. *120*; *20.3*, p. *347*; 22.5.2, p. 485.

P. D. Alekseev, G. I. Baranov, E. P. Kurakina, and K. A. Maltsev
*Formation of Hydrogen Bonding in Doped OH-Group Alkali Halide Crystals by the Action of γ-Radiation
(b) Vol. **120**, No. 2, p. K119 – K121 (1983).
See also 10.2, p. 121; *20.1*, p. *330*; 22.5.2, p. 485.

K. Lenkeit, R. Wedell, and P. Gehrmann
*Nonlinear Effect of Particle Motion under Axial-to-Planar Channeling Transition Conditions
(b) Vol. **120**, No. 2, p. K195 – K199 (1983).
See also 22.1.2, p. 439.

Yu. V. Bulgakov, K. Lenkeit, and R. Wedell
*Measurements of the Ionic Charge of Hydrogen and Helium Ions Moving through Silicon by Means of the Resonance Dechanneling Effect
(b) Vol. **120**, No. 2, p. K201 – K204 (1983).
See also 22.1.2, p. 439.

B. RAUSCHENBACH, A. KOLITSCH, and K. HOHMUTH
*Iron Nitride Phases Formed by Nitrogen Ion Implantation and Thermal Treatment
(a) Vol. **80**, No. 2, p. 471–482 (1983).
See also *1.3*, p. *27*; 1.5, p. 38; 21, p. 362; 21.1.1, p. 385.

C. J. ROSSOUW
A γ- to α-Phase Transition Induced by Heavy Ion Bombardment of a CuAl Alloy
(a) Vol. **80**, No. 2, p. 631–636 (1983).
See also *1.2*, p. *20*; 21, p. 362; 21.1, p. 378.

K. KISHAN RAO and D. B. SIRDESHMUKH
*Effect of X-Ray Irradiation on the Hardness of $NaBrO_3$ Crystals
(a) Vol. **80**, No. 2, p. K117–K119 (1983).
See also *12.2*, p. *144*; 22.8.1, p. 518.

V. A. BYKOVSKII, A. V. MUDRYI, V. P. POSKREBYSHEV, and V. D. TKACHEV
Radiative Recombination Centers in Irradiated Germanium
(a) Vol. **80**, No. 2, p. K133–K136 (1983).
See also *13.4*, p. *191*; *20.3*, p. *347*; 22.1.1, p. 422.

SH. A. VAKHIDOV and A. F. RAKOV
*Radiation Stimulated Recombination Processes in $Y_3Al_5O_{12}$ Crystals
(a) Vol. **80**, No. 2, p. K175–K178 (1983).
See also 13.4, p. 191; 13.5.1, p. 199; *20.3*, p. *348*; 22.8.2, p. 523.

P. N. DASHUK, A. V. KOLOBOV, B. T. KOLOMIETS, and V. M. LYUBIN
*Plasma-Induced Processes in Chalcogenide Vitreous Semiconductors
(a) Vol. **80**, No. 2, p. K187–K192 (1983).
See also *2*, p. *56*; 22, p. 416.

12. Mechanical Properties (Plastic Deformations See 10)

Review Article

A. E. Romanov and V. I. Vladimirov
Disclinations in Solids
(a) Vol. **78**, No. 1, p. 11 – 34 (1983).
See also *1.4*, p. *29*; *10*, p. *97*.

Original Papers and Short Notes

K. Schiele and S. Trimper
*On the Elastic Constants of a Nematic Liquid Crystal
(b) Vol. **118**, No. 1, p. 267 – 274 (1983).
See also *2*, p. *49*.

V. N. Lukerchenko
*On the Relation between the Plastic Strain Rate of a Crystal and Microscopic Characteristics of Dislocation Motion
(a) Vol. **78**, No. 1, p. K9 – K12 (1983).
See also *10*, p. *98*.

A. Y. Belov, V. A. Chamrov, V. L. Indenbom, and J. Lothe
*Elastic Fields of Dislocations Piercing the Interface of an Anisotropic Bicrystal
(b) Vol. **119**, No. 2, p. 565 – 578 (1983).
See also *10*, p. *98*.

12.1 Metals

Original Papers and Short Notes

D. SEN, S. K. SARKAR, S. SENGUPTA, and D. ROY
*Consistency Condition in the Energy Expression of the Pseudopotential Theory of Metals and Unified Study of Alkali Metals
(b) Vol. **115**, No. 2, p. 593–602 (1983).
See also *6*, p. *71*; 21.2, p. 387.

F. ABD EL-SALAM, A. I. EATAH, and A. TAWFIK
Effect of Li Loss on Some Physical Properties of Al–Li Alloy
(a) Vol. **75**, No. 2, p. 379–384 (1983).
See also *1.3*, p. *21*; 10.1, p. 100; 14.1, p. 206; 21, p. 351; 21.2, p. 387.

K. K. RAY and A. K. MALLIK
An Investigation on the Activation Volume for Plastic Flow in Nickel Determined by Micro and Macro Tests
(a) Vol. **75**, No. 2, p. 451–458 (1983).
See also *10.1*, p. *100*; 21.1, p. 365.

A. KORNER and H. P. KARNTHALER
Glide Dislocations on Cube Planes in a Low Stacking-Fault Energy Alloy
(a) Vol. **75**, No. 2, p. 525–535 (1983).
See also *10.1*, p. *100*; 21, p. 351; 21.1, p. 365.

P. DELOBELLE and C. OYTANA
Creep at Low and Average Stress of Several Highly Concentrated Solid Solutions of the Cubic Centered Type
(a) Vol. **75**, No. 2, p. 625–634 (1983).
See also *10.1*, p. *100*; 21.1, p. 365; 21.1.1, p. 380.

K. K. MISRA and A. N. NIGAM
*Development of Stress in Nickel during Hydrogen Diffusion
(a) Vol. **75**, No. 2, p. 635–639 (1983).
See also *9*, p. *93*; 21.1, p. 365.

V. V. ZONINASHVILI, I. A. NASKIDASHVILI, U. MESSERSCHMIDT, V. A. MELIK-SHAKHNAZAROV, N. M. YASTREBOVA, and L. M. KOLESNIKOVA
The Dependence of the Elastic Relaxation on a Bias Stress in Pure Nb in the Temperature Region of the α-Maximum
(a) Vol. **75**, No. 2, p. K205–K208 (1983).
See also *7*, p. *82*; *10.1*, p. *100*; 21, p. 351.

I. KANAZAWA, H. MURAKAMI, and M. TANAKA
*Internal Friction by a Second-Phase Particle in Al–Ge Alloy
(a) Vol. **75**, No. 2, p. K211–K213 (1983).
See also *10.1*, p. *101*; 21, p. 352.

H. Mizubayashi, T. Arai, and S. Okuda
*Low-Temperature Dislocation Pinning and Breakaway Observed in Ta Containing H
(a) Vol. **76**, No. 1, p. 165–177 (1983).
See also *10.1*, p. *101*; 21, p. 352.

H. Misiorek, T. Zakrzewski, and J. Rafałowicz
The Lorenz Function for Aluminium of Different Chemical Purity, Neutron Irradiated or Plastically Deformed
(a) Vol. **76**, No. 1, p. K25–K26 (1983).
See also *8*, p. *87*; *11*, p. *125*; 21, p. 352.

C. W. Lung and L. Y. Xiong
The Dislocation Distribution Function in the Plastic Zone at a Crack Tip
(a) Vol. **77**, No. 1, p. 81–86 (1983).
See also *10.1*, p. *102*; 21, p. 354.

L. G. Khvostantsev and N. A. Nikolaev
Electroresistance of Cerium at Phase Transitions under Hydrostatic Pressure up to 8.0 GPa. *P–T* Diagram of the Cerium Phase Transitions
(a) Vol. **77**, No. 1, p. 161–166 (1983).
See also *1.2*, p. *16*; *14.1*, p. *207*; 21.4, p. 391.

M. Anglada and F. Guiu
*Temperature and Strain Rate Dependence of the Cyclic Flow Stress of Molybdenum Single Crystals
(a) Vol. **77**, No. 1, p. 283–289 (1983).
See also *10.1*, p. *102*; 21, p. 354.

M. Ohta, T. Kanadani, A. Sakakibara, H. Yamada, and M. Yamada
Effects of Grain Boundary on Age Hardening and Reversion in Al–Zn Alloys
(a) Vol. **77**, No. 1, p. K49–K51 (1983).
See also *1.3*, p. *24*; 10.1, p. 102; 21, p. 354; 21.1, p. 369.

T. Soma, Y. Konno, and H.-M. Kagaya
*Local Heine-Abarenkov Potential with Volume-Dependent Parameters of Al
(b) Vol. **117**, No. 2, p. 743–748 (1983).
See also *4*, p. *61*; 21, p. 355.

C. M. van Baal
The Influence of Texture on the X-Ray Determination of Residual Strains in Ground or Worn Surfaces
(a) Vol. **77**, No. 2, p. 521–526 (1983).
See also *1.6*, p. *40*; 10.1, p. 103; 21, p. 355.

N. Kuramoch, H. Mizubayashi, and S. Okuda
*Dislocation Breakaway at Low Temperatures in Nb
(a) Vol. **77**, No. 2, p. 633–642 (1983).
See also *10.1*, p. *103*; 21, p. 355.

M. Ohta, T. Kanadani, A. Sakakibara, H. Yamada, and M. Yamada
Effects of Surface and Thickness of Specimens on the Aging in Al–Zn Alloys
(a) Vol. **78**, No. 1, p. K23–K26 (1983).
See also *1.3*, p. *25*; 10.1, p. 104; 21, p. 357; 21.1, p. 372.

C. Malinowska-Adamska
Dynamics and Thermodynamics of Anharmonic Crystals near the Instability Point
(b) Vol. **118**, No. 2, p. 535–539 (1983).
See also *6*, p. *76*; 8, p. 89; 21.2, p. 387.

T. Soma, Y. Konno, and H.-M. Kagaya
*Compression Effect on the Bulk Modulus of Al
(b) Vol. **118**, No. 2, p. K155–K157 (1983).
See also *4*, p. *62*; 21, p. 357.

O. V. Presnyakova, V. I. Zaitsev, and V. A. Fomchenko
Creep of Aluminium Single Crystals under High Pressure
(a) Vol. **78**, No. 2, p. 377–383 (1983).
See also *10.1*, p. *104*; 21, p. 357.

E. Nembach
Precipitation Hardening Caused by a Difference in Shear Modulus between Particle and Matrix
(a) Vol. **78**, No. 2, p. 571–581 (1983).
See also *10.1*, p. *105*; 21.1, p. 373.

A. R. Yavari
Hardening by Thermal Vacancies and Quenching Strains in Singly-Oriented Pb Single Crystals
(a) Vol. **78**, No. 2, p. 583–588 (1983).
See also *10.1*, p. *105*; 21, p. 357.

Shu Zhen and G. J. Davies
Calculation of the Lennard-Jones *n–m* Potential Energy Parameters for Metals
(a) Vol. **78**, No. 2, p. 595–605 (1983).
See also 1.3, p. 25; *4*, p. *63*; 21, p. 358.

E. Bonetti, E. Evangelista, A. Cavallini, and C. Patuelli
*Internal Friction Peaks and Structural Relaxation on Some Electroless Deposited Amorphous Alloys
(a) Vol. **78**, No. 2, p. 615–621 (1983).
See also *2*, p. *50*; 21.1, p. 373.

V. Navrátil, M. Hamerský, P. Lukáč, V. P. Soldatov, and V. I. Startsev
*Solid Solution Hardening of Cd–Zn Single Crystals at 1.5, 2.8, and 4.2 K
(a) Vol. **78**, No. 2, p. 633–638 (1983).
See also *10.1*, p. *105*; 21, p. 358; 21.1, p. 373.

S. P. Singh and R. P. S. Rathore
Van der Waals' Forces in Face Centred Cubic Cobalt
(b) Vol. **119**, No. 1, p. 343–348 (1983).
See also *4*, p. *63*; 6, p. 77; 21.1, p. 374.

K. Iwasaki
*High-Temperature Internal Friction Peaks of Pure Aluminium
(a) Vol. **79**, No. 1, p. 115–124 (1983).
See also *10.1*, p. *105*; 21, p. 359.

C. F. Nilsen and K. N. Subramanian
*Deformation of α–β Brass with Oriented and Equiaxed Duplex Boundaries
(a) Vol. **79**, No. 1, p. 281 – 292 (1983).
See also **1.4**, p. 33; 21.1, p. 375.

E. M. Gololobov, E. L. Mager, Z. V. Mezhevich, and L. K. Pan
*Elastic Properties and Root-Mean-Square Dynamic Displacement of Niobium Atoms at High Pressures
(b) Vol. **119**, No. 2, p. K139 – K142 (1983).
See also *7*, p. *83*; 21.1, p. 376.

M. Vázquez, W. Fernengel, and H. Kronmüller
The Effect of Tensile Stresses on the Magnetic Properties of $Co_{58}Fe_5Ni_{10}Si_{11}B_{16}$ Amorphous Alloys
(a) Vol. **80**, No. 1, p. 195 – 204 (1983).
See also 2, p. 54; *18.2*, p. *288*; 21.1, p. 377; 21.1.1, p. 385.

C. Malinowska-Adamska
Self-Consistent Calculations of the Lattice Dynamics and Thermodynamics of Anharmonic Noble Metals
(b) Vol. **120**, No. 2, p. 601 – 610 (1983).
See also *6*, p. *81*; 8, p. 92; 21.6, p. 399.

M. Vázquez, W. Fernengel, and H. Kronmüller
*Influence of Applied Stresses on the Magnetic Properties of Annealed $Co_{58}Fe_5Ni_{10}Si_{11}B_{16}$ Amorphous Alloy
(a) Vol. **80**, No. 2, p. 513 – 518 (1983).
See also 2, p. 55; *18.2*, p. *289*; 21.1, p. 378; 21.1.1, p. 385.

A. San Martin, K. Kleinstück, N. H. Quyen, and P. Paufler
*Influence of a Direct Current and a Temperature Gradient on the Creep Rate in V_3Si
(a) Vol. **80**, No. 2, p. K171 – K174 (1983).
See also *14.1*, p. *211*; 21.1, p. 378.

12.2 Non-Metals

Original Papers and Short Notes

B. Subramanyam and Y. V. G. S. Murti
*Elastic Constant Correlations in Ionic Crystals of Rocksalt Structure
(b) Vol. **115**, No. 1, p. K29–K31 (1983).
See also 22.5, p. 473; 22.6, p. 490.

H. Beige, A. Kühnel, and B. Lorenz
Anomalous Behaviour of Nonlinear Dielectric, Elastic, and Electromechanical Coefficients at Ferroelectric Phase Transitions. Rochelle Salt
(a) Vol. **75**, No. 1, p. 129–136 (1983).
See also *14.4.1*, p. *256*; 22.8.1, p. 512.

W. Rehwald, A. Vonlanthen, and W. Meyer
*Single Crystal 2,4-Hexadiynylene-Bis (p-Toluenesulfonate): Elastic Properties and Phase Transitions in the Monomer and Polymer
(a) Vol. **75**, No. 1, p. 219–226 (1983).
See also *1.2*, p. *14*; 7, p. 82; 22.9, p. 525.

H. Koizumi and T. Suzuki
Low-Temperature Plasticity of CsBr and Inertial Effect of Dislocation Motion
(a) Vol. **75**, No. 1, p. 301–309 (1983).
See also *10.2*, p. *109*; 22.5.2, p. 477.

L. Kończewicz and W. Trzeciakowski
Highly Localized Donor States in InSb and Their Pressure Dependence
(b) Vol. **115**, No. 2, p. 359–367 (1983).
See also *13.4*, p. *174*; 14.3, p. 216; 22.2.3, p. 452.

R. K. Singh and N. K. Pandey
Study of Thermophysical and Anharmonic Properties of Fluorite Compounds
(b) Vol. **115**, No. 2, p. 555–562 (1983).
See also *4*, p. *60*; 6, p. 70; 22.5.3, p. 486.

O. G. Alaverdova, M. Ya. Fuks, L. S. Khazan, L. P. Koval, L. A. Matveeva, I. F. Mikhailov, N. N. Soldatenko, and Yu. A. Tkhorik
Coherency of the Interphase Boundary and Elastic Strain in the Epitaxial System Ge/GaAs
(a) Vol. **75**, No. 2, p. 367–371 (1983).
See also *1.5*, p. *35*; *10.2*, p. *109*; 22.1.1, p. 419; 22.2.1, p. 443.

L. Bočánek and J. Musilová
*Modulation Piezoreflectance of Cubic Polycrystals
(a) Vol. **75**, No. 2, p. 555–558 (1983).
See also 13.1, p. 156; *20.1*, p. *314*; 22.1.1, p. 419.

G. V. Berenshtein, A. M. Dyachenko, and I. E. Maronchuk
*The Degradation of Semiconductor Mechanical Properties Induced by Microinclusions
(a) Vol. **75**, No. 2, p. K129–K132 (1983).
See also *10.2*, p. *110*; 22.2.1, p. 443.

S. K. KHANNA
On the Elastic Anisotropy of Ferroelectric Triglycine Selenate
(b) Vol. **116**, No. 1, p. 173–178 (1983).
See also *1.2*, p. *14*; *14.4.1*, p. *256*; 22.9, p. 526.

R. K. SINGH, D. K. NEB, and S. P. SANYAL
Effect of Three-Body Forces on Thermophysical and Anharmonic Properties of Rare Gas Solid Mixtures
(b) Vol. **116**, No. 1, p. 289–297 (1983).
See also *4*, p. *60*; 6, p. 72; 8, p. 86; 23, p. 533.

D. BOSE, A. GHOSH, and A. N. BASU
Perturbation-Theoretical Model Calculation of the Lattice Mechanics of Low Polarisability Ionic Crystals
(b) Vol. **116**, No. 1, p. 313–322 (1983).
See also *6*, p. *72*; 22.5.2, p. 478.

M. HALBWACHS, P. MAZOT, and J. WOIRGARD
*Anelastic Relaxation Phenomena in Plasma-Sprayed $(Al_2O_3)_3MgO$ Spinel with Reference to Diffusion Process
(a) Vol. **76**, No. 1, p. 157–164 (1983).
See also 9, p. 93; *10.2*, p. *110*; 22.8.2, p. 520.

Y. KAWAI and T. OGAWA
*ΔE Effect of a Mn Ferrite Single Crystal at Temperatures between 150 and 300 K
(a) Vol. **76**, No. 1, p. 375–381 (1983).
See also 7, p. 82; *18.3*, p. *291*; 22.8.2, p. 520.

G. N. GAIDUKOV, S. K. MAKSIMOV, and A. A. PODREZOV
*The Effect of the Crystal Anisotropy on the Energy of the Dislocation–Impurity Atmosphere Elastic Interaction in Silicon
(a) Vol. **76**, No. 2, p. 453–459 (1983).
See also *10.2*, p. *111*; 22.1.2, p. 428.

B. BRIDGE, N. D. PATEL, and D. N. WATERS
*On the Elastic Constants and Structure of the Pure Inorganic Oxide Glasses
(a) Vol. **77**, No. 2, p. 655–668 (1983).
See also *2*, p. *49*; 22.6, p. 494.

Y. KAWAI and T. OGAWA
*Domain Wall Motions Due to External Stresses in a Mn–Zn Ferrite Single Crystal
(a) Vol. **77**, No. 2, p. K103–K106 (1983).
See also *18.3*, p. *292*; 22.8.2, p. 521.

R. V. KONAKOVA, YU. A. TKHORIK, I. L. ZAITSEVSKII, P. KORDOS, M. MORVIČ, and J. CERVENAK
*The Influence of Stress on the Fine Structure of Schottky Diode *I–U* Characteristics in the Breakdown Region
(a) Vol. **77**, No. 2, p. K125–K128 (1983).
See also *14.3.4*, p. *246*; 22.2.1, p. 445; 22.2.4, p. 454.

T. Huang and A. I. Ruoff
*Pressure-Induced Phase Transitions of HgTe
(a) Vol. **77**, No. 2, p. K193 – K197 (1983).
See also *1.2*, p. *17*; 22.4.3, p. 468.

B. R. K. Gupta and R. P. Goyal
*A Theoretical Analysis of Interionic Potentials in Rutile-Type Crystals
(b) Vol. **118**, No. 1, p. K7 – K12 (1983).
See also *4*, p. *62*; 22.6, p. 494.

A. I. Prokhvatilov and A. P. Isakina
Lattice Parameters, Thermal Expansion Coefficients, and Vacancy Density in Solid CD_4
(a) Vol. **78**, No. 1, p. 147 – 155 (1983).
See also 1.1, p. 11; 8, p. 88; *10.2*, p. *114*; 22.9, p. 529.

W. Gehlhoff and K. H. Segsa
*Determination of Stresses in Si–SiO_2 and Mechanically Affected Si by EPR
(b) Vol. **118**, No. 2, p. 703 – 712 (1983).
See also *19*, p. *303*; 22.1.2, p. 433; 22.6, p. 495.

G. D. Jain and J. Shanker
Use of Harrison's Potential in Mixed Alkali Halide Crystals
(b) Vol. **119**, No. 1, p. 79 – 82 (1983).
See also *4*, p. *63*; 22.5.4, p. 488.

T. Ito
A Pseudopotential Approach to the Lattice Parameters for GaAs Containing Impurity Elements
(b) Vol. **119**, No. 1, p. 307 – 314 (1983).
See also *10.2*, p. *117*; 22.2, p. 442; 22.2.1, p. 446.

M. Gatulle, M. Fischer, and A. Chevy
*Elastic Constants of the Layered Compounds GaS, GaSe, InSe, and Their Pressure Dependence (I): Experimental Part
(b) Vol. **119**, No. 1, p. 327 – 336 (1983).
See also 22.3, p. 457.

N. Dass and B. M. Mustafa
A Study of the Anderson-Grüneisen Parameter in Sodium Chloride
(b) Vol. **119**, No. 1, p. K27 – K30 (1983).
See also *8*, p. *90*; 22.5.2, p. 483.

T. Soma and H.-Matsuo Kagaya
Binding Force Including Ionic Interactions and Bulk Properties for Tetrahedral Compounds
(b) Vol. **119**, No. 1, p. K41 – K45 (1983).
See also *4*, p. *63*; 22.2, p. 442; 22.4, p. 461.

B. P. Chandra, M. Elyas, and B. Majumdar
Pressure and Temperature Dependence of the Mechanoluminescence of γ-Irradiated Alkali Halide Crystals
(a) Vol. **79**, No. 1, p. 95 – 102 (1983).
See also *20.3*, p. *344*; 22.5.2, p. 483.

V. I. BARBASHOV and YU. B. TKACHENKO
*Dislocation Mobility in Hydrostatically Compressed KBr Crystals
(a) Vol. **79**, No. 1, p. K17–K19 (1983).
See also *10.2*, p. *118*; 22.5.2, p. 483.

A. A. DEMIDENKO, A. P. ZDEBSKII, V. N. PISKOVOI, and A. U. SAVCHUK
*Determination of Nonlinear Elastic Coefficients by Resonant Methods
(a) Vol. **79**, No. 1, p. K73–K76 (1983).
See also *14.4.1*, p. *258*; 22.8.1, p. 516.

R. B. YADAV
*Brillouin Scattering in Single Crystals of Carbon Tetraiodide
(a) Vol. **79**, No. 2, p. 345–350 (1983).
See also 22.5, p. 475.

G. YA. AKIMOV and I. YU. PROKHOROV
*Fracture of Hydrostatically Compressed NaCl Single Crystals
(a) Vol. **79**, No. 2, p. 423–431 (1983).
See also *10.2*, p. *119*; 22.5.2, p. 484.

H. BAHADUR and R. PARSHAD
*Some New Results on Irradiation Characteristics of Synthetic Quartz Crystals and Their Application to Radiation Hardening
(a) Vol. **79**, No. 2, p. 639–644 (1983).
See also *11*, p. *133*; 22.6, p. 496.

A. V. ZAMKOV, I. T. KOKOV, and A. T. ANISTRATOV
*The Acousto-Optical Properties and Photoelasticity of $PbBr_2$ Single Crystals
(a) Vol. **79**, No. 2, p. K177–K180 (1983).
See also *7*, p. *83*; 12.2, p. 144; 22, p. 413.

M. T. DOVE, I. U. HEILMANN, J. K. KJEMS, J. KURITTU, and G. S. PAWLEY
A Neutron Scattering Study of Phonons in Per-Deuterated S-Triazine
(b) Vol. **120**, No. 1, p. 173–181 (1983).
See also *1.2*, p. *19*; *6*, p. *79*; 22.9, p. 531.

I. E. BOLOTOV and V. L. PRILEPO
On the Vacancy Mechanism of the Occurrence of Compressive Stresses during the Crystallization of Amorphous Films
(a) Vol. **80**, No. 1, p. K67–K70 (1983).
See also *1.5*, p. *38*; 3, p. 59; 10.2, p. 120; 22.1.3, p. 441; 22.4.1, p. 465.

A. A. KAMINSKII, I. M. SILVESTROVA, S. E. SARKISOV, and G. A. DENISENKO
Investigation of Trigonal $(La_{1-x}Nd_x)_3Ga_5SiO_{14}$ Crystals (II): Spectral Laser and Electromechanical Properties
(a) Vol. **80**, No. 2, p. 607–620 (1983).
See also *14.4*, p. *254*; 20.1, p. 331; *20.2*, p. *333*; 20.3, p. 347; 22.8.1, p. 518.

K. KISHAN RAO and D. B. SIRDESHMUKH
*Effect of X-Ray Irradiation on the Hardness of $NaBrO_3$ Crystals
(a) Vol. **80**, No. 2, p. K117–K119 (1983).
See also *11*, p. *135*; 22.8.1, p. 518.

I. Manika and J. Teteris
*Compositional Dependence of Photo-Induced Microhardness Changes in As–Se Films
(a) Vol. **80**, No. 2, p. K121 – K123 (1983).
See also *10.2*, p. *121*; 22, p. 415.

S. R. Murthy
*ΔE Effect in Co–Zn Ferrites
(a) Vol. **80**, No. 2, p. K161 – K165 (1983).
See also 18.3, p. 294; 22.8.2, p. 523.

13. Electron States

Review Articles

N. C. Halder and K. C. Phillips
Electronic Density of States of Liquid Metals Studied by Perturbation and Green Function Methods
(b) Vol. **115**, No. 1, p. 9–35 (1983).
See also *2*, p. *43*; 21, p. 349.

K. H. Fischer
Spin Glasses (I)
(b) Vol. **116**, No. 2, p. 357–414 (1983).
See also 2, p. 43; 8, p. 85; 14.1, p. 206; *18.2*, p. *280*; *18.4*, p. *295*; 21.1, p. 363; 21.1.1, p. 379.

K. N. Shrivastava
*Theory of Spin–Lattice Relaxation
(b) Vol. **117**, No. 2, p. 437–458 (1983).
See also *6*, p. *69*; 20.1, p. 312; 22.5, p. 473; 22.8.1, p. 512.

Original Papers and Short Notes

A. M. Taratin and S. A. Vorobev
Electronic Energy Loss of Channeled Negative Pions
(b) Vol. **115**, No. 1, p. 37–46 (1983).
See also *11*, p. *122*; 22.1.2, p. 423.

J. Shanker and R. Sundaraj
Dielectric and Anharmonic Behaviour of Silver Halides, Thallous Halides, Alkaline Earth Fluorides, and Lead Fluoride Crystals
(b) Vol. **115**, No. 1, p. 67–74 (1983).
See also 4, p. 60; *14.4*, p. *250*; 22.5, p. 473; 22.5.1, p. 476; 22.5.3, p. 486.

F. Goedsche, A. Richter, and G. Vojta
Dynamical Susceptibility and Electrical Resistivity of Disordered Spin Systems
(b) Vol. **115**, No. 1, p. 209–218 (1983).
See also *14.1*, p. *206*; *18.2*, p. *280*; 21.1, p. 363.

T. Warmiński, M. Maier, and B. A. Orłowski
*Energy Levels in $Cd_{1-x}Mn_xF_2$ Crystals Obtained from the Auger Spectra
(b) Vol. **115**, No. 1, p. K7–K10 (1983).
See also *17*, p. *269*; 22.5.4, p. 488.

E. A. Pamyatnykh and A. V. Poltavets
*Short-Wavelength Cyclotron Waves in Metals in Quantizing Magnetic Fields
(b) Vol. **115**, No. 1, p. K51–K54 (1983).
See also *19*, p. *299*; 21, p. 349.

M. Wagner
*Non-Radiative Transitions: Fundamental Difficulties of the Adiabatic Base Approach
(b) Vol. **115**, No. 2, p. 427–435 (1983).
See also *14*, p. *204*.

K. H. J. Buschow, H.-J. Eifert, and B. Elschner
*Electronic Structure and Stability of Amorphous Pd–Si Alloys
(b) Vol. **115**, No. 2, p. 455–462 (1983).
See also *2*, p. *44*; 19, p. 299; 21, p. 350; 22.1.2, p. 425.

P. Ugliengo and C. Pisani
*Trigonometric Series Expansion of Projected Densities of States
(b) Vol. **115**, No. 2, p. K79–K81 (1983).

V. M. Dubovik
*Properties of Electron Form Factors in One-Dimensional Crystals
(b) Vol. **115**, No. 2, p. K91–K93 (1983).

A. N. Kushnirenko
*Quantum-Electrodynamic Theory of Solids
(b) Vol. **115**, No. 2, p. K95–K99 (1983).

A. M. Korostil
*A New Approach to the Theory of Tunnelling Relaxation
(b) Vol. **115**, No. 2, p. K117–K119 (1983).
See also *2*, p. *44*; 6, p. 71.

V. Drchal and J. Kudrnovský
*Second-Order Perturbation Treatment of Correlations in Transition Metal Alloys
(b) Vol. **116**, No. 1, p. 119–128 (1983).
See also 21.1, p. 365.

P. Streitenberger
*Generalized Pseudoatom Model of Inhomogeneous Systems and Its Application to a Simple-Metal Surface
(b) Vol. **116**, No. 1, p. 179–190 (1983).
See also *1.6*, p. *39*; 21, p. 352.

S. Ghosh and S. Khan
*Stimulated Brillouin Scattering of Transversely Polarized Electromagnetic Waves in Magnetoactive Moderately Piezoelectric Semiconductor Plasmas
(b) Vol. **116**, No. 1, p. 323–334 (1983).
See also *20*, p. *307*; 22, p. 404.

K. P. Maheshwari and Ganesh Tarey
Excitation of Magnetoacoustic Waves by Two Microwave Beams in InSb
(b) Vol. **116**, No. 1, p. 339–346 (1983).
See also *7*, p. *82*; 18, p. 274; *20*, p. *307*; 22, p. 405.

J. Katriel and G. F. Kventzel
*The Constant-Coupling Approximation: Some Exact Results
(b) Vol. **116**, No. 1, p. K51 – K56 (1983).
See also 2, p. 45; *18.2*, p. *281*.

J. Chojcan, M. Szuszkiewicz, S. Szuszkiewicz, and B. Rozenfeld
*Localization of Electron in Fe–Co Alloys
(b) Vol. **116**, No. 1, p. K57 – K60 (1983).
See also 21.1, p. 366; 21.1.1, p. 380.

V. A. Khodyrev and E. I. Sirotinin
Charged-Particle Stopping in Crystals
(b) Vol. **116**, No. 2, p. 659 – 671 (1983).
See also *11* p. *126*.

J. Zieliński and J. Kowalska
*On the Two-Site Spinless Falicov-Kimball Model with Hybridization
(b) Vol. **116**, No. 2, p. 677 – 683 (1983).
See also 21, p. 353.

J. Dóczi-Réger and P. C. Hemmer
*The Antiferromagnetic Phase of the Ising Model with First- and Second-Neighbour Interactions
(b) Vol. **116**, No. 2, p. K67 – K71 (1983).
See also *18.4*, p. *295*.

W. E. Hagston
*Quantum Theory of Tunnelling
(b) Vol. **116**, No. 2, p. K85 – K89 (1983).
See also *14*, p. *204*.

Ch. Opitz and F. Dübler
Modelling and Illustration of a Penetration Mechanism for Hydrogen into Metal
(a) Vol. **76**, No. 2, p. 505 – 512 (1983).
See also 4, p. 61; *9*, p. *94*; 22.5, p. 473.

E. G. Vyatkin, Yu. M. Filimonov, A. M. Taratin and S. A. Vorobiev
Molecular-Type Channeling of Relativistic Electrons in Crystals
(b) Vol. **117**, No. 1, p. 41 – 50 (1983).
See also *11*, p. *127*; 22.1.1, p. 420.

V. K. Arora and M. Prasad
*Quantum Transport in Quasi-One-Dimensional Systems
(b) Vol. **117**, No. 1, p. 127 – 140 (1983).
See also 14, p. 204.

A. M. Berezhkovskii and A. A. Ovchinnikov
*Temperature Dependence of the Wannier-Stark Level Width in a Semiconductor in a Strong Electric Field
(b) Vol. **117**, No. 1, p. 289 – 299 (1983).
See also *6*, p. *73*; 14.3.4, p. 244; 22.2.3, p. 452.

I. Bose and S. Chatterjee
*Quantum Spin Hamiltonians: A Renormalization Group Study
(b) Vol. **117**, No. 1, p. K31 – K36 (1983).

A. Le Paillier-Malecot
*Study of the Single and Double Groups C_8 and S_8 and Properties of Their Two Isomorphisms
(b) Vol. **117**, No. 2, p. 521 – 529 (1983).
See also *1*, p. *9*.

A. M. Berezhkovskii and A. A. Ovchinnikov
*The Interband Interaction and the Tunnel Level Width of the Wannier-Stark Ladder in a Semiconductor Subject to a Strong Electric Field
(b) Vol. **117**, No. 2, p. 585 – 588 (1983).
See also 22, p. 407.

M. Suzszkiewicz, J. Chojcan, and A. Ostrasz
*Investigation of the Electronic Structure of V–Fe–H Alloys by Positron Annihilation
(b) Vol. **117**, No. 2, p. 645 – 651 (1983).
See also 21.1, p. 370; 21.1.1, p. 382.

O. Bílek, L. Skála, and L. Künne
*Tight-Binding Cluster Model for C, Si, and Ge (I): Analytic Solution for s-States in Diamond Lattice
(b) Vol. **117**, No. 2, p. 675 – 681 (1983).
See also 22.1, p. 417; 22.1.1, p. 420; 22.1.2, p. 431.

E. Dębowska
*Investigation of the Electronic Structure of TiH_x and $Ti_{1-y}V_yH_x$ by Positron Annihilation
(b) Vol. **117**, No. 2, p. 699 – 706 (1983).
See also 21, p. 355.

K. Elk
*Susceptibility of the Periodic Anderson Model: Hartree-Fock Approximation
(b) Vol. **117**, No. 2, p. K87 – K90 (1983).
See also *18.1*, p. *278*; 21, p. 355.

M. Dębowska and A. Baranowski
*Positron Diffusion Constant for Gold
(b) Vol. **117**, No. 2, p. K127 – K129 (1983).
See also 21.6, p. 397.

H. Luchian
*Magnetic Symmetry in Several Cubic Compounds
(b) Vol. **118**, No. 1, p. 31 – 39 (1983).
See also *18*, p. *274*; 22.8.2, p. 522.

L. Künne, L. Skála, and O. Bílek
*Tight-Binding Cluster Model for C, Si, and Ge (II): Size and Shape Dependence of Some Cluster Properties
(b) Vol. **118**, No. 1, p. 173 – 178 (1983).
See also 22.1, p. 417; 22.1.1, p. 420; 22.1.2, p. 432.

H. KÖSTER, JR. and K. HÜBNER
*Statistics and Transport Behaviour of Electrons in SiO_2
(b) Vol. **118**, No. 1, p. 293–301 (1983).
See also *6*, p. *75*; 14.4, p. 252; 22.6, p. 494.

A. K. GHOSH, S. CHAKRABORTY, and A. MANNA
*Density Functional Approach for Metals. The Matheiss Prescription
(b) Vol. **118**, No. 1, p. 373–379 (1983).
See also 21.1, p. 371; 21.1.1, p. 382; 21.6, p. 398.

O. A. C. NUNES
*On the Multiphoton Laser Absorption in Free-Carrier Semiconductors
(b) Vol. **118**, No. 1, p. K25–K29 (1983).
See also 6, p. 75; *20.1*, p. *322*; 22, p. 409.

B. BABIĆ, M. STOJIĆ, and J. KONSTANTINOVIĆ
*Electrical Properties of the Metastable $Mn_{1.11}Al_{0.89}$ Alloy
(a) Vol. **78**, No. 1, p. K77–K80 (1983).
See also *14.1*, p. *208*; 21, p. 357; 21.1, p. 372.

V. N. BAIER, V. M. KATKOV, and V. M. STRAKHOVENKO
Radiation at Planar Channeling of Relativistic Electrons in Thick Crystals
(b) Vol. **118**, No. 2, p. 499–509 (1983).
See also *11*, p. *130*.

H. V. SHARMA and I. SINGH
*Valence Fluctuations and Magnetic–Nonmagnetic Transitions in a Model for Intermediate-Valence Systems
(b) Vol. **118**, No. 2, p. 511–520 (1983).
See also 8, p. 89; 18.1, p. 278; *18.2*, p. *285*; 21.4, p. 392.

J. ZIELIŃSKI and J. KOWALSKA
*Role of Phonon-Induced Hybridization in the Spinless Anderson Model
(b) Vol. **118**, No. 2, p. 805–810 (1983).
See also *6*, p. *76*.

S. LIPIŃSKI
*Effects of Order on the Electronic Structure of Ni_3Fe
(b) Vol. **118**, No. 2, p. 861–869 (1983).
See also 18.2, p. 285; 21.1, p. 372; 21.1.1, p. 383.

U. BEHN
*Quasi-Particle Motion in a Dichotomous Stochastic Potential: Comparison of a Decoupling Approach with the Dynamic CPA
(b) Vol. **118**, No. 2, p. K151–K153 (1983).
See also 14, p. 204.

K. G. CHAKRABORTY and J. F. BARET
*Spin-One Lattice Gas Model for Isotherm-Reversal in Spin-Labelled Lipid Monolayers
(b) Vol. **119**, No. 1, p. 61–71 (1983).
See also *2*, p. *51*; 22.9, p. 529.

M. S. Brodin, V. M. Bandura, and M. G. Matsko
*Direct Luminescence Evidence for the Degenerate Electron–Hole Plasma Formation in $A^{II}B^{VI}$ Semiconductors
(b) Vol. **119**, No. 1, p. 83–87 (1983).
See also 13.5.1, p. 197; *20.3*, p. *343*; 22.4.2, p. 467; 22.4.3, p. 468.

A. V. Skripov, A. P. Stepanov, A. D. Shevchenko, and Z. D. Kovalyuk
*NMR Study of the Charge-Density-Wave State in VSe_2
(b) Vol. **119**, No. 1, p. 401–410 (1983).
See also *19*, p. *303*; 22, p. 411.

D. Chattopadhyay
*Transient Behaviour of Hot Electrons in Quantum-Well Structures of Polar Semiconductors
(b) Vol. **119**, No. 1, p. K77–K81 (1983).
See also 6, p. 78; *14.3*, p. *223*; 22, p. 411.

G. P. Das and P. Chaddah
*Correlation Energy from Compton Scattering and Positron Annihilation Data
(b) Vol. **119**, No. 2, p. 467–472 (1983).
See also 21.2, p. 388.

V. M. Fomin and E. P. Pokatilov
Nonlinear Kinetics and Optical Properties of Band Charge Carriers
(b) Vol. **119**, No. 2, p. 483–492 (1983).
See also 6. p. 78; 14.3.4, p. 247; *20.1*, p. *327*; 22, p. 412.

D. J. González, M. L. Zorita, J. A. Alonso, and M. P. Iñiguez
*Density Functional Pseudopotential Calculation of the Cohesive Properties of Disordered Solid Alloys of Nontransition Metals. Application to Zn–Cd
(b) Vol. **119**, No. 2, p. 589–594 (1983).
See also *1.3*, p. *26*; 21, p. 359; 21.1, p. 375.

A. V. Soldatov, A. N. Gusatinskii, M. G. Karin, and O. A. Sadovskaya
*Energy Position and Participation of the Rare-Earth Element 5p-States in Chemical Interaction
(b) Vol. **119**, No. 2, p. K123–K126 (1983).
See also *20*, p. *311*; 21.4, p. 393.

H.-G. Fritsche
*The Electronic Charge Distribution of Hydrogen Absorbed in Niobium – a Comparison of SW-Xα Cluster Calculations and Neutron Diffraction Experiments
(b) Vol. **119**, No. 2, p. K127–K130 (1983).
See also 21.1, p. 375.

D. Ihle
*Cluster Approach to the Spinless Fermion Model
(b) Vol. **119**, No. 2, p. K169–K173 (1983).
See also *18*, p. *275*.

G. A. Scholz and R. F. Frindt
Electrical Resistivity of Silver Intercalated 2H-TaS_2
(a) Vol. **79**, No. 2, p. 483–488 (1983).
See also *14.1*, p. *210*; *14.3*, p. *225*; 22, p. 412.

J. Filipecki, B. Całusiński, and Z. Mandecki
Investigation of Electron Phenomena in the Metallic Glasses $Fe_xNi_{78-x}B_{10}Si_{12}$ with the Use of Positron Annihilation Methods before and after Their Crystallization
(a) Vol. **79**, No. 2, p. 585–591 (1983).
See also *2*, p. *53*; 21.1, p. 376; 21.1.1, p. 384.

A. I. Rubinshtein
*Image Forces at the Metal Electrode–Electrolyte Solution Interface: Account of Spatial Dispersion of the Solution Permittivity
(b) Vol. **120**, No. 1, p. 65–76 (1983).
See also *14.4*, p. *254*.

U. Pietsch
*Hydrostatic Pressure and Uniaxial Stress Dependence of the Silicon (222) X-Ray Reflection Power
(b) Vol. **120**, No. 1, p. 183–188 (1983).
See also *4*, p. *63*; 22.1.2, p. 437.

J. Šlechta
*Comparison between the Self-Consistent Continued-Fraction and Other Continued-Fraction Methods
(b) Vol. **120**, No. 1, p. 329–339 (1983).
See also *2*, p. *54*.

M. Chaturvedi and V. Srivastava
*Studies on Quantum Percolation
(b) Vol. **120**, No. 1, p. 421–433 (1983).
See also *2*, p. *54*.

C. Rudowicz
*Note on the "Comment on Energy of Magnetic Anisotropy"
(b) Vol. **120**, No. 1, p. K51–K52 (1983).
See also *18*, p. *275*.

M. Asche and H. Kostial
Experimental Indication of the Presence of Loops for Hot Electron Properties in Many-Valley Semiconductors
(b) Vol. **120**, No. 1, p. K83–K85 (1983).
See also *14.3.4*, p. *248*; 22.1.2, p. 437.

L. N. Vasilev, V. M. Grabov, A. V. Golubkov, A. G. Gorobets, V. S. Oskotskii, I. A. Smirnov, and V. V. Tikhonov
Physical Properties and Phase Transitions of the Rare Earth Monosulphides in the Homogeneity Range
(a) Vol. **80**, No. 1, p. 237–244 (1983).
See also *1.3*, p. *26*; *8*, p. *91*; 14.3, p. 226; 18.4, p. 297; 22, p. 413.

I. P. Zvyagin
*The Temperature Dependence of the Conductivity near the Localization Threshold
(b) Vol. **120**, No. 2, p. 503–509 (1983).
See also *14*, p. *204*.

V. A. Kochelap and V. N. Sokolov
*Deformational Phase Transitions and Periodic Structures in a Magnetic Field
(b) Vol. **120**, No. 2, p. 565–575 (1983).
See also *6*, p. *81*; 22, p. 414.

H. Fehske
*Spin Fluctuations in Alloys with Random Transfer
(b) Vol. **120**, No. 2, p. 611–620 (1983).
See also 14, p. 204; 18.1, p. 279; 18.2, p. 288; 21.1, p. 377.

F. Goedsche, R. Richter, H. Wegener, and G. Röpke
Correlation Function Method for the Thermoelectric Power of Metals
(b) Vol. **120**, No. 2, p. 631–644 (1983).
See also *14.1*, p. *210*; *15*, p. *263*; 21, p. 361.

P. Enders
*Structure and Bonding in Cubic IV–VI Crystals (I): Evidence for Covalent Bonding from LCAO Parameters
(b) Vol. **120**, No. 2, p. 735–744 (1983).
See also *4*, p. *64*; 22, p. 414.

O. L. T. de Menezes and A. Troper
*A Model for Inhomogeneous Mixed-Valence f-Semiconductors
(b) Vol. **120**, No. 2, p. 767–774 (1983).
See also 14, p. 205; 21.4, p. 393.

N. A. Nikolov and P. N. Malinov
*Resonances and Averaged Forces in Bounded Semiconductor Plasmas
(b) Vol. **120**, No. 2, p. 775–782 (1983).
See also *14.3.4*, p. *248*.

H. Stramska
*Higher Harmonics Generated by Mobile Carriers in Semiconductors
(b) Vol. **120**, No. 2, p. 783–788 (1983).
See also 14, p. 205; 18, p. 276; 22, p. 415.

N. Ya. Borets, I. Ya. Kanovskii, and V. M. Nitsovich
*Coexistence of Charge and Magnetic Orderings in a Polar Ferromagnet
(b) Vol. **120**, No. 2, p. K167–K171 (1983).
See also *18.2*, p. *289*.

V. Koubele, M. Zvára, and P. Hlídek
Compositional Variations of the Nonresonant Interband Faraday Rotation in $Cd_xHg_{1-x}Te$ Single Crystals
(b) Vol. **120**, No. 2, p. K173–K177 (1983).
See also *18*, p. *276*; *20.1*, p. *330*; 22.4.4, p. 472.

O. Z. Alekperov, V. G. Golubev, V. I. Ivanov-Omskii, and A. Sh. Mekhtiev
*Plasma Shift of Cyclotron Resonance Line in n-GaAs
(b) Vol. **120**, No. 2, p. K179–K181 (1983).
See also *14.3*, p. *227*; 18, p. 276; 22.2.1, p. 447.

13.1 Band Structure

Review Articles

R. E. Hummel
*Differential Reflectometry and Its Application to the Study of Alloys, Ordering, Corrosion, and Surface Properties
(a) Vol. **76**, No. 1, p. 11 – 44 (1983).
See also *1.3*, p. *21*; 10.1, p. 99; 13.3, p. 167; *20.1*, p. *312*. 21, p. 349; 21.1, p. 363; 22.6, p. 490.

R. A. Evarestov and V. P. Smirnov
*Special Points of the Brillouin Zone and Their Use in the Solid State Theory
(b) Vol. **119**, No. 1, p. 9 – 40 (1983).
See also 13.3, p. 167; 13.4, p. 172; 22, p. 403.

Original Papers and Short Notes

O. Ziep
*Estimation of Second-Order Auger Recombination in Lead Chalcogenides
(b) Vol. **115**, No. 1, p. 161 – 170 (1983).
See also *6*, p. *69*; 13.5.2, p. 200; 22.8, p. 502.

E. Krüger
Exchange Interaction in Superconducting Metals
(b) Vol. **115**, No. 1, p. 181 – 185 (1983).
See also *14.2*, p. *212*; 21, p. 349.

P. Rennert, G. Carl, and U.-H. Gläser
*Theoretical Compton Profiles of Magnesium
(b) Vol. **115**, No. 1, p. 219 – 224 (1983).
See also 13.2, p. 166; *20*, p. *307*; 21.3, p. 389.

M. Downey and P. V. Smith
*A Localized Orbital Description of Si
(b) Vol. **115**, No. 1, p. 255 – 266 (1983).
See also 22.1.2, p. 423.

C. W. M. Timmermans, S. O. Cholakh, R. L. van der Woude, and G. Blasse
Some Optical and Electrical Measurements on $Cs_3Bi_2Br_9$ Single Crystals
(b) Vol. **115**, No. 1, p. 267 – 271 (1983).
See also *16*, p. *264*; *20.1*, p. *312*; 22.8, p. 502.

C. Engler and W. Lorenz
*Erratum to: Calculation of Local Densities of States, Local Charges, and Adsorption Energy on Large Semiconductor Substrate with Single Adsorbate
(b) Vol. **115**, No. 1, p. 325 (1983).
See also 4, p. 60; 13.3, p. 167; 22.2.2, p. 448.

G. A. Bordovskii, N. L. Gordeev, A. N. Ermoshkin, V. A. Izvozchikov, and R. A. Evarestov
*Self-Consistent Energy Band Structure of Tetragonal Lead Oxide
(b) Vol. **115**, No. 1, p. K15–K19 (1983).
See also 22.6, p. 490.

S. M. Wasim
*On the Electron Effective Mass in n–$CuInTe_2$
(a) Vol. **75**, No. 1, p. K69–K72 (1983).
See also *14.3*, p. *215*; 22.8, p. 502.

N. M. Ravindra, C. Ance, J. P. Ferraton, A. Donnadieu, and S. Robin
Penn-Like Model Calculations. Application to Hydrogenated Amorphous Silicon Films Prepared by Glow Discharge
(b) Vol. **115**, No. 2, p. 347–352 (1983).
See also *2*, p. *44*; 14.4, p. 250; 22.1.2, p. 425.

K. Takeda, Y. Maeda, E. Ohta, M. Sakata, G. Kido, and N. Miura
*Study of Energy Band Parameters in p-Type $Ge_{0.9}Si_{0.1}$ Alloys
(b) Vol. **115**, No. 2, p. 369–379 (1983).
See also 14.3, p. 216; 20.1, p. 314; 22.1.1, p. 418; 22.1.2, p. 425.

M. Mocker and O. Ziep
*Intrinsic Recombination in Dependence on Doping Concentration and Excitation Level. Application to Lead Chalcogenides
(b) Vol. **115**, No. 2, p. 415–425 (1983).
See also *14.3*, p. *216*; 20.3, p. 335; 22.8, p. 503.

S. Eränen and J. Sinkkonen
*Conduction Electron Self-Energy Effects in Ferromagnetic Semiconductors
(b) Vol. **115**, No. 2, p. 519–527 (1983).
See also 14.3, p. 216; 22, p. 403.

A. P. Litvinchuk and N. I. Vitrikhovskii
*The Variation of the Energy Gap with Composition in the Quaternary Alloy System $ZnTe_{1-2x}S_xSe_x$
(b) Vol. **115**, No. 2, p. K151–K154 (1983).
See also 13.5.1, p. 195; *20.3*, p. *336*; 22.4.4, p. 470.

L. Bočánek and J. Musilová
Modulation Piezoreflectance of Cubic Polycrystals
(a) Vol. **75**, No. 2, p. 555–558 (1983).
See also *12.2*, p. *141*; *20.1*, p. *314*; 22.1.1, p. 419.

A. N. Chakravarti, K. P. Ghatak, K. K. Ghosh, and H. M. Mukherjee
*Effect of Varying the Orientation of a Quantizing Magnetic Field on the Effective Electron Mass in Degenerate Tetragonal Semiconductors
(b) Vol. **116**, No. 1, p. 17–24 (1983).
See also 18, p. 273; 22.3, p. 456.

B. I. Reser
*On the Brillouin Zone Integration
(b) Vol. **116**, No. 1, p. 31–40 (1983).

L. G. Khvostantsev and V. A. Sidorov
Thermoelectric Properties of GeTe at High Hydrostatic Pressure up to 8.5 GPa and Its Valence-Band Structure
(b) Vol. **116**, No. 1, p. 83–89 (1983).
See also 14.3, p. 217; *15*, p. *261*; 22.7, p. 499.

P. V. Smith
*A Localized Orbital Description of Ge and α-Sn
(b) Vol. **116**, No. 1, p. 101–109 (1983).
See also 22.1, p. 417; 22.1.1, p. 419.

M. Scrocco
*Satellite Structure of ZnO and Zn(II) Halides Studied by X-Ray Photoelectron Spectroscopy
(b) Vol. **116**, No. 1, p. 137–144 (1983).
See also 17, p. 269; 22.4, p. 460; 22.5, p. 473.

P. Enders
*Universal Tight-Binding Parameters for Cubic IV–VI Compounds
(b) Vol. **116**, No. 1, p. K15–K18 (1983).
See also 22, p. 405.

Ch. Müller, W. Blau, and P. Ziesche
*Local Partial DOS and Experimental Al K Spectra of Transition Metal Aluminides. Charge Transfer Trend
(b) Vol. **116**, No. 2, p. 561–573 (1983).
See also 20, p. 308; 21.1, p. 367; 21.1.1, p. 381.

D. Tjapkin and V. Milanović
*The Influence of Parameter Nonuniformity on the Two-Dimensional Einstein Relation in Periodic and Non-Periodic Semiconductor Structures
(b) Vol. **116**, No. 2, p. 653–658 (1983).
See also *14.3.2*, p. *233*; 22.2.4, p. 454.

V. N. Ermakov, V. V. Kolomoets, and V. S. Timoshuk
*The Deformation Potentials Ξ_u^X, Ξ_d^X of the X-Valleys of the Germanium Conduction Band
(b) Vol. **116**, No. 2, p. K77–K78 (1983).
See also 22.1.1, p. 419.

P. I. Baranskii, V. V. Kolomoets, and S. S. Korolyuk
*The Non-Parabolicity of the n-Si Conduction Band Caused by Elastic Deformation along the [111] Direction
(b) Vol. **116**, No. 2, p. K109–K112 (1983).
See also *14.3*, p. *218*; 22.1.2, p. 428.

A. I. Basyk, V. F. Kovalenko, and G. P. Peka
*Transfer Mechanisms of Nonequilibrium Charge Carriers in Graded Band-Gap Semiconductors
(a) Vol. **76**, No. 2, p. 429–435 (1983).
See also *20.3*, p. *337*; 22.2.4, p. 454.

A. BENNOUNA, P. Y. TESSIER, M. PRIOL, Q. DANG TRAN, and S. ROBIN
*Far Ultraviolet Photoelectric Study of Thin SnSe Evaporated Films
(b) Vol. **117**, No. 1, p. 51–56 (1983).
See also *17*, p. *270*; 22, p. 406.

G. VALIUKONIS, F. M. GASHIMZADE, D. A. GUSEINOVA, G. KRIVAITĖ, A. M. KULIBEKOV, G. S. ORUDZHEV, and A. ŠILEIKA
*Reflectance and Thermoreflectance Spectra and Energy Band Structure of GeSe Crystals
(b) Vol. **117**, No. 1, p. 81–92 (1983).
See also *20.1*, p. *317*; 22, p. 406.

M. BURD, R. STEARNS, and R. BRAUNSTEIN
*De-Correlation Technique for Separation of Drude Parameters from Wavelength Modulation Spectroscopy Data
(b) Vol. **117**, No. 1, p. 101–107 (1983).
See also *20.1*, p. *317*; 21.1, p. 368; 21.6, p. 397.

N. C. AMARAL, B. MAFFEO, and D. GUENZBURGER
*Self-Consistent Embedded-Cluster Calculations of the Electronic Structure of Alkaline Earth Fluorides in the Hartree-Fock-Slater Approximation
(b) Vol. **117**, No. 1, p. 141–153 (1983).
See also 22.5.3, p. 486.

J. ROSE and R. SCHUCHARDT
*A Theoretical Investigation of the Fermi Surfaces of Bismuth and Antimony
(b) Vol. **117**, No. 1, p. 213–224 (1983).
See also *13.2*, p. *166*; 21.7, p. 401.

M. STORDEUR and H. T. LANGHAMMER
*Anisotropy of Transport Properties Due to Direction-Dependent Band Nonparabolicity
(b) Vol. **117**, No. 1, p. 329–333 (1983).
See also 14.3, p. 219; 15, p. 261; 22, p. 406.

I. S. EDELMAN, G. DUSTMURADOV, and V. P. KONONOV
Polarized Light Absorption of the Magnetic Semiconductor $CdCr_2Se_4$
(b) Vol. **117**, No. 1, p. 351–354 (1983).
See also *20.1*, p. *318*; 22.8, p. 505.

R. A. EVARESTOV, V. A. LOVCHIKOV, and I. I. TUPITSIN
*Hartree-Fock Exchange and LCAO Approximation in the Band Structure Calculations of Solids
(b) Vol. **117**, No. 1, p. 417–427 (1983).
See also 22, p. 407.

N. V. VOLKENSHTEIN, E. V. GALOSHINA, M. E. KOST, and T. S. SHUBINA
*Magnetic Properties of the Sc–H System
(b) Vol. **117**, No. 1, p. K47–K52 (1983).
See also *18.1*, p. *278*; 21.1, p. 368.

H. VAN CONG
*Band-Gap Narrowing in n-Type Moderately Doped Silicon at 300 K
(b) Vol. 117, No. 2, p. 575–584 (1983).
See also *13.4*, p. *179*; 22.1.2, p. 430.

TH. HARBICH and G. MAHLER
*Bistability of the Electron-Hole Plasma Density in a Strong Electromagnetic Field
(b) Vol. 117, No. 2, p. 635–644 (1983).
See also *13.5*, p. *193*; 22.2.1, p. 444.

K. P. GHATAK and A. N. CHAKRAVARTI
*Carrier Statistics in Superlattices of Small-Gap Semiconductors in the Presence of a Quantizing Magnetic Field
(b) Vol. 117, No. 2, p. 707–715 (1983).
See also 22.2.1, p. 444; 22.2.3, p. 452.

K. R. ALLAKHVERDIEV, T. G. MAMEDOV, E. YU. SALAEV, and I. K. EFENDIEVA
*The Fundamental Absorption Spectra of $TlInSe_2$ Crystals under Pressure
(b) Vol. 117, No. 2, p. K109–K111 (1983).
See also *20.1*, p. *320*; 22.8, p. 506.

C. RINCÓN, L. GONZÁLEZ, and G. SÁNCHEZ PÉREZ
*Temperature Dependence of the Energy Gap in $CuInSe_2$
(b) Vol. 117, No. 2, p. K123–K126 (1983).
See also *20.1*, p. *320*; 22.8, p. 506.

T. L. TANSLEY, D. F. NEELY, and C. P. FOLEY
Optical Dispersion in Zinc Oxide
(a) Vol. 77, No. 2, p. 491–496 (1983).
See also *20.1*, p. *320*; 22.4, p. 460.

B. PISTOULET, D. COQUILLAT, J. C. TEDENAC, G. BRUN, and M. MAURIN
*Hole Effective Mass and Impurity Levels in Undoped AgTlTe. Optimum Thermoelectric Figure of Merit
(a) Vol. 77, No. 2, p. 669–677 (1983).
See also *13.4*, p. *181*; 14.3, p. 221; 15, p. 262; 22.8, p. 506.

A. S. ALEKSANDROV, V. F. ELESIN, and M. P. KAZEKO
The Influence of Defects on the Low-Temperature Resistivity of Intermetallic Compounds with A-15 Structure
(a) Vol. 77, No. 2, p. 725–732 (1983).
See also 13.4, p. 181; *14.2*, p. *213*.

A. K. AHUJA, S. AULUCK, and S. K. JOSHI
*Density of States and Optical Conductivity of Ferromagnetic Nickel
(b) Vol. 118, No. 1, p. 105–111 (1983).
See also 14.1, p. 208; 21.1, p. 371.

A. N. ERMOSHKIN, R. A. EVARESTOV, S. A. KUCHINSKII, and V. K. ZAKHAROV
*The Quasimolecular Approach to the Electronic Structure Calculations for Silver and Copper Halides
(b) Vol. 118, No. 1, p. 191–203 (1983).
See also 13.4, p. 182; 22.5.1, p. 476.

G. Paasch and H. Übensee
*Carrier Density near the Semiconductor–Insulator Interface. Local Density Approximation for Non-Isotropic Effective Mass
(b) Vol. **118**, No. 1, p. 255 – 266 (1983).
See also *13.3*, p. *169*; *14.3.2*, p. *234*; 22.1.1, p. 420; 22.1.2, p. 432.

P. M. Tomchuk, V. A. Shenderovskii, P. N. Gorley, and N. P. Gavaleshko
Hall Coefficient and Magnetoresistance of Semiconductors with Anisotropic Energy Dispersion
(b) Vol. **118**, No. 1, p. 433 – 439 (1983).
See also 13.4, p. 182; *14.3*, p. *221*; 22.8, p. 507.

L. M. Kanskaya, S. I. Kokhanovskii, R. P. Seysyan, and Al. L. Efros
*Direct Determination of Indium Antimonide Energy Band Parameters from Diamagnetic Exciton Spectra
(b) Vol. **118**, No. 1, p. 447 – 452 (1983).
See also *13.5.1*, p. *197*; 20.1, p. 322; 22.2.3, p. 453.

S. Maciejewski
*Cooper Pair Condensation Energy Dependence on Electron Structure
(b) Vol. **118**, No. 1, p. K3 – K5 (1983).
See also *14.2*, p. *213*; 21, p. 356.

H. Englisch
*There is No Energy Interval Lying in the Spectra of All One-Dimensional Periodic Alloys
(b) Vol. **118**, No. 1, p. K17 – K19 (1983).
See also 21, p. 356.

C. Rincón and J. González
*Residual Optical Absorption below the Band Gap in $CuInSe_2$
(b) Vol. **118**, No. 1, p. K21 – K24 (1983).
See also *20.1*, p. *322*; 22.8, p. 507.

B. Jagannatha Rao and A. R. Chetal
*On the Chemical Shifts of the K-Absorption Discontinuity of Copper in Some of Its Systems
(b) Vol. **118**, No. 2, p. 551 – 555 (1983).
See also *20*, p. *309*; 21.1, p. 372; 22, p. 409.

A. K. Ahuja, S. Auluck, and S. K. Joshi
*Application of Special-Directions Method to the Optical Properties of Copper and Silver
(b) Vol. **118**, No. 2, p. 575 – 586 (1983).
See also 20.1, p. 323; 21.1, p. 372; 21.6, p. 398.

H. J. Fischbeck
On the Translation Mode of the Kink Lattice of a Quasi-One-Dimensional Peierls-Fröhlich System with a Nearly Half-Filled Band
(b) Vol. **118**, No. 2, p. 595 – 603 (1983).
See also *1.2*, p. *18*; *6*, p. *76*; 7, p. 83.

NGUYEN VAN HUONG
*Free-Carrier Absorption in n-PbSe at Low Temperatures Due to Electron–Electron Scattering
(b) Vol. **118**, No. 2, p. 631–640 (1983).
See also *20.1*, p. *323*; 22, p. 409.

F. LÓPEZ-AGUILAR and J. COSTA-QUINTANA
*Renormalized Atom Model (RAM) Applied to the Heavy Rare-Earth Sesquioxides in Cubic Crystal Structure
(b) Vol. **118**, No. 2, p. 779–787 (1983).
See also 22.6, p. 495.

K. HRICOVÍNI, P. MIKUŠÍK, and J. KREMPASKÝ
Photoemission Studies of $Pd_{80-x}Si_{20}Co_x$ Amorphous Alloys
(b) Vol. **118**, No. 2, p. 823–827 (1983).
See also *2*, p. *50*; *17*, p. *270*; 21.1, p. 372.

A. N. CHAKRAVARTI, K. P. GHATAK, K. K. GHOSH, and H. M. MUKHERJEE
Effect of Crystal-Field Splitting on the Einstein Relation in Tetragonal Semiconductors
(b) Vol. **118**, No. 2, p. 843–852 (1983).
See also *14.3*, p. *222*; 22.3, p. 457.

K. SIERAŃSKI and J. SZATKOWSKI
*Intraband Absorption Due to Alloy Scattering in Lead Chalcogenide Semiconductors
(b) Vol. **118**, No. 2, p. 871–879 (1983).
See also *20.1*, p. *324*; 22, p. 410.

R. TOVAR BARRADAS, J. GONZÁLEZ, and C. RINCÓN
Impurity States near the Fundamental Absorption Edge in $p\text{-CuInS}_2$
(b) Vol. **118**, No. 2, p. K103–K106 (1983).
See also 13.4, p. 184; *20.1*, p. *324*; 22.8, p. 508.

V. HEERA, G. SEIFERT, and P. ZIESCHE
*Semirelativistic SCF-SW-Xα Calculations of Uranium Compounds. Electronic Structure and Chemical Bond
(b) Vol. **118**, No. 2, p. K107–K112 (1983).
Erratum
(b) Vol. **120**, No. 2, p. K227 (1983).
See also 4, p. 62; 21.5, p. 395.

G. DRÄGER, R. WIPPERMANN, O. BRÜMMER, and A. ŠIMŮNEK
*Polarized ZnK- and OK-Emission Bands of ZnO
(b) Vol. **118**, No. 2, p. K113–K116 (1983).
See also *20*, p. *309*; 22.4, p. 460.

P. MUTH
*On the Resistivity Behaviour of Some Weak-Ferromagnetic Nickel-Based Alloys
(b) Vol. **118**, No. 2, p. K117–K120 (1983).
See also *14.1*, p. *208*; 21.1, p. 373.

I. I. Gegusin and V. N. Datsyuk
*APW Approach to the Momentum Matrix Element Calculation
(b) Vol. **118**, No. 2, p. K121 – K124 (1983).
See also *20.1*, p. *324*.

P. Muth
*A Critical Concentration of Ni–Mn Alloys at about 6 at% Mn
(b) Vol. **118**, No. 2, p. K137 – K139 (1983).
See also *14.1*, p. *208*; 21.1, p. 373.

D. Weaire and J. P. Kermode
*Electronic States for Incommensurate Lattice Potentials
(b) Vol. **118**, No. 2, p. K163 – K166 (1983).

S. G. Abdullaeva, N. T. Mamedov, and G. S. Orudzhev
*Band Structure of $TlGaSe_2$
(b) Vol. **119**, No. 1, p. 41 – 48 (1983).
See also 22.3.1, p. 459.

P. K. Katana
*Two-Photon Interband Absorption of Light in Semiconductors in External Parallel Electric and Magnetic Fields
(b) Vol. **119**, No. 1, p. 89 – 98 (1983).
See also 14.3.4, p. 247; 18, p. 275; *20.1*, p. *325*; 22, p. 411.

S. F. Kolodeznaya and E. V. Kozlov
Calculation of the Properties and the Structure of Ordering Alloys of the Cu–Au System
(b) Vol. **119**, No. 1, p. 131 – 139 (1983).
See also *1.3*, p. *25*; 21.1, p. 374; 21.6, p. 398.

V. I. Ivanov-Omskii, A. Sh. Mekhtiev, S. A. Rustambekova, and E. N. Ukraintsev
*Plasma Reflection and Hole Optical Effective Mass in p-HgTe
(b) Vol. **119**, No. 1, p. 159 – 169 (1983).
See also *20.1*, p. *325*; 22.4.3, p. 469.

P. Bhattacharya and A. R. Chetal
*Study of Cu K Edge in Some Copper (II) Octahedral Complexes
(b) Vol. **119**, No. 1, p. 179 – 183 (1983).
See also *20*, p. *310*; 22.9, p. 530.

O. Goede and D. Hennig
*Isovalent-Impurity Clusters in Koster-Slater Approximation
(b) Vol. **119**, No. 1, p. 261 – 268 (1983).
See also *13.4*, p. *185*; 22.2.2, p. 450; 22.4.1, p. 464.

E. Ojala
*K and L Absorption Spectra of Aluminium from APW Calculation
(b) Vol. **119**, No. 1, p. 269 – 277 (1983).
See also *20*, p. *310*; 21, p. 358.

O. ZIEP and M. MOCKER
*New Arguments to the Problem of Screening in Auger Recombination in Semiconductors
(b) Vol. **119**, No. 1, p. 299–306 (1983).
See also 22.2.3, p. 453; 22.8, p. 508.

A. V. MALAKHOVSKII, T. P. MOROZOVA, and V. I. YUZVAK
Magneto-Optical Properties of $RbMnF_3$ in the Paramagnetic State
(b) Vol. **119**, No. 1, p. 411–417 (1983).
See also *18*, p. *275*; *20.1*, p. *326*; 22.5, p. 475.

V. HEERA, G. SEIFERT, and P. ZIESCHE
*Semirelativistic SCF-SW-X_α Calculations of Uranium Compounds. Charge Density Differences at the Uranium Nucleus
(b) Vol. **119**, No. 1, p. K1–K4 (1983).
See also 21.5, p. 395.

P. CHANDRASEKHARAM, D. RAJA REDDY, and B. K. REDDY
*Optical Properties of $(ZnSe)_x(CdTe)_{1-x}$ Single Crystals
(b) Vol. **119**, No. 1, p. K5–K9 (1983).
See also *20.1*, p. *326*; 22.4.4, p. 471.

S. K. PLACHKOVA, G. CH. CHRISTAKUDIS, and L. V. VASSILEV
*Density-of-States Effective Mass in the System $(GeTe)_{1-x}(AgSbTe_2)_x$
(b) Vol. **119**, No. 1, p. K73–K76 (1983).
See also *14.3*, p. *223*; 15, p. 262; 22, p. 411.

T. SOMA and H.-MATSUO KAGAYA
Local Heine-Abarenkov Model Potential for III–V and II–VI Covalent Compounds
(b) Vol. **119**, No. 2, p. 547–553 (1983).
See also *4*, p. *63*; 22.2, p. 442; 22.4, p. 461.

K. KÖHLER, H. P. GESERICH, TH. WOLF, and G. MÜLLER-VOGT
*Optical Properties of TiN Single Crystals
(b) Vol. **119**, No. 2, p. 605–610 (1983).
See also *20.1*, p. *327*; 22, p. 412.

K. H. GOETZ and W. RICHTER
*Interaction of Free Hole Intraband Excitations with the A_1-Phonon in Highly Doped (p-Type) Tellurium
(b) Vol. **119**, No. 2, p. 631–638 (1983).
See also *6*, p. *79*; 20.1, p. 327; 22.1.3, p. 440.

J. FLOHRER and M. PORSCH
*Effect of Valence Band Splitting on the Biexciton Binding Energy in Wurtzite-Type Semiconductors
(b) Vol. **119**, No. 2, p. 639–649 (1983).
See also *13.5.1*, p. *198*; 22.4.1, p. 464.

P. BHATTACHARYA and A. R. CHETAL
*Chemical Shift and EXAFS Studies of Some Copper Complexes
(b) Vol. **119**, No. 2, p. 735–740 (1983).
See also 20, p. 310; 22.8.1, p. 517.

S. K. Decheva-Plachkova
*On the Fundamental Energy Gap in the System $(GeTe)_{1-x}(AgSbTe_2)_x$
(b) Vol. **119**, No. 2, p. K97–K100 (1983).
See also 1.2, p. 19; 22.8, p. 509.

S. M. Rafique, N. R. Mitra, and P. L. Srivastava
*On Choosing a Set of Eigenvalues and Wave Functions in Harrison's First Principle Method
(b) Vol. **119**, No. 2, p. K113–K116 (1983).
See also *14.1*, p. *210*; 21, p. 359.

R. Požėra, A. Sutkus, and R. Tolutis
*The Influence of Hydrostatic Pressure on the Effective Concentration and Cyclotron Mobility of Electrons in Semiconducting $Bi_{1-x}Sb_x$ Alloys
(b) Vol. **120**, No. 1, p. 165–171 (1983).
See also 14.3, p. 226; 22.7.1, p. 501.

H. J. Fischbeck
*Self-Consistent Kink Lattice in a Quasi-One-Dimensional Peierls-Fröhlich System with a Nearly One-Third-Filled Band (I): Analytical Solution
(b) Vol. **120**, No. 1, p. 249–262 (1983).

J. Szatkowski and K. Sierański
Theory of Free-Carrier Magnetoabsorption in Mixed Zincblende Narrow-Gap Semiconductors
(b) Vol. **120**, No. 1, p. 263–272 (1983).
See also *20.1*, p. *329*; 22.4.4, p. 472.

P. Rennert, G. Carl, and W. Hergert
*Spin-Polarized Compton Profile of Nickel
(b) Vol. **120**, No. 1, p. 273–281 (1983).
See also 13.2, p. 166; 18.2, p. 287; *20*, p. *311*; 21.1, p. 376.

V. V. Nemoshkalenko, A. E. Krasovskii, V. N. Antonov, Vl. N. Antonov, U. Fleck, H. Wonn and P. Ziesche
The Relativistic Linear Muffin-Tin Orbital Method. Application to Au
(b) Vol. **120**, No. 1, p. 283–296 (1983).
See also 21.6, p. 399.

C. Regnaut, E. Fusco, and J. P. Badiali
Structure and Thermodynamics of the Liquid Noble Metals from the Generalized OPW and Resonant Model Approaches
(b) Vol. **120**, No. 1, p. 373–382 (1983).
See also *2*, p. *54*; 8, p. 91; 21.1, p. 376; 21.6, p. 399.

N. B. Mustafaev and M. G. Shakhtakhtinskii
*Density of Electron States in Thin Bi Films
(b) Vol. **120**, No. 1, p. K23–K26 (1983).
See also 21.7, p. 401.

B. G. S. Doman
*Electron Energy Levels in a Thin Wire in a Magnetic Field
(b) Vol. **120**, No. 1, p. K79–K82 (1983).
See also 21.7, p. 402.

P. B. ALLEN
*A Tetrahedron Method for Doubly Constrained Brillouin Zone Integrals. Application to Silicon Optic Phonon Decay
(b) Vol. **120**, No. 2, p. 529–538 (1983).
See also *6*, p. *80*; 22.1.2, p. 439.

A. C. SHARMA, N. M. RAVINDRA, S. AULUCK, and V. K. SRIVASTAVA
*Temperature-Dependent Effective Masses in III–V Compound Semiconductors
(b) Vol. **120**, No. 2, p. 715–721 (1983).
See also 22.2, p. 442.

M. ELLNER, M. EL-BORAGY, and B. PREDEL
On the Formation of Palladium-Rich Metallic Glasses in Binary and Ternary Systems with Copper, Silver, and Gold as One Component and Phosphorus and Arsenic as Further Component
(a) Vol. **80**, No. 2, p. 519–533 (1983).
See also *2*, p. *55*; 21.1, p. 378; 21.6, p. 399.

13.2 Fermi Surfaces

Original Papers and Short Notes

T. Nautiyal and S. Auluck
*Fermi Surface of Noble Metals Using the Eight-Cone Model
(b) Vol. **115**, No. 1, p. 127–130 (1983).
See also 21.1, p. 363; 21.6, p. 396.

P. Rennert, G. Carl, and U.-H. Gläser
*Theoretical Compton Profiles of Magnesium
(b) Vol. **115**, No. 1, p. 219–224 (1983).
See also *13.1*, p. *155*; *20*, p. *307*; 21.3, p. 389.

V. A. Maslov, I. I. Naumov, and V. E. Panin
*On the Problem of Electron Phase Stability
(b) Vol. **115**, No. 1, p. 317–324 (1983).
See also *1.3*, p. *21*; 21.1, p. 363.

J. S. Brown
*Some Exchange-Correlation Approximations Applied to Stopping Powers of Slow Ions in Metals
(b) Vol. **115**, No. 2, p. K141–K146 (1983).
See also *17*, p. *269*; 21, p. 351; 21.6, p. 396.

K.-H. Steinmetz, H.-P. Fischer, M. Martin, and E. Nembach
*Positron-Annihilation Measurements of the Fermi Surface of Copper-Base Solid Solutions with Gallium, Germanium, and Nickel
(b) Vol. **116**, No. 2, p. 575–583 (1983).
See also *1.3*, p. *22*; 21.1, p. 367.

J. Rose and R. Schuchardt
*A Theoretical Investigation of the Fermi Surfaces of Bismuth and Antimony
(b) Vol. **117**, No. 1, p. 213–224 (1983).
See also *13.1*, p. *158*; 21.7, p. 401.

M. S. Wartak
Cyclotron Resonance in Metals with *g*-Factor Anisotropy
(b) Vol. **118**, No. 1, p. K1–K2 (1983).
See also *19*, p. *302*; 21, p. 356.

P. Rennert, G. Carl, and W. Hergert
Spin-Polarized Compton Profile of Nickel
(b) Vol. **120**, No. 1, p. 273–281 (1983).
See also *13.1*, p. *164*; 18.2, p. 287; 20, p. 311; 21.1, p. 376.

13.3 Surface and Interface States

Review Articles

R. E. Hummel
Differential Reflectometry and Its Application to the Study of Alloys, Ordering, Corrosion, and Surface Properties
(a) Vol. **76**, No. 1, p. 11–44 (1983).
See also *1.3*, p. *21*; 10.1, p. 99; 13.1, p. 155; *20.1*, p. *312*; 21, p. 349; 21.1, p. 363; 22.6, p. 490.

R. A. Evarestov and V. P. Smirnov
Special Points of the Brillouin Zone and Their Use in the Solid State Theory
(b) Vol. **119**, No. 1, p. 9–40 (1983).
See also *13.1*, p. *155*; 13.4, p. 172; 22, p. 403.

Original Papers and Short Notes

A. I. Bobrysheva, S. I. Beryl, V. T. Zyukov, S. A. Moskalenko, and E. P. Pokatylov
*The One- and Two-Photon Absorption Due to Surface State Excitons
(b) Vol. **115**, No. 1, p. 153–160 (1983).
See also *13.5.1*, p. *194*; 22.2.1, p. 443.

C. Engler and W. Lorenz
Erratum to: Calculation of Local Densities of States, Local Charges, and Adsorption Energy on Large Semiconductor Substrate with Single Adsorbate
(b) Vol. **115**, No. 1, p. 325 (1983).
See also 4, p. 60; *13.1*, p. *155*; 22.2.2, p. 448.

I. Bartoš, M. Kolář, and G. Paasch
*Modified Thomas-Fermi Approximation for Systems with Finite Surface Barriers
(b) Vol. **115**, No. 2, p. 437–442 (1983).
See also 22, p. 403.

G. Paasch
*Current Density near the Semiconductor–Insulator Interface
(b) Vol. **115**, No. 2, p. 471–475 (1983).
See also *14.3.2*, p. *233*; 22, p. 403.

K. Sen, R. S. Srivastava, D. P. Joshi, and V. K. Goyal
Mobility and Its Effect on the Performance of Polycrystalline Solar Cells
(a) Vol. **75**, No. 2, p. 657–661 (1983).
See also *14.3.3*, p. *237*; *16*, p. *264*; 22.1.2, p. 426.

F. Fenske
*Simple Improvement of the Lindner High-Frequency MOS Capacitance Approximation
(a) Vol. **75**, No. 2, p. K141–K144 (1983).
See also *14.3.3*, p. *237*; 22.1.2, p. 426.

T. G. Abdel-Malik, A. A. Aly, A. M. Abdeen, and H. A. Motaweh
*The Influence of the Surface Barrier on the Trajectory of Sputtered Particles
(b) Vol. **116**, No. 1, p. 169–172 (1983).
See also 11, p. 124; 21.1, p. 365.

E. V. Anufriev and V. A. Gurtov
*Accumulation and Discharging of the Hole Charge in Oxide Layers of MIS Structures
(a) Vol. **76**, No. 1, p. 107–111 (1983).
See also 11, p. 125; *14.3.3*, p. *237*; 22.6, p. 492.

A. G. Kazanskii, V. F. Kiselev, E. A. Silaev, and V. S. Vavilov
*On the Role of Surface Phenomena in the Staebler-Wronski Effect
(a) Vol. **76**, No. 1, p. 337–343 (1983).
See also 2, p. 46; *14.3.2*, p. *233*; 16, p. 265; 22.1.2, p. 427.

V. Kumar and S. B. Iyer
*Characterization of Surface States in MOS Capacitors by a Modified DLTS Technique
(a) Vol. **76**, No. 2, p. 637–640 (1983).
See also 14.3.3, p. 238; 22.1.2, p. 428.

D. Flamm and E.-H. Weber
Adsorption of Oxygen on GaAs (001) Surfaces by UPS
(a) Vol. **76**, No. 2, p. K163–K166 (1983).
See also *17*, p. *269*; 22.2.1, p. 444.

W. Lorenz, C. Engler, A. Nechwatal, and G. Adler
*Quantumchemical Studies of Chemisorption at Anion or Cation Sites of the GaP (110) Surface. Coverage Dependence of Adsorption Energy and Equilibrium in Cases of Positive and Negative Partial Charge Injection
(b) Vol. **117**, No. 1, p. 203–212 (1983).
See also 22.2.2, p. 449.

F. Bechstedt, R. Enderlein, and D. Reichardt
*Inverse Dielectric Function for a Semi-Infinite Solid
(b) Vol. **117**, No. 1, p. 261–270 (1983).
See also *14.4*, p. *251*; 22, p. 406.

O. Hardouin Duparc and V. R. Velasco
*Influence of a Thin Overlayer on the Surface Brillouin Scattering. Cubic Crystals
(b) Vol. **117**, No. 1, p. K37–K41 (1983).
See also 21, p. 354.

I. Thurzo, K. Gmucová, J. Červenák, and T. Lalinský
GaAs–Anodic Oxide Interface Examination by Deep-Level Transient Current Spectroscopy
(a) Vol. **77**, No. 1, p. 323–330 (1983).
See also *14.3.2*, p. *234*; 22.2.1, p. 444.

A. Hübler and D. Marton
*Energy Dependence of the Si–SiO_2 Interface State Cross-Section Measured by the TSC Method
(a) Vol. **77**, No. 1, p. K17–K21 (1983).
See also 22.1.2, p. 430; 22.6, p. 493.

J. WOLLBRANDT, U. BRÜCKNER, and E. LINKE
*Investigations of Mechanically Induced Excited States on Cleavage Planes of Ionic Crystals (I): Method and Apparatus for Charge and Electron Emission Measurements
(a) Vol. **77**, No. 2, p. 545–552 (1983).
See also *14.4*, p. *251*; 17, p. 270; 22.5.2, p. 480.

R. N. LITOVSKII, V. S. LYSENKO, A. N. NAZAROV, T. E. RUDENKO, S. B. KASCHIEVA, and N. R. NEDEV
The Influence of Defect Surface Layers on the Capacitive Properties of MOS Structures
(a) Vol. **77**, No. 2, p. 699–707 (1983).
See also *14.3.3*, p. *239*; 22.1.2, p. 431.

G. PAASCH and H. ÜBENSEE
*Carrier Density near the Semiconductor–Insulator Interface. Local Density Approximation for Non-Isotropic Effective Mass
(b) Vol. **118**, No. 1, p. 255–266 (1983).
See also 13.1, p. 160; *14.3.2*, p. *234*; 22.1.1, p. 420; 22.1.2, p. 432.

F. BECHSTEDT, R. ENDERLEIN, and D. REICHARDT
*Electronic Relaxation Effects in Core Level Spectra of Surfaces and Interfaces
(b) Vol. **118**, No. 1, p. 327–336 (1983).
See also 14.4, p. 252; 21.6, p. 398; 22.1.2, p. 432; 22.2.1, p. 445; 22.6, p. 494.

G. D. BAGRATISHVILI, R. B. DZHANELIDZE, and D. A. JISHIASHVILI
*Nonstoichiometric Germanium Nitride. A New Insulating Material for MIS Microelectronics (I): The Choice of the Insulator and Deposition
(a) Vol. **78**, No. 1, p. 115–123 (1983).
See also *14.3.2*, p. *234*; 14.3.3, p. 239; 22, p. 409.

E. P. POKALILOV and S. I. BERIL
*Electron–Phonon Interaction in Periodic Two-Layer Structures
(b) Vol. **118**, No. 2, p. 567–573 (1983).
See also *6*, p. *76*; 22.2, p. 442.

G. PAASCH, T. FIEDLER, M. KOLÁŘ, and I. BARTOŠ
*Quasi-Classical Quantization for Potentials with an Infinite Surface Barrier. Subbands in Inversion Layers
(b) Vol. **118**, No. 2, p. 641–652 (1983).
See also 22, p. 410.

K. HÜBNER, H. KÖSTER, JR., B. DERLICH, and W. ECKE
*DLTS Investigations of Si–SiO_2 Interface States of Electron-Beam Irradiated MOS Structures
(b) Vol. **118**, No. 2, p. K133–K136 (1983).
See also *11*, p. *130*; 14.3.2, p. 234; 22.1.2, p. 433; 22.6, p. 495.

G. D. BAGRATISHVILI, R. B. DZHANELIDZE, and D. A. JISHIASHVILI
*Nonstoichiometric Germanium Nitride – A New Insulating Material for MIS Microelectronics (II): Composition, Properties, and the Interface with a Semiconductor
(a) Vol. **78**, No. 2, p. 391–400 (1983).
See also *14.3.2*, p. *234*; 14.3.3, p. 240; 22, p. 410.

K. Kawabata
Intense Exoelectron Emission from Fine Particles of Al_2O_3 and Its Dependence on the Conductivity of the Substrate Material
(a) Vol. **78**, No. 2, p. K127–K130 (1983).
See also 11, p. 131; *17*, p. *270*; 22.6, p. 495.

B. Raykov
Density Functional Theory of Metal Surface Properties
(b) Vol. **119**, No. 1, p. 437–444 (1983).
See also 21, p. 358.

P. Gluck
*Polarization Forces between Macroscopic Bodies
(b) Vol. **119**, No. 2, p. K131–K133 (1983).
See also 4, p. 63.

F. Fenske
*Influence of Lateral Charge Inhomogeneities in MOS Structures on the Determination of the Interface State Density Distribution from HF CV Curves
(a) Vol. **79**, No. 2, p. 417–421 (1983).
See also *14.3.3*, p. *241*; 22.1.2, p. 436.

A. A. Frantsuzov and A. P. Kovchavtsev
*The Determination of Surface Potential Fluctuation Values and Spatial Distribution of Traps in Insulators from the Temperature Dependence of the $(G_p/\omega) - \omega$ Curves of MOS Structures
(a) Vol. **79**, No. 2, p. 503–511 (1983).
See also *14.3.3*, p. *241*; 22.1.2, p. 436.

D. V. Korbutyak and V. G. Litovchenko
*Expansion of the Electron–Hole Liquid Plasma at GaAs Surface
(b) Vol. **120**, No. 1, p. 87–97 (1983).
See also 13.5.1, p. 198; *20.3*, p. *346*; 22.2.1, p. 446.

Ch. Schnittler, J. Kelber, G. Paasch, I. Bartoš and H. Übensee
*A Modified Local Density Approximation for Potentials with a Finite Step
(b) Vol. **120**, No. 1, p. 297–309 (1983).
See also 22, p. 413.

S. G. Davison and K. W. Sulston
**N*-Interface Problem Revisited
(b) Vol. **120**, No. 1, p. 415–420 (1983).

M. Mondal and K. P. Ghatak
*Effect of Size Quantization on the Electronic Heat Capacity in Ultrathin Films of Semiconductors
(b) Vol. **120**, No. 1, p. K63–K68 (1983).
See also *8*, p. *91*; 22.2.1, p. 447.

P. Bogdanski, R. Carin, O. Gardais et J. Vizot
*Caractérisation de structures MIS (Al–Si_3N_4–GaAs) par différentes méthodes capacitives
(Characterisation of MIS Structures (Al–Si_3N_4–GaAs) by Different Capacitive Methods)
(a) Vol. **80**, No. 1, p. 185–193 (1983).
See also *14.3.2*, p. *234*; 22.2.1, p. 447.

V. A. Sablikov and V. B. Sandomirskii
*The Photoacoustic Effect in Semiconductors
(b) Vol. **120**, No. 2, p. 471–480 (1983).
See also 8, p. 92; *14.3*, p. *227*; 22, p. 414.

M. Bhattacharya, S. Mukhopadhyay, and T. K. Mitra
*Study of a Bound Surface Optical Polaron by a Non-Perturbative Method
(b) Vol. **120**, No. 2, p. 555–564 (1983).
See also *13.5.3*, p. *202*; 22.2.1, p. 447; 22.4.1, p. 465; 22.5.2, p. 485.

D. S. Gupta, M. M. Chandra, and V. Kumar
*Direct Measurement of Interface State Electron Capture Cross-Section in MOS System by DLTS Technique
(a) Vol. **80**, No. 2, p. K209–K212 (1983).
See also *14.3.2*, p. *235*; 22.1.2, p. 439.

13.4 Impurity and Defects States

Review Articles

R. A. Evarestov and V. P. Smirnov
Special Points of the Brillouin Zone and Their Use in the Solid State Theory
(b) Vol. **119**, No. 1, p. 9–40 (1983).
See also *13.1*, p. *155*; 13.3, p. 167; 22, p. 403.

E. G. Sieverts
Classification of Defects in Silicon after Their g-Values
(b) Vol. **120**, No. 1, p. 11–29 (1983).
See also *10.2*, p. *108*; 19, p. 299; 22.1.2, p. 423.

Original Papers and Short Notes

B. I. Craig
*The Electronic Structure of a Hydrogen Impurity in Aluminium (III): The Energy and the Metal–Impurity Interaction
(b) Vol. **115**, No. 1, p. 53–61 (1983).
See also 21, p. 349.

E. A. Radzhabov
*A Model of Titanium Luminescence Centers in Lithium and Sodium Fluorides
(b) Vol. **115**, No. 1, p. 83–88 (1983).
See also 20.1, p. 312; *20.3*, p. *334*; 22.5.2, p. 477.

W. Korczak, M. Paradowski, and M. Subotowicz
EPR Linewidth Studies of Gd^{3+}-Doped $La_xCe_{1-x}F_3$ Single Crystals
(b) Vol. **115**, No. 1, p. 89–94 (1983).
See also *19*, p. *299*; 22.5, p. 473.

M. Stutzmann, J. Stuke, and H. Dersch
*Electron Spin Resonance of Doped Glow-Discharge Amorphous Germanium
(b) Vol. **115**, No. 1, p. 141–151 (1983).
See also 2, p. 43; 14.3, p. 215; *19*, p. *299*; 22.1.1, p. 418; 22.1.2, p. 423.

H. Kahnt and F. Schirrmeister
Electric Field Dependent Conductivity in Amorphous Semiconductors
(b) Vol. **115**, No. 1, p. 171–179 (1983).
See also *2*, p. *43*; *14.3.4*, p. *243*; 22, p. 403.

Fat Duen Ho
Approximate Wave Functions for the F^+-Centers in $LiNbO_3$ and $LiTaO_3$
(b) Vol. **115**, No. 1, p. 299–303 (1983).
See also *10.2*, p. *108*; *20.1*, p. *313*; 22.8.1, p. 512.

A. Ferreira da Silva, M. Fabbri, and I. C. da Cunha Lima
Electron Hopping Energy Influence on the Specific Heat of Phosphorus-Doped Silicon
((b) Vol. **115**, No. 1, p. 311–316 (1983).
See also *8*, p. *85*; 22.1.2, p. 423.

E. A. RADZHABOV
*Luminescence Centers in LiF:Li_2O
(b) Vol. **115**, No. 1, p. K25–K28 (1983).
See also *20.3*, p. *334*; 22.5.2, p. 477.

P. HARI, K. PURANDAR, J. LAKSHMANA RAO, and S. V. J. LAKSHMAN
*Optical Absorption Spectrum of Co^{2+} in Lithium Ammonium Sulphate Single Crystals
(b) Vol. **115**, No. 1, p. K45–K50 (1983).
See also *20.1*, p. *313*; 22.8.1, p. 512.

G. M. NIFTIEF, B. G. TAGIEV, and G. K. ASLANOV
Injection and Thermodepolarization Currents in $YbGa_2Se_4$ Single Crystals
(a) Vol. **75**, No. 1, p. 59–65 (1983).
See also *14.3.4*, p. *243*; 22.8, p. 502.

B. R. SETHI, P. K. GOYAL, O. P. SHARMA, and P. C. MATHUR
High-Field DC Conductivity in n-Type CdS Single Crystals Annealed in Molten Indium
(a) Vol. **75**, No. 1, p. 83–89 (1983).
See also *14.3*, p. *215*; 22.4.1, p. 462.

P. L. KUKK and A. J. ERM
Luminescent Centre Formation Processes in Activated CdS Phosphors
(a) Vol. **75**, No. 1, p. 91–100 (1983).
See also *10.2*, p. *108*; *20.3*, p. *334*; 22.4.1, p. 462.

D. MERGEL
*Recombination at Dislocations in Silicon
(a) Vol. **75**, No. 1, p. 255–262 (1983).
See also *16*, p. *264*; 22.1.2, p. 424.

K. ŽĎÁNSKÝ, L. KRATĚNA, M. MATYÁŠ, JR., and J. ZELINKA
*Absolute Energy of the Nitrogen-Related Electron Trap in Gallium Phosphide
(a) Vol. **75**, No. 1, p. 335–339 (1983).
See also *14.3.3*, p. *236*; 22.2.2, p. 448.

A. A. KAMINSKII, N. R. AGAMALYAN, L. P. KOZEEVA, V. F. NESTERENKO, and A. A. PAVLYUK
New Data on Stimulated Emission of Nd^{3+} Ions in Disordered Crystals with Scheelite Structure
(a) Vol. **75**, No. 1, p. K1–K4 (1983).
See also *20.3*, p. *335*; 22.8, p. 502.

K. IRMSCHER, H. KLOSE, and K. MAASS
*Electric Field Enhanced Electron Emission from Gold Acceptor Level and A-Centre in Silicon
(a) Vol. **75**, No. 1, p. K25–K28 (1983).
See also 14.3.4, p. 243; 22.1.2, p. 425.

G. BLASSE
*Energy Transfer from Ce^{3+} to Eu^{3+} in (Y, Cd)F_3
(a) Vol. **75**, No. 1, p. K41–K43 (1983).
See also *20.3*, p. *335*; 22.8, p. 502.

D. Nesheva and E. Vateva
*Some Peculiarities of High Temperature Sensitization in CdS
(a) Vol. **75**, No. 1, p. K45 – K48 (1983).
See also *16*, p. *264*; 22.4.1, p. 462.

H. Lemke
*Ionenpaare zwischen Chrom und flachen Akzeptoren in Silizium
(Ion Pairs between Chromium and Shallow Acceptors in Silicon)
(a) Vol. **75**, No. 1, p. K49 – K51 (1983).
See also 14.3, p. 215; 22.1.2, p. 425.

C. N. Duckworth, A. W. Brinkman, and J. Woods
*Deep Level Transient Spectroscopy in TiO_2:Nb
(a) Vol. **75**, No. 1, p. K99 – K102 (1983).
See also 14.3.3, p. 236; 22.6, p. 490.

L. Kończewicz and W. Trzeciakowski
*Highly Localized Donor States in InSb and Their Pressure Dependence
(b) Vol. **115**, No. 2, p. 359 – 367 (1983).
See also 12.2, p. 141; 14.3, p. 216; 22.2.3, p. 452.

H. Köster, Jr., I. N. Iassievich, and K. Hübner
*Thermally Stimulated Excitation of Electrons from Deep Traps of Insulators in Electric Fields
(b) Vol. **115**, No. 2, p. 409 – 414 (1983).
See also *6*, p. *70*; 14.3.4, p. 243; 22.6, p. 491.

W. Gehlhoff and K. H. Segsa
*EPR of Iron–Boron Centres in Silicon
(b) Vol. **115**, No. 2, p. 443 – 453 (1983).
See also *19*, p. *299*; 22.1.2, p. 425.

A. N. Nikitina, I. S. Osadko, I. A. Vasilieva, and L. A. Yanovskaya
*Vibronic and Electron–Phonon Coupling in Spectra of Substituted Polyenes in Rigid Matrices
(b) Vol. **115**, No. 2, p. 493 – 498 (1983).
See also *6*, p. *70*; 20.1, p. 314; 20.3, p. 335; 22.9, p. 525.

D. Verity, J. Nahum, and F. J. Bryant
*Donor-Related Edge Luminescence in Ion-Implanted ZnTe
(b) Vol. **115**, No. 2, p. 499 – 508 (1983).
See also *20.3*, p. *335*; 22.4.3, p. 468.

V. S. Grunin, I. B. Patrina, and Z. N. Zonn
*Impurities and Electron Localization in V_2O_5 Crystals
(b) Vol. **115**, No. 2, p. 545 – 553 (1983).
See also *19*, p. *300*; 22.6, p. 491.

I. Gyémánt and B. Vasvári
*Cluster Phase Shifts and Virtual Bond States for V, Mn, and Fe Impurities in Aluminium
(b) Vol. **115**, No. 2, p. K83 – K86 (1983).
See also 21, p. 350.

P. L. KUKK and A. J. ERM
Isochronal Annealing of Quenched CdS Phosphors
(a) Vol. **75**, No. 2, p. 393–399 (1983).
See also *10.2*, p. *109*; *20.3*, p. *336*; 22.4.1, p. 462.

H. LEMKE
*Eigenschaften einiger Störstellenkomplexe von Gold in Silizium
(Properties of Some Impurity Complexes of Gold in Silicon)
(a) Vol. **75**, No. 2, p. 473–482 (1983).
See also 14.3, p. 216; 22.1.2, p. 425.

H. PFLEIDERER and B. RAUSCHER
Capacitance of Amorphous Silicon pin Solar Cells
(a) Vol. **75**, No. 2, p. 537–545 (1983).
See also 2, p. 44; *14.3.1*, p. *230*; *14.3.3*, p. *237*; 16, p. 264; 22.1.2, p. 426.

F. SCHAUER, O. ZMEŠKAL, and S. NEŠPŮREK
*A New Approach to the Analysis of Steady-State Space-Charge-Limited Currents Using Their Activation Energies. An Application to a-As_2Se_3
(a) Vol. **75**, No. 2, p. 591–594 (1983).
See also 2, p. 45; *14.3.4*, p. *244*; 22, p. 404.

V. D. AKHMETOV, V. V. BOLOTOV, L. S. SMIRNOV, and V. A. KHARCHENKO
*The Defect Accumulation and Irradiation-Induced Lattice Strain in Si at High-Dose Neutron Irradiation
(a) Vol. **75**, No. 2, p. 601–606 (1983).
See also 11, p. 124; *20.1*, p. *314*; 22.1.2, p. 426.

V. M. GARMASH, A. A. KAMINSKII, M. I. POLYAKOV, S. E. SARKISOV, and A. A. FILIMONOV
Luminescence and Stimulated Emission of Nd^{3+} Ions in LaMg $Al_{11}O_{19}$ Crystals in the $^4F_{3/2} \rightarrow {}^4I_{11/2}$ and $^4F_{3/2} \rightarrow {}^4I_{13/2}$ Transitions
(a) Vol. **75**, No. 2, p. K111–K116 (1983).
See also *20.2*, p. *332*; *20.3*, p. *336*; 22.8, p. 503.

G. RUDLOF and H. GLAEFEKE
*Analysis of First-Order Glow Curves by the Balarin-Zetzsche Method in Differential Form
(a) Vol. **75**, No. 2, p. K169–K173 (1983).
See also 22, p. 404.

H. NEUMANN, R. D. TOMLINSON, N. AVGERINOS, and E. NOWAK
Electrical Properties of $CuInSe_2$ Single Crystals Grown by the Vertical Bridgman Technique
(a) Vol. **75**, No. 2, p. K199–K203 (1983).
See also *14.3*, p. *217*; 22.8, p. 504.

V. B. SHARAN
Absorption, Fluorescence, and Excitation Spectra of Thallium-Activated Rubidium Bromide
(b) Vol. **116**, No. 1, p. 41–47 (1983).
See also *20.1*, p. *315*; *20.3*, p. *336*; 22.5, p. 473.

W. PASTUSIAK
*A Microscopic Interpretation of the Spin–Lattice Coupling Coefficients for S-State $4f^7$ Ions
(b) Vol. **116**, No. 1, p. 63–70 (1983).
See also *19*, p. *300*; 22.5.3, p. 486.

N. ISHII, M. KUMEDA, and T. SHIMIZU
*Defects States in Amorphous Silicon
(b) Vol. **116**, No. 1, p. 91–100 (1983).
See also *2*, p. *45*; 10.2, p. 110; 22.1.2, p. 426.

YU. P. DROZHOV
Nonlinear Susceptibility of Impurity Centres in Semiconductors
(b) Vol. **116**, No. 1, p. 111–118 (1983).
See also *20*, p. *307*; 22, p. 404.

K. SAKATA
*Study of Electronic Structure of Reduced TiO_2 and $V_xTi_{1-x}O_2$ Crystals by ESCA and Optical Absorption
(b) Vol. **116**, No. 1, p. 145–153 (1983).
See also 17, p. 269; *20.1*, p. *315*; 22.6.1, p. 498.

M. MOCKER and M. BEILER
*Theory of Intrinsic Recombination at Zero Temperature in Small Gap Semiconductors. Estimations for PbSSe
(b) Vol. **116**, No. 1, p. 205–215 (1983).
See also *14.3*, p. *217*; 20.3, p. 336; 22.8, p. 504.

H. LEMKE
*Energieniveaus und Bindungsenergien von Ionenpaaren in Silizium
(Energy Levels and Binding Energies of Ion Pairs in Silicon)
(a) Vol. **76**, No. 1, p. 223–234 (1983).
See also 14.3, p. 217; 22.1.2, p. 427.

A. N. GEORGOBIANI, M. V. GLUSHKOV, E. S. LOGOZINSKAYA, ZH. A. PUKHLII, I. M. TIGINYANU, and I. A. SHCHERBAKOV
*Radiative Recombination in γ-La_2S_3 Single Crystals
(a) Vol. **76**, No. 1, p. 311–317 (1983).
See also *20.3*, p. *337*; 22.3, p. 456.

D. V. GITSU, I. N. GRINCHESHEN, and N. S. POPOVICH
The Electric and Photoelectric Properties of $TlSbSe_2$ Crystals in the Region of Impurity Conduction
(a) Vol. **76**, No. 1, p. K5–K7 (1983).
See also *14.3*, p. *217*; *16*, p. *265*; 22.8, p. 504.

L. N. CHRISTOPHOROV and V. N. KHARKYANEN
*Theory of Interimpurity Transitions in Condensed Medium
(b) Vol. **116**, No. 2, p. 415–425 (1983).
See also 22, p. 405.

V. G. Baryshevskii, S. A. Kuten, and V. I. Rapoport
*Quadrupole Splitting of Hydrogen-Like Atom Levels in Crystals
(b) Vol. **116**, No. 2, p. 489–500 (1983).
Erratum
(b) Vol. **119**, No. 2, p. 755 (1983).
See also *19*, p. *301*; 22.6, p. 492.

R. Meisels and F. Kuchar
*Far-Infrared Photoconductivity Spectroscopy of Shallow Acceptors in InSb
(b) Vol. **116**, No. 2, p. 557–560 (1983).
See also *16*, p. *265*; 22.2.3, p. 452.

M. Satarić and R. Žakula
*On the Role of the Soliton Mechanism in the Electron Transfer along the Quasi-One-Dimensional Molecular Structures
(b) Vol. **116**, No. 2, p. 617–632 (1983).
See also 22, p. 405.

V. I. Sugakov and Yu. D. Shtepa
*Spin–Lattice Relaxation Theory of Excited Triplet-State Impurities in Molecular Crystals
(b) Vol. **116**, No. 2, p. 633–638 (1983).
See also 22.9, p. 527.

B. I. Craig
*Structure of Muons Trapped in Aluminium
(b) Vol. **116**, No. 2, p. K91–K94 (1983).
See also *10.1*, p. *101*; 21, p. 353.

T. G. Abdel-Malik, A. A. Aly, A. M. Abdeen, and H. M. El-Labany
*Electric and Photoelectric Investigations of β-Nickel Phthalocyanine Thin Films
(a) Vol. **76**, No. 2, p. 651–659 (1983).
See also *14.3.1*, p. *231*; 16, p. 265; 22.9, p. 527.

B. Pödör, J. Pfeifer, L. Csontos, N. Nádor, and F. Deák
*Electrical Properties of Donors in Gallium Phosphide
(a) Vol. **76**, No. 2, p. 695–704 (1983).
See also *14.3*, p. *218*; 22.2.2, p. 449.

T. G. Abdel-Malik
*Effect of Oxygen on the Electrical Properties of β-Zinc Phthalocyanine Single Crystals
(a) Vol. **76**, No. 2, p. K159–K162 (1983).
See also *14.3*, p. *218*; 22.9, p. 527.

H. Lemke
*Über die chemische Natur von Rekombinationszentren in Si-Leistungsbauelementen
(On the Chemical Nature of Recombination Centres in Si Power Devices)
(a) Vol. **76**, No. 2, p. K193–K196 (1983).
See also *14.3.3*, p. *238*; 22.1.2, p. 429.

J. Kočka, O. Štika, and J. Stuchlík
*Comments on the Interpretation of DLTS Spectra of a-Si:H
(a) Vol. **76**, No. 2, p. K217–K220 (1983).
See also 2, p. 47; *14.3.2*, p. *233*; 22.1.2, p. 429.

N. A. Poklonski and V. F. Stelmakh
*Screening of Electrostatic Fields in Crystalline Semiconductors by Electrons Hopping over Defects
(b) Vol. **117**, No. 1, p. 93–99 (1983).
See also *14.3*, p. *218*; 22, p. 406.

H. Zimmermann, R. Boyn, and N. Nagel
Photoinduced Absorption Spectra of ZnS:Fe Crystals
(b) Vol. **117**, No. 1, p. 229–239 (1983).
See also *20.1*, p. *318*; 22.4.1, p. 463.

J. Gryko
*Optical Conductivity in the Impurity Band of Doped Semiconductors
(b) Vol. **117**, No. 1, p. 313–322 (1983).
See also *14.3*, p. *219*; 20.1, p. 318; 22.1.1, p. 420.

I. Mertig, E. Mrosan, R. Zeller, P. H. Dederichs, and P. Ziesche
*Electronic Properties of Dilute Copper Alloys (I): Theory
(b) Vol. **117**, No. 1, p. 335–344 (1983).
See also 21.1, p. 368.

N. T. Bagraev
*Electron Localization on Defects and Optical Nuclear Polarization in Disordered and Semimagnetic Semiconductors
(b) Vol. **117**, No. 1, p. 367–379 (1983).
See also *20*, p. *308*; 22.1.2, p. 429.

K. D. Chaudhuri, A. Malik, S. Kukreti, and P. C. Mathur
*Impurity Band Conduction in Single Crystals of p-Type GaSb
(b) Vol. **117**, No. 1, p. 407–416 (1983).
See also *14.3*, p. *219*; 22.2.3, p. 452.

W. Siegel, E. Ziegler, and G. Kühnel
*Impurity Parameters of Undoped and Zn-Doped p-Type GaP with Two Acceptor Levels Determined by Hall Measurements
(a) Vol. **77**, No. 1, p. 167–173 (1983).
See also *14.3*, p. *219*; 22.2.2, p. 449.

M. Balarin
*Zum Einfluß einer Temperaturabhängigkeit des Frequenzfaktors auf die Form und Breite von Glow-Peaks
(On the Influence of a Temperature Dependence of the Frequency Factor on the Shope and Width of Glow Peaks)
(a) Vol. **77**, No. 1, p. 181–194 (1983).
See also 22, p. 407.

S. I. GOLOVKOVA, A. M. GURVICH, A. I. KRAVCHENKO, V. V. MIKHAILIN, and A. N. VASILEV
*Enhancing and Quenching of Intrinsic Luminescence and Characteristic Features of Calcium Tungstate Phosphorescence in the Presence of Lanthanides
(a) Vol. **77**, No. 1, p. 375–380 (1983).
See also *20.3*, p. *339*; 22.8.1, p. 514.

A. NIKLAS and W. JELEŃSKI
*X-Ray Luminescence of YAG:Nd^{3+}
(a) Vol. **77**, No. 1, p. 393–398 (1983).
See also *20.3*, p. *339*; 22.8.2, p. 521.

S. V. MOHARIL
A Phenomenological Model for Supralinearity and Sensitization in LiF-TLD 100
(a) Vol. **77**, No. 1, p. 413–420 (1983).
See also *20.3*, p. *339*; 22.5.2, p. 479.

V. N. BRUDNYI, A. I. POTAPOV, YU. V. RUD, and M. SERGINOV
Electrical Properties of Electron-Irradiated GaAs and $ZnSiAs_2$ Solid Solutions
(a) Vol. **77**, No. 1, p. K33–K36 (1983).
See also *11*, p. *128*; *14.3*, p. *219*; 22.2.1, p. 444; 22.8, p. 505.

A. S. POPOV, D. B. KUSHEV, and SH. CHADRAABAL
*A Model for Photoconductivity of Fe-Doped GaP
(a) Vol. **77**, No. 1, p. K53–K57 (1983).
See also *16*, p. *265*; 22.2.2, p. 449.

J. JIMENEZ, M. A. GONZALEZ, V. CARBAYO, and J. BONNAFE
*On the Thermal Evolution of the Photoconductivity in Bulk Fe-Doped Semi-Insulating InP
(a) Vol. **77**, No. 1, p. K69–K73 (1983).
See also *16*, p. *266*; 22.2.2, p. 449.

B. VENGALIS and K. VALATSKA
Phase Segregation in Tellurium-Saturated GeTe
(b) Vol. **117**, No. 2, p. 471–476 (1983).
See also *1.2*, p. *16*; 8, p. 88; 14.3, p. 220; 20.1, p. 319; 22.3, p. 457.

Z. DZIUBA
*Impurity Band Conduction in HgTe (I): Experimental Data
(b) Vol. **117**, No. 2, p. 531–540 (1983).
See also *14.3*, p. *220*; 22.4.3, p. 468.

H. VAN CONG
*Band-Gap Narrowing in n-Type Moderately Doped Silicon at 300 K
(b) Vol. **117**, No. 2, p. 575–584 (1983).
See also *13.1*, p. *159*; 22.1.2, p. 430.

H. ZIMMERMANN and R. BOYN
*On the Configuration Coordinate Diagram of ZnS:Fe
(b) Vol. **117**, No. 2, p. 611–618 (1983).
See also 6, p. 74; 20.1, p. 320; 20.3, p. 339; 22.4.1, p. 463.

I. Mertig, E. Mrosan, R. Zeller, and P. H. Dederichs
*Electronic Properties of Dilute Copper Alloys (II): Residual Resistivity
(b) Vol. **117**, No. 2, p. 619–623 (1983).
See also 21.1, p. 370.

R. Borromei, E. Cavalli, and G. Ingletto
*Electronic Spectrum of the Co(II) Ion in the Columbite $CoNb_2O_6$ Host Lattice at 4.2 K
(b) Vol. **117**, No. 2, p. 733–742 (1983).
See also *20.1*, p. *320*; 22.8, p. 506.

J. Zieliński and J. Kowalska
*On the Virtual Bound State Model with Phonons
(b) Vol. **117**, No. 2, p. K105–K108 (1983).
See also 13.5.3, p. 201.

S. V. J. Lakshman and A. Sundar Jacob
*Absorption Spectrum of VO^{2+} in $LiKSO_4$ Single Crystals
(b) Vol. **117**, No. 2, p. K119–K122 (1983).
See also 19, p. 301; *20.1*, p. *320*; 22.8.1, p. 514.

O. A. Golikova and S. Samatov
*Conduction Mechanism of Some "Quasi-Amorphous" Semiconductors on the Basis of Boron
(a) Vol. **77**, No. 2, p. 449–453 (1983).
See also *14.3*, p. *220*; 15, p. 262; 22, p. 408.

S. P. Kuila
*Analysis of the 215 °C Glow Peak in Li-Doped KCl Single Crystals
(a) Vol. **77**, No. 2, p. 477–482 (1983).
See also *20.3*, p. *340*; 22.5.2, p. 480.

I. E. Türe, F. Poulin, A. W. Brinkman, and J. Woods
*Electron Traps and Deep Levels in Cadmium Selenide
(a) Vol. **77**, No. 2, p. 535–544 (1983).
See also *14.3.3*, p. *239*; 22.4.2, p. 467.

P. Gaworzewski and K. Schmalz
*Oxygen-Related Donors Formed at 600 °C in Silicon in Dependence on Oxygen and Carbon Content
(a) Vol. **77**, No. 2, p. 571–582 (1983).
See also *14.3*, p. *220*; 22.1.2, p. 431.

J. Claverie, G. Campet, D. Conte, G. Le Flem, and P. Hagenmuller
Influence of Covalent Bonding on the Photoelectrochemical Properties of Some Perovskite-Type Related Compounds
(a) Vol. **77**, No. 2, p. 603–609 (1983).
See also *4*, p. *61*; *16*, p. *266*; 22.6.1, p. 498.

A. J. Cohen and G. G. Janezic
*Relationships among Trapped Hole and Trapped Electron Centers in Oxidized Soda–Silica Glasses of High Purity
(a) Vol. **77**, No. 2, p. 619–624 (1983).
See also 2, p. 48; *20.1*, p. *321*; 22.6.1, p. 498.

B. Pistoulet, D. Coquillat, J. C. Tedenac, G. Brun, and M. Maurin
*Hole Effective Mass and Impurity Levels in Undoped AgTlTe. Optimum Thermoelectric Figure of Merit
(a) Vol. **77**, No. 2, p. 669–677 (1983).
See also *13.1*, p. *159*; 14.3, p. 221; 15, p. 262; 22.8, p. 506.

Sh. Chadraabal, A. S. Popov, and D. B. Kushev
*Photoconductivity in n-Type InP:Fe
(a) Vol. **77**, No. 2, p. 709–714 (1983).
See also *16*, p. *266*; 22.2.2, p. 450.

A. S. Aleksandrov, V. F. Elesin, and M. P. Kazeko
The Influence of Defects on the Low-Temperature Resistivity of Intermetallic Compounds with A-15 Structure
(a) Vol. **77**, No. 2, p. 725–732 (1983).
See also 13.1, p. 159; *14.2*, p. *213*.

O. Breitenstein and T. Wosinski
*Scanning-DLTS Investigation of the EL 2 Level in Plastically Deformed GaAs
(a) Vol. **77**, No. 2, p. K107–K110 (1983).
See also *10.2*, p. *113*; 14.3, p. 221; 22.2.1, p. 444.

T. Taguchi
Crystal Growth and Neutral-Acceptor Bound-Exciton Emission of ZnCdTe by THM with Te Solvent
(a) Vol. **77**, No. 2, p. K115–K119 (1983).
See also *3*, p. *58*; 13.5.1, p. 196; *20.3*, p. *341*; 22.4.4, p. 471.

I. Sildos, O. Sild, and I. Dolindo
*Manifestation of Spin–Rotation Coupling in Intersystem Crossing in a Rotating Impurity Molecule
(b) Vol. **118**, No. 1, p. 25–29 (1983).
See also 20.1, p. 321; 22.5.2, p. 480.

A. S. Agabekyan
*Resonant Coherent Energy Transfer. Role of Initial Conditions
(b) Vol. **118**, No. 1, p. 51–56 (1983).
See also 20, p. 309; 22, p. 408.

S. Asano, N. Yamashita et Y. Ogawa
*Luminescence et structure vibrationnelle des luminophores CaS:Ce^{3+} et CaSe:Ce^{3+} (Luminescence and Vibrational Structure of the Luminophors CaS:Cr^{3+} and CaSe:Cr^{3+})
(b) Vol. **118**, No. 1, p. 89–100 (1983).
See also *20.3*, p. *341*; 22.4.1, p. 464; 22.4.2, p. 467.

V. Ya. Prinz and S. N. Rechkunov
*Influence of a Strong Electric Field on the Carrier Capture by Nonradiative Deep-Level Centers in GaAs
(b) Vol. **118**, No. 1, p. 159–166 (1983).
See also *14.3.4*, p. *246*; 22.2.1, p. 445.

A. N. ERMOSHKIN, R. A. EVARESTOV, S. A. KUCHINSKII, and V. K. ZAKHAROV
The Quasimelecular Approach to the Electronic Structure Calculations for Silver and Copper Halides
(b) Vol. **118**, No. 1, p. 191 – 203 (1983).
See also *13.1*, p. *159*; 22.5.1, p. 476.

R. KRÜSSMANN, H. VOLLMER, and R. LABUSCH
*The Activitation Energy of Phosphorus Donors in Silicon-Rich Silicon–Germanium Alloys
(b) Vol. **118**, No. 1, p. 275 – 282 (1983).
See also 20.1, p. 321; 22.1.1, p. 421; 22.1.2, p. 432.

Z. DZIUBA
*Impurity Band Conduction in HgTe (II): Theoretical Analysis
(b) Vol. **118**, No. 1, p. 319 – 326 (1983).
See also *14.3*, p. *221*; 22.4.3, p. 468.

A. E. NIKIFOROV, S. YU. SHASHKIN, M. L. LEVITAN, and T. H. AGAMALYAN
*Cooperative Jahn-Teller Orbital Ordering in $KCuF_3$ and K_2CuF_4 Crystals
(b) Vol. **118**, No. 1, p. 419 – 426 (1983).
See also 22.5.4, p. 488.

P. M. TOMCHUK, V. A. SHENDEROVSKII, P. N. GORLEY, and N. P. GAVALESHKO
Hall Coefficient and Magnetoresistance of Semiconductors with Anisotropic Energy Dispersion
(b) Vol. **118**, No. 1, p. 433 – 439 (1983).
See also 13.1, p. 160; *14.3*, p. *221*; 22.8, p. 507.

B. G. TAGIEV, G. M. NIFTIEV, and S. A. ABUSHOV
*Intrashell Yb^{3+} Transitions in Gallium Sulphide Single Crystals
(b) Vol. **118**, No. 1, p. K13 – K16 (1983).
See also *20.3*, p. *341*; 22.3, p. 457.

B. R. BULKA
*On the Electronic Structure of Interacting Solitons with Impurities in Polyacetylene
(b) Vol. **118**, No. 1, p. K57 – K59 (1983).
See also 22.9, p. 528.

N. ISHII, M. KUMEDA, and T. SHIMIZU
*Defect States in Amorphous Germanium
(b) Vol. **118**, No. 1, p. K65 – K68 (1983).
See also 2, p. 49; 19, p. 302; 22.1.1, p. 421.

R. CAPELLETTI, O. CARRANZA TORRES, J. GARCÍA SOLÉ, and M. MANFREDI
On the 430 nm Emission Band in the NaCl:Pb^{++} System
(a) Vol. **78**, No. 1, p. 35 – 39 (1983).
See also *10.2*, p. *114*; *20.3*, p. *342*; 22.5.2, p. 481.

J. Wollbrandt, U. Brückner, and E. Linke
*Investigations of Mechanically Induced Excited States on Cleavage Planes of Ionic Crystals (II): Investigations of the Mechanisms of Electron Emission from Cleaved Alkali Halide Crystals and of the Influence of Surface Charges
(a) Vol. **78**, No. 1, p. 163–168 (1983).
See also 14.4, p. 252; 17, p. 270; 22.5.2, p. 481.

J. Tyczkowski and M. Kryszewski
*Poole-Frenkel Centres in Dielectrics. Kind and Space Density in Plasma-Polymerized Organosilicon Films
(a) Vol. **78**, No. 1, p. 259–265 (1983).
See also 14.3.1, p. 231; *14.3.4*, p. *246*; 22.9, p. 529.

D. Verity, D. Shaw, F. J. Bryant, and C. G. Scott
*DLTS Investigation of Electron Traps in As-Grown and Cd-Fired CdS
(a) Vol. **78**, No. 1, p. 267–275 (1983).
See also *14.3.4*, p. *247*; 22.4.1, p. 464.

A. A. Kaminskii, B. P. Sobolev, and T. V. Uvarova
Stimulated Emission of Ho^{3+} in $BaTm_2F_8$ and Tm^{3+} Ions in $BaYb_2F_8$:Er^{3+} Crystals
(a) Vol. **78**, No. 1, p. K13–K17 (1983).
See also *20.2*, p. *332*; *20.3*, p. *342*; 22.8, p. 507.

C. De Blasi, C. Manfredotti, G. Micocci, and A. Tepore
Thermoluminescence of n-GaS
(a) Vol. **78**, No. 1, p. K31–K34 (1983).
See also *20.3*, p. *342*; 22.3, p. 457.

O. B. Tagiev, G. M. Niftiev, and I. M. Askerov
*The Frenkel Thermal-Field Emission in $EuGa_2S_4$ Single Crystals
(a) Vol. **78**, No. 1, p. K43–K46 (1983).
See also *14.3.4*, p. *247*; 22.8, p. 507.

S. J. Pearton and J. M. Kahn
*Dislocations in Germanium: Effects of Plasma Hydrogenation
(a) Vol. **78**, No. 1, p. K65–K69 (1983).
See also *10.2*, p. *115*; 22.1.1, p. 421.

S. W. Biernacki
*Splitting of the 3d Levels of Iron Group Impurities in Crystals with Zincblende Structure
(b) Vol. **118**, No. 2, p. 525–533 (1983).
See also *20.1*, p. *323*; 22.4.2, p. 467.

B. Rambabu, C. Ramasastry, and B. V. R. Chowdari
Cobalt Centres in Ammonium Chloride: Optical Absorption Studies
(b) Vol. **118**, No. 2, p. 541–549 (1983).
See also *10.2*, p. *115*; *20.1*, p. *323*; 22.5, p. 475.

R. U. Bauer, J. R. Niklas, and J.-M. Spaeth
*Two Types of F Centres in Alkaline Earth Fluoro-Halides (I): Superhyperfine Interactions and Electronic Structure
(b) Vol. **118**, No. 2, p. 557–566 (1983).
See also *19*, p. *302*; 22.5.4, p. 488.

R. Englisch
*Variational Calculations for Ground-State Energies of Donor–Acceptor Pairs
(b) Vol. **118**, No. 2, p. 605–612 (1983).

P. Gosar
*Many-Electron Hopping
(b) Vol. **118**, No. 2, p. 853–860 (1983).
See also *14.3*, p. *222*.

Ch. Opitz, H. Müller, and H. Miessner
EHT Calculations for Interstitial Hydrogen Bonding in Nickel
(b) Vol. **118**, No. 2, p. K85–K87 (1983).
See also *4*, p. *62*; *10.1*, p. *104*; 21.1, p. 372.

R. Tovar Barradas, J. González, and C. Rincón
Impurity States near the Fundamental Absorption Edge in p-$CuInS_2$
(b) Vol. **118**, No. 2, p. K103–K106 (1983).
See also 13.1, p. 161; *20.1*, p. *324*; 22.8, p. 508.

S. M. Wasim, G. Marcano, and G. Sánchez Pérez
*Electrical Properties of $CuGaTe_2$
(a) Vol. **78**, No. 2, p. 423–430 (1983).
See also *14.3*, p. *222*; 22.8, p. 508.

P. I. Baranskii, V. M. Babich, N. P. Baran, Yu. P. Dotsenko, V. B. Kovalchuk, and V. A. Shershel
*Investigation of Formation Conditions of Thermal Donors-I and -II in Oxygen-Containing n-Type Silicon within the Temperature Range 400 to 800 °C
(a) Vol. **78**, No. 2, p. 733–739 (1983).
See also *14.3*, p. *223*; 19, p. 303; 22.1.2, p. 433.

P. Kurek and W. Bogusz
*Electrical Conductivity of $Li_{16-2x}Zn_x(GeO_4)_4$
(a) Vol. **78**, No. 2, p. K117–K119 (1983).
See also 22.8.1, p. 516.

K. Schmalz and P. Gaworzewski
On the Effect of Ambients on the Formation of Oxygen-Related Donors in Cs-Si
(a) Vol. **78**, No. 2, p. K141–K145 (1983).
See also *10.2*, p. *117*; 14.3, p. 223; 22.1.2, p. 434.

B. G. Tagiev, G. M. Niftiev, and S. A. Abushov
*Recombination and Trapping Levels in GaS:Mn Single Crystals
(a) Vol. **78**, No. 2, p. K183–K187 (1983).
See also *16*, p. *266*; 22.3, p. 457.

B. G. Tagiev, G. M. Niftiev, G. K. Aslanov, I. M. Askerov, and F. N. Sultanov
*Effect of Ytterbium on the Electrophysical Properties of Ga_2Se_3 Single Crystals
(a) Vol. **78**, No. 2, p. K189–K193 (1983).
See also *14.3*, p. *223*; 22.3, p. 457.

V. Ziraps
Ionic and Electronic Processes during Frenkel Defect Annealing in KBr Crystals above LHeT
(b) Vol. **119**, No. 1, p. 49–59 (1983).
See also *10.2*, p. *117*; 11, p. 131; 20.1, p. 325; 22.5.2, p. 482.

J. R. Niklas, R. U. Bauer, and J.-M. Spaeth
*Two Types of F Centres in Alkaline Earth Fluoro-Halides (II): Double ENDOR Experiments
(b) Vol. **119**, No. 1, p. 171–178 (1983).
See also *19*, p. *303*; 22.5.4, p. 488.

L. Stauffer, B. Stébé, and G. Munschy
*Optical Absorption by an Exciton Bound to an Ionized Donor Impurity in Anisotropic Semiconductors
(b) Vol. **119**, No. 1, p. 193–198 (1983).
See also *13.5.1*, p. *197*; 20.1, p. 325; 22.4.1, p. 464.

K.-P. Oboth and K. Schmitt
*Effective Dynamical Coupling Constants of the A-Band of Tl^+ Centers in NaCl, KCl, KBr, and KI
(b) Vol. **119**, No. 1, p. 215–219 (1983).
See also *20.1*, p. *325*; 22.5.2, p. 482.

I. Mertig, E. Mrosan, R. Zeller, and P. H. Dederichs
*Electronic Properties of Dilute Copper Alloys (III): Influence on the de Haas-van Alphen Effect
(b) Vol. **119**, No. 1, p. 251–260 (1983).
See also 21.1, p. 374.

O. Goede and D. Hennig
*Isovalent-Impurity Clusters in Koster-Slater Approximation
(b) Vol. **119**, No. 1, p. 261–268 (1983).
See also *13.1*, p. *162*; 22.2.2, p. 450; 22.4.1, p. 464.

S. Benci, F. Fermi, and G. Lenzi
Dimer Center Luminescence in Low-Doped KI:Tl^+ Crystals
(b) Vol. **119**, No. 1, p. 337–342 (1983).
See also *20.3*, p. *344*; 22.5.2, p. 482.

S. Mahajan and S. Prakash
*Electronic Structure of Impurities in Dilute Alloys of Al
(b) Vol. **119**, No. 1, p. 381–389 (1983).
See also 14.1, p. 209; 21, p. 358.

D. M. Daraselia and D. L. Japaridze
*Determination of the Structure of the Co^{2+} Paramagnetic Centers in α-$LiIO_3$ by the RFDS Method
(b) Vol. **119**, No. 1, p. K57 – K58 (1983).
See also *19*, p. *303*; 22.8.1, p. 516.

T. Mohammed-Brahim, A. Friant, and J. Mellet
*Characterization of Deep Energy Levels in Mercuric Iodide
(a) Vol. **79**, No. 1, p. 71 – 82 (1983).
See also 16, p. 266; 22.5, p. 475.

A. V. Dvurechenskii, I. A. Ryazantsev, L. S. Smirnov, H. Klose, and M. Rieth
Amorphous Silicon–Phosphorus Layers
(a) Vol. **79**, No. 1, p. 83 – 86 (1983).
See also 2, p. 51; *14.3.1*, p. *231*; 16, p. 267; 19, p. 304; 22.1.2, p. 434.

M. Suezawa, Y. Sasaki, and K. Sumino
*Dependence of Photoluminescence on Temperature in Dislocated Silicon Crystals
(a) Vol. **79**, No. 1, p. 173 – 181 (1983).
See also *20.3*, p. *344*; 22.1.2, p. 435.

F. Campabadal, V. Milián, and X. Aymerich-Humet
Trap-Assisted Tunneling in MIS and Schottky Structures
(a) Vol. **79**, No. 1, p. 223 – 236 (1983).
See also *14.3.3*, p. *240*; *14.3.4*, p. *247*; 22.6, p. 496.

K. Somaiah and V. Hari Babu
Thermoluminescence of Thallium-Doped BaFCl
(a) Vol. **79**, No. 1, p. 237 – 242 (1983).
See also 20.1, p. 326; *20.3*, p. *344*; 22.5, p. 475.

R. Chen, D. J. Huntley, and G. W. Berger
*Analysis of Thermoluminescence Date Dominated by Second-Order Kinetics
(a) Vol. **79**, No. 1, p. 251 – 261 (1983).
See also 20.3, p. 344.

Y. D. Yao, S. Arajs, and E. E. Anderson
*Deviations from Matthiessen's Rule in the Nickel–Chromium System
(a) Vol. **79**, No. 1, p. 327 – 330 (1983).
See also *14.1*, p. *209*; 18.2, p. 286; 21.1, p. 375.

R. Pickenhain, B. Rheinländer, and S. Brehme
*P_{Ga} Antisite Defect in Zn-Implanted $GaAs_{0.1}P_{0.9}$ LED
(a) Vol. **79**, No. 1, p. K61 – K63 (1983).
See also *20.3*, p. *344*; 22.2.4, p. 455.

S. P. Kathuria and C. M. Sunta
Fractional Order of Kinetics and Its Variation with Trap Occupancy in LiF TLD-100
(a) Vol. **79**, No. 1, p. K77 – K80 (1983).
See also *20.3*, p. *345*; 22.5.2, p. 483.

S. V. Moharil
*Reply to "Fractional Order of Kinetics and Its Variation with Trap Occupancy in LiF TLD-100" by S. P. Kathuria and C. M. Sunta
(a) Vol. **79**, No. 1, p. K81 – K84 (1983).
See also *20.3*, p. *345*; 22.5.2, p. 483.

V. E. Arkhincheev and S. K. Savvinikh
*Semiempirical Theory of Shallow Short-Range Electron Traps in Silicon in a Magnetic Field
(b) Vol. **119**, No. 2, p. 459 – 466 (1983).
See also 22.1.2, p. 435.

P. A. Alekseev, E. A. Goremychkin, B. Lippold, E. Mühle, and I. P. Sadikov
*Crystal Field Splitting in $NdAl_3$ Studied by Inelastic Neutron Scattering
(b) Vol. **119**, No. 2, p. 651 – 658 (1983).
See also 21.4, p. 392.

V. K. Jain, V. P. Seth, R. K. Malhotra, and S. K. Yadav
*Electron Paramagnetic Resonance of Mn^{2+} and VO^{2+} in $Ba_2Zn(HCOO)_6 \cdot 4\,H_2O$ Single Crystals
(b) Vol. **119**, No. 2, p. 659 – 666 (1983).
See also *19*, p. *304*; 22.8.1, p. 517.

M. Ortuño
*Impurity Potentials in Diamond
(b) Vol. **119**, No. 2, p. 709 – 713 (1983).
See also 22.1, p. 417.

M. Höhne
Gold-Related EPR Centres of Low Symmetry in Silicon
(b) Vol. **119**, No. 2, p. K117 – K121 (1983).
See also *19*, p. *304*; 22.1.2, p. 436.

I. D. Henning and H. Thomas
*Process-Induced Defect States in GaAsP Light Emitting Diodes
(a) Vol. **79**, No. 2, p. 567 – 574 (1983).
See also 14.3.3, p. 241; 20.3, p. 345; 22.2.4, p. 455.

L. Z. Mirets, G. P. Peka, S. I. Radautsan, and E. V. Russu
Photoelectric Investigation of Recombination Centre Parameters in InP(Fe)
(a) Vol. **79**, No. 2, p. 593 – 597 (1983).
See also *16*, p. *267*; 22.2.2, p. 450.

S. V. Moharil
Track Interaction Model for Supralinearity and Sensitization in LiF-TLD 100
(a) Vol. **79**, No. 2, p. 599 – 605 (1983).
See also 10.2, p. 119; *20.3*, p. *345*; 22.5.2, p. 484.

Yu. A. Vidadi, A. V. Ragimov, B. A. Mamedov, S. A. Guseinov, A. A. Medzhidov, and B. I. Liogonkii
On the Contribution of Naphthoxyl-Type Paramagnetic Centres to the Electric Conductivity of Oligonaphthols
(a) Vol. **79**, No. 2, p. 645–650 (1983).
See also *14.3*, p. *225*; 22.9, p. 531.

A. J. Szadkowski and A. Lubomirska-Wittlin
*Imperfection Photoconductivity of $Cd_{1-x}Mn_xTe$
(a) Vol. **79**, No. 2, p. K127–K129 (1983).
See also *16*, p. *267*; 22.4.4, p. 471.

H. Feichtinger and R. Czaputa
*On the Question of Multiple Charge States of Interstitial Fe and Cr in Silicon
(a) Vol. **79**, No. 2, p. K143–K146 (1983).
See also 22.1.2, p. 436.

K. D. Glinchuk, N. M. Litovchenko, Z. A. Salnik, and S. I. Skryl
*Effect of Heat Treatment on the Minority Carrier Lifetime in Oxygen-Containing Silicon
(a) Vol. **79**, No. 2, p. K159–K163 (1983).
See also *10.2*, p. *119*; 22.1.2, p. 436.

P. Petelenz
*Optical Detrapping in Molecular Crystals (I): Vibrational Structure of Detrapping Transitions
(b) Vol. **120**, No. 1, p. 31–40 (1983).
See also 20.1, p. 328; 22.9, p. 531.

L. N. Kantorovich
*Multipole Theory of the Polarization of Solids by Point Defects (I): Dipole Approximation
(b) Vol. **120**, No. 1, p. 77–86 (1983).
See also *14.4*, p. *254*.

J. Kowalska and J. Zieliński
*Spinless Anderson Model with a Local Phonon Mode
(b) Vol. **120**, No. 1, p. 99–104 (1983).
See also *6*, p. *79*.

V. Hizhnyakov, G. Liidya, V. Nagirnyi, T. Soovik, and S. Zazubovich
*Effect of Magnetic Field on the Decay Kinetics and Polarization of the A_T Emission of KCl:Ga at 0.38 K
(b) Vol. **120**, No. 1, p. 105–115 (1983).
See also *20.3*, p. *346*; 22.5.2, p. 484.

V. Nagirnyi, T. Soovik, S. Zazubovich, and N. Janson
Polarized A_T Emission of In^+ Centres in KI:In
(b) Vol. **120**, No. 1, p. 117–122 (1983).
See also *20.3*, p. *346*; 22.5.2, p. 484.

M. Stutzmann and J. Stuke
*Temperature Dependence of ESR Spectra of Doped Amorphous Germanium
(b) Vol. **120**, No. 1, p. 225–234 (1983).
See also 2, p. 54; *19*, p. *305*; 22.1.1, p. 421.

H.-J. Unger
*Localization of Electronic States in Disordered Tight-Binding Systems (I): Derivation and Study of the Model
(b) Vol. **120**, No. 1, p. 401–413 (1983).
See also *2*, p. *54*; 22, p. 413.

M. N. Sapozhnikov and V. I. Alekseev
Study of Optical Relaxation and Inhomogeneous Distribution Function of Impurity Organic Molecules by Site-Selective Laser Spectroscopy
(b) Vol. **120**, No. 1, p. 435–444 (1983).
See also *6*, p. *80*; 20.1, p. 329; *20.3*, p. *346*; 22.9, p. 531.

L. E. Kholodenkov and A. G. Makhanek
Laser-Induced Reversible Deformation of Europium Stark Levels in Fluorite
(b) Vol. **120**, No. 1, p. K5–K8 (1983).
See also *20.1*, p. *329*; *20.3*, p. *346*; 22.5.3, p. 487.

A. N. Georgobiani, A. V. Mikulyonok, I. G. Stoyanova, and I. M. Tiginyanu
*Non-Equilibrium Carrier Radiative Recombination in Indium Phosphide Single Crystals
(a) Vol. **80**, No. 1, p. 109–118 (1983).
See also *20.3*, p. *346*; 22.2.2, p. 451.

G. Kühnel, W. Siegel, and E. Ziegler
*Electrical and Photoelectrical Properties of Semi-Insulating GaP
(a) Vol. **80**, No. 1, p. 159–164 (1983).
See also *14.3*, p. *226*; 16, p. 267; 22.2.2, p. 451.

A. Maciel, S. G. Fries, J. A. H. da Jornada, R. P. Livi, and E. R. Fraga
*Study of the Nuclear Quadrupole Interaction in Tl_5Te_3
(a) Vol. **80**, No. 1, p. 245–248 (1983).
See also *1.3*, p. *27*; 22.7, p. 499.

G. Micocci, A. Rizzo, A. Tepore, and F. Zuanni
*Isothermal Currents in InSe, GaSe, and GaS Single Crystals
(a) Vol. **80**, No. 1, p. 263–267 (1983).
See also *14.3*, p. *226*; 22.3, p. 458.

V. V. Borimskii, K. D. Glinchuk, N. M. Litovchenko, and Z. A. Salnik
*Annealing Behaviour of Oxygen-Induced Recombination Centres in Silicon
(a) Vol. **80**, No. 1, p. 343–348 (1983).
See also 22.1.2, p. 438.

G. M. Guseinov, G. M. Niftiev, and M. G. Shakhtakhtinskii
*Effect of External Factors on the Thermally Stimulated Conductivity of $GdIn_3S_6$ Single Crystals
(a) Vol. **80**, No. 1, p. K21–K24 (1983).
See also 14.3, p. 227; 22.8, p. 510.

A. N. Georgobiani, A. N. Gruzintsev, S. I. Radautsan, and I. M. Tiginyanu
*The Contribution of Iodine Impurity to the Formation of Centers in $CdGa_2S_4$ Single Crystals
(a) Vol. **80**, No. 1, p. K45 – K48 (1983).
See also **16**, p. 268; 20.1, p. 329; 22.8, p. 510.

L. Bragina, V. Hizhnyakov, G. Liidya, V. Nagirnyi, T. Soovik, S. Zazubovich, and N. Janson
*Effect of Magnetic Field on the Decay Kinetics and Polarization of the A_T Emission of KCl:Tl and KI:In Crystals at 0.4 to 0.7 K
(b) Vol. **120**, No. 2, p. 463 – 470 (1983).
See also *20.3*, p. *347*; 22.5.2, p. 485.

G. Gildenblat
*Oscillations of Local Magnetization in the Non-Ideal Antiferromagnetic Chain
(b) Vol. **120**, No. 2, p. 519 – 527 (1983).
See also 18.4, p. 298.

D. Suisky
*Multiphonon Structure of Optical Transitions to Impurity Resonance States. The Influence of Anharmonicity
(b) Vol. **120**, No. 2, p. 645 – 654 (1983).
See also **6**, p. 81; *20.1*, p. *330*; 22.4, p. 461.

K. C. Liu, Kai-Shue Lam, and Th. F. George
*Localized Magnetization Arising from Spin-Dependent Impurity Potentials in the Wolff Model
(b) Vol. **120**, No. 2, p. 695 – 702 (1983).
See also *18*, p. *276*.

C. Pedrini, H. Chermette, A. B. Goldberg, D. S. McClure, and B. Moine
*Optical Absorption and Electronic Structure of NaCl:Ag^+. A Multiple Scattering Xα Study of the $(AgCl_6)^{5-}$ Cluster
(b) Vol. **120**, No. 2, p. 753 – 766 (1983).
See also *20.1*, p. *330*; 22.5.2, p. 485.

A. Ferreira da Silva, R. Micnas, I. C. da Cunha Lima, and R. Kishore
*The Impurity Conduction of n-Type CdS: A Two Band Approach
(b) Vol. **120**, No. 2, p. K101 – K106 (1983).
See also *14.3*, p. *227*; 22.4.1, p. 465.

B. Susła, R. Czajka, and J. Raułuszkiewicz
*Electronic Excitation of Ni^{2+} and Co^{2+} by Tunneling Electrons
(b) Vol. **120**, No. 2, p. K129 – K134 (1983).
See also *14.3.3*, p. *242*; 19, p. 305; 20.1, p. 330; 22.6, p. 497.

J. M. Khalifeh
Asymptotic Elastic Energy in Simple Metals
(b) Vol. **120**, No. 2, p. K161 – K166 (1983).
See also *10.1*, p. *107*; 21, p. 361.

H. ZIMMERMANN und R. BOYN
*Excitation Behaviour of Blue ($^1G_4 \rightarrow {}^3H_6$) and Infrared ($^3F_4 \rightarrow {}^3H_6$) Photoluminescence of ZnS:Tm, Ag Crystals
(b) Vol. **120**, No. 2, p. K189–K193 (1983).
See also *20.3*, p. *347*; 22.4.1, p. 465.

K. ASAMI, and M. ISHIGURO
*Pressure Effects on the Absorption and Emission of F_A Centers in KCl
(b) Vol. **120**, No. 2, p. K205–K209 (1983).
See also *10.2*, p. *121*; 20.1, p. 330; 20.3, p. 347; 22.5.2, p. 485.

B. I. CRAIG
*Low-Temperature Trapping of Positive Muons in Metals
(b) Vol. **120**, No. 2, p. K211–K214 (1983).
See also *10.1*, p. *107*; 21, p. 362.

A. T. OZA
*Electrical Conduction in Some Inclusion Compounds of Iodine
(a) Vol. **80**, No. 2, p. 573–579 (1983).
See also *14.3*, p. *228*; 22.9, p. 532.

S. K. MISRA, J. BANDET, G. BACQUET, and T. E. MCENALLY
Electron Spin Resonance Studies of Defect Centers Containing Boron Atoms in Natural Danburite Crystals
(a) Vol. **80**, No. 2, p. 581–588 (1983).
See also *19*, p. *306*; 22.8.1, p. 518.

B. P. TYAGI and K. SEN
*On the Resistivity of Polycrystalline Silicon
(a) Vol. **80**, No. 2, p. 679–684 (1983).
See also *14.3.4*, p. *249*; 22.1.2, p. 439.

V. A. BYKOVSKII, A. V. MUDRYI, V. P. POSKREBYSHEV, and V. D. TKACHEV
*Radiative Recombination Centers in Irradiated Germanium
(a) Vol. **80**, No. 2, p. K133–K136 (1983).
See also 11, p. 135; *20.3*, p. *347*; 22.1.1, p. 422.

SH. A. VAKHIDOV and A. F. RAKOV
Radiation Stimulated Recombination Processes in $Y_3Al_5O_{12}$ Crystals
(a) Vol. **80**, No. 2, p. K175–K178 (1983).
See also *11*, p. *135*; 13.5.1, p. 199; *20.3*, p. *348*; 22.8.2, p. 523.

P. NOVÁK, M. MARYŠKO, S. KRUPIČKA, J. KUB, J. ŠIMŠOVA, J. ČERMÁK, and M. NEVŘÍVA
*Noncubic Anisotropy in YIG:Mn^{3+} Thin Films
(a) Vol. **80**, No. 2, p. K213–K215 (1983).
See also *18*, p. *277*; 22.8.2, p. 524.

13.5 Elementary Excitations (Phonons See 6)

Original Papers and Short Notes

H. Yoshida and S. Shionoya
Luminescence of High Density Electron–Hole Plasma in CdS and CdSe in a Wide Temperature Range
(b) Vol. **115**, No. 1, p. 203–208 (1983).
See also *20.3*, p. *334*; 22.4.1, p. 462; 22.4.2, p. 466.

V. Schulz and L. Merten
Dispersion Curves of Stimulated Raman Effect in Piezoelectric Cubic Crystals
(b) Vol. **115**, No. 1, p. 225–233 (1983).
See also *6*, p. *69*; *20*, p. *307*; 22.2.2, p. 448.

L. S. Brizhik and A. S. Davydov
*Soliton Excitations in One-Dimensional Molecular Systems
(b) Vol. **115**, No. 2, p. 615–630 (1983).

H. Weinert
*Decay Behaviour of High-Excitation Luminescence in GaP:N (I): Model for the Decay of Two Coexisting Phases
(b) Vol. **116**, No. 2, p. 525–532 (1983).
See also *20.3*, p. *337*; 22.2.2, p. 449.

V. N. Myasnikov, S. V. Marisova, and A. N. Lipovchenko
*Optical Properties of Polaritons in Thin Crystal Films
(b) Vol. **117**, No. 1, p. 109–116 (1983).
See also 20, p. 308; 20.3, p. 338; 22.9, p. 527.

L. Wendler
Effects of a Transition Layer on Slab Polaritons
(b) Vol. **117**, No. 1, p. 241–246 (1983).
See also *6*, p. *73*; *20*, p. *308*.

H. Weinert, R. Bindemann, and K. Kreher
*Decay Behaviour of the High-Excitation Luminescence in GaP:N (II): Experimental Results on the Densities of the Coexisting Phases and the Relevant Recombination Processes
(b) Vol. **117**, No. 1, p. 247–253 (1983).
See also *20.3*, p. *338*; 22.2.2, p. 449.

R. Schwabe, I. Streit, F. Thuselt, and K. Unger
*Experimental Investigations of Electron–Hole Plasma in Tellurium-Doped GaP
(b) Vol. **117**, No. 2, p. 625–634 (1983).
See also *20.3*, p. *339*; 22.2.2, p. 449.

TH. HARBICH and G. MAHLER
*Bistability of the Electron–Hole Plasma Density in a Strong Electromagnetic Field
(b) Vol. **117**, No. 2, p. 635–644 (1983).
See also *13.1*, p. *159*; 22.2.1, p. 444.

R. ZIMMERMANN
*Thermodiffusion in Electron–Hole Plasmas Revised
(b) Vol. **117**, No. 2, p. K135–K140 (1983).
See also 22, p. 408.

L. WENDLER
*Effects of a Transition Layer on Nonlinear Surface Polaritons
(b) Vol. **118**, No. 1, p. 205–210 (1983).
See also 14.4, p. 252.

M. RÖSLER, F. THUSELT, and R. ZIMMERMANN
*Theory of Electron–Hole Liquid in Doped Semiconductors. Application to GaP
(b) Vol. **118**, No. 1, p. 303–317 (1983).
See also 22.2.2, p. 450.

K. REGIŃSKI
*Phonon Polaritons in the Presence of a Magnetic Field
(b) Vol. **118**, No. 2, p. 587–593 (1983).
See also *6*, p. *76*; 18, p. 275; 20, p. 309; 22.5.2, p. 481.

G. D. SHIBARSHINA, S. A. MOSKALENKO, and P. I. KHADZHI
*The Polariton Soliton Formation Due to the Effect of the Dipole Momentum Saturation
(b) Vol. **119**, No. 1, p. 153–158 (1983).
See also 13.5.1, p. 197; *20.1*, p. *325*; 22.4.1, p. 464.

L. WENDLER and E. JÄGER
Phonon-Polaritons in Bilayer Systems. Surface Phonon-Polaritons
(b) Vol. **120**, No. 1, p. 235–247 (1983).
See also *6*, p. *80*; *20.1*, p. *328*; 22.2.1, p. 446; 22.2.2, p. 451; 22.5.3, p. 487.

L. MERTEN and R. KOTZOTT
Excitation of Longitudinal Optical Phonons by Stimulated Raman Effect in Piezoelectric Cubic Crystals
(b) Vol. **120**, No. 2, p. 481–490 (1983).
See also *6*, p. *80*; *20*, p. *311*; 22, p. 414.

13.5.1 Excitons

Review Article

V. Kh. Brinkenstein, V. A. Benderskii, and P. G. Filippov
*Fluorescence of Molecular Crystals at High Levels of Optical Pumping
(b) Vol. **117**, No. 1, p. 9–39 (1983).
See also 6, p. 69; 20.2, p. 332; *20.3*, p. *334*; 22.9, p. 525.

Original Papers and Short Notes

A. A. Manenkov, G. N. Mikhailova, A. M. Prokhorov, A. V. Sidorin, S. Yu. Sokolov, and S. G. Tikhodeev
*Interaction of High Power CO_2 Laser Radiation with a Nonequilibrium Carrier System in Ge at Low Temperatures
(b) Vol. **115**, No. 1, p. 75–81 (1983).
Erratum
(b) Vol. **117**, No. 2, p. 756 (1983).
See also *20.3*, p. *334*; 22.1.1, p. 418.

A. S. Davydov and A. V. Zolotariuk
*Autolocalized Collective Excitations in Molecular Chains with Cubic Anharmonicity
(b) Vol. **115**, No. 1, p. 115–125 (1983).
See also *6*, p. *69*.

A. I. Bobrysheva, S. I. Beryl, V. T. Zyukov, S. A. Moskalenko, and E. P. Pokatylov
*The One- and Two-Photon Absorption Due to Surface State Excitons
(b) Vol. **115**, No. 1, p. 153–160 (1983).
See also *13.3*, p. *167*; 22.2.1, p. 443.

M. P. Lisitsa, S. F. Terekhova, and N. A. Onishchenko
*Two-Beam Interferometry in the Exciton Spectroscopy of CdS
(b) Vol. **115**, No. 1, p. 187–194 (1983).
See also *20.1*, p. *312*; 22.4.1, p. 462.

T. Takase, E. Morikawa, and M. Kotani
*Triplet Exciton Trapping in p-Terphenyl Crystals Studied by Delayed Fluorescence and Phosphorescence
(b) Vol. **115**, No. 1, p. 293–298 (1983).
See also *20.3*, p. *334*; 22.9, p. 525.

H. Rossmann, F. Henneberger, and J. Voigt
*Memory Effect in the Excitonic Transmission of CdS
(b) Vol. **115**, No. 1, p. K63–K67 (1983).
See also 22.4.1, p. 462.

A. V. Malyavkin
Optically Detected Magnetic Resonance in $Zn_{1-x}Mn_xTe$
(b) Vol. **115**, No. 2, p. 353–358 (1983).
See also *19*, p. *299*; *20.3*, p. 335; 22.4.4, p. 470.

E. Ya. Glushko and N. A. Efremov
*Diffusion of Frenkel Excitons in Anisotropic Crystals
(b) Vol. **115**, No. 2, p. 529–538 (1983).
See also 22.9, p. 525.

M. S. Brodin, N. A. Davydova, and I. Yu. Shablii
*Relation of Recombination Radiation Spectra to the Concentration of Structure Defects in CdS Single Crystals
(b) Vol. **115**, No. 2, p. 641–647 (1983).
See also 20.3, p. 336; 22.4.1, p. 462.

S. V. Marisova, E. N. Myasnikov, and A. N. Lipovchenko
*Phonon-Assisted Decay of Excitons in Anthracene Crystals
(b) Vol. **115**, No. 2, p. 649–654 (1983).
See also *20.1*, p. *314*; 22.9, p. 525.

B. M. Nitsovich and G. M. Pestryakov
*Temperature Behaviour of the Integral Intensity of Exciton Absorption
(b) Vol. **115**, No. 2, p. K87–K90 (1983).
See also 20.1, p. 314.

A. P. Litvinchuk and N. I. Vitrikhovskii
The Variation of the Energy Gap with Composition in the Quaternary Alloy System $ZnTe_{1-2x}S_xSe_x$
(b) Vol. **115**, No. 2, p. K151–K154 (1983).
See also *13.1*, p. *156*; *20.3*, p. *336*; 22.4.4, p. 470.

Hoang Ngoc Cam, Nguyen Van Hieu, and Ha Vinh Tan
On the Theory of the Nonlinear Acousto-Optical Effect in Semiconductors
(b) Vol. **116**, No. 1, p. 25–29 (1983).
See also *7*, p. *82*; *20*, p. *307*; 22.2, p. 442; 22.4, p. 460.

A. Haufe, K. Henneberger, J. Röseler, and H.-J. Wünsche
*Parameters of Bound Multiexciton Complexes in Direct Polar Semiconductors
(b) Vol. **116**, No. 1, p. 191–203 (1983).
See also 20.3, p. 336; 22.4.1, p. 463; 22.4.2, p. 466.

R. Baumert, I. Broser, J. Gutowski, and A. Hoffmann
*Forbidden Luminescence and Resonance Raman Scattering of Bound Exciton States in CdS
(b) Vol. **116**, No. 1, p. 261–267 (1983).
See also *20.3*, p. *337*; 22.4.1, p. 463.

I. T. Godmanis, A. N. Trukhin, and K. Hübner
*Exciton–Phonon Interaction in Crystalline and Vitreous SiO_2
(b) Vol. **116**, No. 1, p. 279–287 (1983).
See also *6*, p. *72*; 20.1, p. 315; 22.1.1, p. 419; 22.6, p. 492.

B. S. Tošić, G. S. Davidović-Ristovski, and Lj. M. Ristovski
*Excitons in a Finite Polymer Chain
(b) Vol. **116**, No. 1, p. K1–K4 (1983).
See also 22.9, p. 526.

V. Agekyan, R. Bindemann, R. Schwabe, Yu. Stepanov, and I. Streit
*Luminescence of Excitons Bound by Disorder Potential Fluctuations in $CdS_{1-x}Se_x$: Spatial Diffusion, Excitation Level Dependence, and Kinetics
(b) Vol. **116**, No. 1, p. K43 – K46 (1983).
See also *20.3*, p. *337*; 22.4.4, p. 470.

M. Ya. Valakh, A. P. Litvinchuk, and G. G. Tarasov
Effect of the Pumping Intensity on the Resonant Secondary Emission Spectra of Optically Aligned Hot Excitons in ZnTe Crystals
(b) Vol. **116**, No. 2, p. K61 – K65 (1983).
Erratum
(b) Vol. **118**, No. 2, p. K167 (1983).
See also *6*, p. *73*; *20.3*, p. *337*; 22.4.3, p. 468.

Nguyen The Khoi, J. Ginter, and A. Twardowski
Magnetooptical Study of the Fundamental Absorption Edge in $Cd_{1-x}Mn_xTe$
(b) Vol. **117**, No. 1, p. 67 – 74 (1983).
See also *18*, p. *274*; *20*, p. *308*; 22.4.4, p. 470.

B. J. Schmid and P. Reineker
*On the Theory of Singlet and Triplet Charge-Transfer Excitons
(b) Vol. **117**, No. 1, p. 117 – 126 (1983).
See also 22.9, p. 527.

Yu. P. Gnatenko, Z. D. Kovalyuk, P. A. Skubenko, and Yu. I. Zhirko
*Emission of Free and Bound Excitons in Layered GaSe Crystals
(b) Vol. **117**, No. 1, p. 283 – 287 (1983).
See also *20.3*, p. *338*; 22.3, p. 457.

P. Wiśniewski and M. Nawrocki
*Composition Dependence of the Energy Gap in $Cd_{1-x}Mn_xSe$
(b) Vol. **117**, No. 1, p. K43 – K45 (1983).
See also 20.1, p. 318; 22.4.4, p. 470.

V. Železný, L. Jastrabík, and L. Dobiášová
*The Temperature Dependence of Emission Spectra in α-9,10-Dichloroanthracene
(b) Vol. **117**, No. 2, p. K95 – K97 (1983).
See also *20.3*, p. *340*; 22.9, p. 528.

M. S. Brodin, I. V. Blonskii, and B. M. Nitsovich
*Anisotropic Exciton Scattering by Bending Waves in Layer Crystals
(b) Vol. **117**, No. 2, p. K99 – K103 (1983).
See also 20.1, p. 320; 22.5, p. 474.

T. Taguchi
Crystal Growth and Neutral-Acceptor Bound-Exciton Emission of ZnCdTe by THM with Te Solvent
(a) Vol. **77**, No. 2, p. K115 – K119 (1983).
See also *3*, p. *58*; 13.4, p. 181; *20.3*, p. *341*; 22.4.4, p. 471.

A. V. Efanov and M. V. Entin
Optical Orientation and Polarized Luminescence in Silicon
(b) Vol. **118**, No. 1, p. 63–72 (1983).
See also *20.3*, p. *341*; 22.1.2, p. 432.

T. Amand, A. Cornet, M. Pugnet, J. Collet, M. Brousseau, and B. S. Razbirin
*Picosecond Study of the Exciton Gas Cooling
(b) Vol. **118**, No. 1, p. 113–119 (1983).
See also 20.3, p. 341; 22.4.2, p. 467; 22.5, p. 474.

J. M. Hvam, G. Blattner, M. Reuscher, and C. Klingshirn
*The Biexciton Levels and Nonlinear Optical Transitions in ZnO
(b) Vol. **118**, No. 1, p. 179–189 (1983).
See also *20.3*, p. *341*; 22.4, p. 460.

L. M. Kanskaya, S. I. Kokhanovskii, R. P. Seysyan, and Al. L. Efros
*Direct Determination of Indium Antimonide Energy Band Parameters from Diamagnetic Exciton Spectra
(b) Vol. **118**, No. 1, p. 447–452 (1983).
See also *13.1*, p. *160*; 20.1, p. 322; 22.2.3, p. 453.

N. R. Nurtdinov, R. Stegmann, and M. Griepentrog
*The Influence of the Variation of the Chemical Composition and of the Nitrogen Doping Level on the Energy Spectra of the Bound Excitons in $GaAs_{1-x}P_x$:N
(b) Vol. **118**, No. 2, p. 613–621 (1983).
See also 10.2, p. 115; *20.3*, p. *342*; 22.2.4, p. 455.

J. Kalinovski and R. Signerski
*Exciton-Enhanced Double Injection Currents in Tetracene Crystals
(b) Vol. **118**, No. 2, p. K147–K150 (1983).
See also *14.3.4*, p. *247*; 20.3, p. 343; 22.9, p. 529.

M. S. Brodin, V. M. Bandura, and M. G. Matsko
Direct Luminescence Evidence for the Degenerate Electron–Hole Plasma Formation in $A^{II}B^{VI}$ Semiconductors
(b) Vol. **119**, No. 1, p. 83–87 (1983).
See also *13*, p. *151*; *20.3*, p. *343*; 22.4.2, p. 467; 22.4.3, p. 468.

A. I. Bobrysheva and S. A. Moskalenko
*Biexcitons in Crystals with Large Ortho–Para-Exciton Splitting
(b) Vol. **119**, No. 1, p. 141–146 (1983).
See also 22, p. 411.

G. D. Shibarshina, S. A. Moskalenko, and P. I. Khadzhi
The Polariton Soliton Formation Due to the Effect of the Dipole Momentum Saturation
(b) Vol. **119**, No. 1, p. 153–158 (1983).
See also *13.5*, p. *193*; *20.1*, p. *325*; 22.4.1, p. 464.

L. Stauffer, B. Stébé, and G. Munschy
*Optical Absorption by an Exciton Bound to an Ionized Donor Impurity in Anisotropic Semiconductors
(b) Vol. **119**, No. 1, p. 193–198 (1983).
See also *13.4*, p. *185*; 20.1, p. 325; 22.4.1, p. 464.

J. Puls, F. Henneberger, and J. Voigt
*Excited States of Bound Excitons in Wurtzite-Type Semiconductors
(b) Vol. **119**, No. 1, p. 291—298 (1983).
See also *20.3*, p. *343*; 22.4, p. 461.

E. Fiorino and R. Del Sole
Some Remarks on the Theory of the Dielectric Response of Crystals and on the Lorentz-Lorenz Formula
(b) Vol. **119**, No. 1, p. 315—326 (1983).
See also *20.1*, p. *326*.

I. J. Lalov, K. T. Stoychev, and P. S. Penchev
*Non-Linear Optical Susceptibilities of Molecular Crystals in the Region of Vibronic Excitations (I): The Three-Dimensional Case
(b) Vol. **119**, No. 1, p. 419—427 (1983).
See also 6, p. 77; 20.1, p. 326; 22, p. 411.

R. Baltramiejūnas, E. Kuokštis, and A. Žukauskas
*Self-Heating of Dense Exciton Gas in CdTe Single Crystals
(b) Vol. **119**, No. 2, p. 453—457 (1983).
See also 20.3, p. 345; 22.4.3, p. 469.

V. G. Plekhanov, V. V. Shepelev, and A. U. Grinfelds
*Spectroscopic Manifestation of Self-Trapping Excitons in Ionic Insulators
(b) Vol. **119**, No. 2, p. 493—502 (1983).
See also 6, p. 78; 20.3, p. 345; 22.5.2, p. 483.

J. Flohrer and M. Porsch
*Effect of Valence Band Splitting on the Biexciton Binding Energy in Wurtzite-Type Semiconductors
(b) Vol. **119**, No. 2, p. 639—649 (1983).
See also *13.1*, p. *163*; 22.4.1, p. 464.

B. G. Tagiev, G. M. Niftiev, S. A. Abushov, and A. O. Khalilov
Optical Properties of $InSe_{1-x}S_x$ Single Crystals in the Region of the Fundamental Absorption Edge
(b) Vol. **119**, No. 2, p. K143—K148 (1983).
See also 14.3.4, p. 247; *20.1*, p. *328*; *20.3*, p. *345*; 22.3.1, p. 459.

D. V. Korbutyak and V. G. Litovchenko
Expansion of the Electron–Hole Liquid Plasma at GaAs Surface
(b) Vol. **120**, No. 1, p. 87—97 (1983).
See also *13.3*, p. *170*; *20.3*, p. *346*; 22.2.1, p. 446.

G. O. Müller, S. L. Karpenko, R. P. Seisyan, and M. A. Yakobson
Influence of Doping on Excitonic Luminescence Spectra in CdS
(b) Vol. **120**, No. 1, p. K93—K95 (1983).
See also *20.3*, p. *346*; 22.4.1, p. 465.

I. J. Lalov, K. T. Stoychev, and S. P. Penchev
*Non-Linear Optical Susceptibilities of Molecular Crystals in the Region of Vibronic Excitations (II): Totally Symmetric Vibrations
(b) Vol. **120**, No. 2, p. 671–678 (1983).
See also 6, p. 81; 20.1, p. 330; 22, p. 414.

A. Zaręba and A. J. Nadolny
*Photoconductivity Originated from Dissociation of Excitons in $Cd_{1-x}Mn_xTe$
(b) Vol. **120**, No. 2, p. K97–K100 (1983).
See also *16*, p. *268*; 22.4.4, p. 472.

G. Strinati
*Alternative Derivation of the Coupled Set of Differential Equations for Excitons in Semiconductors with Degenerate Bands
(b) Vol. **120**, No. 2, p. K115–K118 (1983).

Sh. A. Vakhidov and A. F. Rakov
Radiation Stimulated Recombination Processes in $Y_3Al_5O_{12}$ Crystals
(a) Vol. **80**, No. 2, p. K175–K178 (1983).
See also *11*, p. *135*; 13.4, p. 191; *20.3*, p. *348*; 22.8.2, p. 523.

13.5.2 Plasmons

Original Papers and Short Notes

O. Ziep
Estimation of Second-Order Auger Recombination in Lead Chalcogenides
(b) Vol. **115**, No. 1, p. 161 – 170 (1983).
See also *6*, p. *69*; *13.1*, p. *155*; 22.8, p. 502.

E. Kühet and R. Wedel
Energy Losses in Crystalline and Amorphous Solids in the Intermediate Energy Region
(b) Vol. **116**, No. 2, p. 585 – 595 (1983).
See also *11*, p. *126*; 22.1.2, p. 428.

S. Ghosh and S. Khan
*Parametric Conversion of an Electron Plasma Wave into an Electromagnetic Wave in Longitudinally Magnetized n-Type Piezoelectric Semiconductor Plasmas
(b) Vol. **119**, No. 1, p. 363 – 373 (1983).
See also 18, p. 275; 22.2.3, p. 453.

G. Irmer, V. V. Toporov, B. H. Bairamov, and J. Monecke
Determination of the Charge Carrier Concentration and Mobility in n-GaP by Raman Spectroscopy
(b) Vol. **119**, No. 2, p. 595 – 603 (1983).
See also 6, p. 79; *14.3*, p. *224*; *20.1*, p. *327*; 22.2.2, p. 450.

13.5.3 Polarons

Original Papers and Short Notes

F. M. Peeters and J. T. Devreese
*Induced Polarization Charge Density around an Electron in a Polar Semiconductor
(b) Vol. **115**, No. 1, p. 285–291 (1983).

F. M. Peeters and J. T. Devreese
*The $\frac{3}{2}\,kT$ Problem in the Low-Temperature Polaron Mobility Theories
(b) Vol. **115**, No. 2, p. 539–543 (1983).
See also *14.3*, p. *216*; 22, p. 403.

M. D. Smolin, A. F. Andreeva, and V. A. Ogorodnik
Electroconduction and Thermoelectricity of Terbium Oxide Films
(b) Vol. **115**, No. 2, p. 587–592 (1983).
See also *14.3.1*, p. *229*; *15*, p. *261*; 22.6, p. 491.

D. Ihle and B. Lorenz
*Theory of the Verwey Transition in the Mixed Valence Compound Eu_3S_4
(b) Vol. **116**, No. 2, p. 539–546 (1983).
See also *6*, p. *72*; 18.2, p. 282; 22, p. 405.

J. Sinkkonen and P. Kuivalainen
*Many Electron Magnetic Polaron in Magnetic Semiconductors
(b) Vol. **116**, No. 2, p. K73–K76 (1983).
See also *18.2*, p. *282*; 22, p. 405.

P. S. Riseborough
*Quantum Tunneling between Inequivalent Locations. Small Polaron Motion
(b) Vol. **117**, No. 1, p. 381–394 (1983).

J. Zieliński and J. Kowalska
On the Virtual Bound State Model with Phonons
(b) Vol. **117**, No. 2, p. K105–K108 (1983).
See also *13.4*, p. *180*.

M. Saleh
Drift Mobility Measurements on Single Crystals of Orthorhombic Benzophenone Using the Electron Bombardment Technique
(a) Vol. **78**, No. 2, p. K137–K140 (1983).
See also *14.3*, p. *223*; 22.9, p. 529.

B. V. Egorov and B. B. Timofeev
*The Theory of Magnetic Polarons in Antiferromagnetic Semiconductors
(b) Vol. **119**, No. 1, p. 99–112 (1983).
See also *18.4*, p. *297*; 22, p. 411.

B. H. Bairamov, A. Heinrich, G. Irmer, V. V. Toporov, and E. Ziegler
Raman Study of the Phonon Halfwidths and the Phonon–Plasmon Coupling in ZnO
(b) Vol. **119**, No. 1, p. 227–234 (1983).
See also *6*, p. *77*; *20.1*, p. *325*; 22.4, p. 460.

M. Weger and O. Entin-Wohlman
*The Ground State Energy of Small Polaron Gas
(b) Vol. **120**, No. 1, p. 49–54 (1983).

K. Arisawa and M. Saitoh
*Second-Order Cumulant Expansion of the Free Energy and the Effective Masses of a Polaron at Finite Temperatures
(b) Vol. **120**, No. 1, p. 361–372 (1983).

M. Bhattacharyva, S. Mukhopadhyay, and T. K. Mitra
*Study of a Bound Surface Optical Polaron by a Non-Perturbative Method
(b) Vol. **120**, No. 2, p. 555–564 (1983).
See also *13.3*, p. *171*; 22.2.1, p. 447; 22.4.1, p. 465; 22.5.2, p. 485.

13.5.4 Magnons

Original Papers and Short Notes

M. Bancewicz and A. Stasch
*Theoretical Investigation of the Resonant Magnon–Phonon Interaction in FeF_2 in High Magnetic Field
(b) Vol. **115**, No. 2, p. 563–574 (1983).
See also *6*, p. *71*; 18, p. 273; 22, p. 404.

V. Ya. Antonchenko, A. S. Davydov, and A. V. Zolotariuk
Solitons and Proton Motion in Ice-Like Structures
(b) Vol. **115**, No. 2, p. 631–640 (1983).
See also *4*, p. *60*; *6*, p. *71*; 22.6, p. 491.

T. G. Petrova and E. S. Syrkin
*Effect of Uniaxial Long-Range Exchange Interaction on the Properties of Surface Spin Waves in a Heisenberg Ferromagnet
(b) Vol. **116**, No. 2, p. 475–481 (1983).
See also *18.2*, p. *282*.

K. M. Häussler, J. Brandmüller, L. Merten, H. Finsterhölzl, and A. Lehmeyer
*Observation of Dispersion on Magnetic Polaritons
(b) Vol. **117**, No. 1, p. 225–228 (1983).
See also 18, p. 274; *20.1*, p. *318*; 22.5, p. 474.

A. Jezierski
*Application of CPA to the Magnon Hamiltonian
(b) Vol. **118**, No. 1, p. 73–80 (1983).
See also 18.2, p. 284; 21, p. 355.

J. M. Wesselinowa
*Dynamic Behaviour of the s–d Model
(b) Vol. **120**, No. 2, p. 585–593 (1983).
See also *18.2*, p. *288*; 22.8, p. 510.

14. Electrical Properties. Transport Phenomena

Original Papers and Short Notes

M. WAGNER
*Non-Radiative Transitions: Fundamental Difficulties of the Adiabatic Base Approach
(b) Vol. **115**, No. 2, p. 427–435 (1983).
See also *13*, p. *147*.

M. L. KALRA
*Role of the Electron Scattering Processes in the Hubbard Model in the Insulator-to-Metal Transition
(b) Vol. **115**, No. 2, p. 575–585 (1983).
See also 22.6, p. 491.

W. E. HAGSTON
Quantum Theory of Tunnelling
(b) Vol. **116**, No. 2, p. K85–K89 (1983).
See also *13*, p. *148*.

V. K. ARORA and M. PRASAD
Quantum Transport in Quasi-One-Dimensional Systems
(b) Vol. **117**, No. 1, p. 127–140 (1983).
See also *13*, p. *148*.

K. BIEDRZYCKI, J. CHRZANOWSKI, and B. SUJAK
Some Electrical and Optical Effects Observed in Polycrystalline Ice at 200 K
(a) Vol. **77**, No. 1, p. K95–K98 (1983).
See also *1.2*, p. *16*; 20, p. 308; 22.6, p. 493.

L. SKÁLA
*On the $e^{-\alpha t}$ Damping of the Coherent Memory Functions
(b) Vol. **118**, No. 2, p. K89–K92 (1983).

U. BEHN
Quasi-Particle Motion in a Dichotomous Stochastic Potential: Comparison of a Decoupling Approach with the Dynamic CPA
(b) Vol. **118**, No. 2, p. K151–K153 (1983).
See also *13*, p. *150*.

I. P. ZVYAGIN
*The Temperature Dependence of the Conductivity near the Localization Threshold
(b) Vol. **120**, No. 2, p. 503–509 (1983).
See also *13*, p. *153*.

H. FEHSKE
Spin Fluctuations in Alloys with Random Transfer
(b) Vol. **120**, No. 2, p. 611–620 (1983).
See also *13*, p. *153*; 18.1, p. 279; 18.2, p. 288; 21.1, p. 377.

A. Kumar and P. Padhi
*Phonon Conductivity of Ice Single Crystals
(b) Vol. **120**, No. 2, p. 679–683 (1983).
See also *6*, p. *81*; 22.6, p. 497.

O. L. T. de Menezes and A. Troper
A Model for Inhomogeneous Mixed-Valence f-Semiconductors
(b) Vol. **120**, No. 2, p. 767–774 (1983).
See also *13*, p. *153*; 21.4, p. 393.

H. Stramska
Higher Harmonics Generated by Mobile Carriers in Semiconductors
(b) Vol. **120**, No. 2, p. 783–788 (1983).
See also *13*, p. *153*; 18, p. 276; 22, p. 415.

14.1 Metals. Semi-Metals

Review Article

K. H. Fischer
Spin Glasses (I)
(b) Vol. **116**, No. 2, p. 357–414 (1983).
See also 2, p. 43; 8, p. 85; 13, p. 146; *18.2*, p. *280*; *18.4*, p. *295*; 21.1, p. 363; 21.1.1, p. 379.

Original Papers and Short Notes

F. Goedsche, A. Richter, and G. Vojta
*Dynamical Susceptibility and Electrical Resistivity of Disordered Spin Systems
(b) Vol. **115**, No. 1, p. 209–218 (1983).
See also 13, p. 146; *18.2*, p. *280*; 21.1, p. 363.

T. A. Jenkins and S. Arajs
*Electrical Resistivity of Some Directionally-Solidified Aluminium-Based Eutectic Alloys
(a) Vol. **75**, No. 1, p. 73–82 (1983).
See also 21, p. 349; 21.1, p. 363; 21.1.1, p. 379.

U. Dillner
*Analysis of Resistivity, Hall Effect, and Magnetoresistance Measurements of Thin Bismuth Films by Means of an Anisotropic Two-Carrier Model
(a) Vol. **75**, No. 1, p. 227–233 (1983).
See also *14.3.1*, p. *229*; 21.7, p. 401.

W. Kleinn
*Erratum zu: Restwiderstandsbestimmung aus Strom–Spannungs-Kurven kurzer freitragender Elektrotransportproben epitaktischer Goldschichten
(Erratum to: Determination of the Residual Resistivity from Current–Voltage Characteristics of Short Support-Free Samples of Epitaxial Gold Films)
(a) Vol. **75**, No. 1, p. 347 (1983).
See also *1.5*, p. *35*; 21.6, p. 396.

F. Abd El-Salam, A. I. Eatah, and A. Tawfik
Effect of Li Loss on Some Physical Properties of Al–Li Alloy
(a) Vol. **75**, No. 2, p. 379–384 (1983).
See also *1.3*, p. *31*; 10.1, p. 100; 12.1, p. 137; 21, p. 351; 21.2, p. 387.

H. Gruber and E. Krautz
*Magnetoresistance and Conductivity in the Binary System Titanium–Oxygen (II): Semiconductive Titanium Oxides
(a) Vol. **75**, No. 2, p. 511–518 (1983).
See also 8, p. 86; *14.3*, p. *216*; 15, p. 261; 18, p. 273; 22.6, p. 491.

V. M. Pan, I. Ya. Dekhtyar, M. E. Osinovskii, S. E. Litvin, B. G. Nikitin, M. M. Nishenko, and M. P. Voronko
Thermodynamical and Structural Properties of Amorphous $Fe_{80}B_{20}$ and $Fe_{80}P_{13}C_7$ Alloys
(a) Vol. **76**, No. 1, p. 197–206 (1983).
See also *2*, p. *45*; 5, p. 65; *8*, p. 87; *18.2.1*, p. *290*; 21.1.1, p. 380.

P. Czarnecki and G. Fritzkowski
*Transverse Magnetoresistance of Amorphous $Fe_{40-x}Ni_{40}Cr_xB_{20}$ Alloys in Magnetic Fields up to 20 T
(a) Vol. **76**, No. 1, p. K93 – K96 (1983).
See also *2*, p. *46*; 18.2, p. 282; 21.1, p. 366; 21.1.1, p. 380.

J. Richter and G. Schubert
*The Electrical Resistivity of Short-Range Ordered Alloys and Its Temperature Dependence
(b) Vol. **116**, No. 2, p. 597 – 605 (1983).
See also *1.3*, p. *23*; 21, p. 352.

S. N. Kaul
*Two-Current Conduction Model and Anisotropic Magnetoresistance in Transition Metal–Metalloid Glasses
(b) Vol. **116**, No. 2, p. K99 – K104 (1983).
See also *2*, p. *46*; 21.1, p. 367; 21.1.1, p. 381.

E. Balanzat and M. Halbwachs
A Comparative Analysis of the Ordering Kinetic Laws for the Derivation of the Vacancy Mobility Parameters in Concentrated Alloys
(a) Vol. **76**, No. 2, p. 667 – 673 (1983).
See also 9, p. 94; *10.1*, p. *102*; 21.1, p. 367; 21.6, p. 397.

M. Sorohan
*Magnetoresistance of Polycrystalline Ni Wire under Hydrostatic Pressure
(b) Vol. **117**, No. 1, p. K1 – K4 (1983).
See also 18.2, p. 283; 21.1, p. 368.

V. V. Kechin and A. I. Likhter
*Hall Effect and Magnetoresistance of Antimony up to 5 GPa
(b) Vol. **117**, No. 1, p. K13 – K17 (1983).
See also 21.7, p. 401.

L. G. Khvostantsev and N. A. Nikolaev
*Electroresistance of Cerium at Phase Transitions under Hydrostatic Pressure up to 8.0 GPa. *P–T* Diagram of the Cerium Phase Transitions
(a) Vol. **77**, No. 1, p. 161 – 166 (1983).
See also *1.2*, p. *16*; 12.1, p. 138; 21.4, p. 391.

K. T-Y. Kung, B. X. Liu, and M-A. Nicolet
Study of Ni–Nb System by Ion Mixing
(a) Vol. **77**, No. 1, p. 355 – 359 (1983).
See also 1.5 p. 37; *2*, p. *48*; 21, p. 354; 21.1, p. 369.

M. A. Angadi and P. V. Ashrit
*The Effect of Deposition Parameters on the Electrical Properties of Thin Ytterbium Films
(a) Vol. **77**, No. 2, p. 685 – 692 (1983).
See also 21.4, p. 391.

Z. GIERAK, J.W. MOROŃ, and A. LEHR
Recovery of Structural Defects in Molybdenum
(a) Vol. **77**, No. 2, p. 775–783 (1983).
See also *10.1*, p. *103*; 21, p. 355.

H. RATAJCZAK, P. SLANČO, T. TIMA, A. ZENTKO, E. DEMKOVICSOVÁ, and M. KONČ
*Electrical Resistivity and Hall Effect in Thin Amorphous UFe Films
(a) Vol. **77**, No. 2, p. 785–791 (1983).
See also *2*, p. *49*; 18.2, p. 284; 21.1.1, p. 382.

H. K. PHAN and T. H. NGUYEN
*Measurement of the Electrothermal Nonlinearity and Thermal Relaxation Times of Thin Metal Films
(a) Vol. **77**, No. 2, p. K135–K138 (1983).
See also *8*, p. *88*; 21.1, p. 371.

W. KLEINN
*Determination of Residual Resistivity and Observation of Resistance Recovery from Current–Voltage Curves of Short Epitaxial Gold Thin Films
(a) Vol. **77**, No. 2, p. K149–K153 (1983).
See also *1.5*, p. *37*; 21.6, p. 398.

L. SCHWARTZ and A. PAJA
*On the Electrical Conductivity of Disordered Alloys in Muffin-Tin Model
(b) Vol. **118**, No. 1, p. 101–104 (1983).
See also 21, p. 356.

A. K. AHUJA, S. AULUCK, and S. K. JOSHI
Density of States and Optical Conductivity of Ferromagnetic Nickel
(b) Vol. **118**, No. 1, p. 105–111 (1983).
See also *13.1*, p. *159*; 21.1, p. 371.

B. BABIĆ, M. STOJIĆ, and J. KONSTANTINOVIĆ
*Electrical Properties of the Metastable $Mn_{1.11}Al_{0.89}$ Alloy
(a) Vol. **78**, No. 1, p. K77–K80 (1983).
See also *13*, p. *150*; 21, p. 357; 21.1, p. 372.

S. ANZAI, S. OHTA, A. YOSHINO, J. NISHIO, and M. HATORI
*Charge-Density-Wave-Like Transition in V_3Te_4
(b) Vol. **118**, No. 2, p. K99–K102 (1983).
See also 15, p. 262; *18.1*, p. *279*; 22, p. 410.

P. MUTH
*On the Resistivity Behaviour of Some Weak-Ferromagnetic Nickel-Based Alloys
(b) Vol. **118**, No. 2, p. K117–K120 (1983).
See also *13.1*, p. *161*; 21.1, p. 373.

P. MUTH
*A Critical Concentration of Ni–Mn Alloys at about 6 at% Mn
(b) Vol. **118**, No. 2, p. K137–K139 (1983).
See also *13.1*, p. *162*; 21.1, p. 373.

A. R. Ali and G. Said
On the Recovery and Annealing Kinetics of Cold-Worked Ni–4.5 at% Cu Alloy
(a) Vol. **78**, No. 2, p. 463–467 (1983).
See also *10.1*, p. *104*; 21.1, p. 373.

I. Lupsa and E. Burzo
*On the Resistive Anomaly in NiCuSi and NiCuZn Alloys near the Curie Temperature
(a) Vol. **78**, No. 2, p. K155–K158 (1983).
See also 18.2, p. 285; 21.1, p. 373.

S. Mahajan and S. Prakash
Electronic Structure of Impurities in Dilute Alloys of Al
(b) Vol. **119**, No. 1, p. 381–389 (1983).
See also *13.4*, p. *185*; 21, p. 358.

E. Canessa and J. Vignolo
*Electrical Resistivities of the Liquid Alkali Metals
(b) Vol. **119**, No. 1, p. 429–436 (1983).
See also 2, p. 51; 21.2, p. 388.

E. Canessa and J. Vignolo
*The Resistivity of Liquid Mercury at 150 °C
(b) Vol. **119**, No. 1, p. K83–K85 (1983).
See also *2*, p. *51*; 21, p. 358.

P. M. Boerrigter, A. Lodder, and J. Molenaar
*Calculated Residual Resistivities in Al and Cu Alloys
(b) Vol. **119**, No. 1, p. K91–K96 (1983).
See also 21, p. 358; 21.1, p. 374.

L. V. Nomerovannaya and V. A. Rassokhin
*Anomalies in the Optical and Electrical Properties of Cr–Al Alloys
(a) Vol. **79**, No. 1, p. 87–93 (1983).
See also 20.1, p. 326; 21, p. 359; 21.1, p. 374.

O. Schneeweiss, T. Zemčík, T. Žák, and S. Mager
Atomic Structure and Magnetic Properties of the Pseudobinary Alloys Fe_3(Al, Si)
(a) Vol. **79**, No. 1, p. 125–129 (1983).
See also *1.3*, p. *26*; *5*, p. *67*; 18.2, p. 286; 21.1.1, p. 384.

Y. D. Yao, S. Arajs and E. E. Anderson
*Deviations from Matthiessen's Rule in the Nickel–Chromium System
(a) Vol. **79**, No. 1, p. 327–330 (1983).
See also *13.4*, p. *186*; 18.2, p. 286; 21.1, p. 375.

M. Tamaki, H. Matsui, G. Matsumoto, and T. Kirihara
*Critical Composition of Antiferromagnetic Ordering in UN_xC_{1-x}
(a) Vol. **79**, No. 1, p. K93–K98 (1983).
See also *18.4*, p. *297*; 21.5, p. 395.

S. M. Rafique, N. R. Mitra, and P. I. Srivastava
*On Choosing a Set of Eigenvalues and Wave Functions in Harrison's First Principle Method
(b) Vol. **119**, No. 2, p. K113–K116 (1983).
See also *13.1*, p. *164*; 21, p. 359.

G. A. Scholz and R. F. Frindt
*Electrical Resistivity of Silver Intercalated 2H-TaS_2
(a) Vol. **79**, No. 2, p. 483–488 (1983).
See also *14.3*, p. *225*; 22, p. 412.

D. V. Gitsu and F. M. Muntyanu
*Commutation Effect of Transport Phenomena in Antimony–Bismuth Alloys in Quantizing Magnetic Fields
(b) Vol. **120**, No. 1, p. K45–K49 (1983).
See also *15*, p. *263*; 21.7, p. 402.

P. Meisterle and W. Pfeiler
*Short-Range Ordering in the Matrix of a Two-Phase AgAl Alloy
(a) Vol. **80**, No. 1, p. K37–K40 (1983).
See also *1.3*, p. *27*; 21, p. 361.

W. Schumacher
On the Influence of the Magnetic Ordering on Thermodynamic and Electrical Properties in the Hubbard Model
(b) Vol. **120**, No. 2, p. 621–629 (1983).
See also *8*, p. *92*; 18.4, p. 298.

F. Goedsche, R. Richter, H. Wegener, and G. Röpke
*Correlation Function Method for the Thermoelectric Power of Metals
(b) Vol. **120**, No. 2, p. 631–644 (1983).
See also 13, p. 153; *15*, p. *263*; 21, p. 361.

R. Schumann
Temperature Dependence of the Thermopower of the Periodic Anderson Model
(b) Vol. **120**, No. 2, p. K183–K188 (1983).
See also 8, p. 92; *15*, p. *263*; 21, p. 361.

V. Christoph and A. L. Kuzemskii
*The Influence of the Electron–Phonon Interaction on the Electroconductivity of Disordered Metallic Alloys
(b) Vol. **120**, No. 2, p. K219–K224 (1983).

S. V. Tsivinskii
*The Theory of Electric Resistance of Pure Metals and Metals with a Given Concentration of Impurities, Vacancies, and Dislocations
(a) Vol. **80**, No. 2, p. 439–446 (1983).
See also 10.1, p. 107; 21, p. 362; 21.1, p. 377; 21.1.1, p. 385; 21.2, p. 388; 21.6, p. 399.

P. Svoboda and P. Vašek
*The Generalized Thomson-Voigt Relations for Transport Coefficients of Bismuth
(a) Vol. **80**, No. 2, p. 567–571 (1983).
See also *15*, p. *263*; 21.7, p. 402.

H. Kido, T. Hoshikawa, M. Shimada, and M. Koizumi
Preparation and Magnetic and Electrical Properties of RZnSi
(a) Vol. **80**, No. 2, p. 601 – 605 (1983).
See also *1.3*, p. *27*; *18.1*, p. *279*; 21.1, p. 378; 21.4, p. 394.

A. San Martin, K. Kleinstück, N. H. Quyen, and P. Paufler
*Influence of a Direct Current and a Temperature Gradient on the Creep Rate in V_3Si
(a) Vol. **80**, No. 2, p. K171 – K174 (1983).
See also *12.1*, p. *140*; 21.1, p. 378.

14.2 Superconductivity. Superconducting Materials and Devices

Original Papers and Short Notes

E. Krüger
*Exchange Interaction in Superconducting Metals
(b) Vol. **115**, No. 1, p. 181–185 (1983).
See also 13.1, p. 155; 21, p. 349.

J. Szymaszek, J. J. Wnuk, and C. Sułkowski
*Superconductiong Properties of $LaMo_6S_8$ Thin Films
(a) Vol. **75**, No. 1, p. K9–K13 (1983).

B. Głowacky
*X-Ray Diffraction Pattern and Morphology of Superconducting Nb_3Sn Layer Grown on a Polycrystalline Nb Substrate of (111) Main Texture
(a) Vol. **75**, No. 1, p. K103–K105 (1983).
See also *1.5*, p. *35*.

I. M. Chapnik
*Penetration of Electromagnetic Fields through Thin Foils of Niobium and Indium
(a) Vol. **75**, No. 2, p. K157–K160 (1983).
See also 21, p. 351.

M. Surma
*A New Semiempirical Formula for Superconducting Transition Temperatures of Metals and Alloys
(b) Vol. **116**, No. 2, p. 465–474 (1983).

W. Richter and F. Dettmann
*Flux Quantization in a Superconducting Closed Loop with more than Two Josephson Junctions
(b) Vol. **116**, No. 2, p. K95–K98 (1983).

A. I. Eatah, F. Abd El-Salam, and A. Tawfik
*Effect of Argon on the Superconducting Transition Temperature of Lead–Argon Films
(a) Vol. **76**, No. 2, p. 705–708 (1983).

V. E. Antonov, T. E. Antonova, I. T. Belash, E. G. Ponyatovskii, V. I. Rashupkin, and V. G. Thiessen
The Pd–Ni–H System at High Pressure
(a) Vol. **77**, No. 1, p. 71–79 (1983).
See also *1.3*, p. *23*; 18.2, p. 283; 21, p. 354; 21.1, p. 369.

S. Heinemann
*On the Stationary Behaviour of Magnetic Flux Embraced by Asymmetric Two-Junction SQUIDs
(a) Vol. **77**, No. 1, p. 153–159 (1983).

L. A. SHEBANOV, V. YA. FRITSBERG, and A. P. GAEVSKIS
*Crystallographic Properties and Superconductivity of Solid Solutions of the $BaBi_xPb_{1-x}O_3$ System
(a) Vol. 77, No. 1, p. 369–373 (1983).
See also *1.2*, p. *16*; 22.8.1, p. 514.

V. E. ANTONOV, T. E. ANTONOVA, I. T. BELASH, E. G. PONYATOVSKII, and V. I. RASHUPKIN
*Superconductivity of Pd–Au–H Solid Solution
(a) Vol. 77, No. 1, p. K23–K27 (1983).
See also 1.3, p. 24; 21.6, p. 397.

S. L. DRECHSLER and G. M. VUJIČIĆ
*On the Influence of Tunneling States on the Transition Temperature of Metallic Glassy Superconductors
(b) Vol. 117, No. 2, p. 569–574 (1983).
See also 2, p. 48; 21.1, p. 370.

A. S. ALEKSANDROV, V. F. ELESIN, and M. P. KAZEKO
*The Influence of Defects on the Low-Temperature Resistivity of Intermetallic Compounds with A-15 Structure
(a) Vol. 77, No. 2, p. 725–732 (1983).
See also 13.1, p. 159; 13.4, p. 181.

YU. I. VESNIN, YU. F. ELTSEV, S. P. ZAKOVRYASHIN, and M. A. STARIKOV
*Superconductivity and Decay Phenomena of Nb_3Al–Nb_3Ge Solid Solutions
(a) Vol. 77, No. 2, p. 759–764 (1983).
See also *1.3*, p. *24*.

S. MACIEJEWSKI
*Cooper Pair Condensation Energy Dependence on Electron Structure
(b) Vol. 118, No. 1, p. K3–K5 (1983).
See also *13.1*, p. *160*; 21, p. 356.

V. E. ANTONOV, T. E. ANTONOVA, I. T. BELASH, E. G. PONYATOVSKII, and V. I. RASHUPKIN
*The Pd–Pt–H System: Phase Transformations at High Pressure and Superconductivity
(a) Vol. 78, No. 1, p. 137–146 (1983).
See also *1.2*, p. *17*; 21, p. 356; 21.1, p. 371.

CH. SEIDEL and V. N. PRIGODIN
The Impurity Effect on the Phase Transitions in Quasi-One-Dimensional Conductors
(b) Vol. 118, No. 2, p. 713–724 (1983).
See also *1.2*, p. *18*; *10.2*, p. *115*; 22.9, p. 529.

J. SOSNOWSKI
*On the Flux Jumps in Hard Superconductors
(a) Vol. 79, No. 1, p. 207–212 (1983).
See also 21, p. 359.

V. G. CHUDINOV, V. P. GOGOLIN, B. N. GOSHCHITSKII, and V. I. PROTASOV
Simulation of Collision Cascades in Intermetallic V_2Hf Compounds
(a) Vol. 79, No. 1, p. K57–K60 (1983).
See also *11*, p. *132*; 21, p. 359; 21.1, p. 375.

G. J. Chashoo and G. Dharmadurai
*Role of Thermal Transport in the Non-Equilibrium Behaviour of a Cylindrical Ferromagnetic Superconductor
(b) Vol. **120**, No. 1, p. 397–400 (1983).
See also **21**, p. 360.

Ch. Seidel and V. N. Prigodin
Three-Dimensional CDW Ordering and Impurity Effect in Quasi-One-Dimensional Conductors
(b) Vol. **120**, No. 1, p. K73–K77 (1983).
See also *1.2*, p. *19*; *10.1*, p. *106*.

W. D. Schmidt and S. Heinemann
*On the Storage and Detection of Multiple Flux Quantum States in Thin Film Double SQUIDs
(a) Vol. **80**, No. 1, p. 141–146 (1983).

B. Głowacki and J. Chojcan
Studies on the Condition of Optaining a Single Phase. Superconducting Nb_3Sn Layer by Mössbauer Spectroscopy
(a) Vol. **80**, No. 1, p. K93–K96 (1983).
See also *5*, p. *68*; 9, p. 96; **21**, p. 361.

A. Beckmann, E. Kolley, and W. Kolley
*T_c Formula for Strong-Coupling Superconductivity in Random Narrow-Band Alloys
(b) Vol. **120**, No. 2, p. 577–583 (1983).

M. Schossmann and E. Schachinger
*Thermodynamics of Superconducting Alloys with Localized States in the Energy Gap
(b) Vol. **120**, No. 2, p. 723–734 (1983).

14.3 Semiconductors

Original Papers and Short Notes

M. Singh, J. Leotin, and P. R. Wallace
*Theory of the Diffusivity–Mobility Ratio in Cadmium Arsenide
(b) Vol. **115**, No. 1, p. 105 – 113 (1983).
See also 18, p. 273; 22.3, p. 456.

M. Stutzmann, J. Stuke, and H. Dersch
Electron Spin Resonance of Doped Glow-Discharge Amorphous Germanium
(b) Vol. **115**, No. 1, p. 141 – 151 (1983).
See also 2, p. 43; *13.4*, p. *172*; *19*, p. *299*; 22.1.1, p. 418; 22.1.2, p. 423.

V. M. Ivashchenko, V. V. Mitin, and N. A. Zakhleniuk
*Galvanomagnetic Effects in Semiconductors of p-Ge Type
(b) Vol. **115**, No. 1 p. 245 – 254 (1983).
See also 22.1.1, p. 418.

B. R. Sethi, P. K. Goyal, O. P. Sharma, and P. C. Mathur
*High-Field DC Conductivity in n-Type CdS Single Crystals Annealed in Molten Indium
(a) Vol. **75**, No. 1, p. 83 – 89 (1983).
See also 13.4, p. 173; 22.4.1, p. 462.

R. Stumpe, D. Wagner, and D. Bäuerle
*Influence of Bulk and Interface Properties on the Electric Transport in ABO_3 Perovskites
(a) Vol. **75**, No. 1, p. 143 – 154 (1983).
See also 22.8.1, p. 512.

H. Lemke
Ion Pairs between Chromium and Shallow Acceptors in Silicon
(a) Vol. **75**, No. 1, p. K49 – K51 (1983).
See also *13.4*, p. *174*; 22.1.2, p. 425.

S. M. Wasim
*On the Electron Effective Mass in n-$CuInTe_2$
(a) Vol. **75**, No. 1, p. K69 – K72 (1983).
See also *13.1*, p. *156*; 22.8, p. 502.

V. N. Brudnyi, A. I. Potapov, and Yu. V. Rud
*Electrical Properties of H^+-Irradiated p–$ZnGeAs_2$
(a) Vol. **75**, No. 1, p. K73 – K76 (1983).
See also *11*, p. *123*; 22.8, p. 502.

K. B. Kadyrakunov, E. V. Nidaev, I. E. Tyshchenko, and L. S. Smirnov
*Decomposition of Supersaturated Antimony Solutions in Silicon Created by Pulsed Annealing of Ion-Doped Layers
(a) Vol. **75**, No. 1, p. K107 – K109 (1983).
See also *11*, p. *123*; 22.1.2, p. 425.

L. Kończewicz and W. Trzeciakowski
Highly Localized Donor States in InSb and Their Pressure Dependence
(b) Vol. **115**, No. 2, p. 359–367 (1983).
See also 12.2, p. 141; *13.4*, p. *174*; 22.2.3, p. 452.

K. Takeda, Y. Maeda, E. Ohta, M. Sakata, G. Kido, and N. Miura
Study of Energy Band Parameters in p-Type $Ge_{0.9}Si_{0.1}$ Alloys
(b) Vol. **115**, No. 2, p. 369–379 (1983).
See also *13.1*, p. *156*; 20.1, p. 314; 22.1.1, p. 418; 22.1.2, p. 425.

M. Mocker and O. Ziep
*Intrinsic Recombination in Dependence on Doping Concentration and Excitation Level. Application to Lead Chalcogenides
(b) Vol. **115**, No. 2, p. 415–425 (1983).
See also *13.1*, p. *156*; 20.3, p. 335; 22.8, p. 503.

S. Eränen and J. Sinkkonen
*Conduction Electron Self-Energy Effects in Ferromagnetic Semiconductors
(b) Vol. **115**, No. 2, p. 519–527 (1983).
See also *13.1*, p. *156*; 22, p. 403.

F. M. Peeters and J. T. Devreese
*The $\frac{3}{2}$ kT Problem in the Low-Temperature Polaron Mobility Theories
(b) Vol. **115**, No. 2, p. 539–543 (1983).
See also *13.5.3*, p. *201*; 22, p. 403.

V. K. Arora
*Nonohmic Conductivity for Acoustic-Phonon Scattering
(b) Vol. **115**, No. 2, p. K107–K112 (1983).
See also *6*, p. *71*; 22, p. 404.

A. P. Tyutnev, G. S. Mingaleev, A. V. Vannikov, A. F. Dunaev, V. S. Saenko, and V. S. Likhovidov
*Kinetics of Radiation-Induced Conductivity in Some Polymers
(a) Vol. **75**, No. 2, p. 385–392 (1983).
See also *11*, p. *123*; 22.9, p. 525.

J. Reichel
*The Distinction of Different Kinds of Thermally Induced Defects in Silicon
(a) Vol. **75**, No. 2, p. 459–464 (1983).
See also *10.2*, p. *109*; 22.1.2, p. 425.

H. Lemke
Properties of Some Impurity Complexes of Gold in Silicon
(a) Vol. **75**, No. 2, p. 473–482 (1983).
See also *13.4*, p. *175*; 22.1.2, p. 425.

H. Gruber and E. Krautz
*Magnetoresistance and Conductivity in the Binary System Titanium–Oxygen (II): Semiconductive Titanium Oxides
(a) Vol. **75**, No. 2, p. 511–518 (1983).
See also 8, p. 86; *14.1*, p. *206*; 15, p. 261; 18, p. 273; 22.6, p. 491.

A. N. Anagnostopoulos
*Influence of Composition Faults on the AC Conductivity of $ZnIn_2S_4$ (III)
(a) Vol. **75**, No. 2, p. 595–599 (1983).
See also 22.8, p. 503.

G. J. Ashwell, J. G. Allen, E. P. Goodings, D. A. Kennedy, and I. W. Nowell
Crystal Structure of (DEPA) $(TCNQ)_4$ and Stoichiometric Variation of Some Related TCNQ Salts
(a) Vol. **75**, No. 2, p. 663–667 (1983).
See also *1.1*, p. *11*; 22.9, p. 526.

O. Boehnke, E. Nowak, P. Paufler, and H. Neumann
*On a Slip-Induced Anisotropy of the Electrical Resistance of GaP
(a) Vol. **75**, No. 2, p. K137–K139 (1983).
See also *10.2*, p. *110*; 22.2.2, p. 448.

H. Neumann, R. D. Tomlinson, N. Avgerinos, and E. Nowak
*Electrical Properties of $CuInSe_2$ Single Crystals Grown by the Vertical Bridgman Technique
(a) Vol. **75**, No. 2, p. K199–K203 (1983).
See also 13.4, p. 175; 22.8, p. 504.

L. G. Khvostantsev, and V. A. Sidorov
Thermoelectric Properties of GeTe at High Hydrostatic Pressure up to 8.5 GPa and Its Valence-Band Structure
(b) Vol. **116**, No. 1, p. 83–89 (1983).
See also 13.1, p. 157; *15*, p. *261*; 22.7, p. 499.

M. Mocker and M. Beiler
*Theory of Intrinsic Recombination at Zero Temperature in Small Gap Semiconductors. Estimations for PbSSe
(b) Vol. **116**, No. 1, p. 205–215 (1983).
See also *13.4*, p. *176*; 20.3, p. 336; 22.8, p. 504.

H. Lemke
Energy Levels and Binding Energies of Ion Pairs in Silicon
(a) Vol. **76**, No. 1, p. 223–234 (1983).
See also *13.4*, p. *176*; 22.1.2, p. 427.

M. M. Hafiz, A. A. Ammar, A. I. Al-Adl, and Aboutaleb Mohamed
*Effect of Heat Treatment on the Conduction and Structure of Ge–Se–Te Amorphous Alloys
(a) Vol. **76**, No. 1, p. 319–325 (1983).
See also 1.2, p. 15; *2*, p. *46*; 22.8, p. 504.

D. V. Gitsu, I. N. Grincheshen, and N. S. Popovich
*The Electric and Photoelectric Properties of $TlSbSe_2$ Crystals in the Region of Impurity Conduction
(a) Vol. **76**, No. 1, p. K5–K7 (1983).
See also 13.4, p. 176; *16*, p. *265*; 22.8, p. 504.

Y. Watanabe, J. Takeuchi, and M. Oka
*Microwave Transport Properties around the Structural Phase Transition of MnSe
(a) Vol. **76**, No. 1, p. K13 – K15 (1983).
See also *1.2*, p. *15*; 22.4.2, p. 466.

N. Kamegashira and Y. Miyazaki
*Jahn-Teller Transition in $NdMnO_3$
(a) Vol. **76**, No. 1, p. K39 – K42 (1983).
See also *1.2*, p. *15*; 22.8.1, p. 513.

P. Lošťák, J. Horák, and L. Koudelka
Some Physical Properties and Point Defects in $Bi_{2-x}In_xTe_3$ Mixed Crystals
(a) Vol. **76**, No. 1, p. K71 – K75 (1983).
See also *1.3*, p. *22*; *20.1*, p. *316*; 22.7.1, p. 501.

M. Nagabhooshanam and V. Hari Babu
*Conductivity and Hall Coefficient Studies in Plastically Deformed CdTe Single Crystals
(a) Vol. **76**, No. 1, p. K89 – K92 (1983).
See also *10.2*, p. *111*; 22.4.3, p. 468.

P. I. Baranskii, V. V. Kolomoets, and S. S. Korolyuk
*The Non-Parabolicity of the n-Si Conduction Band Caused by Elastic Deformation along the [111] Direction
(b) Vol. **116**, No. 2, p. K109 – K112 (1983).
See also *13.1*, p. *157*; 22.1.2, p. 428.

B. Gillot and F. Jemmali
*Dependence of Electrical Properties in Iron–Cobalt, Iron–Zinc Ferrites near Stoichiometry on Firing Temperature and Atmosphere
(a) Vol. **76**, No. 2, p. 601 – 608 (1983).
See also *15*, p. *261*; 22.8.2, p. 520.

C. Sanchez, R. Morineau, and J. Livage
*Electrical Conductivity of Amorphous V_2O_5
(a) Vol. **76**, No. 2, p. 661 – 666 (1983).
See also *2*, p. *47*; 22.6, p. 493.

B. Pödör, J. Pfeifer, L. Csontos, N. Nádor, and F. Deák
*Electrical Properties of Donors in Gallium Phosphide
(a) Vol. **76**, No. 2, p. 695 – 704 (1983).
See also *13.4*, p. *177*; 22.2.2, p. 449.

T. G. Abdel-Malik
*Effect of Oxygen on the Electrical Properties of β-Zinc Phthalocyanine Single Crystals
(a) Vol. **76**, No. 2, p. K159 – K162 (1983).
See also *13.4*, p. *177*; 22.9, p. 527.

N. A. Poklonski and V. F. Stelmakh
*Screening of Electrostatic Fields in Crystalline Semiconductors by Electrons Hopping over Defects
(b) Vol. **117**, No. 1, p. 93 – 99 (1983).
See also *13.4*, p. *178*; 22, p. 406.

J. Gryko
*Optical Conductivity in the Impurity Band of Doped Semiconductors
(b) Vol. **117**, No. 1, p. 313–322 (1983).
See also *13.4*, p. *178*; 20.1, p. 318; 22.1.1, p. 420.

M. Stordeur and H. T. Langhammer
Anisotropy of Transport Properties Due to Direction-Dependent Band Nonparabolicity
(b) Vol. **117**, No. 1, p. 329–333 (1983).
See also *13.1*, p. *158*; 15, p. 261; 22, p. 406.

K. D. Chaudhuri, A. Malik, S. Kukreti, and P. C. Mathur
*Impurity Band Conduction in Single Crystals of p-Type GaSb
(b) Vol. **117**, No. 1, p. 407–416 (1983).
See also *13.4*, p. *178*; 22.2.3, p. 452.

K. van Cong, S. Brunet, S. Charar, S. Benet, C. Delseny, and J. C. Martin
*Effective Electric Field in the n-Type Moderately Doped Region of Silicon Devices at 300 K
(b) Vol. **117**, No. 1, p. K23–K26 (1983).
See also 22.1.2, p. 429.

L. V. Vassilev and G. Ch. Christakudis
*On the Calculation of Transport Integrals in Lead Chalcogenides
(b) Vol. **117**, No. 1, p. K53–K56 (1983).
See also 15, p. 261; 22, p. 407.

W. Siegel, E. Ziegler, and G. Kühnel
*Impurity Parameters of Undoped and Zn-Doped p-Type GaP with Two Acceptor Levels Determined by Hall Measurements
(a) Vol. **77**, No. 1, p. 167–173 (1983).
See also *13.4*, p. *178*; 22.2.2, p. 449.

J. Betko and K. Měřínský
*Geometrical Magnetoresistance Effect in Semiconductors with Mixed Conductivity. Determination of the Electron Hall Mobility in Semi-Insulating GaAs
(a) Vol. **77**, No. 1, p. 331–337 (1983).
See also 22.2.1, p. 444.

B. Gillot and F. Jemmali
*Electrical Conduction Mechanism in Al-, Cr-, and Zn-Substituted Magnetites over the Temperature Range from 77 to 1000 K
(a) Vol. **77**, No. 1, p. 339–346 (1983).
See also 22.8.2, p. 521.

V. N. Brudnyi, A. I. Potapov, Yu. V. Rud, and M. Serginov
*Electrical Properties of Electron-Irradiated GaAs and $ZnSiAs_2$ Solid Solutions
(a) Vol. **77**, No. 1, p. K33–K36 (1983).
See also *11*, p. *128*; 13.4, p. 179; 22.2.1, p. 444; 22.8, p. 505.

P. V. Reddy and T. S. Rao
*A Study of the Semiconducting Properties of Mixed Mn–Mg Ferrites
(a) Vol. **77**, No. 1, p. K63–K67 (1983).
See also *15*, p. *261*; 22.8.1, p. 514.

B. Vengalis and K. Valatska
Phase Segregation in Tellurium-Saturated GeTe
(b) Vol. **117**, No. 2, p. 471–476 (1983).
See also *1.2*, p. *16*; 8, p. 88; 13.4, p. 179; 20.1, p. 319; 22.3, p. 457.

Z. Dziuba
*Impurity Band Conduction in HgTe (I): Experimental Data
(b) Vol. **117**, No. 2, p. 531–540 (1983).
See also *13.4*, p. *179*; 22.4.3, p. 468.

H. van Cong, S. Charar, S. Brunet, and J. C. Martin
*Transport Parameters in the n-Type Moderately Doped Region of Silicon Devices at 300 K
(b) Vol. **117**, No. 2, p. K61–K65 (1983).
See also 22.1.2, p. 431.

A. K. Sarychev and A. P. Vinogradov
*Effective Medium Theory for the Magnetoconductivity Tensor of Disordered Materials
(b) Vol. **117**, No. 2, p. K113–K118 (1983).
See also 22.3, p. 457.

O. A. Golikova and S. Samatov
*Conduction Mechanism of Some "Quasi-Amorphous" Semiconductors on the Basis of Boron
(a) Vol. **77**, No. 2, p. 449–453 (1983).
See also *13.4*, p. *180*; 15, p. 262; 22, p. 408.

K. Ram Kumar and M. Satyam
*Minority Carrier Lifetime in Polycrystalline Semiconductors
(a) Vol. **77**, No. 2, p. 467–470 (1983).
See also 22, p. 408.

Z. Korczak and M. Subotowicz
*Structural and Galvanomagnetic Properties of $Pb_{1-x}Mn_xTe$ Single Crystals Grown by the Bridgman-Stockbarger Method
(a) Vol. **77**, No. 2, p. 497–503 (1983).
See also *1.3*, p. *24*; 3, p. 58; 22, p. 408.

D. Panknin, E. Wieser, R. Klabes, and H. Syhre
Dose Dependence of the Flash Lamp Annealing of Arsenic-Implanted Silicon
(a) Vol. **77**, No. 2, p. 553–559 (1983).
See also *11*, p. *128*; 22.1.2, p. 431.

P. Gaworzewski and K. Schmalz
*Oxygen-Related Donors Formed at 600 °C in Silicon in Dependence on Oxygen and Carbon Content
(a) Vol. **77**, No. 2, p. 571–582 (1983).
See also *13.4*, p. *180*; 22.1.2, p. 431.

A. N. ANAGNOSTOPOULOS, C. MANOLIKAS, and D. PAPADOPOULOS
*Comparison of the Structure and the Electric Properties of $ZnIn_2S_4$(III)- and $CdInGaS_4$-Layered Crystals
(a) Vol. **77**, No. 2, p. 595–601 (1983).
See also *1*, p. *10*; 22.8, p. 506.

J. ZHANG and J. C. THUILLIER
*The Effect of Annealing Temperature on the Electron Concentration of $Hg_{0.3}Cd_{0.7}Te$
(a) Vol. **77**, No. 2, p. 649–653 (1983).
See also 22.4.4, p. 471.

B. PISTOULET, D. COQUILLAT, J. C. TEDENAC, G. BRUN, and M. MAURIN
Hole Effective Mass and Impurity Levels in Undoped AgTlTe. Optimum Thermoelectric Figure of Merit
(a) Vol. **77**, No. 2, p. 669–677 (1983).
See also *13.1*, p. *159*; *13.4*, p. *181*; 15, p. 262; 22.8, p. 506.

O. BREITENSTEIN and T. WOSINSKI
Scanning-DLTS Investigation of the EL 2 Level in Plastically Deformed GaAs
(a) Vol. **77**, No. 2, p. K107–K110 (1983).
See also *10.2*, p. *113*; *13.4*, p. *181*; 22.2.1, p. 444.

T. DZIUBA
*Impurity Band Conduction in HgTe (II): Theoretical Analysis
(b) Vol. **118**, No. 1, p. 319–326 (1983).
See also *13.4*, p. *182*; 22.4.3, p. 468.

P. M. TOMCHUK, V. A. SHENDEROVSKII, P. N. GORLEY, and N. P. GAVALESHKO
*Hall Coefficient and Magnetoresistance of Semiconductors with Anisotropic Energy Dispersion
(b) Vol. **118**, No. 1, p. 433–439 (1983).
See also 13.1, p. 160; 13.4, p. 182; 22.8, p. 507.

M. K. AGARWAL, P. D. PATEL, and O. VIJAYAN
*Electrical Studies on $(Mo/W)Se_2$ Single Crystals (I): Electrical Resistivity
(a) Vol. **78**, No. 1, p. 133–136 (1983).
See also 22.7, p. 499.

H. HEDLER and G. GÖTZ
Incorporation of Sb in Si during CW-Laser Stimulated Solid Phase Epitaxy
(a) Vol. **78**, No. 1, p. 237–241 (1983).
See also *1.2*, p. *17*; 11, p. 129; 22.1.2, p. 432.

B. N. MUKASHEV, ŻH. A. KUSAINOV, K. KH. NUSUPOV, and S. ZH. TOKMOLDIN
*Electrical Properties of Shallow Implanted Layers in Silicon
(a) Vol. **78**, No. 1, p. K19–K22 (1983).
See also *11*, p. *129*; 22.1.2, p. 433.

E. W. Kreutz
*Magnetoresistance Thickness Dependence of n-InSb
(a) Vol. 78, No. 1, p. K71 – K75 (1983).
See also 22.2.3, p. 453.

K. O. Kechechyan and A. A. Kirakosyan
*Magneto-Resistance of a Semiconductor with the Dislocation Wall in the Quantum Limit
(b) Vol. 118, No. 2, p. 521 – 524 (1983).
See also *10.2*, p. *115*; 22, p. 409.

A. N. Chakravarti, K. P. Ghatak, K. K. Ghosm, and H. M. Mukherjee
*Effect of Crystal-Field Splitting on the Einstein Relation in Tetragonal Semiconductors
(b) Vol. 118, No. 2, p. 843 – 852 (1983).
See also 13.1, p. 161; 22.3, p. 457.

P. Gosar
*Many-Electron Hopping
(b) Vol. 118, No. 2, p. 853 – 860 (1983).
See also *13.4*, p. *184*.

D. Genzow
*Transverse Dember Effect in Lead Chalcogenides under Linearly Polarized Light Excitation
(b) Vol. 118, No. 2, p. K159 – K162 (1983).
See also 16, p. 266; 22, p. 410.

S. M. Wasim, G. Marcano, and G. Sánchez Pérez
*Electrical Properties of $CuGaTe_2$
(a) Vol. 78, No. 2, p. 423 – 430 (1983).
See also *13.4*, p. *184*; 22.8, p. 508.

A. A. El-Sharkawy, M. S. Zaghloul, and A. M. Sanad
Study of Thermal Properties, Electrical Conductivity, and the Crystallization Process of the Glassy Alloy $SeGe_{20}$
(a) Vol. 78, No. 2, p. 457 – 462 (1983).
See also *3*, p. *59*; 8, p. 89; 22.1.1, p. 421; 22.1.3, p. 440.

M. K. Agarwal, P. D. Patel, and O. Vijayan
*Electrical Studies on $(Mo/W)Se_2$ Single Crystals (II): Temperature Dependence of Hall Effect
(a) Vol. 78, No. 2, p. 469 – 473 (1983).
See also 22.7, p. 499.

A. P. Tyutnev, V. S. Saenko, A. V. Vannikov, and G. S. Mingaleev
*Radiation-Induced Conductivity in Polyethylene
(a) Vol. 78, No. 2, p. 689 – 696 (1983).
See also 11, p. 130; 22.9, p. 529.

P. I. Baranskii, V. M. Babich, N. P. Baran, Yu. P. Dotsenko, V. B. Kovalchuk, and V. A. Shershel
*Investigation of Formation Conditions of Thermal Donors-I and -II in Oxygen-Containing n-Type Silicon within the Temperature Range 400 to 800 °C
(a) Vol. **78**, No. 2, p. 733 – 739 (1983).
See also *13.4*, p. *184*; 19, p. 303; 22.1.2, p. 433.

P. Venugopal Reddy, R. Satyanarayana, and T. Seshagiri Rao
*Electrical Conduction in Nickel Ferrite
(a) Vol. **78**, No. 2, p. K109 – K111 (1983).
See also *15*, p. *262*; 22.8.2, p. 523.

M. Saleh
*Drift Mobility Measurements on Single Crystals of Orthorhombic Benzophenone Using the Electron Bombardment Technique
(a) Vol. **78**, No. 2, p. K137 – K140 (1983).
See also 13.5.3, p. 201; 22.9, p. 529.

K. Schmalz and P. Gaworzewski
On the Effect of Ambients on the Formation of Oxygen-Related Donors in Cs-Si
(a) Vol. **78**, No. 2, p. K141 – K145 (1983).
See also *10.2*, p. *117*; 13.4, p. 184; 22.1.2, p. 434.

B. G. Tagiev, G. M. Niftiev, G. K. Aslanov, I. M. Askerov, and F. N. Sultanov
*Effect of Ytterbium on the Electrophysical Properties of Ga_2Se_3 Single Crystals
(a) Vol. **78**, No. 2, p. K189 – K193 (1983).
See also *13.4*, p. *185*; 22.3, p. 457.

Z. Dziuba and R. Kowalczyk
*The Analysis of the Conductivity Tensor Components in HgTe-Type Semiconductors
(b) Vol. **119**, No. 1, p. K11 – K15 (1983).
See also 22.4.3, p. 469.

J. Horák, P. Lošťák, and A. Montaner
Point Defects in Ga-Doped Bi_2Se_3 Single Crystals
(b) Vol. **119**, No. 1, p. K17 – K21 (1983).
See also *10.2*, p. *117*; 20.1, p. 326; 22.7, p. 499.

S. K. Plachkova, G. Ch. Christakudis, and L. V. Vassilev
*Density-of-States Effective Mass in the System $(GeTe)_{1-x}(AgSbTe_2)_x$
(b) Vol. **119**, No. 1, p. K73 – K76 (1983).
See also *13.1*, p. *163*; 15, p. 262; 22, p. 411.

D. Chattopadhyay
*Transient Behaviour of Hot Electrons in Quantum-Well Structures of Polar Semiconductors
(b) Vol. **119**, No. 1, p. K77 – K81 (1983).
See also 6, p. 78; *13*, p. *151*; 22, p. 411.

V. V. Aristov, Yu. A. Osipyan, R. Scholz, I. I. Snighireva, I. I. Khodos, and S. A. Shevchenko
*The Effect of Annealing on the Defect Structure and Electrical Properties of Deformed Single Crystals of Ge
(a) Vol. **79**, No. 1, p. 47–55 (1983).
See also *10.2*, p. *117*; 22.1.1, p. 421.

M. K. Agarwal, P. D. Patel, and O. Vijayan
*Electrical Studies on (Mo/W)Se_2 Single Crystals (III): Anisotropy in Resistivity
(a) Vol. **79**, No. 1, p. 103–108 (1983).
See also 22.7, p. 499.

A. N. Anagnostopoulos and A. Karoutis
*Cation Mixing and the AC Conductivity of $CdInGaS_4$
(a) Vol. **79**, No. 1, p. K29–K33 (1983).
See also 22.8, p. 509.

G. J. Ashwell and K. Osborne
*Electrical and Structural Characterisation of (DBzBP) $(TCNQ)_4$
(a) Vol. **79**, No. 1, p. K35–K39 (1983).
See also *1.1*, p. *12*; 22.9, p. 530.

J. Grňo, Š. Bederka, and M. Veselý
*Pulse Radiation Annealing of Capless GaAs
(a) Vol. **79**, No. 1, p. K41–K44 (1983).
See also *11*, p. *132*; 22.2.1, p. 446.

V. P. Dobrego and A. K. Fedotov
*Influence of Growth Dislocations on Low Temperature Conductivity in Irradiated p-Germanium
(a) Vol. **79**, No. 1, p. K45–K48 (1983).
See also *10.2*, p. *118*; 11, p. 132; 22.1.1, p. 421.

S. M. Wasim and G. Sánchez Porras
*On the Hole Mobility of $CuGaSe_2$
(a) Vol. **79**, No. 1, p. K65–K68 (1983).
See also 22.8, p. 509.

H. Bergner and V. Brückner
*Dependence of the Ambipolar Diffusion in Silicon on the Carrier Density
(a) Vol. **79**, No. 1, p. K85–K88 (1983).
See also 20.1, p. 327; 22.1.2, p. 435.

P. Chandrasekharam, D. Raja Reddy, and B. K. Reddy
*Electrical Properties of Vapour Phase Grown $(ZnSe)_x(CdTe)_{1-x}$ Single Crystals
(a) Vol. **79**, No. 1, p. K103–K107 (1983).
See also *2*, p. *52*; 22.4.4, p. 471.

G. Irmer, V. V. Toporov, B. H. Bairamov, and J. Monecke
*Determination of the Charge Carrier Concentration and Mobility in n-GaP by Raman Spectroscopy
(b) Vol. **119**, No. 2, p. 595–603 (1983).
See also 6, p. 79; 13.5.2, p. 200; *20.1*, p. *327*; 22.2.2, p. 450.

B. R. SETHI, P. K. GOYAL, O. P. SHARMA, and P. C. MATHUR
*Negative Magnetoresistance in Heavily Doped and Compensated n-Type InSb
(b) Vol. **119**, No. 2, p. 721 – 726 (1983).
See also 22.2.3, p. 453.

L. A. KAZAKEVICH and P. F. LUGAKOV
The Effect of Dislocations on the Interaction Processes of Point and Group Radiation Defects in Silicon
(a) Vol. **79**, No. 2, p. 377 – 380 (1983).
See also *11*, p. *132*; 22.1.2, p. 436.

V. I. KUZNETSOV and P. F. LUGAKOV
Parameters of Defect Clusters in Irradiated Silicon and a Method for Their Determining
(a) Vol. **79**, No. 2, p. 381 – 388 (1983).
See also *11*, p. *132*; 22.1.2, p. 436.

G. J. ASHWELL, J. G. ALLEN, G. H. CROSS, and I. W. NOWELL
*Electrical, Magnetic, and Structural Characterisation of the 1:2 and 1:3 TCNQ Salts of 1,2-Bis(4-Pyridinium)Ethane
(a) Vol. **79**, No. 2, p. 455 – 463 (1983).
See also *1.1*, p. *12*; 15, p. 262; 18.1, p. 279; 22.9, p. 530.

S. A. ABO-NAMOUS, Y. ZAKA, and R. W. FANE
*Dependence of the Electronic and Optical Properties of Unhydrogenated a-Si on Preparation Conditions
(a) Vol. **79**, No. 2, p. 477 – 482 (1983).
See also 16, p. 267; *20.1*, p. *328*; 22.1.2, p. 436.

G. A. SCHOLZ and R. F. FRINDT
*Electrical Resistivity of Silver Intercalated 2H-TaS_2
(a) Vol. **79**, No. 2, p. 483 – 488 (1983).
See also 13, p. 152; *14.1*, p. *210*; 22, p. 412.

A. N. ANAGNOSTOPOULOS and CH. LIOUTAS
*Voltage-Dependent Conductivity along the *c*-Axis of $ZnIn_2S_4$(III)
(a) Vol. **79**, No. 2, p. 513 – 516 (1983).
See also 22.8, p. 509.

YU. A. VIDADI, A. V. RAGIMOV, B. A. MAMEDOV, S. A. GUSEINOV, A. A. MEDZHIDOV, and B. I. LIOGONKII
*On the Contribution of Naphthoxyl-Type Paramagnetic Centres to the Electric Conductivity of Oligonaphthols
(a) Vol. **79**, No. 2, p. 645 – 650 (1983).
See also 13.4, p. 188; 22.9, p. 531.

A. P. TYUTNEV, G. S. MINGALEEV, V. S. SAENKO, and A. V. VANNIKOV
*Radiation Induced Conductivity of Polyethyleneterephthalate and Polystyrene
(a) Vol. **79**, No. 2, p. 651 – 659 (1983).
See also *11*, p. *133*; 22.9, p. 531.

P. G. Rustamov, Ch. I. Abilov, and S. G. Agdamskaya
Thermoelectric Properties of $(PbTe)_{1-x}[(Cr, Fe, Ni)_2Te_3]_x$
(a) Vol. **79**, No. 2, p. K135–K137 (1983).
See also 8, p. 91; *15*, p. *263*; 22, p. 413.

P. I. Baranskii, V. V. Savyak, and Y. V. Simonenko
*Effect of Heat Treatment and Cooling Conditions on the Anisotropy Parameters of Drag TEMF in Neutron-Doped Silicon Crystals
(a) Vol. **79**, No. 2, p. K165–K170 (1983).
See also *15*, p. *263*; 22.1.2, p. 437.

N. M. Khamidullina
*Carrier Mobility in Doped Nondegenerate Semiconductors
(b) Vol. **120**, No. 1, p. 55–63 (1983).
See also 22, p. 413.

R. Požėra, A. Sutkus, and R. Tolutis
The Influence of Hydrostatic Pressure on the Effective Concentration and Cyclotron Mobility of Electrons in Semiconducting $Bi_{1-x}Sb_x$ Alloys
(b) Vol. **120**, No. 1, p. 165–171 (1983).
See also *13.1*, p. *164*; 22.7.1, p. 501.

J. P. Vitton, J. P. Dufour, R. Machet, and J. C. Thuillier
*Magnetoconduction of n-Inversion Layers on $Hg_{1-x}Cd_xTe$
(b) Vol. **120**, No. 1, p. K53–K57 (1983).
See also 22.4.4, p. 472.

G. J. Ashwell, S. Jansen, M. T. Jones, and C. H. L. Kennard
$(Diethyldiphenylphosphonium)^+$ $(7,7,8,8\text{-tetracyano-p-quinodimethanide})_2^-$, a Novel Quasi-One-Dimensional Organic Conductor
(a) Vol. **80**, No. 1, p. 127–134 (1983).
See also *1.1*, p. *12*; 18.1, p. 279; 22.9, p. 532.

G. Kühnel, W. Siegel, and E. Ziegler
*Electrical and Photoelectrical Properties of Semi-Insulating GaP
(a) Vol. **80**, No. 1, p. 159–164 (1983).
See also *13.4*, p. *189*; 16, p. 267; 22.2.2, p. 451.

L. N. Vasilev, V. M. Grabov, A. V. Golubkov, A. G. Gorobets, V. S. Oskotskii, I. A. Smirnov, and V. V. Tikhonov
Physical Properties and Phase Transitions of the Rare Earth Monosulphides in the Homogeneity Range
(a) Vol. **80**, No. 1, p. 237–244 (1983).
See also *1.3*, p. *26*; *8*, p. *91*; 13, p. 152; 18.4, p. 297; 22, p. 413.

G. Micocci, A. Rizzo, A. Tepore, and F. Zuanni
*Isothermal Currents in InSe, GaSe, and GaS Single Crystals
(a) Vol. **80**, No. 1, p. 263–267 (1983).
See also *13.4*, p. *189*; 22.3, p. 458.

L. Hrivnák
*On the Determination of Carrier Mobilities and Densities in Semi-Insulating GaAs
(a) Vol. **80**, No. 1, p. 317–323 (1983).
See also 22.1.2, p. 438.

A. Priemuth and H. Süssmann
*On the Relation between the Carrier Concentration and the Effective Ion Charge of Interstitial Cu at the Electromigration in V_2VI_3 Compounds
(a) Vol. **80**, No. 1, p. K7–K9 (1983).
See also *9*, p. *95*; 22.7, p. 500.

G. M. Guseinov, G. M. Niftiev, and M. G. Shakhtakhtinskii
Effect of External Factors on the Thermally Stimulated Conductivity of $GdIn_3S_6$ Single Crystals
(a) Vol. **80**, No. 1, p. K21–K24 (1983).
See also *13.4*, p. *189*; 22.8, p. 510.

S. K. Plachkova
*Thermoelectric Power in the System $(GeTe)_{1-x}(AgSbTe_2)_x$
(a) Vol. **80**, No. 1, p. K97–K100 (1983).
See also 8, p. 91; *15*, p. *263*; 22.8, p. 510.

G. V. Loseva and G. M. Abramova
*Anomalies of Electric and Magnetic Properties of Ternary Cobalt Sulfides
(a) Vol. **80**, No. 1, p. K109–K111 (1983).
See also *18*, p. *276*; 22.8, p. 510.

V. A. Sablikov and V. B. Sandomirskii
*The Photoacoustic Effect in Semiconductors
(b) Vol. **120** No. 2, p. 471–480 (1983).
See also 8, p. 92; *13.3*, p. *171*; 22, p. 414.

A. Brau, J. P. Farges, A. Filhol, and H. Grassi
*Order–Disorder Phase Transition in $TEA^+(TCNQ)_2^-$
(b) Vol. **120**, No. 2, p. 547–554 (1983).
See also *1*, p. *10*; 8, p. 92; 22.9, p. 532.

O. Uemura, Y. Wada, T. Takahara, and T. Satow
*The Semiconduction in the Molten Tl–$CuTlTe_2$ System
(b) Vol. **120**, No. 2, p. 685–693 (1983).
See also *2*, p. *55*; 15, p. 263; 18, p. 276; 22.8, p. 510.

A. Ferreira da Silva, R. Micnas, I. C. da Cunha Lima, and R. Kishore
*The Impurity Conduction of n-Type CdS: A Two Band Approach
(b) Vol. **120**, No. 2, p. K101–K106 (1983).
See also *13.4*, p. *190*; 22.4.1, p. 465.

O. Z. Alekperov, V. G. Golubev, V. I. Ivanov-Omskii, and A. Sh. Mekhtiev
*Plasma Shift of Cyclotron Resonance Line in n-GaAs
(b) Vol. **120**, No. 2, p. K179–K181 (1983).
See also *13*, p. *154*; 18, p. 276; 22.2.1, p. 447.

C. Manolikas and A. N. Anagnostopoulos
Electrical Effects Associated with the Ordering Process in $CdInGaS_4$ Crystals (I): Electron Microscopy and Electron Diffraction Study
(a) Vol. **80**, No. 2, p. 503–511 (1983).
See also *1.3*, p. *27*; 22.8, p. 510.

H. Süssmann and W. Heiliger
*Seebeck Coefficient and Electrical Conductivity in p-$(Bi_{1-x}Sb_x)_2Te_3$ at Room Temperature
(a) Vol. **80**, No. 2, p. 535–539 (1983).
See also *15*, p. *263*; 22.7.1, p. 501.

A. T. Oza
*Electrical Conduction in Some Inclusion Compounds of Iodine
(a) Vol. 80, No. 2, p. 573–579 (1983).
See also *13.4*, p. *191*; 22.9, p. 532.

J. Tanaka, K. Takahashi, K. Yukino, and S. Horiuchi
*Electrical Conduction of $(La_{0.8}Ca_{0.2})MnO_3$ with Homogeneous Ionic Distribution
(a) Vol. **80**, No. 2, p. 621–630 (1983).
See also *1.3*, p. *27*; 22.8.1, p. 518.

J. K. Jeszka, J. Ulański, and M. Kryszewski
*TTT · TCNQ CT Complex Crystallized in Polymer Matrix
(a) Vol. **80**, No. 2, p. K141–K143 (1983).
See also *3*, p. *59*; 22.9, p. 532.

P. Simova and M. Petrov
*On the Electrical Conductivity in the Nematic Phase of 4-n-Alkyloxybenzoic Acids HOBA, OOBA, and NOBA
(a) Vol. **80**, No. 2, p. K153–K156 (1983).
See also 22.9, p. 532.

L. I. Ryabinkina and G. V. Loseva
*Influence of Non-Stoichiometry and Cation Substitution on the Electrical Properties of α-MnS
(a) Vol. **80**, No. 2, p. K179–K182 (1983).
See also 22, p. 415.

14.3.1 Films

Original Papers and Short Notes

V. E. Borisenko, V. V. Gribkovskii, V. A. Labunov, V. A. Samuilov, and K. D. Yashin
*Incoherent Light Annealing of Phosphorus-Doped Polycrystalline Silicon
(a) Vol. **75**, No. 1, p. 117 – 120 (1983).
See also 20.1, p. 313; 22.1.2, p. 424.

H. Pohlack
*Thin-Film Photodetectors with High Quantum Yields Achieved by Induced Resonance Total Absorption
(a) Vol. **75**, No. 1, p. 179 – 185 (1983).
See also 14.3.3, p. 236; *20.1*, p. *313*; 22.1.2, p. 424.

R. Mach, G. O. Müller, W. Gericke, and J. von Kalben
The Efficiency of Electron Impact Excited Luminescence in ZnS Thin Film Devices
(a) Vol. **75**, No. 1, p. 187 – 194 (1983).
See also *14.3.3*, p. *236*; *20.3*, p. *335*; 22.4.1, p. 462; 22.4.2, p. 466.

U. Dillner
*Analysis of Resistivity, Hall Effect, and Magnetoresistance Measurements of Thin Bismuth Films by Means of an Anisotropic Two-Carrier Model
(a) Vol. **75**, No. 1, p. 227 – 233 (1983).
See also *14.1*, p. *206*; 21.7, p. 401.

C. Gunkel, D. Schalch, and A. Scharmann
*Electrical Conduction in Silicon Nitride Thin Films under 15 keV He^+ Irradiation
(a) Vol. **75**, No. 1, p. K57 – K62 (1983).
See also 2, p. 44; *11*, p. *123*; 22, p. 403.

S. A. Azimov, D. A. Aronov, V. M. Rubinov, and M. Tuichiev
*Quantum Yield of Current Carriers Excited by Electron Beams and Electron Penetration Depth in Thin GaP Films
(a) Vol. **75**, No. 1, p. K63 – K68 (1983).
See also *11*, p. *123*; 22.2.2, p. 448.

M. D. Smolin, A. F. Andreeva, and V. A. Ogorodnik
*Electroconduction and Thermoelectricity of Terbium Oxide Films
(b) Vol. **115**, No. 2, p. 587 – 592 (1983).
See also 13.5.3, p. 201; *15*, p. *261*; 22.6, p. 491.

K. B. Kadyrakunov, E. V. Nidaev, A. E. Plotnikov, L. S. Smirnov, I. G. Melnik, and M. V. Makeev
*Flash Lamp Annealing of Ion-Implanted Polycrystalline Silicon
(a) Vol. **75**, No. 2, p. 483 – 488 (1983).
See also 11, p. 123; 22.1.2, p. 426.

R. Mach, J. van Kalben, W. Gericke, G. O. Müller, and G. U. Reinsperger
ZnSe:Mn in AC Thin Film Electroluminescence Devices
(a) Vol. **75**, No. 2, p. 489–494 (1983).
See also *14.3.3*, p. *237*; *20.3*, p. *336*; 22.4.2, p. 466.

H. Pfleiderer and B. Rauscher
*Capacitance of Amorphous Silicon pin Solar Cells
(a) Vol. **75**, No. 2, p. 537–545 (1983).
See also 2, p. 44; 13.4, p. 175; *14.3.3*, p. *237*; 16, p. 264; 22.1.2, p. 426.

H. Gotoh, K. Sasamoto, S. Kuroda, and M. Kimata
Molecular Beam Epitaxy of AlSb on GaAs and GaSb on AlSb Films
(a) Vol. **75**, No. 2, p. 641–645 (1983).
See also *1.5*, p. *36*; *3*, p. *57*; 22.2.1, p. 443; 22.2.3, p. 452.

D. V. Krishna Sastry and P. Jayarama Reddy
*Hall Effect Measurements in High-Temperature Deposited and Heat Treated Films of InTe and InSe
(a) Vol. **75**, No. 2, p. K183–K185 (1983).
See also 22.3, p. 456.

N. Crisa
*Anomalous Transient-Time Dispersion in Amorphous Solids: Oxadiazole
(b) Vol. **116**, No. 1, p. 269–274 (1983).
See also 22.9, p. 526.

E. A. Silinsh, V. A. Kolesnikov, I. J. Muzikante, and D. R. Balode
Erratum to: On Charge Carrier Photogeneration Mechanisms in Organic Molecular Crystals
(b) Vol. **116**, No. 1, p. 347 (1983).
See also *16*, p. *265*; 22.9, p. 526.

G. Eftekhari, D. de Cogan, and B. Tuck
*Electrical Conduction through Anodic Oxides on InP
(a) Vol. **76**, No. 1, p. 331–336 (1983).
See also *14.3.2*, p. *233*; 22.2.2, p. 448.

D. V. Krishna Sastry and P. Jayarama Reddy
*Influence of Nitrogen Ion Implantation on the Electrical Transport Properties of InTe and InSe Thin Films
(a) Vol. **76**, No. 1, p. 345–350 (1983).
See also 11, p. 125; 22.3, p. 456.

J. Toušková, D. Kindl, J. Toušek, and E. Klier
*Some Electrical Properties of Au_xTe/CdTe Heterojunctions on CdTe Thick Films
(a) Vol. **76**, No. 1, p. 365–374 (1983).
See also *14.3.3*, p. *238*; 22.4.3, p. 468.

H. von Berlepsch and L. Brehmer
*Invalidity of Scaling Property of Surface Potential Kinetics in HDPE
(a) Vol. **76**, No. 1, p. K27–K29 (1983).
See also *14.3.4*, p. *244*; 22.9, p. 526.

F. R. KESSLER, K. DETTMER, and E.-W. RITTERS
*Ge–Al Thin Films for Selective Solar Absorbers
(a) Vol. **76**, No. 2, p. 533–541 (1983).
See also *20.1*, p. *317*; 22.1.1, p. 420.

T. G. ABDEL-MALIK, A. A. ALY, A. M. ABDEEN, and H. M. EL-LABANY
*Electric and Photoelectric Investigations of β-Nickel Phthalocyanine Thin Films
(a) Vol. **76**, No. 2, p. 651–659 (1983).
See also *13.4*, p. *177*; 16, p. 265; 22.9, p. 527.

A. JAKUBOWICZ
*On EBIC versus Time Measurements in Thin Semiconductor Layers with the Scanning Electron Microscope
(a) Vol. **76**, No. 2, p. 729–734 (1983).
See also *14.3.3*, p. *238*; 22, p. 406.

G. P. SRIVASTAVA, K. N. TRIPATHI, and N. K. SEHGAL
*Effect of Laser Irradiation on the Electrical Properties of Polycrystalline Thin Films of Tellurium
(a) Vol. **76**, No. 2, p. K137–K140 (1983).
See also *11*, p. *127*; 22.1.3, p. 440.

J. SZCZYRBOWSKI, A. DIETRICH, and H. HOFFMANN
*Optical and Electrical Properties of RF-Sputtered Indium–Tin Oxide Films
(a) Vol. **78**, No. 1, p. 243–252 (1983).
See also *20.1*, p. *323*; 22.6.1, p. 498.

J. TYCZKOWSKI and M. KRYSZEWSKI
Poole-Frenkel Centres in Dielectrics. Kind and Space Density in Plasma-Polymerized Organosilicon Films
(a) Vol. **78**, No. 1, p. 259–265 (1983).
See also *13.4*, p. *183*; *14.3.4*, p. *246*; 22.9, p. 529.

S. A. AZIMOV, D. A. ARONOV, V. M. RUBINOV, and M. TUICHIEV
*The Influence of Heating on the Characteristics of the Electron-Excited Conduction in Gallium Phosphide Films
(a) Vol. **78**, No. 1, p. K53–K58 (1983).
See also *11*, p. *130*; 22.2.2, p. 450.

J. S. KUMAR, G. NARAYANA, M. CHANDRA SHEKAR, U. V. SUBBA RAO, and V. HARI BABU
*Dielectric and Electrical Properties of SnO_2, Sb_2O_3, and Their Mixed Films
(a) Vol. **78**, No. 2, p. 647–653 (1983).
See also 14.4, p. 253; 22.6, p. 495; 22.6.1, p. 498.

A. V. DVURECHENSKII, I. A. RYAZANTSEV, L. S. SMIRNOV, H. KLOSE, and M. RIETH
*Amorphous Silicon–Phosphorus Layers
(a) Vol. **79**, No. 1, p. 83–86 (1983).
See also 2, p. 51; 13.4, p. 186; 16, p. 267; 19, p. 304; 22.1.2, p. 434.

A. K. SHARMA and P. JAYARAMA REDDY
*Temperature and Thickness Dependence of Electrical Conductivity of Flash-Evaporated InSb Films
(a) Vol. **79**, No. 1, p. 183–188 (1983).
See also 22.2.3, p. 453.

W. FUHS, H. M. WELSCH, and D. C. BOOTH
Optical Quenching of Photoconductivity in a-Si:H Films
(b) Vol. **120**, No. 1, p. 197–205 (1983).
See also 2, p. 53; *16*, p. *267*; 22.1.2, p. 437.

B.-P. KIETIS, R. REKSNYS, and K. TUMKEVIČIUS
*Hall Effect and Acoustoelectric Phenomena in PbTe Films
(a) Vol. **80**, No. 1, p. K57–K58 (1983).
See also *7*, p. *84*; 22, p. 414.

M. NOWAK
*The Photomagnetoelectric Effect and Photoconductivity for Non-Normal Incidence of Radiation
(a) Vol. **80**, No. 2, p. 691–702 (1983).
See also *16*, p. *268*; 22.3, p. 458.

J. HAŁADYJ, J. MICHALSKI, and A. SZYMAŃSKI
*On the Anomalous Field Dependence of Drift Mobility in Thin Organic Films
(a) Vol. **80**, No. 2, p. K137–K139 (1983).
See also *14.3.4*, p. *249*; 22.9, p. 532.

14.3.2 Surfaces and Interfaces

Original Papers and Short Notes

G. Paasch
*Current Density near the Semiconductor–Insulator Interface
(b) Vol. **115**, No. 2, p. 471–475 (1983).
See also *13.3*, p. *167*; 22, p. 403.

V. B. Bikbaev, S. Č. Karpinskas, and J. J. Vaitkus
*Electrical Properties of Mo–Cadmium Chalcogenide Schottky Barrier Diodes
(a) Vol. **75**, No. 2, p. 583–590 (1983).
See also 14.3.3, p. 237; 22.4, p. 460.

G. Eftekhari, D. de Cogan, and B. Tuck
*Electrical Conduction through Anodic Oxides on InP
(a) Vol. **76**, No. 1, p. 331–336 (1983).
See also *14.3.1*, p. *230*; 22.2.2, p. 448.

A. G. Kazanskii, V. F. Kiselev, E. A. Silaev, and V. S. Vavilov
*On the Role of Surface Phenomena in the Staebler-Wronski Effect
(a) Vol. **76**, No. 1, p. 337–343 (1983).
See also 2, p. 46; *13.3*, p. *168*; 16, p. 265; 22.1.2, p. 427.

D. Tjapkin and V. Milanović
*The Influence of Parameter Nonuniformity on the Two-Dimensional Einstein Relation in Periodic and Non-Periodic Semiconductor Structures
(b) Vol. **116**, No. 2, p. 653–658 (1983).
See also *13.1*, p. *157*; 22.2.4, p. 454.

J. Dziesaty, F.-G. Kirscht, M. Mai, and K.-R. Sprung
*Influence of Backside Argon Implantation on the Si–SiO_2 Interface
(a) Vol. **76**, No. 2, p. K105–K108 (1983).
See also *11*, p. *126*; 22.1.2, p. 428.

A. D'Amico, G. Petrocco, G. Stochino, A. Salsano, and G. Luzzi
*Grain Boundaries as Possible Origin of Flicker Noise in GaP Polycrystalline Material
(a) Vol. **76**, No. 2, p. K125–K129 (1983).
See also *1.4*, p. *31*; 22.2.2, p. 449.

J. Kočka, O. Štika, and J. Stuchlík
*Comments on the Interpretation of DLTS Spectra of a-Si:H
(a) Vol. **76**, No. 2, p. K217–K220 (1983).
See also 2, p. 47; *13.4*, p. *178*; 22.1.2, p. 429.

M. Kittler and K.-W. Schröder
Determination of Semiconductor Parameters and of the Vertical Structure of Devices by Numerical Analysis of Energy-Dependent EBIC Measurements
(a) Vol. **77**, No. 1, p. 139–151 (1983).
See also 1.5, p. 36; *14.3.3*, p. *238*; 22.1.2, p. 430.

I. THURZO, K. GMUCOVÁ, J. ČERVENÁK, and T. LALINSKÝ
*GaAs–Anodic Oxide Interface Examination by Deep-Level Transient Current Spectroscopy
(a) Vol. **77**, No. 1, p. 323–330 (1983).
See also 13.3, p. 168; 22.2.1, p. 444.

E. W. KREUTZ
*Size Effects on the Magnetoresistance of p-InSb
(a) Vol. **77**, No. 2, p. 583–594 (1983).
See also 22.2.3, p. 452.

K. N. KANG and S. CRISTOLOVEANU
*On the Role of Exact Boundary Conditions in Phototransport Phenomena: Case of SI–GaAs
(a) Vol. **77**, No. 2, p. K155–K159 (1983).
See also *16*, p. *266*; 22.2.1, p. 445.

G. PAASCH and H. ÜBENSEE
*Carrier Density near the Semiconductor–Insulator Interface. Local Density Approximation for Non-Isotropic Effective Mass
(b) Vol. **118**, No. 1, p. 255–266 (1983).
See also 13.1, p. 160; *13.3*, p. *169*; 22.1.1, p. 420; 22.1.2, p. 432.

G. D. BAGRATISHVILI, R. B. DZHANELIDZE, and D. A. JISHIASHVILI
*Nonstoichiometric Germanium Nitride. A New Insulating Material for MIS Microelectronics (I): The Choice of the Insulator and Deposition
(a) Vol. **78**, No. 1, p. 115–123 (1983).
See also *13.3*, p. *169*; 14.3.3, p. 239; 22, p. 409.

K. HÜBNER, H. KÖSTER, JR., B. DERLICH, and W. ECKE
DLTS Investigations of Si–SiO_2 Interface States of Electron-Beam Irradiated MOS Structures
(b) Vol. **118**, No. 2, p. K133–K136 (1983).
See also *11*, p. *130*; *13.3*, p. *169*; 22.1.2, p. 433; 22.6, p. 495.

G. D. BAGRATISHVILI, R. B. DZHANELIDZE, and D. A. JISHIASHVILI
*Nonstoichiometric Germanium Nitride — A New Insulating Material for MIS Microelectronics (II): Composition, Properties, and the Interface with a Semiconductor
(a) Vol. **78**, No. 2, p. 391–400 (1983).
See also *13.3*, p. *169*; 14.3.3, p. 240; 22, p. 410.

J. KOSHY
*Electrical Properties of Junctions between Sputter Deposited Silicon Films and Monocrystalline Silicon
(a) Vol. **79**, No. 1, p. K21–K24 (1983).
See also *14.3.3*, p. *240*; 22.1.2, p. 435.

P. BOGDANSKI, R. CARIN, O. GARDAIS et J. VIZOT
*Caractérisation de structures MIS (Al–Si_3N_4–GaAs) par différentes méthodes capacitives (Characterisation of MIS Structures (Al–Si_3N_4–GaAs) by Different Capacitive Methods)
(a) Vol. **80**, No. 1, p. 185–193 (1983).
See also *13.3*, p. *171*; 22.2.1, p. 447.

M. Kusaka, M. Hirai, and S. Okazaki
*Change in Capacitance of InP MIS Diodes Caused by Piezoelectric Effect
(a) Vol. **80**, No. 1, p. K25 – K28 (1983).
See also *14.3.3*, p. *242*; 22.2.2, p. 451.

I. Mojzes, R. Veresegyházy, V. Obodnikov, Yu. Vaulin, and S. Vasilev
*Mapping the Redistributed Components of the Au–Ge/GaAs System after Scanning Laser Annealing
(a) Vol. **80**, No. 1, p. K33 – K36 (1983).
See also 14.3.4, p. 248; 22.2.1, p. 447.

C. H. Ling
*Resistivity Change in Ti/Au Bimetal Films Annealed in the Temperature Range 300 to 400 °C
(a) Vol. **80**, No. 1, p. K59 – K62 (1983).
See also *9*, p. *96*; 21.1, p. 377; 21.6, p. 399.

E. I. Goldman, Yu. V. Markin, P. S. Sulzhenko, and A. G. Zhdan
*Grain Boundary Tunneling Conductivity of Semiconductor Poly- and Bicrystals
(a) Vol. **80**, No. 2, p. 423 – 429 (1983).
See also 1.4, p. 34; 22, p. 415.

D. S. Gupta, M. M. Chandra, and V. Kumar
*Direct Measurement of Interface State Electron Capture Cross-Section in MOS System by DLTS Technique
(a) Vol. **80**, No. 2, p. K209 – K212 (1983).
See also *13.3*, p. *171*; 22.1.2, p. 439.

14.3.3 Devices. Junctions

(Contact Problems See 14.3.4)

Review Articles

N. D. Stojadinović and S. D. Ristić
*Failure Physics of Integrated Circuits and Relationship to Reliability
(a) Vol. **75**, No. 1, p. 11–48 (1983).
See also 22.1.2, p. 423.

T. D. Dzhafarov
Photostimulated Diffusion in Semiconductors
(a) Vol. **79**, No. 1, p. 11–45 (1983).
See also *9*, p. *93*; 11, p. 122; **14.3.4**, p. 243; 16, p. 264; 22, p. 403.

Original Papers and Short Notes

Yen-chu Wang and Zheng-an Qiu
*Velocity Overshoot of Electrons in GaAs with Space Charge and Non-Uniform Field
(a) Vol. **75**, No. 1, p. 67–72 (1983).
See also 22.2.1, p. 443.

H. Pohlack
Thin-Film Photodetectors with High Quantum Yields Achieved by Induced Resonance Total Absorption
(a) Vol. **75**, No. 1, p. 179–185 (1983).
See also *14.3.1*, p. *229*; *20.1*, p. *313*; 22.1.2, p. 424.

R. Mach, G. O. Müller, W. Gericke, and J. von Kalben
*The Efficiency of Electron Impact Excited Luminescence in ZnS Thin Film Devices
(a) Vol. **75**, No. 1, p. 187–194 (1983).
See also 14.3.1, p. 229; *20.3*, p. *335*; 22.4.1, p. 462; 22.4.2, p. 466.

J. P. Gourret and J. Paille
*Breakdown Voltage Modeling in Mesa Power Devices
(a) Vol. **75**, No. 1, p. 207–217 (1983).

D. P. Joshi and K. Sen
*Effective Lifetime for Polycrystalline Solar Cells
(a) Vol. **75**, No. 1, p. 311–315 (1983).
See also *16*, p. *264*; 22.1.2, p. 424.

K. Žďánský, L. Kraténa, M. Matyáš, Jr., and J. Zelinka
*Absolute Energy of the Nitrogen-Related Electron Trap in Gallium Phosphide
(a) Vol. **75**, No. 1, p. 335–339 (1983).
See also *13.4*, p. *173*; 22.2.2, p. 448.

C. N. Duckworth, A. W. Brinkman, and J. Woods
Deep Level Transient Spectroscopy in TiO_2:Nb
(a) Vol. **75**, No. 1, p. K99–K102 (1983).
See also *13.4*, p. *174*; 22.6, p. 490.

K.-H. Bäther, W. Brückner, W. Moldenhauer, and H.-P. Brückner
*Microscopic Degradation Model for ZnO Varistors
(a) Vol. **75**, No. 2, p. 465–472 (1983).
See also *1.4*, p. *29*; 22.4, p. 460.

R. Mach, J. van Kalben, W. Gericke, G. O. Müller, and G. U. Reinsperger
*ZnSe: Mn in AC Thin Film Electroluminescence Devices
(a) Vol. **75**, No. 2, p. 489–494 (1983).
See also 14.3.1, p. 230; *20.3*, p. *336*; 22.4.2, p. 466.

H. Pfleiderer and B. Rauscher
*Capacitance of Amorphous Silicon pin Solar Cells
(a) Vol. **75**, No. 2, p. 537–545 (1983).
See also 2, p. 44; 13.4, p. 175; *14.3.1*, p. *230*; 16, p. 264; 22.1.2, p. 426.

G. Oelgart and H. Scholz
*Determination of Electron Depth–Dose Function for Kilovolt Electrons on GaAsP
(a) Vol. **75**, No. 2, p. 547–553 (1983).
See also *11*, p. *124*; 22.2.4, p. 454.

V. B. Bikbaev, S. Č. Karpinskas, and J. J. Vaitkus
Electrical Properties of Mo–Cadmium Chalcogenide Schottky Barrier Diodes
(a) Vol. **75**, No. 2, p. 583–590 (1983).
See also *14.3.2*, p. *233*; 22.4, p. 460.

A. D. Franklin and H. J. de Bruin
*The Fourier Analysis of Impedance Spectra for Electroded Solid Electrolytes
(a) Vol. **75**, No. 2, p. 647–656 (1983).
See also 22.8.2, p. 519.

K. Sen, R. S. Srivastava, D. P. Joshi, and V. K. Goyal
*Mobility and Its Effect on the Performance of Polycrystalline Solar Cells
(a) Vol. **75**, No. 2, p. 657–661 (1983).
See also 13.3, p. 167; *16*, p. *264*; 22.1.2, p. 426.

F. Fenske
*Simple Improvement of the Lindner High-Frequency MOS Capacitance Approximation
(a) Vol. **75**, No. 2, p. K141–K144 (1983).
See also *13.3*, p. *167*; 22.1.2, p. 426.

M. I. Mitkowa and W. Fallmann
*Some Properties of Silver Photodiffused Chalcogenide Glasses
(a) Vol. **75**, No. 2, p. K145–K148 (1983).
See also *2*, p. *45*; 20.1, p. 315; 22.8, p. 503.

E. V. Anufriev and V. A. Gurtov
*Accumulation and Discharging of the Hole Charge in Oxide Layers of MIS Structures
(a) Vol. **76**, No. 1, p. 107–111 (1983).
See also 11, p. 125; *13.3*, p. *168*; 22.6, p. 492.

N. Stojadinović, S. Dimitrijev, S. Mijalković, and Z. Živić
*Reliability of n-Channel and p-Channel MOSTs in CMOS Integrated Circuits
(a) Vol. **76**, No. 1, p. 357 – 364 (1983).
See also **14.3.4**, p. **244**.

J. Toušková, D. Kindl, J. Toušek, and E. Klier
*Some Electrical Properties of $Au_xTe/CdTe$ Heterojunctions on CdTe Thick Films
(a) Vol. **76**, No. 1, p. 365 – 374 (1983).
See also *14.3.1*, p. *230*; 22.4.3, p. 468.

S. Nojima
*Isolation Characteristics in Selectively O^+ and Cr^+ Implanted GaAs
(a) Vol. **76**, No. 1, p. K9 – K12 (1983).
See also *11*, p. *125*; 22.2.1, p. 443.

A. Chovet, S. Cristoloveanu, Y. Daher, and V. K. Malyutenko
*Useful Relations for Highly Sensitive Sensors Based on Carrier Concentration
(a) Vol. **76**, No. 1, p. K43 – K48 (1983).
See also 22.1.1, p. 419; 22.2.3, p. 452.

D. Dădârlat, Rodica M. Cândea, and M. Bârlea
*Majority Carrier Concentration and Quasi-Fermi Level for $Pb_{1-x}Sn_xTe$ DH Diode Lasers
(a) Vol. **76**, No. 1, p. K61 – K65 (1983).
See also *20.2*, p. *332*; 22.8, p. 504.

V. Kumar and S. B. Iyer
Characterization of Surface States in MOS Capacitors by a Modified DLTS Technique
(a) Vol. **76**, No. 2, p. 637 – 640 (1983).
See also *13.3*, p. *168*; 22.1.2, p. 428.

A. Jakubowicz
*On EBIC versus Time Measurements in Thin Semiconductor Layers with the Scanning Electron Microscope
(a) Vol. **76**, No. 2, p. 729 – 734 (1983).
See also *14.3.1*, p. *231*; 22, p. 406.

H. Lemke
*Über die chemische Natur von Rekombinationszentren in Si-Leistungsbauelementen (On the Chemical Nature of Recombination Centres in Si Power Devices)
(a) Vol. **76**, No. 2, p. K193 – K196 (1983).
See also *13.4*, p. *177*; 22.1.2, p. 429.

M. Kittler and K.-W. Schröder
*Determination of Semiconductor Parameters and of the Vertical Structure of Devices by Numerical Analysis of Energy-Dependent EBIC Measurements
(a) Vol. **77**, No. 1, p. 139 – 151 (1983).
See also 1.5, p. 36; 14.3.2, p. 233; 22.1.2, p. 430.

A. U. Shareef, K. K. Saraf, and A. P. Srivastava
*Spontaneous Electric Current from Polyvinyl Alcohol Induced by the First Heating Run
(a) Vol. **77**, No. 1, p. 381 – 386 (1983).
See also 22.9, p. 528.

V. S. Lysenko, R. N. Litovskii, M. M. Lokshin, and M. F. Sherbakova
*Double Injection Current and Field Effect in SOS Diodes
(a) Vol. **77**, No. 2, p. 443–448 (1983).
See also *14.3.4*, p. *245*; 22, p. 408.

B. Rheinländer, G. Oelgart, R. Reulke, H. Haefner, and R. Stegmann
*Degradation of Yellow Light-Emitting $GaAs_{0.1}P_{0.9}$:N Diodes (I): Implanted Diodes
(a) Vol. **77**, No. 2, p. 527–534 (1983).
See also 14.3.4, p. 245; *20.3*, p. *340*; 22.2.4, p. 454.

I. E. Türe, F. Poulin, A. W. Brinkman, and J. Woods
*Electron Traps and Deep Levels in Cadmium Selenide
(a) Vol. **77**, No. 2, p. 535–544 (1983).
See also *13.4*, p. *180*; 22.4.2, p. 467.

R. N. Litovskii, V. S. Lysenko, A. N. Nazarov, T. E. Rudenko, S. B. Kaschieva, and N. R. Nedev
*The Influence of Defect Surface Layers on the Capaictive Properties of MOS Structures
(a) Vol. **77**, No. 2, p. 699–707 (1983).
See also 13.3, p. 169; 22.1.2, p. 431.

M. M. Chandra and M. Prasad
*Schottky Barrier Characteristics at Low Temperatures
(a) Vol. **77**, No. 2, p. 715–719 (1983).
See also 22.1.2, p. 431.

J. C. Lou, M. S. Lin, J. I. Chyi, M. D. Lin, and S. Sche
*Electrical and Optical Characteristics of $GaAs_{0.6}P_{0.4}$ LEDs Fabricated by Zn Semi-Closed Diffusion Method
(a) Vol. **77**, No. 2, p. 741–747 (1983).
See also 14.3.4, p. 245; *20.3*, p. *340*; 22.2.4, p. 454.

I. L. Baginskii, E. G. Kostsov, and E. E. Meerson
*Pulse Technique for Flat-Band Voltage Measurements in MIS Structures
(a) Vol. **77**, No. 2, p. K99–K102 (1983).
See also 22.1.2, p. 431.

B. Rheinländer
*Diffusion-Limited Carrier Lifetime in GaP:N LED
(a) Vol. **77**, No. 2, p. K111–K114 (1983).
See also *20.3*, p. *340*; 22.2.2, p. 450.

P. Müller
*Contactless Determination of the Conductivity in Semi-Insulators
(a) Vol. **78**, No. 1, p. 41–51 (1983).
See also *14.3.4*, p. *246*; 22.2.1, p. 445; 22.5.1, p. 476.

G. D. Bagratishvili, R. B. Dzhanelidze, and D. A. Jishiashvili
Nonstoichiometric Germanium Nitride. A New Insulating Material for MIS Microelectronics (I): The Choice of the Insulator and Deposition
(a) Vol. **78**, No. 1, p. 115–123 (1983).
See also *13.3*, p. *169*; *14.3.2*, p. *234*; 22, p. 409.

G. Oelgart, B. Rheinländer, R. Reulke, H. Haefner, and R. Stegmann
*Degradation of Yellow Light-Emitting $GaAs_{0.1}P_{0.9}$:N Diodes (II): Diffused Diodes
(a) Vol. **78**, No. 1, p. 169–176 (1983).
See also **14.3.4**, p. **246**; *20.3*, p. *342*; **22.2.4**, p. **455**.

M. Katayama, H. Morimoto, and K. Sugawara
*Electron Irradiation Effects on Amorphous Silicon Solar Cells
(a) Vol. **78**, No. 1, p. K5–K8 (1983).
See also **11**, p. **129**; *16*, p. *266*; **22.1.2**, p. **433**.

G. D. Bagratishvili, R. B. Dzhanelidze, and D. A. Jishiashvili
Nonstoichiometric Germanium Nitride – A New Insulating Material for MIS Microelectronics (II): Composition, Properties, and the Interface with a Semiconductor
(a) Vol. **78**, No. 2, p. 391–400 (1983).
See also *13.3*, p. *169*; *14.3.2*, p. *234*; **22**, p. **410**.

R. Gleichmann, H. Blumtritt, and J. Heydenreich
New Morphological Types of CuSi Precipitates in Silicon and Their Electrical Effects
(a) Vol. **78**, No. 2, p. 527–538 (1983).
See also *10.2*, p. *116*; **22.1.2**, p. **433**.

N. N. Gerasimenko, A. E. Gershinskii, A. Yu. Surtaev, S. A. Sokolov, E. I. Cherepov, B. I. Fomin, and G. V. Timofeeva
*Silicon–Nickel Silicide Schottky Barriers Formed by Ion Mixing
(a) Vol. **78**, No. 2, p. K151–K154 (1983).
See also **11**, p. **131**; **22.1.2**, p. **434**.

H. Tsuya and F. Shimura
Transient Behaviour of Intrinsic Gettering in CZ Silicon Wafers
(a) Vol. **79**, No. 1, p. 199–206 (1983).
See also *1.6*, p. *41*; *10.2*, p. *118*; **22.1.2**, p. **435**.

F. Campabadal, V. Milián, and X. Aymerich-Humet
*Trap-Assisted Tunneling in MIS and Schottky Structures
(a) Vol. **79**, No. 1, p. 223–236 (1983).
See also **13.4**, p. **186**; *14.3.4*, p. *247*; **22.6**, p. **496**.

G. S. Popkirov
*An Electrochemical Evidence of the O_2 Photodesorption from n-TiO_2 Surface
(a) Vol. **79**, No. 1, p. 323–326 (1983).
See also *1.6*, p. *41*; **22.6**, p. **496**.

J. Koshy
*Electrical Properties of Junctions between Sputter Deposited Silicon Films and Monocrystalline Silicon
(a) Vol. **79**, No. 1, p. K21–K24 (1983).
See also *14.3.2*, p. *234*; **22.1.2**, p. **435**.

H. K. Phan, L. H. Phu, and C. V. Chiem
*Stress Effects on the Switching Voltage of Metal–SnO_2–Si(n)–Si(p^+) Devices
(a) Vol. **79**, No. 1, p. K25 – K28 (1983).
See also *14.3.4*, p. *247*; 22.1.2, p. 435.

J. Walachová and D. Nohavica
*Determination of p–n Junctions in InP/GaInAsP Laser Heterostructures Using a Profilometer
(a) Vol. **79**, No. 1, p. K69 – K72 (1983).
See also 22.2, p. 442.

J. Muto and Y. Tsunekawa
*Photovoltaic Behaviour of Rhodamine 6G Schottky-Type Cells
(a) Vol. **79**, No. 1, p. K109 – K112 (1983).
See also *16*, p. *267*; 22.9, p. 530.

F. Fenske
*Influence of Lateral Charge Inhomogeneities in MOS Structures on the Determination of the Interface State Density Distribution from HF CV Curves
(a) Vol. **79**, No. 2, p. 417 – 421 (1983).
See also *13.3*, p. *170*; 22.1.2, p. 436.

A. A. Frantsuzov and A. P. Kovchavtsev
*The Determination of Surface Potential Fluctuation Values and Spatial Distribution of Traps in Insulators from the Temperature Dependence of the $(G_p/\omega) - \omega$ Curves of MOS Structures
(a) Vol. **79**, No. 2, p. 503 – 511 (1983).
See also *13.3*, p. *170*; 22.1.2, p. 436.

I. D. Henning and H. Thomas
Process-Induced Defect States in GaAsP Light Emitting Diodes
(a) Vol. **79**, No. 2, p. 567 – 574 (1983).
See also *13.4*, p. *187*; 20.3, p. 345; 22.2.4, p. 455.

B. M. Wilamowski
*The Punch-Through Transistor with MOS Controled Gate
(a) Vol. **79**, No. 2, p. 631 – 637 (1983).
See also 14.3.4, p. 248; 22, p. 412.

E. E. Velmre and B. P. Freidin
*Numerical Modeling of the Electrothermal Transient Process in Diode Structures Based on Direct-Gap Semiconductors
(a) Vol. **79**, No. 2, p. K131 – K134 (1983).
See also 8, p. 90; 22.2.1, p. 446.

G. Kaden
*Zum Einfluß des Oberflächenpotentials bei Photoinjektion von der Metallelektrode in MOS-Strukturen
(On the Influence of the Surface Potential on the Photoinjection from the Metal Electrode in MOS Structures)
(a) Vol. **79**, No. 2, p. K187 – K191 (1983).
See also 22.1.2, p. 437.

J. GONG, C. M. VAN VLIET, D. SCHOENFELD, and G. BOSMAN
*Noise from Hole-Initiated Photoavalanche Processes
(a) Vol. **79**, No. 2, p. K201 – K203 (1983).
See also *14.3.4*, p. *248*; 16, p. 267; 22.1.2, p. 437.

S. S. SIMEONOV and M. D. IVANOVITCH
*Determination of Quasi-Fermi Level at the Metal–Semiconductor Interface in a Forward-Biased Schottky Barrier Diode
(a) Vol. **80**, No. 1, p. 79 – 84 (1983).
See also 22.1.2, p. 437.

K. LÖHNERT and E. KUBALEK
*On the Degradation of Electroluminescence Efficiency in Gallium Phosphide Green Light Emitting Diodes
(a) Vol. **80**, No. 1, p. 173 – 183 (1983).
See also *20.3*, p. *346*; 22.2.2, p. 451.

M. KUSAKA, M. HIRAI, and S. OKAZAKI
*Change in Capacitance of InP MIS Diodes Caused by Piezoelectric Effect
(a) Vol. **80**, No. 1, p. K25 – K28 (1983).
See also *14.3.2*, p. *235*; 22.2.2, p. 451.

B. SUSŁA, R. CZAJKA, and J. RAUŁUSZKIEWICZ
*Electronic Excitation of Ni^{2+} and Co^{2+} by Tunneling Electrons
(b) Vol. **120**, No. 2, p. K129 – K134 (1983).
See also *13.4*, p. *190*; 19, p. 305; 20.1, p. 330; 22.6, p. 497.

W. HOERSTEL, W. KRAAK, and A. F. RUDOLPH
*The Influence of Hydrostatic Pressure on the *I–U* Characteristic of $Pb_{1-x}Sn_xTe$ Diodes
(a) Vol. **80**, No. 2, p. K193 – K197 (1983).
See also 22, p. 416.

14.3.4 High-Field Phenomena, Space-Charge, Inhomogeneities, Injected Carriers (Electroluminescence See 20.3; Junctions See 14.3.3)

Review Article

T. D. DZHAFAROV
Photostimulated Diffusion in Semiconductors
(a) Vol. **79**, No. 1, p. 11 – 45 (1983).
See also *9*, p. *93*; 11, p. 122; 14.3.3, p. 236; 16, p. 264; 22, p. 403.

Original Papers and Short Notes

H. KAHNT and F. SCHIRRMEISTER
*Electric Field Dependent Conductivity in Amorphous Semiconductors
(b) Vol. **115**, No. 1, p. 171 – 179 (1983).
See also *2*, p. *43*; 13.4, p. 172; 22, p. 403.

G. M. NIFTIEV, B. G. TAGIEV, and G. K. ASLANOV
*Injection and Thermodepolarization Currents in $YbGa_2Se_4$ Single Crystals
(a) Vol. **75**, No. 1, p. 59 – 65 (1983).
See also 13.4, p. 173; 22.8, p. 502.

H.-J. FITTING and J. BOYDE
*Monte-Carlo Calculation of Electron Attenuation in SiO_2
(a) Vol. **75**, No. 1, p. 137 – 142 (1983).
See also *6*, p. *70*; 22.6, p. 490.

M. K. HAN, W. A. ANDERSON, P. SUNG, and R. LAHRI
*Photogenerated Carrier Conduction Mechanisms for Schottky a-Si:H Solar Cells
(a) Vol. **75**, No. 1, p. 283 – 288 (1983).
See also *16*, p. *264*; 22.1.2, p. 424.

K. IRMSCHER, H. KLOSE, and K. MAASS
Electric Field Enhanced Electron Emission from Gold Acceptor Level and A-Centre in Silicon
(a) Vol. **75**, No. 1, p. K25 – K28 (1983).
See also *13.4*, p. *173*; 22.1.2, p. 425.

H. KÖSTER, JR., I. N. IASSIEVICH, and K. HÜBNER
Thermally Stimulated Excitation of Electrons from Deep Traps of Insulators in Electric Fields
(b) Vol. **115**, No. 2, p. 409 – 414 (1983).
See also *6*, p. *70*; *13.4*, p. *174*; 22.6, p. 491.

W. WESTERA
*Electro-Acoustic Effects in CdSe
(a) Vol. **75**, No. 2, p. 405 – 408 (1983).
See also *7*, p. *82*; 22.4.2, p. 466.

H. KAWAKAMI, S. EGUCHI, M. IKEDA, S. MATSUMOTO, and T. NIIMI
*Negative Differential Resistances in Ferromagnetic Semiconductor $CdCr_2Se_4$ Single Crystals
(a) Vol. **75**, No. 2, p. 433–439 (1983).
See also 22.8.2, p. 519.

F. SCHAUER, O. ZMEŠKAL, and S. NEŠPŮREK
*A New Approach to the Analysis of Steady-State Space-Charge-Limited Currents Using Their Activation Energies. An Application to a-As_2Se_3
(a) Vol. **75**, No. 2, p. 591–594 (1983).
See also 2, p. 45; *13.4*, p. *175*; 22, p. 404.

A. L. KARTUZHANSKII, V. M. ULANOV, and A. F. YURCHENKO
*Latent-Image and Fog Formation in Silver Halide Microcrystals by the Action of a Sequence of Light and Electric Pulses
(a) Vol. **75**, No. 2, p. 669–673 (1983).
See also 16, p. 265; 22.5.1, p. 476.

S. SARKAR
*Second Harmonic Suppression in Delayed Domain Mode of GaAs Samples
(a) Vol. **75**, No. 2, p. K161–K163 (1983).
See also 22.2.1, p. 443.

N. STOJADINOVIĆ, S. DIMITRIJEV, S. MIJALKOVIĆ, and Z. ŽIVIĆ
Reliability of n-Channel and p-Channel MOSTs in CMOS Integrated Circuits
(a) Vol. **76**, No. 1, p. 357–364 (1983).
See also *14.3.3*, p. *238*.

H. VON BERLEPSCH and L. BREHMER
*Invalidity of Scaling Property of Surface Potential Kinetics in HDPE
(a) Vol. **76**, No. 1, p. K27–K29 (1983).
See also *14.3.1*, p. *230*; 22.9, p. 526.

A. M. BELYANTSEV and M. N. DROZDOV
*On $I-U$ Characteristics of Semiconductor Structures with Inhomogeneous Mobility Profile
(a) Vol. **76**, No. 2, p. 721–728 (1983).
See also 22, p. 406.

A. M. BEREZHKOVSKII and A. A. OVCHINNIKOV
Temperature Dependence of the Wannier-Stark Level Width in a Semiconductor in a Strong Electric Field
(b) Vol. **117**, No. 1, p. 289–299 (1983).
See also *6*, p. *73*; *13*, p. *148*; 22.2.3, p. 452.

H. KLOSE, A. MERTENS, R. REETZ, and NG. TANG
*Ion Beam Contacting of Silicon (I): Aluminium in p-Silicon
(a) Vol. **77**, No. 1, p. 233–239 (1983).
See also *11*, p. *128*; 22.1.2, p. 430.

R. P. Bhardwaj, J. K. Quamara, K. K. Nagpaul, and B. L. Sharma
The Non-Isothermal Depolarization Current Studies in Kapton-H Thermoelectrets
(a) Vol. **77**, No. 1, p. 347–354 (1983).
Erratum
(a) Vol. **78**, No. 2, p. 759 (1983).
See also *14.4*, p. *251*; 22.9, p. 527.

G. Juška and K. Arlauskas
*Features of Hot Carriers in Amorphous Selenium
(a) Vol. **77**, No. 1, p. 387–391 (1983).
See also *2*, p. *48*; 22.1.3, p. 440.

D. A. Aronov and V. V. Rubinov
*On the Minimum Coefficient of Light Absorption by Free Carriers under the Conditions of Exclusion in "Pure" Semiconductors with Non-Injecting Contacts
(a) Vol. **77**, No. 1, p. K41–K44 (1983).
See also *20.1*, p. *319*; 22.1.1, p. 420.

H. Stramska
The Magneto-Optical Effect under a Strong Electric Field in Semiconductors
(b) Vol. **117**, No. 2, p. 541–547 (1983).
See also *18*, p. *274*; *20*, p. *309*; 22, p. 407.

A. M. Belyantsev and M. N. Drozdov
*The Effect of Spatial Variation of a High-Frequency Field on the Instability Occurrence in Inhomogeneous Semiconductors
(a) Vol. **77**, No. 2, p. 437–442 (1983).
See also 22, p. 408.

V. S. Lysenko, R. N. Litovskii, M. M. Lokshin, and M. F. Sherbakova
*Double Injection Current and Field Effect in SOS Diodes
(a) Vol. **77**, No. 2, p. 443–448 (1983).
See also *14.3.3*, p. *239*; 22, p. 408.

B. Rheinländer, G. Oelgart, R. Reulke, H. Haefner, and R. Stegmann
Degradation of Yellow Light-Emitting $GaAs_{0.1}P_{0.9}$:N Diodes (I): Implanted Diodes
(a) Vol. **77**, No. 2, p. 527–534 (1983).
See also *14.3.3*, p. *239*; *20.3*, p. *340*; 22.2.4, p. 454.

J. C. Lou, M. S. Lin, J. I. Chyi, M. D. Lin, and S. Sche
Electrical and Optical Characteristics of $GaAs_{0.6}P_{0.4}$ LEDs Fabricated by Zn Semi-Closed Diffusion Method
(a) Vol. **77**, No. 2, p. 741–747 (1983).
See also *14.3.3*, p. *239*; *20.3*, p. *340*; 22.2.4, p. 454.

V. P. Gribkovskii, A. A. Gladyshchuk, V. V. Zubritskii, V. V. Parashchuk, and G. P. Yablonskii
*Streamer Discharges in Semiconductors
(a) Vol. **77**, No. 2, p. 765–774 (1983).
See also 20.3, p. 340; 22.4.1, p. 463.

N. P. Kalmykova, B. T. Kolomiets, V. Kh. Shpunt, and E. A. Smorgonskaya
*On the Influence of the Non-Homogeneous Structure on the Current–Voltage Behaviour of Chalcogenide Vitreous Semiconductors
(a) Vol. **77**, No. 2, p. 803–808 (1983).
See also *2*, p. *49*; 22, p. 408.

R. V. Konakova, Yu. A. Tkhorik, I. L. Zaitsevskii, P. Kordos, M. Morvič, and J. Cervenak
*The Influence of Stress on the Fine Structure of Schottky Diode I–U Characteristics in the Breakdown Region
(a) Vol. **77**, No. 2, p. K125–K128 (1983).
See also *12.2*, p. *142*; 22.2.1, p. 445; 22.2.4, p. 454.

O. Yu. Borkovskaya, N. L. Dmitruk, R. V. Konakova, V. I. Fineberg, and V. I. Shakhovtsov
*The Radiation Effects on the Metal–GaP Interface
(a) Vol. **77**, No. 2, p. K129–K133 (1983).
See also *11*, p. *129*; 22.2.2, p. 450.

G. O. Müller and R. Mach
*On the Mechanism of Electron Impact Excited Luminescence Devices
(a) Vol. **77**, No. 2, p. K179–K183 (1983).
See also *20.3*, p. *341*; 22.4.1, p. 464.

Th. Jung and E. Nebauer
Interdiffusion Profiles of AuGe/n-GaAs Ohmic Contacts Studied by AES
(a) Vol. **77**, No. 2, p. K203–K206 (1983).
See also *9*, p. *94*; 22.2.1, p. 445.

V. Ya. Prinz and S. N. Rechkunov
*Influence of a Strong Electric Field on the Carrier Capture by Nonradiative Deep-Level Centers in GaAs
(b) Vol. **118**, No. 1, p. 159–166 (1983).
See also *13.4*, p. *181*; 22.2.1, p. 445.

P. Müller
*Contactless Determination of the Conductivity in Semi-Insulators
(a) Vol. **78**, No. 1, p. 41–51 (1983).
See also *14.3.3*, p. *239*; 22.2.1, p. 445; 22.5.1, p. 476.

G. Oelgart, B. Rheinländer, R. Reulke, H. Haefner, and R. Stegmann
Degradation of Yellow Light-Emitting $GaAs_{0.1}P_{0.9}$:N Diodes (II): Diffused Diodes
(a) Vol. **78**, No. 1, p. 169–176 (1983).
See also *14.3.3*, p. *240*; *20.3*, p. *342*; 22.2.4, p. 455.

J. Tyczkowski and M. Kryszewski
*Poole-Frenkel Centres in Dielectrics. Kind and Space Density in Plasma-Polymerized Organosilicon Films
(a) Vol. **78**, No. 1, p. 259–265 (1983).
See also *13.4*, p. *183*; 14.3.1, p. 231; 22.9, p. 529.

D. Verity, D. Shaw, F. J. Bryant, and C. G. Scott
*DLTS Investigation of Electron Traps in As-Grown and Cd-Fired CdS
(a) Vol. **78**, No. 1, p. 267 – 275 (1983).
See also *13.4*, p. *183*; 22.4.1, p. 464.

O. B. Tagiev, G. M. Niftiev, and I. M. Askerov
*The Frenkel Thermal-Field Emission in $EuGa_2S_4$ Single Crystals
(a) Vol. **78**, No. 1, p. K43 – K46 (1983).
See also *13.4*, p. *183*; 22.8, p. 507.

J. Kalinowski and R. Signerski
*Exciton-Enhanced Double Injection Currents in Tetracene Crystals
(b) Vol. **118**, No. 2, p. K147 – K150 (1983).
See also *13.5.1*, p. *197*; 20.3, p. 343; 22.9, p. 529.

P. K. Katana
Two-Photon Interband Absorption of Light in Semiconductors in External Parallel Electric and Magnetic Fields
(b) Vol. **119**, No. 1, p. 89 – 98 (1983).
See also *13.1*, p. *162*; 18, p. 275; *20.1*, p. *325*; 22, p. 411.

T. M. Gasymov, A. A. Katanov, and M. M. Babaev
*Thermopower and Thermomagnetic Effects in Kane Semiconductors under Conditions of Electron and Phonon Heating and Their Mutual Drag
(b) Vol. **119**, No. 1, p. 391 – 399 (1983).
See also 6, p. 77; *15*, p. *262*; 22, p. 411.

F. Campabadal, V. Milián, and X. Aymerich-Humet
*Trap-Assisted Tunneling in MIS and Schottky Structures
(a) Vol. **79**, No. 1, p. 223 – 236 (1983).
See also 13.4, p. 186; *14.3.3*, p. *240*; 22.6, p. 496.

H. K. Phan, L. H. Phu, and C. V. Chiem
*Stress Effects on the Switching Voltage of Metal–SnO_2–Si(n)–Si(p^+) Devices
(a) Vol. **79**, No. 1, p. K25 – K28 (1983).
See also *14.3.3*, p. *241*; 22.1.2, p. 435.

G. M. Guseinov, G. M. Niftiev, and M. G. Shakhtakhtinskii
*Effect of Illumination on Current–Voltage Characteristics of $GdIn_3S_6$ Single Crystals
(a) Vol. **79**, No. 1, p. K53 – K55 (1983).
See also 22.8, p. 509.

V. M. Fomin and E. P. Pokatilov
Nonlinear Kinetics and Optical Properties of Band Charge Carriers
(b) Vol. **119**, No. 2, p. 483 – 492 (1983).
See also 6, p. 78; 13, p. 151; *20.1*, p. *327*; 22, p. 412.

B. G. Tagiev, G. M. Niftiev, S. A. Abushov, and A. O. Khalilov
Optical Properties of $InSe_{1-x}S_x$ Single Crystals in the Region of the Fundamental Absorption Edge
(b) Vol. **119**, No. 2, p. K143 – K148 (1983).
See also 13.5.1, p. 198; *20.1*, p. *328*; *20.3*, p. *345*; 22.3.1, p. 459.

B. M. Wilamowski
The Punch-Through Transistor with MOS Controled Gate
(a) Vol. **79**, No. 2, p. 631—637 (1983).
See also *14.3.3*, p. *241*; 22, p. 412.

J. Gong, C. M. van Vliet, D. Schoenfeld, and G. Bosman
*Noise from Hole-Initiated Photoavalanche Processes
(a) Vol. **79**, No. 2, p. K201—K203 (1983).
See also *14.3.3*, p. *242*; 16, p. 267; 22.1.2, p. 437.

M. Asche and H. Kostial
*Experimental Indication of the Presence of Loops for Hot Electron Properties in Many-Valley Semiconductors
(b) Vol. **120**, No. 1, p. K83—K85 (1983).
See also *13*, p. *152*; 22.1.2, p. 437.

J. C. Bernède
*Influence du métal de l'anode sur l'existence d'une résistance différentielle négative dans des structures MIM
(Influence of the Anode Metal on the Existence of a Negative Differential Resistance)
(a) Vol. **80**, No. 1, p. 85—93 (1983).
See also 22.6, p. 497.

H. Müller, P. Haasen, and W. Schröter
Thermopower of Plastically Deformed Cadmium Telluride
(a) Vol. **80**, No. 1, p. 223—230 (1983).
See also *10.2*, p. *120*; *15*, p. *263*; 22.4.3, p. 469.

R. P. Bhardwaj, J. K. Quamara, K. K. Nagpaul, and B. L. Sharma
*Thermally Stimulated Spontaneous Currents from Kapton Polyimide–Metal Contacts
(a) Vol. **80**, No. 1, p. 325—332 (1983).
See also *14.4*, p. *254*; 22.9, p. 532.

I. Mojzes, R. Veresegyházy, V. Obodnikov, Yu. Vaulin, and S. Vasilev
Mapping the Redistributed Components of the Au–Ge/GaAs System after Scanning Laser Annealing
(a) Vol. **80**, No. 1, p. K33—K36 (1983).
See also *14.3.2*, p. *235*; 22.2.1, p. 447.

N. A. Nikolov and P. N. Malinov
*Resonances and Averaged Forces in Bounded Semiconductor Plasmas
(b) Vol. **120**, No. 2, p. 775—782 (1983).
See also *13*, p. *153*.

R. P. Bhardwaj, J. K. Quamara, K. K. Nagpaul, and B. L. Sharma
*Field-Induced Thermally Stimulated Currents in Kapton-H Polyimide Films
(a) Vol. **80**, No. 2, p. 559—566 (1983).
See also *14.4*, p. *254*; 22.9, p. 532.

B. P. TYAGI and K. SEN
*On the Resistivity of Polycrystalline Silicon
(a) Vol. **80**, No. 2, p. 679–684 (1983).
See also *13.4*, p. *191*; 22.1.2, p. 439.

K. HERRMANN, J. W. TOMM, A. JALYSCHKO, and W. BREMSER
*On the Behaviour of the Threshold Current Densities in Dual-Wavelength $Pb_{1-x}Sn_xTe$ Injection Lasers
(a) Vol. **80**, No. 2, p. K113–K116 (1983).
See also *20.2*, p. *333*; 22, p. 415.

J. HAŁADYJ, J. MICHALSKI, and A. SZYMAŃSKI
*On the Anomalous Field Dependence of Drift Mobility in Thin Organic Films
(a) Vol. **80**, No. 2, p. K137–K139 (1983).
See also *14.3.1*, p. *232*; 22.9, p. 532.

K. G. NIKIFOROV, V. P. BUZHOR, S. I. RADAUTSAN, and V. E. TEZLEVAN
*Switching Effect Based on N-Shaped Negative Differential Resistance in n-$CdCr_2Se_4$ Single Crystals
(a) Vol. **80**, No. 2, p. K149–K151 (1983).
See also 22.8, p. 511.

B. RHEINLÄNDER, R. PICKENHAIN, H. HAEFNER, and G. OELGART
*Degradation of LPE-GaP:N LED
(a) Vol. **80**, No. 2, p. K203–K207 (1983).
See also *20.3*, p. *348*; 22.2.2, p. 451.

14.4 Dielectrics

Original Papers and Short Notes

J. SHANKER and R. SUNDARAJ
*Dielectric and Anharmonic Behaviour of Silver Halides, Thallous Halides, Alkaline Earth Fluorides, and Lead Fluoride Crystals
(b) Vol. **115**, No. 1, p. 67 – 74 (1983).
See also **4**, p. **60**; **13**, p. **146**; **22.5**, p. **473**; **22.5.1**, p. **476**; **22.5.3**, p. **486**.

W. MEDYCKI
*Mean Charge Depth in Non-Metallized Electrets
(a) Vol. **75**, No. 1, p. K21 – K24 (1983).
See also **22.9**, p. **525**.

N. M. RAVINDRA, C. ANCE, J. P. FERRATON, A. DONNADIEU, and S. ROBIN
Penn-Like Model Calculations. Application to Hydrogenated Amorphous Silicon Films Prepared by Glow Discharge
(b) Vol. **115**, No. 2, p. 347 – 352 (1983).
See also *2*, p. *44*; **13.1**, p. **156**; **22.1.2**, p. **425**.

A. G. FOKIN
*The Use of an Iteration Method for Calculating the Potential Energy of Nonhomogeneous Dielectrics
(b) Vol. **115**, No. 2, p. 509 – 518 (1983).

G. P. KRAMAR, YA. I. PANOVA, and V. V. PASSYNKOV
*The Influence of Microstructure on the Dielectric Spectra of Ferrites
(a) Vol. **76**, No. 1, p. 95 – 100 (1983).
See also **22.8.2**, p. **519**.

H. SODOLSKI
*Depolarization Current in a Dielectric with Deep Trapped Charge Layers
(a) Vol. **76**, No. 1, p. 303 – 309 (1983).
See also **22.9**, p. **526**.

J. W. BRIGHTWELL, C. N. BUCKLEY, L. S. MILLER, and B. RAY
*Electrical Transport and Phase Stability in Silver Iodide–Cadmium Iodide
(a) Vol. **76**, No. 1, p. 391 – 398 (1983).
See also *1.2*, p. *15*; **22.5.4**, p. **488**.

W. STARK and R. DANZ
*Investigations of the Thermally Stimulated Change of the Surface Potential of FEP Electrets
(a) Vol. **76**, No. 1, p. K101 – K103 (1983).
See also **22.9**, p. **526**.

S. A. Y. AL-ISMAIL and C. A. HOGARTH
*Temperature Dependence of Voltage Controlled Negative Resistance in an Electroformed Cu–SiO_x–Cu Structure
(a) Vol. **76**, No. 2, p. 559 – 563 (1983).
See also **22.6**, p. **493**.

F. GLASSE
*The Influence of Thermal Measurement Conditions on the Results of Dielectric Measurements in PLZT Ceramic with Diffused Phase Transition
(a) Vol. **76**, No. 2, p. K121 – K123 (1983).
See also 22.8, p. 505.

K. SOMAIAH and V. HARI BABU
*Ionic Conductivity of BaFCl
(b) Vol. **117**, No. 1, p. 75 – 79 (1983).
See also *10.2*, p. *112*; 22.5, p. 474.

F. BECHSTEDT, R. ENDERLEIN, and D. REICHARDT
*Inverse Dielectric Function for a Semi-Infinite Solid
(b) Vol. **117**, No. 1, p. 261 – 270 (1983).
See also *13.3*, p. *168*; 22, p. 406.

R. P. BHARDWAJ, J. K. QUAMARA, K. K. NAGPAUL, and B. L. SHARMA
*The Non-Isothermal Depolarization Current Studies in Kapton-H Thermoelectrets
(a) Vol. **77**, No. 1, p. 347 – 354 (1983).
Erratum
(a) Vol. **78**, No. 2, p. 759 (1983).
See also 14.3.4, p. 245; 22.9, p. 527.

H. B. LAL and O. P. SRIVASTAVA
*Electrical Transport Studies in Copper Mercuric Iodide
(a) Vol. **77**, No. 1, p. 405 – 412 (1983).
See also 10.2, p. 112; *15*, p. *261*; 22.5.4, p. 488.

I. V. STASYUK and S. S. KOTSUR
*The Microscopic Theory of the Gyration and Electrogyration in Dielectric Crystals
(b) Vol. **117**, No. 2, p. 557 – 568 (1983).
See also *20*, p. *309*; 22, p. 407.

H. KLIEM and G. ARLT
*A Relation between Dielectric Polarization Currents and Distribution Functions
(b) Vol. **117**, No. 2, p. K81 – K85 (1983).

J. WOLLBRANDT, U. BRÜCKNER, and E. LINKE
Investigations of Mechanically Induced Excited States on Cleavage Planes of Ionic Crystals (I): Method and Apparatus for Charge and Electron Emission Measurements
(a) Vol. **77**, No. 2, p. 545 – 552 (1983).
See also *13.3*, p. *169*; 17, p. 270; 22.5.2, p. 480.

P. K. C. PILLAI, G. K. NARULA, A. K. TRIPATHI, and R. G. MENDIRATTA
*A Study of Absorption Currents in Polycarbonate: Polypropylene Blend
(a) Vol. **77**, No. 2, p. 693 – 698 (1983).
See also 22.9, p. 528.

H. SODOLSKI
*DC Conductivity and Electric Relaxation Currents in a Polyester Polymer
(a) Vol. **77**, No. 2, p. 749 – 758 (1983).
See also 22.9, p. 528.

R. K. Chaudhary, A. V. Krishna Rao, and K. V. Rao
*Dielectric Properties of Sm of Gd Doped CaF_2 Single Crystals
(a) Vol. **77**, No. 2, p. K161–K165 (1983).
See also 22.5.3, p. 486.

L. Wendler
*Effects of a Transition Layer on Nonlinear Surface Polaritons
(b) Vol. **118**, No. 1, p. 205–210 (1983).
See also *13.5*, p. *193*.

U. Pilatus and F. Fischer
*Thermally Stimulated Depolarization of AsO_2^- Centres in Alkali Halide Crystals
(b) Vol. **118**, No. 1, p. 217–221 (1983).
See also *10.2*, p. *113*; 22.5.2, p. 480.

J. Prakash and Rahul
*A Method for the Evaluation of Dielectric Relaxation Parameters in Electronic Excited State
(b) Vol. **118**, No. 1, p. 229–237 (1983).
See also 10.2, p. 113; 22.5.2, p. 480.

H. Köster, Jr. and K. Hübner
Statistics and Transport Behaviour of Electrons in SiO_2
(b) Vol. **118**, No. 1, p. 293–301 (1983).
See also *6*, p. *75*; *13*, p. *150*; 22.6, p. 494.

F. Bechstedt, R. Enderlein, and D. Reichardt
Electronic Relaxation Effects in Core Level Spectra of Surfaces and Interfaces
(b) Vol. **118**, No. 1, p. 327–336 (1983).
See also *13.3*, p. *169*; 21.6, p. 398; 22.1.2, p. 432; 22.2.1, p. 445; 22.6, p. 494.

J. Poźniak and G. Berg
Theoretical Analysis of the Dipole Polarization and Depolarization in Me^{++}-Doped NaCl-Type Alkali Halide Crystals (II): Next-Nearest Neighbour Approach
(a) Vol. **78**, No. 1, p. 69–76 (1983).
See also *10.2*, p. *114*; 22.5.2, p. 481.

J. Wollbrandt, U. Brückner, and E. Linke
Investigations of Mechanically Induced Excited States on Cleavage Planes of Ionic Crystals (II): Investigations of the Mechanisms of Electron Emission from Cleaved Alkali Halide Crystals and of the Influence of Surface Charges
(a) Vol. **78**, No. 1, p. 163–168 (1983).
See also *13.4*, p. *183*; 17, p. 270; 22.5.2, p. 481.

H. Mashiyama
Modified Phase Modulation-Only Approximation of Incommensurate Phases in K_2SeO_4-Type Crystals
(b) Vol. **118**, No. 2, p. 757–767 (1983).
See also *1.2*, p. *18*; 8, p. 89; 22.8.1, p. 515.

R. R. Nigmatullin and R. M. Hill
*A Relationship between the Universal Law of Susceptibility Response and the Correlation Weakening Principle
(b) Vol. **118**, No. 2, p. 769 – 778 (1983).

J. S. Kumar, G. Narayana, M. Chandra Shekar, U. V. Subba Rao, and V. Hari Babu
Dielectric and Electrical Properties of SnO_2, Sb_2O_3, and Their Mixed Films
(a) Vol. **78**, No. 2, p. 647 – 653 (1983).
See also *14.3.1*, p. *231*; 22.6, p. 495; 22.6.1, p. 498.

A. Torres, J. Jimenez, V. Carbayo et J. A. de Saja
*Courants de dépolarisation stimulés thermiquement dans le sulfate de cérium (III) enneahydraté
(Thermally Stimulated Depolarization Currents in Cerium Sulfate Hydrates)
(a) Vol. **78**, No. 2, p. 671 – 677 (1983).
See also 22.8.2, p. 522.

J. Shanker and G. D. Jain
*Analysis of the Dielectric Properties of Mixed Alkali Halide Crystals
(b) Vol. **119**, No. 1, p. 185 – 191 (1983).
See also 20, p. 310; 22.5.4, p. 489.

J. W. Brightwell, L. S. Miller, A. Munday, and B. Ray
The Silver Iodide–Thallium Iodide Pseudo-Binary System
(a) Vol. **79**, No. 1, p. 293 – 300 (1983).
See also *1.2*, p. *19*; 22.5, p. 475; 22.5.1, p. 476.

A. G. Fokin
*Macroscopical Dielectric Permittivities of Nonhomogeneous Media
(b) Vol. **119**, No. 2, p. 741 – 754 (1983).

A. D. Reddy, S. G. Sathyanarayan, and G. S. Sastry
*Thermally Stimulated Depolarisation Current Studies on Some Hydrogen-Bonded Crystals
(a) Vol. **79**, No. 2, p. 575 – 583 (1983).
See also 22.8.1, p. 517.

B. Gross, H. von Seggern, and D. A. Berkley
*Long Term Behaviour of Radiation-Induced Currents in Fluorinated Ethylene Prophylene Copolymer
(a) Vol. **79**, No. 2, p. 607 – 615 (1983).
See also *11*, p. *133*; 22.9, p. 530.

L. Komitov and A. G. Petrov
Quasihomeotropic, Reversely Pretilted Nematic Liquid Crystal Layers. Electrooptical Behaviour
(a) Vol. **79**, No. 2, p. 623 – 629 (1983).
See also *2*, p. *53*; 20.1, p. 328; 22.9, p. 531.

G. Pretzsch
*Charge State of PTFE Thermoelectrets
(a) Vol. **79**, No. 2, p. K139 – K142 (1983).
See also 22.9, p. 531.

P. Fedorko and F. Masin
*DC Conductivity and Optical Absorption Spectra of Polyacetylene Doped with HSO_3F
(a) Vol. **79**, No. 2, p. K155 – K158 (1983).
See also *20.1*, p. *328*; 22.9, p. 531.

A. I. Rubinshtein
*Image Forces at the Metal Electrode–Electrolyte Solution Interface: Account of Spatial Dispersion of the Solution Permittivity
(b) Vol. **120**, No. 1, p. 65 – 76 (1983).
See also *13*, p. *152*.

L. N. Kantorovich
*Multipole Theory of the Polarization of Solids by Point Defects (I): Dipole Approximation
(b) Vol. **120**, No. 1, p. 77 – 86 (1983).
See also *13.4*, p. *188*.

A. Alegria, J. M. Barandiarán, and J. Colmenero
Dielectric Relaxation at the Glass Transition as a Free Volume Process. A Single Relaxation Time Approach
(b) Vol. **120**, No. 1, p. 349 – 360 (1983).
See also *2*, p. *54*; 22.9, p. 531.

R. P. Bhardwaj, J. K. Quamara, K. K. Nagpaul, and B. L. Sharma
*Thermally Stimulated Spontaneous Currents from Kapton Polyimide–Metal Contacts
(a) Vol. **80**, No. 1, p. 325 – 332 (1983).
See also *14.3.4*, p. *248*; 22.9, p. 532.

H. Jain
Haven Ratio for Sodium in Synthetic α-Quartz
(a) Vol. **80**, No. 1, p. K15 – K20 (1983).
See also *9*, p. *96*; 10.2, p. 120; 22.6, p. 497.

K. Lal and P. Thoma
*Study of Microstructural Changes in Silicon Single Crystals Subject to High Electric Fields by Using a High Resolution X-Ray Diffraction Method
(a) Vol. **80**, No. 2, p. 491 – 502 (1983).
See also *1.1*, p. *13*; 22.1.2, p. 439.

R. P. Bhardwaj, J. K. Quamara, K. K. Nagpaul, and B. L. Sharma
*Field-Induced Thermally Stimulated Currents in Kapton-H Polyimide Films
(a) Vol. **80**, No. 2, p. 559 – 566 (1983).
See also *14.3.4*, p. *248*; 22.9, p. 532.

A. A. Kaminskii, I. M. Silvestrova, S. E. Sarkisov, and G. A. Denisenko
*Investigation of Trigonal $(La_{1-x}Nd_x)_3Ga_5SiO_{14}$ Crystals (II): Spectral Laser and Electromechanical Properties
(a) Vol. **80**, No. 2, p. 607 – 620 (1983).
See also 12.2, p. 144; 20.1, p. 331; *20.2*, p. *333*; 20.3, p. 347; 22.8.1, p. 518.

Yu. Yu. Sidorin and Yu. A. Zakharov
*The Light Stimulated Charge Carrier Transfer in Silver Azide
(a) Vol. **80**, No. 2, p. K157 – K160 (1983).
See also *16*, p. *268*; 22, p. 415.

14.4.1 Ferroelectrics

Original Papers and Short Notes

V. Kalesinskas, J. Grigas, R. Jankevičius, and A. Audzionis
*Soft Mode in the Microwave Dielectric Spectra of the SbSI–BiSI System
(b) Vol. **115**, No. 1, p. K11–K13 (1983).
See also 6, p. 70; 22.8, p. 502.

H. Beige, A. Kühnel, and B. Lorenz
*Anomalous Behaviour of Nonlinear Dielectric, Elastic, and Electromechanical Coefficients at Ferroelectric Phase Transitions. Rochelle Salt
(a) Vol. **75**, No. 1, p. 129–136 (1983).
See also 12.2, p. 141; 22.8.1, p. 512.

W. Kucharczyk and P. Górski
*Measurement of Nonlinear Electrooptic Effects in KDP Crystals
(a) Vol. **75**, No. 1, p. K87–K90 (1983).
See also *20.1*, p. *313*; 22.8.1, p. 512.

I. P. Aleksandrova, S. Grande, Yu. N. Moskvich, A. I. Krieger, and V. A. Koptsik
Second-Order Quadrupole Shift in the Incommensurate Structures of Rb_2ZnBr_4 and Rb_2ZnCl_4
(b) Vol. **115**, No. 2, p. 603–613 (1983).
See also *1.2*, p. *14*; *19*, p. *300*; 22.8, p. 503.

O. Kersten, A. Rost, and G. Schmidt
*Dielectric Dispersion of Relaxor Ferroelectrics (SBN 75 and PLZT 8/65/35)
(a) Vol. **75**, No. 2, p. 495–500 (1983).
See also 22.8, p. 503.

A. Sternberg, M. Zvára, and P. Hlídek
*Influence of Permanent Magnetic Field on Dielectric Properties of Transparent Ferroelectric PLZT Ceramics
(a) Vol. **75**, No. 2, p. K179–K182 (1983).
See also 18, p. 273; 22.8.1, p. 513.

J. M. Wesselinowa
*Spin–Phonon Interactions in the Ising Model with a Transverse Field
(b) Vol. **116**, No. 1, p. 71–82 (1983).
See also *6*, p. *71*; 22.8.1, p. 513.

S. K. Khanna
*On the Elastic Anisotropy of Ferroelectric Triglycine Selenate
(b) Vol. **116**, No. 1, p. 173–178 (1983).
See also *1.2*, p. *14*; 12.2, p. 142; 22.9, p. 526.

L. M. Avkhutskii, R. L. Davidovich, L. A. Zemnukhova, P. S. Gordienko, V. Urbonavičius, and J. Grigas
*Peculiarities of Phase Transitions and Physical Properties in $(NH_4)_2SbF_5$
(b) Vol. **116**, No. 2, p. 483–488 (1983).
See also 8, p. 87; 19, p. 301; 22.8, p. 505.

S. PAČESOVÁ, B. BŘEZINA, and L. JASTRABÍK
Optical Properties of K_2SeO_4 in the Vicinity of Phase Transitions
(b) Vol. **116**, No. 2, p. 645–652 (1983).
See also *6*, p. *73*; *20.1*, p. *316*; 22.8.1, p. 513.

G. BORCHARDT, J. VON CIEMINSKI, and G. SCHMIDT
*Switching Properties of SBN Single Crystals
(a) Vol. **76**, No. 2, p. K141–K145 (1983).
See also 22.8.1, p. 513.

Z. CZAPLA, O. CZUPIŃSKI, L. SOBCZYK, J. MRÓZ, and H. PYKACZ
*Dielectric Properties and Ferroelectric Phase Transitions in Mixed $Rb_x(NH_4)_{1-x}HSeO_4$ Crystals
(a) Vol. **76**, No. 2, p. K155–K157 (1983).
See also *1.2*, p. *16*; 22.8.1, p. 513.

A. K. BATRA, S. C. MATHUR, and A. MANSINGH
*Dielectric Dispersion in Triglycine Sulphate Single Crystals
(a) Vol. **77**, No. 1, p. 399–404 (1983).
See also 22.9, p. 528.

V. M. MUKHORTOV, YU. I. GOLOVKO, V. A. ALESHIN, E. V. SVIRIDOV, VL. M. MUKHORTOV, V. P. DUDKEVICH, and E. G. FESENKO
*Ferroelectric Phase Transition in Heteroepitaxial Films of (Ba, Sr) TiO_3
(a) Vol. **77**, No. 1, p. K37–K40 (1983).
See also 1.5, p. 37; 22.8.1, p. 514.

V. N. KASHCHEEV
Phonon Relaxation Due to Anharmonism and Central Peak in Structural Phase Transitions (I): Soft Optic Modes and Acoustic Phonons in Cubic Crystals
(b) Vol. **117**, No. 2, p. 483–492 (1983).
See also *6*, p. *74*.

M. D. KAPLAN
The Dynamical Electron–Phonon Mode Coupling in Jahn-Teller Crystals with Antiferroelastic Ordering
(b) Vol. **118**, No. 1, p. 81–87 (1983).
See also *6*, p. *74*; 7, p. 83.

V. N. KASHCHEEV
*Central Peak in Ising Model with Transverse Field (II)
(b) Vol. **118**, No. 1, p. 427–432 (1983).

V. V. ARISTOV, L. S. KOKHANCHIK, K.-P. MEYER, and H. BLUMTRITT
*Scanning Electron Microscopic Investigation of Peculiarities of the $BaTiO_3$ Ferroelectric Domain Contrast
(a) Vol. **78**, No. 1, p. 229–236 (1983).
See also 22.8.1, p. 515.

V. M. Mukhortov, Yu. I. Golovko, V. A. Aleshin, E. V. Sviridov, Vl. M. Mukhortov, V. P. Dudkevich, and E. G. Fesenko
*(Ba, Sr)TiO_3 Heteroepitaxial Films
(a) Vol. 78, No. 1, p. 253 – 257 (1983).
See also 1.5, p. 37; 22.8.1, p. 515.

T. Kala
*Contribution to the Study of Tetragonal and Rhombohedral Phase Coexistence in the $PbZrO_3$–$PbTiO_3$ System
(a) Vol. 78, No. 1, p. 277 – 282 (1983).
See also *1.2*, p. *17*; 22.8.1, p. 515.

A. L. Aleksandrovskii and A. I. Nagaev
*Electrooptic Diffraction in Polydomain Barium–Sodium Niobate Crystals
(a) Vol. 78, No. 2, p. 431 – 438 (1983).
See also *20*, p. *310*; 22.8.1, p. 516.

R. Jakubas and L. Sobczyk
*Dielectric Relaxation Study of the Incommensurate Phase of $[(CH_3)_4N]_2 \cdot CuBr_4$ near the Transition Temperature
(a) Vol. 78, No. 2, p. K147 – K150 (1983).
See also 22.9, p. 529.

I. Henning, E. Hegenbarth, and B. Brezina
*Specific Heat of a K_2SeO_4 Single Crystal in Temperature Range 0.6 to 7 K
(a) Vol. 78, No. 2, p. K167 – K170 (1983).
See also *8*, p. *89*; 22.8.1, p. 516.

A. S. T. Pires and F. C. Sá Barreto
*Solitons in Ferroelectric Liquid Crystal
(b) Vol. 119, No. 1, p. 121 – 130 (1983).
See also *2*, p. *51*; 22.9, p. 530.

S. Różański
*Determination of Pitch and Chiral Smectic C DOBAMBC
(a) Vol. 79, No. 1, p. 309 – 312 (1983).
See also 2, p. 51; 22.9, p. 530.

A. A. Demidenko, A. P. Zdebsyii, V. N. Piskovoi, and A. U. Savchuk
*Determination of Nonlinear Elastic Coefficients by Resonant Methods
(a) Vol. 79, No. 1, p. K73 – K76 (1983).
See also *12.2*, p. *144*; 22.8.1, p. 516.

R. Styrkowiec
*The Hampered, Sideways Motion of a Single Domain Wall in Rochelle Salt Crystals
(a) Vol. 79, No. 1, p. K99 – K102 (1983).
See also 22.8.1, p. 517.

V. N. Kashcheev
Phonon Relaxation Due to Anharmonism and Central Peak in Structural Phase Transitions (II): Soft Optic Modes in Uniaxial Ferroelectrics and Soft Acoustic Modes in Cubic Crystals
(b) Vol. **119**, No. 2, p. 507–514 (1983).
See also *6*, p. *78*.

I. Muševič, B. Žekš, R. Blinc, Th. Rasing, and P. Wyder
Dielectric Study of the Modulated Smectic C–Uniform Smectic C Transition in a Magnetic Field
(b) Vol. **119**, No. 2, p. 727–733 (1983).
See also 2, p. 52; 22.9, p. 530.

V. S. Gorelik, B. S. Umarov, and M. Umarov
On the Connection between Isofrequency Temperature Dependences of Inelastic Light Scattering and Dielectric Anomalies in Lithium Niobate Crystals
(b) Vol. **120**, No. 1, p. 131–136 (1983).
See also *6*, p. *79*; *20.1*, p. *328*; 22.8.1, p. 517.

J. Berdowski
*The Influence of Isotope Effects on Anomalous Changes of Dielectric and Electric Properties of DADA Crystals in the Range of the Phase Transition
(a) Vol. **80**, No. 1, p. 65–72 (1983).
See also *1.2*, p. *19*; 22.8.1, p. 517.

A. Aresti, F. Meloni, G. G. Pegna, and M. Serra
*Microwave Radiometric Study of Ferro–Paraelectric Phase Transitions in KNO_3
(a) Vol. **80**, No. 1, p. 119–126 (1983).
See also *1.2*, p. *19*; 22.8.1, p. 518.

G. I. Rosenman, V. I. Pechorskii, Yu. L. Chepelev, E. I. Boikova, and L. E. Issakova
*Exoemission of Pyroelectrics
(b) Vol. **120**, No. 2, p. 667–670 (1983).
See also *17*, p. *271*; 22.8.1, p. 518.

E. Bărbulescu, D. Barb, A. Bărbulescu, and D. Geana
*Computer Calculated F–P Diagram for the $BaTiO_3$–$PbTiO_3$ System
(a) Vol. **80**, No. 2, p. 431–437 (1983).
See also *1.2*, p. *20*; 22.8.1, p. 518.

R. B. Lal, M. D. Aggarwal, R. L. Kroes, and W. R. Wilcox
A New Technique of Solution Crystal Growth
(a) Vol. **80**, No. 2, p. 547–552 (1983).
See also *3*, p. *59*; 22.9, p. 532.

L. A. Zemnukhova, R. L. Davidovich, P. S. Gordienko, J. Grigas, A. N. Kovrianov, S. I. Kuznetsov, T. A. Kaidalova, and V. Urbonavičius
*Investigation of Phase Transitions in K_2SbF_5
(a) Vol. **80**, No. 2, p. 553–558 (1983).
See also *1.2*, p. *20*; 8, p. 92; 19, p. 306; 22.8, p. 511.

F. Vasiliu, P. Gr. Lucuta, and F. Constantinescu
*Structure–Sintering Temperature Relationship in PZT Solid Solutions at the Morphotropic Phase Boundary
(a) Vol. **80**, No. 2, p. 637–642 (1983).
See also *1.3*, p. *28*; 22.8, p. 511.

15. Thermoelectric and Thermomagnetic Properties

Original Papers and Short Notes

M. D. SMOLIN, A. F. ANDREEVA, and V. A. OGORODNIK
*Electroconduction and Thermoelectricity of Terbium Oxide Films
(b) Vol. **115**, No. 2, p. 587–592 (1983).
See also 13.5.3, p. 201; *14.3.1*, p. *229*; 22.6, p. 491.

H. GRUBER and E. KRAUTZ
Magnetoresistance and Conductivity in the Binary System Titanium–Oxygen (II): Semiconductive Titanium Oxides
(a) Vol. **75**, No. 2, p. 511–518 (1983).
See also 8, p. 86; *14.1*, p. *206*; *14.3*, p. *216*; 18, p. 273; 22.6, p. 491.

L. G. KHVOSTANTSEV and V. A. SIDOROV
*Thermoelectric Properties of GeTe at High Hydrostatic Pressure up to 8.5 GPa and Its Valence-Band Structure
(b) Vol. **116**, No. 1, p. 83–89 (1983).
See also 13.1, p. 157; 14.3, p. 217; 22.7, p. 499.

B. GILLOT and F. JEMMALI
*Dependence of Electrical Properties in Iron–Cobalt, Iron–Zinc Ferrites near Stoichiometry on Firing Temperature and Atmosphere
(a) Vol. **76**, No. 2, p. 601–608 (1983).
See also *14.3*, p. *218*; 22.8.2, p. 520.

M. STORDEUR and H. T. LANGHAMMER
Anisotropy of Transport Properties Due to Direction-Dependent Band Nonparabolicity
(b) Vol. **117**, No. 1, p. 329–333 (1983).
See also *13.1*, p. *158*; 14.3, p. 219; 22, p. 406.

L. V. VASSILEV and G. CH. CHRISTAKUDIS
On the Calculation of Transport Integrals in Lead Chalcogenides
(b) Vol. **117**, No. 1, p. K53–K56 (1983).
See also *14.3*, p. *219*; 22, p. 407.

H. B. LAL and O. P. SRIVASTAVA
*Electrical Transport Studies in Copper Mercuric Iodide
(a) Vol. **77**, No. 1, p. 405–412 (1983).
See also 10.2, p. 112; *14.4*, p. *251*; 22.5.4, p. 488.

P. V. REDDY and T. S. RAO
*A Study of the Semiconducting Properties of Mixed Mn–Mg Ferrites
(a) Vol. **77**, No. 1, p. K63–K67 (1983).
See also *14.3*, p. *219*; 22.8.1, p. 514.

A. K. SHARMA
*Thickness Dependence of the Thermoelectric Power of Tellurium Films
(a) Vol. **77**, No. 1, p. K81–K85 (1983).
See also 22.1.3, p. 440.

O. A. Golikova and S. Samatov
Conduction Mechanism of Some "Quasi-Amorphous" Semiconductors on the Basis of Boron
(a) Vol. **77**, No. 2, p. 449–453 (1983).
See also *13.4*, p. *180*; *14.3*, p. *220*; 22, p. 408.

B. Pistoulet, D. Coquillat, J. C. Tedenac, G. Brun, and M. Maurin
Hole Effective Mass and Impurity Levels in Undoped AgTlTe. Optimum Thermoelectric Figure of Merit
(a) Vol. **77**, No. 2, p. 669–677 (1983).
See also *13.1*, p. *159*; *13.4*, p. *181*; 14.3, p. 221; 22.8, p. 506.

P. Vašek and P. Svoboda
*Thermopower of Bismuth in Low Longitudinal Magnetic Fields
(b) Vol. **118**, No. 1, p. 405–408 (1983).
See also 21.7, p. 401.

A. K. Sharma and P. Jayarama Reddy
*Thermoelectric Power in InAs Films Obtained by Flash Evaporation
(a) Vol. **78**, No. 1, p. K27–K29 (1983).
See also 22.2.1, p. 445.

S. Anzai, S. Ohta, A. Yoshino, J. Nishio, and M. Hatori
Charge-Density-Wave-Like Transition in V_3Te_4
(b) Vol. **118**, No. 2, p. K99–K102 (1983).
See also *14.1*, p. *208*; *18.1*, p. *279*; 22, p. 410.

P. Venugopal Reddy, R. Satyanarayana, and T. Seshagiri Rao
*Electrical Conduction in Nickel Ferrite
(a) Vol. **78**, No. 2, p. K109–K111 (1983).
See also *14.3*, p. *223*; 22.8.2, p. 523.

R. Schumann and K. Elk
*On Thermopower and Thermal Conductivity of the Periodic Anderson Model
(b) Vol. **119**, No. 1, p. 221–225 (1983).
See also *8*, p. *90*; 21, p. 358.

T. M. Gasymov, A. A. Katanov, and M. M. Babaev
*Thermopower and Thermomagnetic Effects in Kane Semiconductors under Conditions of Electron and Phonon Heating and Their Mutual Drag
(b) Vol. **119**, No. 1, p. 391–399 (1983).
See also 6, p. 77; *14.3.4*, p. *247*; 22, p. 411.

S. K. Plachkova, G. Ch. Christakudis, and L. V. Vassilev
Density-of-States Effective Mass in the System $(GeTe)_{1-x}(AgSbTe_2)_x$
(b) Vol. **119**, No. 1, p. K73–K76 (1983).
See also *13.1*, p. *163*; *14.3*, p. *223*; 22, p. 411.

G. J. Ashwell, J. G. Allen, G. H. Cross, and I. W. Nowell
Electrical, Magnetic, and Structural Characterisation of the 1:2 and 1:3 TCNQ Salts of 1,2-Bis(4-Pyridinium)Ethane
(a) Vol. **79**, No. 2, p. 455–463 (1983).
See also *1.1*, p. *12*; *14.3*, p. *225*; 18.1, p. 279; 22.9, p. 530.

P. G. Rustamov, Ch. I. Abilov, and S. G. Agdamskaya
*Thermoelectric Properties of $(PbTe)_{1-x}[(Cr, Fe, Ni)_2Te_3]_x$
(a) Vol. **79**, No. 2, p. K135–K137 (1983).
See also 8, p. 91; 14.3, p. 226; 22, p. 413.

P. I. Baranskii, V. V. Savyak, and Y. V. Simonenko
*Effect of Heat Treatment and Cooling Conditions on the Anisotropy Parameters of Drag TEMF in Neutron-Doped Silicon Crystals
(a) Vol. **79**, No. 2, p. K165–K170 (1983).
See also *14.3*, p. *226*; 22.1.2, p. 437.

D. V. Gitsu and F. M. Muntyanu
*Commutation Effect of Transport Phenomena in Antimony–Bismuth Alloys in Quantizing Magnetic Fields
(b) Vol. **120**, No. 1, p. K45–K49 (1983).
See also *14.1*, p. *210*; 21.7, p. 402.

H. Müller, P. Haasen, and W. Schröter
*Thermopower of Plastically Deformed Cadmium Telluride
(a) Vol. **80**, No. 1, p. 223–230 (1983).
See also *10.2*, p. *120*; 14.3.4, p. 248; 22.4.3, p. 469.

S. K. Plachkova
*Thermoelectric Power in the System $(GeTe)_{1-x}(AgSbTe_2)_x$
(a) Vol. **80**, No. 1, p. K97–K100 (1983).
See also 8, p. 91; *14.3*, p. *227*; 22.8, p. 510.

F. Goedsche, R. Richter, H. Wegener, and G. Röpke
*Correlation Function Method for the Thermoelectric Power of Metals
(b) Vol. **120**, No. 2, p. 631–644 (1983).
See also 13, p. 153; *14.1*, p. *210*; 21, p. 361.

O. Uemura, Y. Wada, T. Takahara, and T. Satow
The Semiconduction in the Molten Tl–$CuTlTe_2$ System
(b) Vol. **120**, No. 2, p. 685–693 (1983).
See also *2*, p. *55*; *14.3*, p. *227*; 18, p. 276; 22.8, p. 510.

R. Schumann
*Temperature Dependence of the Thermopower of the Periodic Anderson Model
(b) Vol. **120**, No. 2, p. K183–K188 (1983).
See also 8, p. 92; 14.1, p. 210; 21, p. 361.

H. Süssmann and W. Heiliger
*Seebeck Coefficient and Electrical Conductivity in p-$(Bi_{1-x}Sb_x)_2Te_3$ at Room Temperature
(a) Vol. **80**, No. 2, p. 535–539 (1983).
See also *14.3*, p. *228*; 22.7.1, p. 501.

P. Svoboda and P. Vašek
The Generalized Thomson-Voigt Relations for Transport Coefficients of Bismuth
(a) Vol. **80**, No. 2, p. 567–571 (1983).
See also *14.1*, p. *210*; 21.7, p. 402.

16. Photoconductivity. Photovoltaic Effects

Review Article

T. D. Dzhafarov
Photostimulated Diffusion in Semiconductors
(a) Vol. **79**, No. 1, p. 11–45 (1983).
See also *9*, p. *93*; 11, p. 122; 14.3.3, p. 236; 14.3.4, p. 243; 22, p. 403.

Original Papers and Short Notes

C. W. M. Timmermans, S. O. Cholakh, R. L. van der Woude, and G. Blasse
Some Optical and Electrical Measurements on $Cs_3Bi_2Br_9$ Single Crystals
(b) Vol. **115**, No. 1, p. 267–271 (1983).
See also 13.1, p. 155; *20.1*, p. *312*; 22.8, p. 502.

D. Mergel
*Recombination at Dislocations in Silicon
(a) Vol. **75**, No. 1, p. 255–262 (1983).
See also *13.4*, p. *173*; 22.1.2, p. 424.

M. K. Han, W. A. Anderson, P. Sung, and R. Lahri
*Photogenerated Carrier Conduction Mechanisms for Schottky a-Si:H Solar Cells
(a) Vol. **75**, No. 1, p. 283–288 (1983).
See also *14.3.4*, p. *243*; 22.1.2, p. 424.

D. P. Joshi and K. Sen
*Effective Lifetime for Polycrystalline Solar Cells
(a) Vol. **75**, No. 1, p. 311–315 (1983).
See also *14.3.3*, p. *236*; 22.1.2, p. 424.

D. Nesheva and E. Vateva
*Some Peculiarities of High Temperature Sensitization in CdS
(a) Vol. **75**, No. 1, p. K45–K48 (1983).
See also *13.4*, p. *174*; 22.4.1, p. 462.

H. Pfleiderer and B. Rauscher
Capacitance of Amorphous Silicon pin Solar Cells
(a) Vol. **75**, No. 2, p. 537–545 (1983).
See also 2, p. 44; 13.4, p. 175; *14.3.1*, p. *230*; *14.3.3*, p. *237*; 22.1.2, p. 426.

K. Sen, R. S. Srivastava, D. P. Joshi, and V. K. Goyal
*Mobility and Its Effect on the Performance of Polycrystalline Solar Cells
(a) Vol. **75**, No. 2, p. 657–661 (1983).
See also 13.3, p. 167; *14.3.3*, p. *237*; 22.1.2, p. 426.

A. L. Kartuzhanskii, V. M. Ulanov, and A. F. Yurchenko
Latent-Image and Fog Formation in Silver Halide Microcrystals by the Action of a Sequence of Light and Electric Pulses
(a) Vol. **75**, No. 2, p. **669–673** (1983).
See also *14.3.4*, p. *244*; 22.5.1, p. 476.

E. A. Silinsh, V. A. Kolesnikov, I. J. Muzikante, and D. R. Balode
*Erratum to: On Charge Carrier Photogeneration Mechanisms in Organic Molecular Crystals
(b) Vol. **116**, No. 1, p. 347 (1983).
See also 14.3.1, p. 230; 22.9, p. 526.

A. G. Kazanskii, V. F. Kiselev, E. A. Silaev, and V. S. Vavilov
On the Role of Surface Phenomena in the Staebler-Wronski Effect
(a) Vol. **76**, No. 1, p. **337–343** (1983).
See also 2, p. **46**; *13.3*, p. *168*; *14.3.2*, p. *233*; 22.1.2, p. 427.

D. V. Gitsu, I. N. Grincheshen, and N. S. Popovich
*The Electric and Phootelectric Properties of $TlSbSe_2$ Crystals in the Region of Impurity Conduction
(a) Vol. **76**, No. 1, p. K5–K7 (1983).
See also 13.4, p. 176; *14.3*, p. *217*; 22.8, p. 504.

K. H. Herrmann and K.-P. Möllmann
*Characteristic Steps in the Relaxation Curves of Field-Quenched Persistent Photoconductivity of $Pb_{0.78}Sn_{0.22}Te$:In
(a) Vol. **76**, No. 1, p. K67–K70 (1983).
See also 22.8, p. 504.

R. Meisels and F. Kuchar
*Far-Infrared Photoconductivity Spectroscopy of Shallow Acceptors in InSb
(b) Vol. **116**, No. 2, p. 557–560 (1983).
See also *13.4*, p. *177*; 22.2.3, p. 452.

T. G. Abdel-Malik, A. A. Aly, A. M. Abdeen, and H. M. El-Labany
Electric and Photoelectric Investigations of β-Nickel Phthalocyanine Thin Films
(a) Vol. **76**, No. 2, p. 651–659 (1983).
See also *13.4*, p. *177*; *14.3.1*, p. *231*; 22.9, p. 527.

V. N. Semkin, R. M. Vlasova, and G. Yu. Yashin
*Absence of 3.39 μm Photoconductivity in the Quasi-One-Dimenisonal Organic Semiconductor $Cs_2(TCNQ)_3$
(a) Vol. **77**, No. 1, p. 421–425 (1983).
See also 22.9, p. 528.

A. S. Popov, D. B. Kushev, and Sh. Chadraabal
*A Model for Photoconductivity of Fe-Doped GaP
(a) Vol. **77**, No. 1, p. K53–K57 (1983).
See also *13.4*, p. *179*; 22.2.2, p. 449.

J. JIMENEZ, M. A. GONZALEZ, V. CARBAYO, and J. BONNAFE
*On the Thermal Evolution of the Photoconductivity in Bulk Fe-Doped Semi-Insulating InP
(a) Vol. **77**, No. 1, p. K69–K73 (1983).
See also *13.4*, p. *179*; 22.2.2, p. 449.

J. CLAVERIE, G. CAMPET, D. CONTE, G. LE FLEM, and P. HAGENMULLER
*Influence of Covalent Bonding on the Photoelectrochemical Properties of Some Perovskite-Type Related Compounds
(a) Vol. **77**, No. 2, p. 603–609 (1983).
See also *4*, p. *61*; 13.4, p. 180; 22.6.1, p. 498.

SH. CHADRAABAL, A. S. POPOV, and D. B. KUSHEV
*Photoconductivity in n-Type InP:Fe
(a) Vol. **77**, No. 2, p. 709–714 (1983).
See also *13.4*, p. *181*; 22.2.2, p. 450.

K. N. KANG and S. CRISTOLOVEANU
*On the Role of Exact Boundary Conditions in Phototransport Phenomena: Case of SI–GaAs
(a) Vol. **77**, No. 2, p. K155–K159 (1983).
See also *14.3.2*, p. *234*; 22.2.1, p. 445.

M. KATAYAMA, H. MORIMOTO, and K. SUGAWARA
*Electron Irradiation Effects on Amorphous Silicon Solar Cells
(a) Vol. **78**, No. 1, p. K5–K8 (1983).
See also 11, p. 129; *14.3.3*, p. *240*; 22.1.2, p. 433.

F. CALLENS and W. MAENHOUT-VAN DER VORST
*The Influence of Triazaindolisine on the Microwave Photoconductivity of AgBr Emulsion Grains
(a) Vol. **78**, No. 1, p. K47–K52 (1983).
See also 22.5.1, p. 476.

D. GENZOW
Transverse Dember Effect in Lead Chalcogenides under Linearly Polarized Light Excitation
(b) Vol. **118**, No. 2, p. K159–K162 (1983).
See also *14.3*, p. *222*; 22, p. 410.

B. G. TAGIEV, G. M. NIFTIEV, and S. A. ABUSHOV
*Recombination and Trapping Levels in GaS:Mn Single Crystals
(a) Vol. **78**, No. 2, p. K183–K187 (1983).
See also *13.4*, p. *184*; 22.3, p. 457.

T. MOHAMMED-BRAHIM, A. FRIANT, and J. MELLET
*Characterization of Deep Energy Levels in Mercuric Iodide
(a) Vol. **79**, No. 1, p. 71–82 (1983).
See also *13.4*, p. *186*; 22.5, p. 475.

A. V. Dvurechenskii, I. A. Ryazantsev, L. S. Smirnov, H. Klose, and M. Rieth
Amorphous Silicon–Phosphorus Layers
(a) Vol. **79**, No. 1, p. 83–86 (1983).
See also 2, p. 51; 13.4, p. 186; *14.3.1*, p. *231*; 19, p. 304; 22.1.2, p. 434.

J. Muto and Y. Tsunekawa
*Photovoltaic Behaviour of Rhodamine 6G Schottky-Type Cells
(a) Vol. **79**, No. 1, p. K109–K112 (1983).
See also *14.3.3*, p. *241*; 22.9, p. 530.

A. V. Efanov and M. V. Entin
*Photogalvanic Effect in Crystal with Dislocations
(b) Vol. **119**, No. 2, p. 473–481 (1983).
See also 22, p. 412.

S. A. Abo-Namous, Y. Zaka, and R. W. Fane
Dependence of the Electronic and Opitcal Properties of Unhydrogenated a-Si on Preparation Conditions
(a) Vol. **79**, No. 2, p. 477–482 (1983).
See also *14.3*, p. *225*; *20.1*, p. *328*; 22.1.2, p. 436.

L. Z. Mirets, G. P. Peka, S. I. Radautsan, and E. V. Russu
*Photoelectric Investigation of Recombination Centre Parameters in InP(Fe)
(a) Vol. **79**, No. 2, p. 593–597 (1983).
See also 13.4, p. 187; 22.2.2, p. 450.

A. J. Szadkowski and A. Lubomirska-Wittlin
*Imperfection Photoconductivity of $Cd_{1-x}Mn_xTe$
(a) Vol. **79**, No. 2, p. K127–K129 (1983).
See also *13.4*, p. *188*; 22.4.4, p. 471.

J. Gong, C. M. van Vliet, D. Schoenfeld, and G. Bosman
Noise from Hole-Initiated Photoavalanche Processes
(a) Vol. **79**, No. 2, p. K201–K203 (1983).
See also *14.3.3*, p. *242*; *14.3.4*, p. *248*; 22.1.2, p. 437.

W. Fuhs, H. M. Welsch, and D. C. Booth
*Optical Quenching of Photoconductivity in a-Si:H Films
(b) Vol. **120**, No. 1, p. 197–205 (1983).
See also 2, p. 53; 14.3.1, p. 232; 22.1.2, p. 437.

D. Della Sala, F. Evangelisti, P. Fiorini, and C. Giovannella
*Influence of Oxygen on the Photoconductivity of Hydrogenated Amorphous Silicon
(b) Vol. **120**, No. 1, p. K59–K62 (1983).
See also 2, p. 54; 22.1.2, p. 437.

G. Kühnel, W. Siegel, and E. Ziegler
Electrical and Photoelectrical Properties of Semi-Insulating GaP
(a) Vol. **80**, No. 1, p. 159–164 (1983).
See also *13.4*, p. *189*; *14.3*, p. *226*; 22.2.2, p. 451.

A. N. Georgobiani, A. N. Gruzintsev, S. I. Radautsan, and I. M. Tiginyanu
*The Contribution of Iodine Impurity to the Formation of Centers in $CdGa_2S_4$ Single Crystals
(a) Vol. **80**, No. 1, p. K45 – K48 (1983).
See also *13.4*, p. *190*; 20.1, p. 329; 22.8, p. 510.

K. H. Herrmann and K.-P. Möllmann
*Curie Temperature as a Critical Temperature for Dielectric, Galvanomagnetic, and Photoelectric Phenomena in Strongly Doped $Pb_{1-x}Sn_xTe$
(a) Vol. **80**, No. 1, p. K101 – K104 (1983).
See also 22, p. 414.

A. Zaręba and A. J. Nadolny
*Photoconductivity Originated from Dissociation of Excitons in $Cd_{1-x}Mn_xTe$
(b) Vol. **120**, No. 2, p. K97 – K100 (1983).
See also *13.5.1*, p. *199*; 22.4.4, p. 472.

K. H. Herrmann, K.-P. Möllmann, and M. Wendt
*Photoeffects in Strongly Gallium-Doped Lead Telluride above Critical Temperature
(a) Vol. **80**, No. 2, p. 541 – 546 (1983).
See also 22, p. 415.

M. Nowak
*The Photomagnetoelectric Effect and Photoconductivity for Non-Normal Incidence of Radiation
(a) Vol. **80**, No. 2, p. 691 – 702 (1983).
See also *14.3.1*, p. *232*; 22.3, p. 458.

Yu. Yu. Sidorin and Yu. A. Zakharov
*The Light Stimulated Charge Carrier Transfer in Silver Azide
(a) Vol. **80**, No. 2, p. K157 – K160 (1983).
See also *14.4*, p. *255*; 22, p. 415.

17. Emission of Electrons and Ions

Original Papers and Short Notes

T. WARMIŃSKI, M. MAIER, and B. A. ORŁOWSKI
*Energy Levels in $Cd_{1-x}Mn_xF_2$ Crystals Obtained from the Auger Spectra
(b) Vol. **115**, No. 1, p. K7–K10 (1983).
See also *13*, p. *146*; 22.5.4, p. 488.

J. S. BROWN
*Some Exchange-Correlation Approximations Applied to Stopping Powers of Slow Ions in Metals
(b) Vol. **115**, No. 2, p. K141–K146 (1983).
See also *13.2*, p. *166*; 21, p. 351; 21.6, p. 396.

G. NEUBERT and S. ROGASCHEWSKI
*Determination of the Electron Backscattering Coefficient for Thin Films Using Double Layers
(a) Vol. **75**, No. 2, p. K191–K193 (1983).
See also *11*, p. *124*.

M. SCROCCO
Satellite Structure of ZnO and Zn(II) Halides Studied by X-Ray Photoelectron Spectroscopy
(b) Vol. **116**, No. 1, p. 137–144 (1983).
See also *13.1*, p. *157*; 22.4, p. 460; 22.5, p. 473.

K. SAKATA
Study of Electronic Structure of Reduced TiO_2 and $V_xTi_{1-x}O_2$ Crystals by ESCA and Optical Absorption
(b) Vol. **116**, No. 1, p. 145–153 (1983).
See also *13.4*, p. *176*; *20.1*, p. *315*; 22.6.1, p. 498.

D. SCHULZE, J. FINSTER, E. HENSEL, W. SKORUPA, and U. KREISSIG
*XPS Studies of SiO_2 Surface Layers Formed by Oxygen Ion Implantation into Silicon
(a) Vol. **76**, No. 1, p. K21–K24 (1983).
See also *1.6*, p. *39*; 22.6, p. 492.

H.-J. FITTING and R. TECHNOW
Electron Backscattering at Various Angles of Incidence
(a) Vol. **76**, No. 2, p. K151–K154 (1983).
See also *11*, p. *127*; 21, p. 353; 22.1.2, p. 428.

D. FLAMM and E.-H. WEBER
*Adsorption of Oxygen on GaAs (001) Surfaces by UPS
(a) Vol. **76**, No. 2, p. K163–K166 (1983).
See also 13.3, p. 168; 22.2.1, p. 444.

A. Bennouna, P. Y. Tessier, M. Priol, Q. Dang Tran, and S. Robin
*Far Ultraviolet Photoelectric Study of Thin SnSe Evaporated Films
(b) Vol. **117**, No. 1, p. 51–56 (1983).
See also *13.1*, p. *158*; 22, p. 406.

Y. Momose, M. Ohkubo, and J. Gunji
*Interaction of Mechanically Rubbed Aluminium with Polar Organic Vapors Found by Measurements of Optically Stimulated Exoelectron Emission under the Influence of Applied Electric Potential
(a) Vol. **77**, No. 1, p. K45–K48 (1983).
See also 21, p. 354.

J. Wollbrandt, U. Brückner, and E. Linke
Investigations of Mechanically Induçed Excited States on Cleavage Planes of Ionic Crystals (I): Method and Apparatus for Charge and Electron Emission Measurements
(a) Vol. **77**, No. 2, p. 545–552 (1983).
See also *13.3*, p. *169*; *14.4*, p. *251*; 22.5.2, p. 480.

J. Wollbrandt, U. Brückner, and E. Linke
Investigations of Mechanically Induced Excited States on Cleavage Planes of Ionic Crystals (II): Investigations of the Mechanisms of Electron Emission from Cleaved Alkali Halide Crystals and of the Influence of Surface Charges
(a) Vol. **78**, No. 1, p. 163–168 (1983).
See also *13.4*, p. *183*; 14.4, p. 252; 22.5.2, p. 481.

K. Hricovíni, P. Mikušík, and J. Krempaský
*Photoemission Studies of $Pd_{80-x}Si_{20}Co_x$ Amorphous Alloys
(b) Vol. **118**, No. 2, p. 823–827 (1983).
See also *2*, p. *50*; 13.1, p. 161; 21.1, p. 372.

K. Kawabata
*Intense Exoelectron Emission from Fine Particles of Al_2O_3 and Its Dependence on the Conductivity of the Substrate Material
(a) Vol. **78**, No. 2, p. K127–K130 (1983).
See also 11, p. 131; 13.3, p. 170; 22.6, p. 495.

V. Atzrodt and H. Lange
Principal Component Analysis as a Method for Silicide Investigation with Auger Electron Spectroscopy
(a) Vol. **79**, No. 2, p. 489–496 (1983).
See also *1.6*, p. *42*; 21, p. 360; 22.1.2, p. 436.

W. Frentrup, M. Griepentrog, and U. Müller-Jahreis
*The Influence of Cesium Implanted in Silicon on the Secondary Ion Emission
(a) Vol. **79**, No. 2, p. K193–K196 (1983).
See also 11, p. 133; 22.1.2, p. 437.

D. Hasselkamp and A. Scharmann
*Ion-Induced Electron Emission from Carbon
(a) Vol. **79**, No. 2, p. K197–K200 (1983).
See also 22.1.3, p. 440.

G. I. ROSENMAN, V. I. PECHORSKII, YU. L. CHEPELEV, E. I. BOIKOVA, and L. E. ISSAKOVA
*Exoemission of Pyroelectrics
(b) Vol. **120**, No. 2, p. 667–670 (1983).
See also *14.4.1*, p. *259*; 22.8.1, p. 518.

A. B. GERASIMOV, M. M. MERKIN, and A. A. TSERTSVADZE
*Calculation of the Exoelectron Energy Spectrum for NaCl
(b) Vol. **120**, No. 2, p. 711–714 (1983).
See also 22.5.2, p. 485.

L. ČERVINKA, V. CHÁB, and S. KOC
XPS Investigation of the Amorphous $TlSe_x$ System
(a) Vol. **80**, No. 2, p. 685–689 (1983).
See also *2*, p. *56*; *4*, p. *64*; 22.3, p. 458.

17.1 Field Emission Microscope Investigations

Original Papers

S. Surma, M. Błaszkiewicz, R. Męclewski, and W. Gubernator
A Rearranged Structure on a Rhenium Surface
(a) Vol. **80**, No. 1, p. 269–276 (1983).
See also *1.6*, p. *42*; 21.4, p. 393.

D. J. Farnum, O. T. Inal, and R. J. Walko
A Field Ion Microscopy Study of Ion Damage to Tungsten
(a) Vol. **80**, No. 1, p. 287–303 (1983).
See also *11*, p. *134*; 21, p. 361.

18. Magnetic Properties

Original Papers and Short Notes

K. A. Krezhov, M. T. Mikhov, A. M. Kadomtseva, I. B. Krinetskii, and M. M. Lukina
*Magnetic and Neutron Diffraction Study of the Phase Transition in $TmFe_{1-x}Mn_xO_3$
(b) Vol. **115**, No. 1, p. 47–51 (1983).
See also 22.8.2, p. 519.

O. N. Dubey and R. S. Tripathi
*Susceptibility and Specific Heat of Spin Glasses in a Magnetic Field
(b) Vol. **115**, No. 1, p. 63–66 (1983).
See also *8*, p. *85*.

M. Singh, J. Leotin, and P. R. Wallace
Theory of the Diffusivity–Mobility Ratio in Cadmium Arsenide
(b) Vol. **115**, No. 1, p. 105–113 (1983).
See also *14.3*, p. *215*; 22.3, p. 456.

M. Bancewicz and A. Stasch
Theoretical Investigation of the Resonant Magnon–Phonon Interaction in FeF_2 in High Magnetic Field
(b) Vol. **115**, No. 2, p. 563–574 (1983).
See also *6*, p. *71*; *13.5.4*, p. *203*; 22, p. 404.

H. Gruber and E. Krautz
Magnetoresistance and Conductivity in the Binary System Titanium–Oxygen (II): Semiconductive Titanium Oxides
(a) Vol. **75**, No. 2, p. 511–518 (1983).
See also 8, p. 86; *14.1*, p. *206*; *14.3*, p. *216*; 15, p. 261; 22.6, p. 491.

A. Sternberg, M. Zvára, and P. Hlídek
*Influence of Permanent Magnetic Field on Dielectric Properties of Transparent Ferroelectric PLZT Ceramics
(a) Vol. **75**, No. 2, p. K179–K182 (1983).
See also *14.4.1*, p. *256*; 22.8.1, p. 513.

A. N. Chakravarti, K. P. Ghatak, K. K. Ghosh, and H. M. Mukherjee
*Effect of Varying the Orientation of a Quantizing Magnetic Field on the Effective Electron Mass in Degenerate Tetragonal Semiconductors
(b) Vol. **116**, No. 1, p. 17–24 (1983).
See also *13.1*, p. *156*; 22.3, p. 456.

H. Stramska
*Nonlinear Magneto-Optical Phenomena of the Third Harmonic in Semiconductors
(b) Vol. **116**, No. 1, p. 129–135 (1983).
See also *20*, p. *307*; 22, p. 404.

K. P. MAHESHWARI and GANESH TAREY
Excitation of Magnetoacoustic Waves by Two Microwave Beams in InSb
(b) Vol. **116**, No. 1, p. 339–346 (1983).
See also *7*, p. *82*; 13, p. 147; *20*, p. *307*; 22, p. 405.

G. L. WHITTLE
*Magnetic Order in Isoelectronic Co(GaVFe) Alloys
(a) Vol. **76**, No. 1, p. 351–356 (1983).
See also *5*, p. *66*; 21.1, p. 366.

NGUYEN THE KHOI, J. GINTER, and A. TWARDOWSKI
*Magnetooptical Study of the Fundamental Absorption Edge in $Cd_{1-x}Mn_xTe$
(b) Vol. **117**, No. 1, p. 67–74 (1983).
See also 13.5.1, p. 196; *20*, p. *308*; 22.4.4, p. 470.

K. M. HÄUSSLER, J. BRANDMÜLLER, L. MERTEN, H. FINSTERHÖLZL, and A. LEHMEYER
Observation of Dispersion on Magnetic Polaritons
(b) Vol. **117**, No. 1, p. 225–228 (1983).
See also *13.5.4*, p. *203*; *20.1*, p. *318*; 22.5, p. 474.

E. BURZO, I. LUPŞA, and I. COSMA
*On the Magnetic Behaviour of NiCuSi Solid Solutions
(a) Vol. **77**, No. 1, p. 127–132 (1983).
See also 18.2, p. 283; 21.1, p. 369.

H. STRAMSKA
*The Magneto-Optical Effect under a Strong Electric Field in Semiconductors
(b) Vol. **117**, No. 2, p. 541–547 (1983).
See also 14.3.4, p. 245; *20*, p. *309*; 22, p. 407.

B. A. MEN, M. L. LEVITAN, and A. N. MEN
*Influence of the Ground State Symmetry on Qualitative Form of Ordering Phase Diagram
(a) Vol. **77**, No. 2, p. 455–461 (1983).
See also *1.2*, p. *17*; 21, p. 355.

H. LUCHIAN
*Magnetic Symmetry in Several Cubic Compounds
(b) Vol. **118**, No. 1, p. 31–39 (1983).
See also *13*, p. *149*; 22.8.2, p. 522.

K. SOKALSKI and W. WÓJCIK
*Critical Temperature of a Quasi-One-Dimensional Ising Model
(b) Vol. **118**, No. 1, p. 441–446 (1983).

V. V. KELAREV, V. V. CHUEV, A. N. PIROGOV, S. K. SIDOROV, and V. N. DVINYANINOV
*Magnetization of the Rare-Earth Sublattice in Intermetallic Compounds of the Type $R(Co_{1-x}Ni_x)_5$
(b) Vol. **118**, No. 1, p. K75–K78 (1983).
See also 21.1, p. 371; 21.4, p. 392.

G. I. MAKOVETSKII
*Magnetic Properties of the $Cr_{7-x}Fe_xTe_8$ ($x = 0$ to 3) System
(a) Vol. **78**, No. 1, p. K39 – K41 (1983).
See also 22.8, p. 507.

K. REGIŃSKI
Phonon Polaritons in the Presence of a Magnetic Field
(b) Vol. **118**, No. 2, p. 587 – 593 (1983).
See also *6*, p. *76*; *13.5*, p. *193*; 20, p. 309; 22.5.2, p. 481.

S. KATSURA and A. MATSUNO
*Phase Diagram of a Spin Glass of $Eu_pSr_{1-p}S$
(b) Vol. **119**, No. 1, p. 73 – 77 (1983).
See also *2*, p. *51*; 18.1, p. 279; 18.2, p. 286; 22, p. 410.

P. K. KATANA
Two-Photon Interband Absorption of Light in Semiconductors in External Parallel Electric and Magnetic Fields
(b) Vol. **119**, No. 1, p. 89 – 98 (1983).
See also *13.1*, p. *162*; 14.3.4, p. 247; *20.1*, p. *325*; 22, p. 411.

S. GHOSH and S. KHAN
Parametric Conversion of an Electron Plasma Wave into an Electromagnetic Wave in Longitudinally Magnetized n-Type Piezoelectric Semiconductor Plasmas
(b) Vol. **119**, No. 1, p. 363 – 373 (1983).
See also *13.5.2*, p. *200*; 22.2.3, p. 453.

A. V. MALAKHOVSKII, T. P. MOROZOVA, and V. I. YUZVAK
*Magneto-Optical Properties of $RbMnF_3$ in the Paramagnetic State
(b) Vol. **119**, No. 1, p. 411 – 417 (1983).
See also 13.1, p. 163; *20.1*, p. *326*; 22.5, p. 475.

A. A. KISELEV
*Circular Phonons and Optical Magnetization of Crystals
(b) Vol. **119**, No. 2, p. 525 – 530 (1983).
See also *6*, p. *78*; 20, p. 310; 22.5.2, p. 484.

B. LORENZ
Thermodynamic Properties of Uniform and Dimerized Hubbard Chains
(b) Vol. **119**, No. 2, p. 555 – 563 (1983).
See also *8*, p. *90*; 22.9, p. 530.

D. IHLE
*Cluster Approach to the Spinless Fermion Model
(b) Vol. **119**, No. 2, p. K169 – K173 (1983).
See also *13*, p. *151*.

C. RUDOWICZ
*Note on the "Comment on Energy of Magnetic Anisotropy"
(b) Vol. **120**, No. 1, p. K51 – K52 (1983).
See also *13*, p. *152*.

G. V. Loseva and G. M. Abramova
*Anomalies of Electric and Magnetic Properties of Ternary Cobalt Sulfides
(a) Vol. **80**, No. 1, p. K109–K111 (1983).
See also *14.3*, p. *227*; 22.8, p. 510.

O. Uemura, Y. Wada, T. Takahara, and T. Satow
The Semiconduction in the Molten Tl–$CuTlTe_2$ System
(b) Vol. **120**, No. 2, p. 685–693 (1983).
See also *2*, p. *55*; *14.3*, p. *227*; 15, p. 263; 22.8, p. 510.

K. C. Liu, Kai-Shue Lam, and Th. F. George
*Localized Magnetization Arising from Spin-Dependent Impurity Potentials in the Wolff Model
(b) Vol. **120**, No. 2, p. 695–702 (1983).
See also *13.4*, p. *190*.

H. Stramska
Higher Harmonics Generated by Mobile Carriers in Semiconductors
(b) Vol. **120**, No. 2, p. 783–788 (1983).
See also *13*, p. *153*; 14, p. 205; 22, p. 415.

G. Heber
*Some Remarks on Domains and Kinks in Spin Glasses
(b) Vol. **120**, No. 2, p. K111–K114 (1983).

A. H. Nayyar and P. Shukla
*Spin Glass Behaviour in $Eu_xSr_{1-x}S$: Calculation of the Phase Diagram in Mean Field Approximation
(b) Vol. **120**, No. 2, p. K141–K144 (1983).
See also 18.2, p. 289; 22, p. 415.

K. I. Grozdev and S. I. Tzenov
*Effect of Impurities on the Indirect Polarization of Spins in an Anisotropic Double Ising Chain
(b) Vol. **120**, No. 2, p. K157–K159 (1983).

V. Koubele, M. Zvára, and P. Hlídek
*Compositional Variations of the Nonresonant Interband Faraday Rotation in $Cd_xHg_{1-x}Te$ Single Crystals
(b) Vol. **120**, No. 2, p. K173–K177 (1983).
See also 13, p. 153; *20.1*, p. *330*; 22.4.4, p. 472.

O. Z. Alekperov, V. G. Golubev, V. I. Ivanov-Omskii, and A. Sh. Mekhtiev
Plasma Shift of Cyclotron Resonance Line in n-GaAs
(b) Vol. **120**, No. 2, p. K179–K181 (1983).
See also *13*, p. *154*; *14.3*, p. *227*; 22.2.1, p. 447.

R. Piasecki
*Phase Diagrams of a Layered Magnet with Coupled Order Parameters
(b) Vol. **120**, No. 2, p. K215–K218 (1983).

P. Novák, M. Maryško, S. Krupička, J. Kub, J. Šimšova, J. Čermák, and M. Nevřívа
*Noncubic Anisotropy in YIG:Mn^{3+} Thin Films
(a) Vol. **80**, No. 2, p. K213–K215 (1983).
See also *13.4*, p. *191*; 22.8.2, p. 524.

18.1 Paramagnetic Properties

Original Papers and Short Notes

W. Zarek, A. Kasprzyk, and E. Kozubek
*Magnetic and ESR Investigations of $Sm_6(Mn_{1-x}Fe_x)_{23}$ Compounds for $0 < x < 0.55$ in the Paramagnetic Range
(a) Vol. **75**, No. 2, p. 359 – 366 (1983).
See also *19*, p. *300*; 21.1, p. 364; 21.1.1, p. 379.

A. Apostolov, H. Hristov, M. Mikhov, and V. Skumriev
*Magnetic Properties of Bulk Amorphous Alloys Gd_4Co_3, Er_4Co_3, and Sm_4Co_3
(a) Vol. **75**, No. 2, p. 401 – 404 (1983).
See also *2*, p. *44*; 18.2, p. 281; 21.4, p. 390.

S. I. Tyukhtenko, D. A. Stakhov, and G. I. Batalin
*High-Temperature Paramagnetic Susceptibility of the Erbium–Germanium Alloys
(a) Vol. **76**, No. 2, p. K179 – K181 (1983).
See also 21.4, p. 391.

N. V. Volkenshtein, E. V. Galoshina, M. E. Kost, and T. S. Shubina
*Magnetic Properties of the Sc–H System
(b) Vol. **117**, No. 1, p. K47 – K52 (1983).
See also *13.1*, p. *158*; 21.1, p. 368.

O. J. Żogal
Knight Shifts of Protons in Some Rare-Earth Dihydrides
(b) Vol. **117**, No. 2, p. 717 – 723 (1983).
See also *19*, p. *301*; 21.4, p. 391.

K. Elk
*Susceptibility of the Periodic Anderson Model: Hartree-Fock Approximation
(b) Vol. **117**, No. 2, p. K87 – K90 (1983).
See also 13, p. 149; 21, p. 355.

J. Ulner
*Theorie des durch das äußere Magnetfeld induzierten Jahn-Teller-Effektes in orthorhombischen $J = 1$-Paramagneten
(Theory of the Jahn-Teller Effect Induced by an External Magnetic Field in Orthorhombic $J = 1$ Paramagnets)
(b) Vol. **118**, No. 1, p. 389 – 398 (1983).
See also *7*, p. *83*.

H. V. Sharma and I. Singh
Valence Fluctuations and Magnetic–Nonmagnetic Transitions in a Model for Intermediate-Valence Systems
(b) Vol. **118**, No. 2, p. 511 – 520 (1983).
See also 8, p. 89; *13*, p. *150*; *18.2*, p. *285*; 21.4, p. 392.

P. R. Silva and F. C. Sá Barreto
*The Diluted Spin $-\frac{1}{2}$ Ising Model in the Honeycomb Lattice
(b) Vol. **118**, No. 2, p. K93 – K97 (1983).
See also *18.2*, p. *285*.

S. ANZAI, S. OHTA. A. YOSHINO, J. NISHIO, and M. HATORI
*Charge-Density-Wave-Like Transition in V_3Te_4
(b) Vol. **118**, No. 2, p. K99 – K102 (1983).
See also *14.1*, p. *208*; 15, p. 262; 22, p. 410.

J. PEBLER
*Paramagnetic Hyperfine Structure of Iron (57) in $V_{1-x}Fe_xO_2$
(a) Vol. **78**, No. 2, p. 589 – 594 (1983).
See also *5*, p. *67*; 22.6.1, p. 498.

S. KATSURA and A. MATSUNO
Phase Diagram of a Spin Glass of $Eu_pSr_{1-p}S$
(b) Vol. **119**, No. 1, p. 73 – 77 (1983).
See also *2*, p. *51*; *18*, p. *275*; 18.2, p. 286; 22, p. 410.

E. KOLLEY and W. KOLLEY
*Spin Fluctuations in Weak Itinerant Ferromagnets near T_c
(b) Vol. **119**, No. 2, p. K175 – K179 (1983).
See also *18.2*, p. *287*.

G. J. ASHWELL, J. G. ALLEN, G. H. CROSS, and I. W. NOWELL
Electrical, Magnetic, and Structural Characterisation of the 1:2 and 1:3 TCNQ Salts of 1,2-Bis(4-Pyridinium) Ethane
(a) Vol. **79**, No. 2, p. 455 – 463 (1983).
See also *1.1*, p. *12*; *14.3*, p. *225*; 15, p. 262; 22.9, p. 530.

G. J. ASHWELL, S. JANSEN, M. T. JONES, and C. H. L. KENNARD
(Diethyldiphenylphosphonium)$^+$ (7,7,8,8-tetracyano-p-quinodimethanide)$_2^-$, a Novel Quasi-One-Dimensional Organic Conductor
(a) Vol. **80**, No. 1, p. 127 – 134 (1983).
See also *1.1*, p. *12*; 14.3, p. 226; 22.9, p. 532.

H. FEHSKE
Spin Fluctuations in Alloys with Random Transfer
(b) Vol. **120**, No. 2, p. 611 – 620 (1983).
See also *13*, p. *153*; 14, p. 204; 18.2, p. 288; 21.1, p. 377.

H. KIDO, T. HOSHIKAWA, M. SHIMADA, and M. KOIZUMI
*Preparation and Magnetic and Electrical Properties of RZnSi
(a) Vol. **80**, No. 2, p. 601 – 605 (1983).
See also *1.3*, p. *27*; 14.1, p. 211; 21.1, p. 378; 21.4, p. 394.

18.2 Ferromagnetic Properties

Review Articles

K. H. Fischer
*Spin Glasses (I)
(b) Vol. **116**, No. 2, p. 357–414 (1983).
See also 2, p. 43; 8, p. 85; 13, p. 146; 14.1, p. 206; *18.4*, p. *295*; 21.1, p. 363; 21.1.1, p. 379.

V. S. Shpinel
Exchange Interaction and Hyperfine Fields at Nonmagnetic Impurities in Rare-Earth Metals
(b) Vol. **118**, No. 1, p. 11–24 (1983).
See also 5, p. 65; *18.4*, p. *295*; 21.4, p. 390.

Original Papers and Short Notes

F. Goedsche, A. Richter, and G. Vojta
*Dynamical Susceptibility and Electrical Resistivity of Disordered Spin Systems
(b) Vol. **115**, No. 1, p. 209–218 (1983).
See also 13, p. 146; *14.1*, p. *206*; 21.1, p. 363.

A. Bobák
*Curie Temperature of Amorphous Ferromagnets
(b) Vol. **115**, No. 1, p. K21–K24 (1983).
See also *2*, p. *43*.

I. Tomáš, L. Murtinová, and J. Kaczér
*Easy Magnetization Axes in Materials with Combined Cubic and Uniaxial Anisotropies
(a) Vol. **75**, No. 1, p. 121–127 (1983).
See also 18.3, p. 291.

F. Walz, H. J. Blythe, and F. Dworschak
*Annealing Behaviour of Dilute FeCu, FeMn, and FeTi Alloys in the Temperature Range up to Stage III, Following Low-Temperature Electron Irradiation
(a) Vol. **75**, No. 1, p. 235–242 (1983).
See also *10.1*, p. *99*; 11, p. 122; 21.1, p. 364; 21.1.1, p. 379.

F. T. Parker and H. Oesterreicher
*Temperature Dependence of Intrinsic Magnetic Hardness in Some Rare-Earth Pseudobinary Compounds and Amorphous Materials
(a) Vol. **75**, No. 1, p. 273–282 (1983).
See also 2, p. 43; 21.1, p. 364; 21.4, p. 390.

A. S. Lileev and V. P. Menushenkov
*The Crystal Structure and Coercive Force of $SmCo_5$ Sintered Permanent Magnets
(a) Vol. **75**, No. 1, p. 341–346 (1983).
See also *1.3*, p. *21*; 21.1, p. 364; 21.4, p. 390.

S. Tyagi
*Ferromagnetic Resonance in Single Particles of Amorphous $Fe_{81.5}B_{14.5}Si_4$
(a) Vol. **75**, No. 1, p. K29–K32 (1983).
See also *2*, p. *43*; 19, p. 299; 21.1.1, p. 379.

A. Apostolov, H. Hristov, M. Mikhov, and V. Skumriev
Magnetic Properties of Bulk Amorphous Alloys Gd_4Co_3, Er_4Co_3, and Sm_4Co_3
(a) Vol. **75**, No. 2, p. 401–404 (1983).
See also *2*, p. *44*; *18.1*, p. *278*; 21.4, p. 390.

T. J. McKenna, S. J. Campbell, D. H. Chaplin, and G. V. H. Wilson
*Temperature Modulation and Temperature Hysteresis Studies of Dysprosium
(a) Vol. **75**, No. 2, p. 421–432 (1983).
See also *18.4*, p. *295*; 21.4, p. 390.

K. H. J. Buschow, J. H. N. van Vucht, P. G. van Engen, D. B. de Mooij, and A. M. van der Kraan
*Structure, Magnetic and Magneto-Optical Properties, and ^{57}Fe Mössbauer Effect in a New Variety of Heusler Alloy: PtFeSb
(a) Vol. **75**, No. 2, p. 617–623 (1983).
See also *1.3*, p. *22*; 5, p. 65; 20.1, p. 314; 21.1.1, p. 379.

J. R. Faleiro Ferreira and N. P. Silva
*Investigation of Oguchi's Procedure Applied to the Ising Model
(b) Vol. **116**, No. 1, p. 155–162 (1983).

L. Wojtczak and T. Balcerzak
*Reaction Field Approximation for Inhomogeneous Ferromagnets
(b) Vol. **116**, No. 1, p. 217–225 (1983).
See also *2*, p. *45*.

A. V. Zaborov and M. V. Medvedev
*A Binary Ferromagnetic Alloy with Magnetic Moments Dependent on Local Environment (I): The Jaccarino-Walker Model in the Effective Field Approximation
(b) Vol. **116**, No. 1, p. 227–238 (1983).
See also 21.1, p. 366; 21.1.1, p. 380.

J. Katriel and G. F. Kventzel
*The Constant-Coupling Approximation: Some Exact Results
(b) Vol. **116**, No. 1, p. K51–K56 (1983).
See also 2, p. 45; *13*, p. *148*.

F. Walz, H. J. Blythe, and F. Dworschak
*Annealing Behaviour of Dilute <u>Fe</u>Ti, <u>Fe</u>Cu, and <u>Fe</u>Mn Alloys in the Temperature Range above Stage III Following Low-Temperature Electron Irradiation
(a) Vol. **76**, No. 1, p. 267–276 (1983).
See also *10.1*, p. *101*; 11, p. 125; 21.1.1, p. 380.

P. Czarnecki and G. Fritzkowski
Transverse Magnetoresistance of Amorphous $Fe_{40-x}Ni_{40}Cr_xB_{20}$ Alloys in Magnetic Fields up to 20 T
(a) Vol. **76**, No. 1, p. K93–K96 (1983).
See also *2*, p. *46*; *14.1*, p. *207*; 21.1, p. 366; 21.1.1, p. 380.

S. Takahashi and K. Ikeda
*Magnetic Interactions in Plastically Deformed Ni_3Mn Alloy
(a) Vol. **76**, No. 1, p. K97–K100 (1983).
See also 10.1, p. 101; *18.4*, p. *295*; 21.1, p. 366.

T. G. Petrova and E. Syrkin
*Effect of Uniaxial Long-Range Exchange Interaction on the Properties of Surface Spin Waves in a Heisenberg Ferromagnet
(b) Vol. **116**, No. 2, p. 475–481 (1983).
See also *13.5.4*, p. *203*.

A. V. Zaborov and M. V. Medvedev
*A Binary Ferromagnetic Alloy with Magnetic Moments Dependent on Local Environment (II): Neutron Magnetic Elastic Scattering and Hyperfine Field Distribution
(b) Vol. **116**, No. 2, p. 511–523 (1983).
See also 21.1, p. 367; 21.1.1, p. 381.

D. Ihle and B. Lorenz
Theory of the Verwey Transition in the Mixed Valence Compound Eu_3S_4
(b) Vol. **116**, No. 2, p. 539–546 (1983).
See also *6*, p. *72*; *13.5.3*, p. *201*; 22, p. 405.

J. Sinkkonen and P. Kuivalainen
*Many Electron Magnetic Polaron in Magnetic Semiconductors
(b) Vol. **116**, No. 2, p. K73–K76 (1983).
See also *13.5.3*, p. *201*; 22, p. 405.

H. Börner and R. Höhne
*Study of Magnetic After-Effects in the Amorphous Alloy $Fe_{40}Ni_{40}P_{14}B_6$
(a) Vol. **76**, No. 2, p. 499–503 (1983).
See also *2*, p. *47*; 21.1, p. 367; 21.1.1, p. 381.

K. H. J. Buschow and P. G. van Engen
*Note on the Magnetic and Magneto-Optical Properties of Ni_2In Type 3d Transition Metal Compounds
(a) Vol. **76**, No. 2, p. 615–620 (1983).
See also 1.1, p. 11; *20*, p. *308*; 21, p. 353; 21.1, p. 367; 21.1.1, p. 381.

K. Sumiyama, M. Shiga, and Y. Nakamura
Characteristic Martensitic Transformation of the Fe–Pt Disordered Alloys around the γ–α Phase Boundary
(a) Vol. **76**, No. 2, p. 747–751 (1983).
See also *1.2*, p. *16*; 21.1.1, p. 381.

W. Bażela-Wróbel and A. Szytuła
Neutron Diffraction Study of the $Ni_2Mn_{1-x}Cr_xSn$ System
(a) Vol. **76**, No. 2, p. K117 – K120 (1983).
See also *1.3*, p. *23*; 21.1, p. 368.

J. Schneider, J. Kamarad, Z. Arnold, and M. Bitterlich
*The Pressure Dependence of the Curie Temperature of Amorphous Fe–P–Ga and Fe–P–Ge Alloys
(a) Vol. **76**, No. 2, p. K183 – K188 (1983).
See also *2*, p. *47*; 21.1.1, p. 382.

H. A. Brown
*On the Schrödinger Spin-Exchange Ferromagnet
(b) Vol. **117**, No. 1, p. 177 – 184 (1983).

G. A. Petrakovskii and S. S. Aplesnin
*Ground State and Effective Field Distribution of Ising and Heisenberg Disordered Antiferromagnets
(b) Vol. **117**, No. 1, p. 401 – 405 (1983).
See also *18.4*, p. *236*.

M. Sorohan
Magnetoresistance of Polycrystalline Ni Wire under Hydrostatic Pressure
(b) Vol. **117**, No. 1, p. K1 – K4 (1983).
See also *14.1*, p. *207*; 21.1, p. 368.

D. V. Kapor, K. R. Surla, D. I. Surla, J. P. Šetrajčić, and M. J. Škrinjar
*Anisotropy Effects in an Anisotropic Heisenberg Model
(b) Vol. **117**, No. 1, p. K9 – K11 (1983).

Z. Bąk
*A Heisenberg Ferromagnet with Anisotropic Exchange between Nearest Neighbours
(b) Vol. **117**, No. 1, p. K19 – K22 (1983).

N. N. Delyagin, V. I. Nesterov, and S. I. Semenov
*The Oscillations of the Mössbauer Line Shift near T_c in the Ferromagnetic Ni–Rh Alloy
(b) Vol. **117**, No. 1, p. K27 – K29 (1983).
See also *5*, p. *66*; 21.1, p. 368.

V. E. Antonov, T. E. Antonova, I. T. Belash, E. G. Ponyatovskii, V. I. Rashupkin, and V. G. Thiessen
The Pd–Ni–H System at High Pressure
(a) Vol. **77**, No. 1, p. 71 – 79 (1983).
See also *1.3*, p. *23*; 14.2, p. 212; 21, p. 354; 21.1, p. 369.

E. Burzo, I. Lupsa, and I. Cosma
On the Magnetic Behaviour of NiCuSi Solid Solutions
(a) Vol. **77**, No. 1, p. 127 – 132 (1983).
See also *18*, p. *274*; 21.1, p. 369.

G. Vértesy, I. Tomáš, Z. Vértesy, and M. Balaskó
Temperature Hysteresis of Demagnetized Domain Structures
(a) Vol. **77**, No. 1, p. K87 – K90 (1983).
See also 22.8.2, p. 521.

V. Brédimas, C. Papatriantafyllou, and H. Gamari-Seale
*On the Magnetic Structure of the Intermetallic Compounds Gd_2Al and Gd_2Au
(b) Vol. **117**, No. 2, p. 513 – 519 (1983).
See also 18.4, p. 296; 21.4, p. 391.

H. Grimm and H. Kronmüller
*Investigation of Structural Defects in the Amorphous Ferromagnetic Alloy $Fe_{40}Ni_{40}P_{14}B_6$
(b) Vol. **117**, No. 2, p. 663 – 674 (1983).
See also 2, p. 48; *10.1*, p. *103*; 21.1, p. 370; 21.1.1, p. 382.

S. Trimper and A. Kühnel
*On the Phase Transition Temperature for a Heisenberg Ferromagnet
(b) Vol. **117**, No. 2, p. K75 – K79 (1983).
See also *1.2*, p. *17*.

H. Ratajczak, P. Slančo, T. Tima, A. Zentko, E. Demkovicsová, and M. Konč
Electrical Resistivity and Hall Effect in Thin Amorphous UFe Films
(a) Vol. **77**, No. 2, p. 785 – 791 (1983).
See also *2*, p. *49*; *14.1*, p. *208*; 21.1.1, p. 382.

H. Kido, T. Hoshikawa, M. Shimada, and M. Koizumi
*Synthesis and Magnetic Properties of RCuSi (R = Y, Ce, Nd, Sm, Gd, Ho)
(a) Vol. **77**, No. 2, p. K121 – K123 (1983).
See also *18.4*, p. *296*; 21.1, p. 371; 21.4, p. 391.

V. S. Patil, R. G. Pillay, P. N. Tandon, and H. G. Devare
Hyperfine Field Investigations in Some Ru and Rh Based Heusler Alloys
(b) Vol. **118**, No. 1, p. 57 – 61 (1983).
See also *5*, p. *67*; 21, p. 355; 21.1.1, p. 382.

A. Jezierski
Application of CPA to the Magnon Hamiltonian
(b) Vol. **118**, No. 1, p. 73 – 80 (1983).
See also *13.5.4*, p. *203*; 21, p. 355.

I. Tamura, E. F. Sarmento, I. P. Fittipaldi, and T. Kaneyoshi
*Magnetic Properties of the Ising Model with a Free Surface
(b) Vol. **118**, No. 1, p. 409 – 418 (1983).

A. Bobák
*The Behaviour of an Amorphous Ferromagnet with Dilution
(b) Vol. **118**, No. 1, p. K61 – K63 (1983).
See also *2*, p. *49*.

C. M. Mô and P. Moser
Hydrogen in $Fe_{40}Ni_{40}P_{14}B_6$ Amorphous Alloy Studied by Magnetic After Effect, Internal Friction, and Positron Annihilation
(a) Vol. **78**, No. 1, p. 201 – 210 (1983).
See also *2*, p. *49*; 10.1, p. 103; 21.1, p. 371; 21.1.1, p. 383.

R. Höhne, H. Börner, and M. Krötzsch
*On the Annealing Behaviour and Crystallization of the Amorphous Alloy $Fe_{40}Ni_{40}P_{11}B_6$
(a) Vol. **78**, No. 1, p. 211 – 216 (1983).
See also *2*, p. *50*; 21.1, p. 371; 21.1.1, p. 383.

H. V. Sharma and I. Singh
*Valence Fluctuations and Magnetic–Nonmagnetic Transitions in a Model for Intermediate-Valence Systems
(b) Vol. **118**, No. 2, p. 511 – 520 (1983).
See also 8, p. 89; *13*, p. *150*; 18.1, p. 278; 21.4, p. 392.

H. Kronmüller
*The Rôle of Two-Level Systems in Amorphous Metallic Alloys (I): Induced Anisotropy in Amorphous Ferromagnets
(b) Vol. **118**, No. 2, p. 661 – 673 (1983).
See also *2*, p. *50*; 21.1.1, p. 383.

T. Kaneyoshi
*On an Anomalous Behaviour of Spin-Wave Stiffness Constant in Amorphous Ferromagnets
(b) Vol. **118**, No. 2, p. 751 – 755 (1983).
See also *2*, p. *50*.

J. R. Faleiro Ferreira and N. P. Silva
*An Approximate Method for the Free Energy
(b) Vol. **118**, No. 2, p. 829 – 833 (1983).

S. Lipiński
Effects of Order on the Electronic Structure of Ni_3Fe
(b) Vol. **118**, No. 2, p. 861 – 869 (1983).
See also *13*, p. *150*; 21.1, p. 372; 21.1.1, p. 383.

P. R. Silva and F. C. Sá Barreto
*The Diluted Spin $-\frac{1}{2}$ Ising Model in the Honeycomb Lattice
(b) Vol. **118**, No. 2, p. K93 – K97 (1983).
See also *18.1*, p. *278*.

Y. Muraoka, M. Shiga, and Y. Nakamura
*Magnetic Properties and Thermal Expansion of $(Zr_{1-x}Ta_x)Fe_2$
(a) Vol. **78**, No. 2, p. 717 – 722 (1983).
See also *5*, p. *67*; 8, p. 89; 18.4, p. 297; 21.1.1, p. 383.

I. Lupsa and E. Burzo
On the Resistive Anomaly in NiCuSi and NiCuZn Alloys near the Curie Temperature
(a) Vol. **78**, No. 2, p. K155 – K158 (1983).
See also *14.1*, p. *209*; 21.1, p. 373.

E. König
*Variation of the Magnetic Properties of $SmCo_5$ Powders by Water Milling
(a) Vol. **78**, No. 2, p. K159 – K160 (1983).
See also 21.1, p. 374; 21.4, p. 392.

S. Katsura and A. Matsuno
Phase Diagram of a Spin Glass of $Eu_pSr_{1-p}S$
(b) Vol. **119**, No. 1, p. 73 – 77 (1983).
See also *2*, p. *51*; *18*, p. *275*; 18.1, p. 279; 22, p. 410.

P. H. Müller, A. Kasten, and M. Schienle
*Relaxation in the Ferromagnetic Phase of Terbium Arsenate ($TbAsO_4$)
(b) Vol. **119**, No. 1, p. 239 – 249 (1983).
See also 22.8.1, p. 516.

J. M. Szlanta
*Temperature Shift of the Maximum of the Critically Scattered Neutrons
(b) Vol. **119**, No. 1, p. K23 – K25 (1983).
See also 21.1.1, p. 383.

A. F. Siqueira and I. P. Fittipaldi
*On the Phase Transition in the Blume-Capel Model
(b) Vol. **119**, No. 1, p. K31 – K36 (1983).

W. Beck and H. Kronmüller
*On the Coercive Field of the Nearly Non-Magnetostrictive Amorphous Ferromagnetic Alloy $Co_{58}Ni_{10}Fe_5Si_{11}B_{16}$
(a) Vol. **79**, No. 1, p. 109 – 114 (1983).
See also 2, p. 51; 10.1, p. 105; 21.1, p. 375; 21.1.1, p. 383.

O. Schneeweiss, T. Zemčík, T. Žák, and S. Mager
Atomic Structure and Magnetic Properties of the Pseudobinary Alloys Fe_3(Al, Si)
(a) Vol. **79**, No. 1, p. 125 – 129 (1983).
See also *1.3*, p. *26*; *5*, p. *67*; 14.1, p. 209; 21.1.1, p. 384

D. K. Chakrabarty, A. Bandyopadhyay, S. B. Patil, and S. N. Shringi
*Magnetic and Mössbauer Studies of Some Mixed Cobaltites of the Lanthanides and Alkali Metals
(a) Vol. **79**, No. 1, p. 213 – 222 (1983).
See also *5*, p. *67*; 22.8, p. 508.

Y. D. Yao, S. Arajs, and E. E. Anderson
Deviations from Matthiessen's Rule in the Nickel–Chromium System
(a) Vol. **79**, No. 1, p. 327 – 330 (1983).
See also *13.4*, p. *186*; *14.1*, p. *209*; 21.1, p. 375.

A. V. Zaborov and M. V. Medvedev
*A Binary Magnetic Alloy with Competing Exchange Interactions and Magnetic Moments Dependent on Local Environment
(b) Vol. **119**, No. 2, p. 621 – 630 (1983).
See also *18.4*, p. *297*.

Z. Z. AKSELROD, M. BUDZYNSKI, B. A. KOMISSAROVA, L. N. KRYUKOVA, G. K. RYASNY, and A. A. SOROKIN
*Hyperfine Magnetic Field at ^{181}Ta in YFe_2
(b) Vol. **119**, No. 2, p. 667–673 (1983).
See also 21.1.1, p. 384; 21.4, p. 392.

P. R. SILVA and F. C. SÁ BARRETO
*Thermodynamics of the Diluted Transverse Ising Model: A Mean Field Type Treatment
(b) Vol. **119**, No. 2, p. K105–K108 (1983).

S. ENGELS and W. MÖRKE
*Silica Supported Nickel Clusters: The Particle Size Dependence of FMR Intensity and Magnetization
(b) Vol. **119**, No. 2, p. K109–K112 (1983).
See also *19*, p. *304*; 21.1, p. 375.

E. KOLLEY and W. KOLLEY
Spin Fluctuations in Weak Itinerant Ferromagnets near T_c
(b) Vol. **119**, No. 2, p. K175–K179 (1983).
See also *18.1*, p. *279*.

W. BECK and H. KRONMÜLLER
*The Influence of Surface Roughness on the Coercive Field of the Nearly Non-Magnetostrictive Amorphous Ferromagnetic Alloy $Co_{58}Ni_{10}Fe_5Si_{11}B_{16}$
(a) Vol. **79**, No. 2, p. 389–394 (1983).
See also 1.6, p. 41; 2, p. 53; 21.1, p. 376; 21.1.1, p. 384.

J. ŠTERNBERK, E. KRATOCHVÍLOVÁ, J. HŘEBÍK, and A. GEMPERLE
*Coercivity and Microstructure of Low-Alloy Cr–Mo Steel
(a) Vol. **79**, No. 2, p. 523–529 (1983).
See also *1.4*, p. *34*; 21.1.1, p. 384.

R. Z. SADYKHOV, L. M. VALIEV, D. A. GUSEINOV, and A. O. ISMAILOV
*Magnetic Properties of the $Fe_xZn_{1-x}Cr_2S_4$ $(0 < x < 1)$ System
(a) Vol. **79**, No. 2, p. K147–K150 (1983).
See also *18.4*, p. *297*; 22.8, p. 509.

K. WITAŃSKI
*Theoretical Calculations of the Curie Temperature for $Gd_{1-x}RE_xAl_2$ Alloys
(b) Vol. **120**, No. 1, p. 149–154 (1983).
See also 21.4, p. 393.

J. A. PLASCAK
Thermodynamical Properties of the Diluted Transverse Ising Model with Anisotropic Coupling Constants
(b) Vol. **120**, No. 1, p. 215–224 (1983).

P. RENNERT, G. CARL, and W. HERGERT
Spin-Polarized Compton Profile of Nickel
(b) Vol. **120**, No. 1, p. 273–281 (1983).
See also *13.1*, p. *164*; 13.2, p. 166; *20*, p. *311*; 21.1, p. 376.

D. Fraitová
*An Analytical Theory of FMR in Bulk Metals (Parallel Configuration) (I): Dispersion Relations
(b) Vol. **120**, No. 1, p. 341 – 348 (1983).
See also *19*, p. *305*.

A. F. Sadreev, I. S. Sandalov, and A. N. Podmarkov
*The Spin-Wave Theory of Electron and s–f-Exchange Magnetostriction in the Rare Earth Metals
(b) Vol. **120**, No. 1, p. 383 – 395 (1983).
See also 21.4, p. 393.

W. Korneta and Z. Pytel
*Autocorrelation Time in Anisotropic Semi-Infinite Ferromagnets
(b) Vol. **120**, No. 1, p. K69 – K72 (1983).

D. E. G. Williams and A. Jezierski
*The Spin Wave Stiffness Constant in $Pt_x(Mn_{0.5}Cr_{0.5})_{1-x}$ Alloys
(b) Vol. **120**, No. 1, p. K87 – K91 (1983).
See also 21, p. 360.

I. Tomáš and Yu. A. Kuzin
*Internal Forces in Magnetic Bubble Lattices
(a) Vol. **80**, No. 1, p. 43 – 50 (1983).

G. L. Whittle and A. M. Stewart
Mössbauer Effect Study of $Fe_{85-x}Ni_xB_{15}$ Metallic Glasses
(a) Vol. **80**, No. 1, p. 101 – 107 (1983).
See also 2, p. 54; *5*, p. *68*; 21.1, p. 377; 21.1.1, p. 385.

M. Vázquez, W. Fernengel, and H. Kronmüller
*The Effect of Tensile Stresses on the Magnetic Properties of $Co_{58}Fe_5Ni_{10}Si_{11}B_{16}$ Amorphous Alloys
(a) Vol. **80**, No. 1, p. 195 – 204 (1983).
See also 2, p. 54; 12.1, p. 140; 21.1, p. 377; 21.1.1, p. 385.

Y. Shimada
*Fabrication and Some Magnetization Characteristics of Amorphous Cores with Multilayered Structure
(a) Vol. **80**, No. 1, p. K87 – K91 (1983).
See also 21.1, p. 377.

J. M. Wesselinowa
*Dynamic Behaviour of the s–d Model
(b) Vol. **120**, No. 2, p. 585 – 593 (1983).
See also *13.5.4*, p. *203*; 22.8, p. 510.

H. Fehske
Spin Fluctuations in Alloys with Random Transfer
(b) Vol. **120**, No. 2, p. 611 – 620 (1983).
See also *13*, p. *153*; 14, p. 204; 18.1, p. 279; 21.1, p. 377.

D. Fraitová

*An Analytical Theory of FMR in Bulk Metals (Parallel Configuration) (II): Penetration Depths
(b) Vol. **120**, No. 2, p. 659–665 (1983).
See also *19*, p. *305*.

A. H. Nayyar and P. Shukla

*Spin Glass Behaviour in $Eu_xSr_{1-x}S$: Calculation of the Phase Diagram in Mean Field Approximation
(b) Vol. **120**, No. 2, p. K141–K144 (1983).
See also *18*, p. *276*; 22, p. 415.

N. Ya. Borets, I. Ya. Kanovskii, and V. M. Nitsovich

*Coexistence of Charge and Magnetic Orderings in a Polar Ferromagnet
(b) Vol. **120**, No. 2, p. K167–K171 (1983).
See also *13*, p. *153*.

M. Vázquez, W. Fernengel, and H. Kronmüller

*Influence of Applied Stresses on the Magnetic Properties of Annealed $Co_{58}Fe_5Ni_{10}Si_{11}B_{16}$ Amorphous Alloy
(a) Vol. **80**, No. 2, p. 513–518 (1983).
See also 2, p. 55; 12.1, p. 140; 21.1, p. 378; 21.1.1, p. 385.

M. Jurczyk

*Magnetic Properties of Substituted $Pr_2(Co_{1-x}M_x)_{17}$ Compounds (M = Fe, Mn, and Cr)
(a) Vol. **80**, No. 2, p. 657–662 (1983).
See also *1.3*, p. *28*; 21.1, p. 378; 21.4, p. 394.

18.2.1 Ferromagnetic Films

Original Papers and Short Notes

Y. Iwama and Y. Takeno
*Formation Process of MnBi Thin Films by Williams Method
(a) Vol. **76**, No. 1, p. 75–83 (1983).
See also *3*, p. *57*; 22.7, p. 499.

L. Wojtczak, A. Urbaniak-Kucharczyk, J. Mielnicki, and B. Mrygon
The Magnetic Contribution to Specific Heat in Thin Ferromagnetic Films
(a) Vol. **76**, No. 1, p. 113–120 (1983).
See also *8*, p. *86*.

V. M. Pan, I. Ya. Dekhtyar, M. E. Osinovskii, S. E. Litvin, B. G. Nikitin, M. M. Nishenko, and M. P. Voronko
*Thermodynamical and Structural Properties of Amorphous $Fe_{80}B_{20}$ and $Fe_{80}P_{13}C_7$ Alloys
(a) Vol. **76**, No. 1, p. 197–206 (1983).
See also *2*, p. *45*; 5, p. 65; 8, p. 87; 14.1, p. 206; 21.1.1, p. 380.

Yu. I. Bespyatykh, I. E. Dikshtein, S. V. Meriakri, and V. V. Tarasenko
*Spectrum of Magnetization Oscillations in a Uniaxial Ferromagnetic Plate
(b) Vol. **116**, No. 2, p. 607–616 (1983).

J. Baszyński
*Measurements of Magnetic Parameters of YBiFe Garnet Films by Ferrimagnetic Resonance and Ferrimagnetic Antiresonance
(a) Vol. **76**, No. 2, p. K189–K191 (1983).
See also *18.3*, p. *292*; 22.8.2, p. 520.

A. A. Alekhin
*The Bloch Lines Dynamics in Permalloy Films
(a) Vol. **77**, No. 1, p. 51–61 (1983).
See also 21.1, p. 368; 21.1.1, p. 382.

K. Sokalski and W. Wójcik
*A Simple Model of Thin Magnetic Films
(b) Vol. **117**, No. 2, p. K71–K73 (1983).

L. Půst
*Influence of Magnetostriction on FMR in Thin Garnet (111) Films
(a) Vol. **78**, No. 2, p. 741–749 (1983).
See also 19, p. 303; 22.8.2, p. 522.

18.3 Ferrimagnetic Properties

Original Papers and Short Notes

I. Tomáš, L. Murtinová, and J. Kaczér
Easy Magnetization Axes in Materials with Combined Cubic and Uniaxial Anisotropies
(a) Vol. **75**, No. 1, p. 121 – 127 (1983).
See also *18.2*, p. *280*.

L. M. Dedukh, V. S. Gornakov, and V. I. Nikitenko
*One-Direction Motion of Bloch Lines during Their Nonlinear Oscillation
(a) Vol. **75**, No. 2, p. K117 – K119 (1983).
See also 22.8.2, p. 519.

V. K. Vlasko-Vlasov and L. S. Uspenskaya
*Dynamic Changes of the Domain Structure Period in Magnetic Dielectric
(a) Vol. **75**, No. 2, p. K125 – K127 (1983).
See also 22.8.2, p. 519.

C. Pasnicu, D. Condurache, and E. Luca
*Influence of Substitution and Addition of Calcium on Magnetic Properties of $Ni_{0.245}Zn_{0.755}Fe_2O_4$ Ferrite
(a) Vol. **76**, No. 1, p. 145 – 150 (1983).
See also 22.8.2, p. 520.

Y. Kawai and T. Ogawa
*ΔE Effect of a Mn Ferrite Single Crystal at Temperatures between 150 and 300 K
(a) Vol. **76**, No. 1, p. 375 – 381 (1983).
See also 7, p. 82; *12.2*, p. *142*; 22.8.2, p. 520.

V. G. Kleparskii and I. Pintér
*Direct Observation of Domain Nucleation in Garnet Films
(a) Vol. **76**, No. 1, p. K1 – K4 (1983).
See also 22.8.2, p. 520.

D. Adlienė, L. Pranevičius, and G. Zubauskas
Acoustic Emission Initiated by Ion Implantation in Bubble Garnet Films
(a) Vol. **76**, No. 1, p. K35 – K37 (1983).
See also *7*, p. *82*; *11*, p. *125*; 22.8.2, p. 520.

Z. Drzazga
*Magnetic Anisotropy of $ErCo_{4.5}Cu$ Single Crystals
(a) Vol. **76**, No. 2, p. 647 – 649 (1983).
See also 21.4, p. 391.

B. Banerjee and S. Lahiry
*Superparamagnetism in γ-Mn_2O_3–α-Fe_2O_3–α-Mn_2O_3 System
(a) Vol. **76**, No. 2, p. 683 – 694 (1983).
See also *18.4*, p. *295*; 22.6, p. 493; 22.8.2, p. 520.

J. Baszyński
*Measurements of Magnetic Parameters of YBiFe Garnet Films by Ferrimagnetic Resonance and Ferrimagnetic Antiresonance
(a) Vol. **76**, No. 2, p. K189–K191 (1983).
See also *18.2.1*, p. *290*; 22.8.2, p. 520.

P. C. M. Gubbens, W. Ras, A. M. van der Kraan, and K. H. J. Buschow
*Magnetization and ^{161}Dy Mössbauer Effect Study of $DyMn_2$ and Dy_6Mn_{23} and Their Ternary Hydrides
(b) Vol. **117**, No. 1, p. 277–282 (1983).
See also *5*, p. *66*; 21.4, p. 391.

B. Y. Yousif and R. G. Bowers
*Spin Dependence and the Interrelation between Staggered and Uniform Field Properties in the Exact Solution of One-Dimensional Mixed-Spin Ising Ferrimagnets
(b) Vol. **117**, No. 1, p. 395–400 (1983).

U. Heinecke, C. Cruz, E. Wieser, and C. Bernhardt
*Structural Transformations and Connected Changes of Magnetic Properties by Mechanical Treatment of $SrFe_{12}O_{19}$
(a) Vol. **77**, No. 1, p. 225–232 (1983).
See also *1.2*, p. *16*; 5, p. 66; 22.8.2, p. 521.

S. Ligenza, M. Łukasiak, Z. Kucharski, and J. Suwalski
Study of the $3d^5$ Shell Deformation of Fe^{3+} Ions in $ZnFe_2O_4$ by High-Field Mössbauer Spectroscopy
(b) Vol. **117**, No. 2, p. 465–470 (1983).
See also *5*, p. *66*; 22.8.2, p. 521.

G. P. Kramar and Ya. I. Panova
*Resonance and Relaxation of Domain Walls in Polycrystalline Ferrites
(a) Vol. **77**, No. 2, p. 483–489 (1983).
See also 22.8.2, p. 521.

O. P. Smirnov and V. P. Plakhty
*On the Role of Intrasublattice Interactions in YIG with Diamagnetically Substituted a-Sublattice
(a) Vol. **77**, No. 2, p. 505–511 (1983).
See also 22.8.2, p. 521.

Y. Kawai and T. Ogawa
*Domain Wall Motions Due to External Stresses in a Mn–Zn Ferrite Single Crystal
(a) Vol. **77**, No. 2, p. K103–K106 (1983).
See also *12.2*, p. *142*; 22.8.2, p. 521.

L. Pranevičius, A. Sakalas, and G. Zubauskas
*Ion Beam Induced Mixing in Bubble Garnet Films
(a) Vol. **77**, No. 2, p. K171–K172 (1983).
See also *11*, p. *129*; 22.8.2, p. 522.

A. G. Arias and F. Calderón
*Effect of Some Additives in Nickel–Zinc Ferrite Losses
(a) Vol. **77**, No. 2, p. K185 – K188 (1983).
See also 22.8.2, p. 522.

N. L. Pakhomova, L. M. Kasimenko, and V. Christoph
*Ferromagnetic Intra-Sublattice Exchange in Multicomponent Ferrites
(a) Vol. **78**, No. 1, p. 111 – 114 (1983).
See also 22.8.2, p. 522.

M. Wurlitzer
*Richter Type Magnetic After-Effect in YIG at Helium Temperature
(a) Vol. **78**, No. 1, p. K87 – K91 (1983).
See also 10.2, p. 115; 22.8.2, p. 522.

H. Hülsing and A. Kasten
*Magnetic Properties of Dysprosium Oxisulphate ($Dy_2O_2SO_4$)
(b) Vol. **118**, No. 2, p. 725 – 741 (1983).
See also *18.4*, p. *296*; 22.8.1, p. 515.

G. Dehe and U. Behn
*Contribution of the Direct B–B Interaction to the Supertransferred Hyperfine Field at Diamagnetic Antimony in Oxidic Spinels Determined by ^{121}Sb Mössbauer Spectroscopy
(a) Vol. **78**, No. 2, p. 485 – 488 (1983).
See also *5*, p. *67*; 22.8.2, p. 522.

P. Novotný
*Bubble Film Characterization by Differential Susceptibility Measurements
(a) Vol. **78**, No. 2, p. K101 – K104 (1983).
See also 22.8.2, p. 522.

V. V. Kelarev, V. V. Chuev, A. N. Pirogov, and S. K. Sidorov
*Anisotropy and Exchange Effects in Heavy Rare-Earth–Cobalt Compounds of the RCo_5 Type
(a) Vol. **79**, No. 1, p. 57 – 66 (1983).
See also *1.3*, p. *25*; 21.1, p. 374; 21.4, p. 392.

G. Dehe and J. Suwalski
*^{121}Sb and ^{57}Fe Mössbauer Effect in Sb Substituted Lithium Ferrite in the Presence of a Magnetic Field
(b) Vol. **119**, No. 2, p. K155 – K158 (1983).
See also *5*, p. *68*; 22.8.2, p. 523.

Yu. G. Chukalkin, V. R. Shtirts, and B. N. Goshchitskii
Amorphization of Neutron-Irradiated Yttrium Iron Garnet
(a) Vol. **79**, No. 2, p. 361 – 366 (1983).
See also 2, p. 53; *11*, p. *132*; 22.8.2, p. 523.

E. Gmelin, N. Lenge, and H. Kronmüller
Vacancy Effect on the Transition Temperature of Magnetite
(a) Vol. **79**, No. 2, p. 465 – 475 (1983).
See also *8*, p. *90*; 22.6, p. 496.

V. K. Vlasko-Vlasov and L. S. Uspenskaya
*Direct Experimental Observation of a Localized Wave of Magnetization in a Ferrimagnetic
(b) Vol. **120**, No. 1, p. K1 – K4 (1983).
See also 22.8.2, p. 523.

R. Pfranger, D. Płusa, and B. Wysłocki
*Magnetic Domain Structure in Polycrystalline $DyFe_2$ Compounds
(a) Vol. **80**, No. 1, p. 73 – 78 (1983).
See also 21.1.1, p. 384; 21.4, p. 393.

Z. Drzazga and T. Mydlarz
*Magnetic Properties of the $ErCo_{5.5-x}Cu_x$ Alloys
(a) Vol. **80**, No. 1, p. 403 – 409 (1983).
Se also 21.4, p. 393.

M. Wurlitzer
*The Influence of Irradiation on Richter-Type Magnetic After-Effects in YIG:Si at Helium Temperature
(a) Vol. **80**, No. 2, p. 457 – 461 (1983).
See also 22.8.2, p. 523.

S. R. Murthy
ΔE Effect in Co–Zn Ferrites
(a) Vol. **80**, No. 2, p. K161 – K165 (1983).
See also *12.2*, p. *145*; 22.8.2, p. 523.

18.4 Antiferromagnetic Properties

Review Article

K. H. Fischer
*Spin Glasses (I)
(b) Vol. **116**, No. 2, p. 357 – 414 (1983).
See also 2, p. 43; 8, p. 85; 13, p. 146; 14.1, p. 206; *18.2*, p. *280*; 21.1, p. 363; 21.1.1, p. 379.

V. S. Shpinel
*Exchange Interaction and Hyperfine Fields at Nonmagnetic Impurities in Rare-Earth Metals
(b) Vol. **118**, No. 1, p. 11 – 24 (1983).
See also 5, p. 65; *18.2*, p. *280*; 21.4, p. 390.

Original Papers and Short Notes

M. Ya. Flyagin and A. N. Men
*Calculation of Magnetic Hyperfine Interactions in the Rare-Earth Orthoferrites with Model Covalency Parameter
(b) Vol. **115**, No. 1, p. 277 – 284 (1983).
See also *5*, p. *65*; 22.8.2, p. 519.

G. D. Khattak
Specific Heat of Uranium Dioxide (UO_2) between 0.3 and 50 K
(a) Vol. **75**, No. 1, p. 317 – 321 (1983).
See also *6*, p. *70*; *8*, p. *85*; 22.6, p. 490.

T. J. McKenna, S. J. Campbell, D. H. Chaplin, and G. V. H. Wilson
*Temperature Modulation and Temperature Hysteresis Studies of Dysprosium
(a) Vol. **75**, No. 2, p. 421 – 432 (1983).
See also *18.2*, p. *281*; 21.4, p. 390.

S. Takahashi and K. Ikeda
*Magnetic Interactions in Plastically Deformed Ni_3Mn Alloy
(a) Vol. **76**, No. 1, p. K97 – K100 (1983).
See also 10.1, p. 101; *18.2*, p. *282*; 21.1, p. 366.

J. Dóczi-Réger and P. C. Hemmer
*The Antiferromagnetic Phase of the Ising Model with First- and Second-Neighbour Interactions
(b) Vol. **116**, No. 2, p. K67 – K71 (1983).
See also *13*, p. *148*.

B. Banerjee and S. Lahiry
*Superparamagnetism in γ-Mn_2O_3–α-Fe_2O_3–α-Mn_2O_3 System
(a) Vol. **76**, No. 2, p. 683 – 694 (1983).
See also *18.3*, p. *291*; 22.6, p. 493; 22.8.2, p. 520.

M. A. SEMARY, M. F. MOSTAFA, and M. M. ABDEL-KADER
*Magnetic Susceptibility of the Canted Spin Compounds $(CH_2)_n(NH_3)_2FeCl_4$; $n = 4, 5$, and 7
(a) Vol. **76**, No. 2, p. K131 – K136 (1983).
See also 22.9, p. 527.

D. DĂDÂRLAT, T. PETRIŞOR, A. GIURGIU, and I. POP
*Localization Effect in CrRh Alloys
(b) Vol. **117**, No. 1, p. 155 – 159 (1983).
See also 21.1, p. 368; 21.4, p. 391.

J. BRANDMÜLLER, A. LEHMEYER, K. M. HÄUSSLER, and L. MERTEN
*Electromagnetic Wave Propagation in the Uniaxial Antiferromagnetic Crystals FeF_2, CoF_2, and MnF_2
(b) Vol. **117**, No. 1, p. 323 – 328 (1983).
See also *20.1*, p. *318*; 22.5, p. 474.

G. A. PETRAKOVSKII and S. S. APLESNIN
*Ground State and Effective Field Distribution of Ising and Heisenberg Disordered Antiferromagnets
(b) Vol. **117**, No. 1, p. 401 – 405 (1983).
See also *18.2*, p. *283*.

S. V. TARASENKO and O. G. MEDVEDOVSKAYA
*Orthoferrite Dynamical Properties in the Vicinity of the Tricritical Point
(b) Vol. **117**, No. 2, p. 459 – 463 (1983).

V. BRÉDIMAS, C. PAPATRIANTAFYLLOU, and H. GAMARI-SEALE
On the Magnetic Structure of the Intermetallic Compounds Gd_2Al and Gd_2Au
(b) Vol. **117**, No. 2, p. 513 – 519 (1983).
See also *18.2*, p. *284*; 21.4, p. 391.

H. KIDO, T. HOSHIKAWA, M. SHIMADA, and M. KOIZUMI
*Synthesis and Magnetic Properties of RCuSi (R = Y, Ce, Nd, Sm, Gd, Ho)
(a) Vol. **77**, No. 2, p. K121 – K123 (1983).
See also *18.2*, p. *284*; 21.1, p. 371; 21.4, p. 391.

A. LEHMEYER, L. MERTEN, and K. M. HÄUSSLER
*On the Reflectivity for Oblique Incidence of the Uniaxial Antiferromagnetic Crystals FeF_2, CoF_2, and MnF_2
(b) Vol. **118**, No. 2, p. 675 – 679 (1983).
See also *20.1*, p. *324*; 22.5, p. 475.

H. HÜLSING and A. KASTEN
*Magnetic Properties of Dysprosium Oxisulphate ($Dy_2O_2SO_4$)
(b) Vol. **118**, No. 2, p. 725 – 741 (1983).
See also *18.3*, p. *293*; 22.8.1, p. 515.

D. FRUCHART, B. MALAMAN, G. LE CAER et B. ROQUES
*Structures magnétiques de FeGe monoclinique
(Magnetic Studies of Monoclinic FeGe)
(a) Vol. **78**, No. 2, p. 555 – 569 (1983).
See also 5, p. 67; 21.1.1, p. 383.

Y. Muraoka, M. Shiga, and Y. Nakamura
*Magnetic Properties and Thermal Expansion of $(Zr_{1-x}Ta_x)Fe_2$
(a) Vol. **78**, No. 2, p. 717–722 (1983).
See also *5*, p. *67*; 8, p. 89; *18.2*, p. *285*; 21.1.1, p. 383.

V. P. Miroshkin and Ya. I. Panova
*Temperature Dependence and Frequency Dispersion Mechanisms of the Magnetic Permeability of Monocrystalline Mn–Zn Ferrite
(a) Vol. **78**, No. 2, p. K179–K182 (1983).
See also 22.8.2, p. 523.

B. V. Egorov and B. B. Timofeev
*The Theory of Magnetic Polarons in Antiferromagnetic Semiconductors
(b) Vol. **119**, No. 1, p. 99–112 (1983).
See also *13.5.3*, p. *201*; 22, p. 411.

W. Schumacher
*On Incommensurate Phases in the Magnetic Phase Diagram of the Hubbard Model
(b) Vol. **119**, No. 1, p. 235–238 (1983).

M. Tamaki, H. Matsui, G. Matsumoto, and T. Kirihara
*Critical Composition of Antiferromagnetic Ordering in UN_xC_{1-x}
(a) Vol. **79**, No. 1, p. K93–K98 (1983).
See also *14.1*, p. *209*; 21.5, p. 395.

M. Schienle, A. Kasten, and P. H. Müller
*Magnetic Relaxation in Dysprosium Phosphate ($DyPO_4$)
(b) Vol. **119**, No. 2, p. 611–620 (1983).
See also 22.8.1, p. 517.

A. V. Zaborov and M. V. Medvedev
*A Binary Magnetic Alloy with Competing Exchange Interactions and Magnetic Moments Dependent on Local Environment
(b) Vol. **119**, No. 2, p. 621–630 (1983).
See also *18.2*, p. *286*.

J. Richter
*Disorder and Competing Interactions in Spin Systems: Finite-Cluster Calculations
(b) Vol. **119**, No. 2, p. K159–K163 (1983).
See also *2*, p. *52*; 21, p. 359.

R. Z. Sadykhov, L. V. Valiev, D. A. Guseinov, and A. O. Ismailov
*Magnetic Properties of the $Fe_xZn_{1-x}Cr_2S_4$ $(0 < x < 1)$ System
(a) Vol. **79**, No. 2, p. K147–K150 (1983).
See also *18.2*, p. *287*; 22.8, p. 509.

L. N. Vasilev, V. M. Grabov, A. V. Golubkov, A. G. Gorobets, V. S. Oskotskii, I. A. Smirnov, and V. V. Tikhonov
Physical Properties and Phase Transitions of the Rare Earth Monosulphides in the Homogeneity Range
(a) Vol. **80**, No. 1, p. 237–244 (1983).
See also *1.3*, p. *26*; *8*, p. *91*; 13, p. 152; 14.3, p. 226; 22, p. 413.

G. GILDENBLAT
Oscillations of Local Magnetization in the Non-Ideal Antiferromagnetic Chain
(b) Vol. **120**, No. 2, p. 519 – 527 (1983).
See also *13.4*, p. *190*.

W. SCHUMACHER
On the Influence of the Magnetic Ordering on Thermodynamic and Electrical Properties in the Hubbard Model
(b) Vol. **120**, No. 2, p. 621 – 629 (1983).
See also *8*, p. *92*; 14.1, p. 210.

19. Magnetic Resonance

Review Article

E. G. SIEVERTS
Classification of Defects in Silicon after Their g-Values
(b) Vol. **120**, No. 1, p. 11–29 (1983).
See also *10.2*, p. *108*; 13.4, p. 172; 22.1.2, p. 423.

Original Papers and Short Notes

W. KORCZAK, M. PARADOWSKI, and M. SUBOTOWICZ
*EPR Linewidth Studies of Gd^{3+}-Doped $La_xCe_{1-x}F_3$ Single Crystals
(b) Vol. **115**, No. 1, p. 89–94 (1983).
See also 13.4, p. 172; 22.5, p. 473.

M. STUTZMANN, J. STUKE, and H. DERSCH
*Electron Spin Resonance of Doped Glow-Discharge Amorphous Germanium
(b) Vol. **115**, No. 1, p. 141–151 (1983).
See also 2, p. 43; *13.4*, p. *172*; 14.3, p. 215; 22.1.1, p. 418; 22.1.2, p. 423.

K. N. SHRIVASTAVA
*Nuclear Relaxation in Antiferromagnets at Low Temperatures
(b) Vol. **115**, No. 1, p. K37–K40 (1983).
See also 22.8, p. 502.

E. A. PAMYATNYKH and A. V. POLTAVETS
*Short-Wavelength Cyclotron Waves in Metals in Quantizing Magnetic Fields
(b) Vol. **115**, No. 1, p. K51–K54 (1983).
See also *13*, p. *146*; 21, p. 349.

S. TYAGI
Ferromagnetic Resonance in Single Particles of Amorphous $Fe_{81.5}B_{14.5}Si_4$
(a) Vol. **75**, No. 1, p. K29–K32 (1983).
See also *2*, p. *43*; *18.2*, p. *281*; 21.1.1, p. 379.

A. V. MALYAVKIN
*Optically Detected Magnetic Resonance in $Zn_{1-x}Mn_xTe$
(b) Vol. **115**, No. 2, p. 353–358 (1983).
See also 13.5.1, p. 194; *20.3*, p. *335*; 22.4.4, p. 470.

W. GEHLHOFF and K. H. SEGSA
*EPR of Iron–Boron Centres in Silicon
(b) Vol. **115**, No. 2, p. 443–453 (1983).
See also *13.4*, p. *174*; 22.1.2, p. 425.

K. H. J. BUSCHOW, H.-J. EIFERT, and B. ELSCHNER
Electronic Structure and Stability of Amorphous Pd–Si Alloys
(b) Vol. **115**, No. 2, p. 455–462 (1983).
See also *2*, p. *44*; *13*, p. *147*; 21, p. 350; 22.1.2, p. 425.

A. V. Mitin and N. V. Polyakov
γ-Magnetic Resonance in the Presence of Electronic Spin Relaxation
(b) Vol. **115**, No. 2, p. 477 – 484 (1983).
See also *5*, p. *65*.

V. S. Grunin, I. B. Patrina, and Z. N. Zonn
*Impurities and Electron Localization in V_2O_5 Crystals
(b) Vol. **115**, No. 2, p. 545 – 553 (1983).
See also *13.4*, p. *174*; 22.6, p. 491.

I. P. Aleksandrova, S. Grande, Yu. N. Moskvich, A. I. Krieger, and V. A. Koptsik
*Second-Order Quadrupole Shift in the Incommensurate Structures of Rb_2ZnBr_4 and Rb_2ZnCl_4
(b) Vol. **115**, No. 2, p. 603 – 613 (1983).
See also *1.2*, p. *14*; 14.4.1, p. 256; 22.8, p. 503.

V. K. Jain, V. P. Seth, and S. K. Yadav
*EPR of VO^{2+} in $ZnSeO_4 \cdot 6H_2O$ Single Crystals
(b) Vol. **115**, No. 2, p. K113 – K115 (1983).
See also 22.8.1, p. 512.

N. Guskos, J. Kuriata, I. H. Salikhov, N. I. Silkin, and S. Yagadin
*EPR Study of Eu^{2+} in PbF_2 Single Crystals
(b) Vol. **115**, No. 2, p. K133 – K135 (1983).
See also 22.5.3, p. 486.

J. Kuriata and N. Guskos
*Superhyperfine Structure in the EPR of PbF_2:Eu^{2+} Single Crystals
(b) Vol. **115**, No. 2, p. K137 – K140 (1983).
See also 22.5.3, p. 486.

W. Zarek, A. Kasprzyk, and E. Kozubek
*Magnetic and ESR Investigations of $Sm_6(Mn_{1-x}Fe_x)_{23}$ Compounds for $0 < x < 0.55$ in the Paramagnetic Range
(a) Vol. **75**, No. 2, p. 359 – 366 (1983).
See also *18.1*, p. *278*; 21.1, p. 364; 21.1.1, p. 379.

W. Pastusiak
*A Microscopic Interpretation of the Spin–Lattice Coupling Coefficients for S-State $4f^7$ Ions
(b) Vol. **116**, No. 1, p. 63 – 70 (1983).
See also *13.4*, p. *176*; 22.5.3, p. 486.

V. L. Ermakov and D. Ya. Osokin
*Theory of Coherent Quasi-Steady States in NQR Multipulse Experiments
(b) Vol. **116**, No. 1, p. 239 – 248 (1983).

W. Gunsser and K. Rohwer
*Determination of the Correlation between the Crystal Field Axis System and the Crystallographic Axes in Chromium-Doped β-Ga_2O_3 by EPR
(b) Vol. **116**, No. 1, p. 275 – 278 (1983).
See also 22.6, p. 492.

N. Takeuchi, K. Inabe, and S. Nakamura
*Paramagnetic Species in NaCl:Ca, Cu Crystals Irradiated at RT
(b) Vol. **116**, No. 1, p. K5–K8 (1983).
See also *11*, p. *124*; 22.5.2, p. 478.

Yu. N. Moskvich, A. M. Polyakov, S. P. Prozorov, V. E. Zobov, and A. V. Ponomarenko
*The Spin–Lattice Relaxation of the Moving Two-Spin Group in Multipulse Spin Locking
(b) Vol. **116**, No. 1, p. K35–K38 (1983).
See also 22.8, p. 504.

L. M. Avkhutskii, R. L. Davidovich, L. A. Zemnukhova, P. S. Gordienko, V. Urbonavičius, and J. Grigas
Peculiarities of Phase Transitions and Physical Properties in $(NH_4)_2SbF_5$
(b) Vol. **116**, No. 2, p. 483–488 (1983).
See also 8, p. 87; *14.4.1*, p. *256*; 22.8, p. 505.

V. G. Baryshevskii, S. A. Kuten, and V. I. Rapoport
*Quadrupole Splitting of Hydrogen-Like Atom Levels in Crystals
(b) Vol. **116**, No. 2, p. 489–500 (1983).
Erratum
(b) Vol. **119**, No. 2, p. 755 (1983).
See also *13.4*, p. *177*; 22.6, p. 492.

Ph. Colomban and D. Vivien
*EPR Study of Ordering in Stoichiometric β-Aluminate
(a) Vol. **76**, No. 2, p. 565–574 (1983).
See also 22.8.1, p. 513.

M. Heming, G. Lehmann, K. Recker, and F. Wallrafen
*EPR and Luminescence of Mn^{2+} in Distorted Tetrahedral Symmetry in $CaZnF_4$ Crystals
(b) Vol. **117**, No. 1, p. 271–276 (1983).
See also 3, p. 58; *20.3*, p. *338*; 22.5, p. 474.

E. A. Harris and J. W. Tucker
*EPR Transitions for Ions with $S = 3/2$ in the Presence of Small Distortions of the Cubic Crystal Field
(b) Vol. **117**, No. 1, p. 301–311 (1983).

D. Gerritzen and D. Siebert
*Second Order Shift of ^{71}Ga NMR in the Two Different Sites of a $CdGa_2S_4$ Single Crystal
(b) Vol. **117**, No. 1, p. K5–K7 (1983).
See also *1*, p. *9*; 22.8, p. 505.

O. J. Żogał
*Knight Shifts of Protons in Some Rare-Earth Dihydrides
(b) Vol. **117**, No. 2, p. 717–723 (1983).
See also 18.1, p. 278; 21.4, p. 391.

S. V. J. Lakshman and A. Sundar Jacob
Absorption Spectrum of VO^{2+} in $LiKSO_4$ Single Crystals
(b) Vol. **117**, No. 2, p. K119–K122 (1983).
See also *13.4*, p. *180*; *20.1*, p. *320*; 22.8.1, p. 514.

R. HRABAŃSKI
*EPR of Mn^{2+} in $[Mg(H_2O)_6]$ $SnCl_6$ Single Crystals
(a) Vol. **77**, No. 2, p. K143 – K147 (1983).
See also 22.8, p. 506.

C. JÄGER and M. MÜLLER
*Crystallization in a Fluorozirconate Glass: Determination of the ^{19}F Chemical Shift in β-$BaZrF_6$
(a) Vol. **77**, No. 2, p. K167 – K170 (1983).
See also *2*, p. *49*; 22.8, p. 506.

H. J. FUCHS, W. KARTHE, and S. BÖRNER
On the Kinetics of a Paramagnetic Oxygen Centre in CaF_2
(b) Vol. **118**, No. 1, p. 211 – 215 (1983).
See also 9, p. 94; *10.2*, p. *113*; 22.5.3, p. 487.

K. HÜBNER, A. V. SHENDRIK, and A. M. PRAULINSH
Decomposition of SiO_x Films Due to Internal Stress
(b) Vol. **118**, No. 1, p. 239 – 244 (1983).
See also *1.5*, p. *37*; *4*, p. *61*; 20.1, p. 321; 22.6, p. 494.

B. RAMBABU, C. RAMASASTRY, and B. V. R. CHOWDARI
Radiation Damage Studies of Co^{2+} and Ni^{2+} Doped NH_4Cl Crystals
(b) Vol. **118**, No. 1, p. 381 – 387 (1983).
See also *10.2*, p. *114*; *20.1*, p. *322*; 22.5, p. 474.

M. S. WARTAK
*Cyclotron Resonance in Metals with *g*-Factor Anisotropy
(b) Vol. **118**, No. 1, p. K1 – K2 (1983).
See also 13.2, p. 166; 21, p. 356.

A. MANOOGIAN and B. W. CHAN
*Electron Spin Resonance and Electron Nuclear Double Resonance of $^{53}Cr^{3+}$ in RbIn Alum
(b) Vol. **118**, No. 1, p. K31 – K34 (1983).
See also 22.8, p. 507.

F. CALLENS, P. MATTHYS, and E. BOESMAN
*ESR of a New Sulphur Centre in KCl
(b) Vol. **118**, No. 1, p. K35 – K37 (1983).
See also *10.2*, p. *114*; 22.5.2, p. 480.

N. ISHII, M. KUMEDA, and T. SHIMIZU
Defect States in Amorphous Germanium
(b) Vol. **118**, No. 1, p. K65 – K68 (1983).
See also 2, p. 49; *13.4*, p. *182*; 22.1.1, p. 421.

R. U. BAUER, J. R. NIKLAS, and J.-M. SPAETH
*Two Types of F Centres in Alkaline Earth Fluoro-Halides (I): Superhyperfine Interactions and Electronic Structure
(b) Vol. **118**, No. 2, p. 557 – 566 (1983).
See also *13.4*, p. *184*; 22.5.4, p. 488.

W. GEHLHOFF and K. H. SEGSA
*Determination of Stresses in Si–SiO_2 and Mechanically Affected Si by EPR
(b) Vol. **118**, No. 2, p. 703–712 (1983).
See also *12.2*, p. *143*; 22.1.2, p. 433; 22.6, p. 495.

L. BOTTYÁN, D. L. BEKE, and K. TOMPA
*Diffusion-Induced Quadrupole Relaxation of ^{27}Al Nuclei in Dilute Al–Ti, Al–Cr, Al–Mn, and Al–Cu Alloys at High Temperatures
(b) Vol. **118**, No. 2, p. 835–842 (1983).
See also *9*, p. *95*; 10.1, p. 104; 21, p. 357; 21.1, p. 372.

H. OPYRCHAŁ, K. D. NIERZEWSKI, and H. DRULIS
*Effect of γ-Irradiation on EPR Spectra of Eu^{2+} Doped KCl and NaCl Crystals
(b) Vol. **118**, No. 2, p. K125–K127 (1983).
See also 10.2, p. 116; *11*, p. *130*; 22.5.2, p. 482.

P. I. BARANSKII, V. M. BABICH, N. P. BARAN, YU. P. DOTSENKO, V. B. KOVALCHUK, and V. A. SHERSHEL
Investigation of Formation Conditions of Thermal Donors-I and -II in Oxygen-Containing n-Type Silicon within the Temperature Range 400 to 800 °C
(a) Vol. **78**, No. 2, p. 733–739 (1983).
See also *13.4*, p. *184*; *14.3*, p. *223*; 22.1.2, p. 433.

L. PŮST
Influence of Magnetostriction on FMR in Thin Garnet (111) Films
(a) Vol. **78**, No. 2, p. 741–749 (1983).
See also *18.2.1*, p. *290*; 22.8.2, p. 522.

J. R. NIKLAS, R. U. BAUER, and J.-M. SPAETH
*Two Types of F Centres in Alkaline Earth Fluoro-Halides (II): Double ENDOR Experiments
(b) Vol. **119**, No. 1, p. 171–178 (1983).
See also *13.4*, p. *185*; 22.5.4, p. 488.

A. V. SKRIPOV, A. P. STEPANOV, A. D. SHEVCHENKO, and Z. D. KOVALYUK
NMR Study of the Charge-Density-Wave State in VSe_2
(b) Vol. **119**, No. 1, p. 401–410 (1983).
See also *13*, p. *151*; 22, p. 411.

S. SASTRY, K. V. LINGAM, and M. J. RAO
*EPR of Cu(II) Doped in Trisethylenediamine Cadmium Chloride
(b) Vol. **119**, No. 1, p. K51–K55 (1983).
See also 22.9, p. 530.

D. M. DARASELIA and D. L. JAPARIDZE
*Determination of the Structure of the Co^{2+}Paramagnetic Centers in α-$LiIO_3$ by the RFDS Method
(b) Vol. **119**, No. 1, p. K57–K58 (1983).
See also *13.4*, p. *186*; 22.8.1, p. 516.

C. S. Sunandana
*Low Temperature ESR Studies on $KHSO_4$
(b) Vol. **119**, No. 1, p. K59 – K64 (1983).
See also 22.8.1, p. 516.

P. Muralt, E. Voit, and R. Kind
*Cl-NQR Study of the Low Temperature Phase of Rb_2ZnCl_4
(b) Vol. **119**, No. 1, p. K65 – K67 (1983).
See also *1.2*, p. *18*; 22.8, p. 508.

A. V. Dvurechenskii, I. A. Ryazantsev, L. S. Smirnov, H. Klose, and M. Rieth
Amorphous Silicon–Phosphorus Layers
(a) Vol. **79**, No. 1, p. 83 – 86 (1983).
See also 2, p. 51; 13.4, p. 186; *14.3.1*, p. *231*; 16, p. 267; 22.1.2, p. 434.

Z. G. Akhvlediani, I. G. Akhvlediani, and T. L. Kalabegishvili
Effect of Combined Mechanical Load and Irradiation on Hydrogen Stabilization in LiF:OH
(b) Vol. **119**, No. 2, p. 503 – 506 (1983).
See also *10.2*, p. *119*; 20.1, p. 327; 22.5.2, p. 483.

V. K. Jain, V. P. Seth, R. K. Malhotra, and S. K. Yadav
Electron Paramagnetic Resonance of Mn^{2+} and VO^{2+} in $Ba_2Zn(HCOO)_6 \cdot 4\,H_2O$ Single Crystals
(b) Vol. **119**, No. 2, p. 659 – 666 (1983).
See also *13.4*, p. *187*; 22.8.1, p. 517.

A. Morimoto, S. Oozora, M. Kumeda, and T. Shimizu
Annealing Behaviour of Hydrogenated Amorphous Silicon–Nitrogen Alloy Films Prepared by Sputtering
(b) Vol. **119**, No. 2, p. 715 – 720 (1983).
See also *2*, p. *52*; *10.2*, p. *119*; 20.1, p. 327; 22, p. 412.

O. S. Torosyan
Hyperfine Structure of the EPR Spectra of Mn^{4+} Ions in a Strong Trigonal Crystal Field
(b) Vol. **119**, No. 2, p. K101 – K104 (1983).
See also 22.8.1, p. 517.

S. Engels and W. Mörke
*Silica Supported Nickel Clusters: The Particle Size Dependence of FMR Intensity and Magnetization
(b) Vol. **119**, No. 2, p. K109 – K112 (1983).
See also *18.2*, p. *287*; 21.1, p. 375.

M. Höhne
*Gold-Related EPR Centres of Low Symmetry in Silicon
(b) Vol. **119**, No. 2, p. K117 – K121 (1983).
See also 13.4, p. 187; 22.1.2, p. 436.

A. N. Rodionov, R. I. Kalendarev, A. V. Shendrik, and Yu. R. Zakis
*Photoinduced Paramagnetic States of Yellow Arsenic
(a) Vol. **79**, No. 2, p. K151 – K154 (1983).
See also *2*, p. *53*; 22.1, p. 417.

M. Stutzmann and J. Stuke
*Temperature Dependence of ESR Spectra of Doped Amorphous Germanium
(b) Vol. **120**, No. 1, p. 225 – 234 (1983).
See also 2, p. 54; *13.4*, p. *189*; 22.1.1, p. 421.

D. Fraitová
*An Analytical Theory of FMR in Bulk Metals (Parallel Configuration) (I): Dispersion Relations
(b) Vol. **120**, No. 1, p. 341 – 348 (1983).
See also *18.2*, p. *288*.

K. Budde and H. Richtering
*Einbau und Diffusion von Cu(I)-Ionen in CsI anhand von NMR-Messungen (Incorporation and Diffusion of Cu(I)-Ions in CsI by NMR Measurements)
(b) Vol. **120**, No. 2, p. 595 – 599 (1983).
See also *9*, p. *96*; 10.2, p. 120; 22.5.2, p. 485.

D. Fraitová
*An Analytical Theory of FMR in Bulk Metals (Parallel Configuration) (II); Penetration Depths
(b) Vol. **120**, No. 2, p. 659 – 665 (1983).
See also *18.2*, p. *289*.

H. J. A. Koopmans, M. M. A. Perik, B. Nieuwenhuijse, and P. J. Gellings
*The Interpretation of Broadened Gd^{3+} EPR Spectra of Polycrystalline Ceramics
(b) Vol. **120**, No. 2, p. 745 – 751 (1983).
See also 22.8.1, p. 518.

B. Susła, R. Czajka, and J. Raułuszkiewicz
Electronic Excitation of Ni^{2+} and Co^{2+} by Tunneling Electrons
(b) Vol. **120**, No. 2, p. K129 – K134 (1983).
See also *13.4*, p. *190*; *14.3.3*, p. *242*; 20.1, p. 330; 22.6, p. 497.

J. Kuriata, J. M. Baker, T. Rewaj, and N. Guskos
*Temperature Dependence of the "Intrinsic" $\bar{b}_2$ Parameter for $CdF_2:Eu^{2+}$
(b) Vol. **120**, No. 2, p. K135 – K139 (1983).
See also 22.5.3, p. 487.

W. Pastusiak
*The Relativistic Field Lattice Contributions to the Spin–Lattice Coupling Coefficients of the s-State $4f^7$ Ions
(b) Vol. **120**, No. 2, p. K151 – K155 (1983).
See also 22.5.3, p. 487.

L. A. Zemnukhova, R. L. Davidovich, P. S. Gordienko, J. Grigas, A. N. Kovrianov, S. I. Kuznetsov, T. A. Kaidalova, and V. Urbanovičius
Investigation of Phase Transitions in K_2SbF_5
(a) Vol. 80, No. 2, p. 553–558 (1983).
See also *1.2*, p. *20*; 8, p. 92; *14.4.1*, p. *259*; 22.8, p. 511.

S. K. Misra, J. Bandet, G. Bacquet, and T. E. McEnally
*Electron Spin Resonance Studies of Defect Centers Containing Boron Atoms in Natural Danburite Crystals
(a) Vol. 80, No. 2, p. 581–588 (1983).
See also 13.4, p. 191; 22.8.1, p. 518.

20. Optical Properties

Original Papers and Short Notes

P. Rennert, G. Carl, and U.-H. Gläser
*Theoretical Compton Profiles of Magnesium
(b) Vol. **115**, No. 1, p. 219–224 (1983).
See also *13.1*, p. *155*; 13.2, p. 166; 21.3, p. 389.

V. Schulz and L. Merten
*Dispersion Curves of Stimulated Raman Effect in Piezoelectric Cubic Crystals
(b) Vol. **115**, No. 1, p. 225–233 (1983).
See also *6*, p. *69*; 13.5, p. 192; 22.2.2, p. 448.

D. P. Mahapatra and H. C. Padhi
*The Compton Profiles of TiB_2 and VB_2
(b) Vol. **115**, No. 2, p. K101–K105 (1983).
See also 22, p. 404.

W. Thulke, R. Frahm, R. Haensel, and P. Rabe
*X-Ray Study of Niobium Disulfide Intercalated with Copper
(a) Vol. **75**, No. 2, p. 501–509 (1983).
See also *1.3*, p. *21*; *4*, p. *60*; 22.8, p. 503.

Hoang Ngo Cam, Nguyen Van Hieu, and Ha Vinh Tan
On the Theory of the Nonlinear Acousto-Optical Effect in Semiconductors
(b) Vol. **116**, No. 1, p. 25–29 (1983).
See also *7*, p. *82*; 13.5.1, p. 195; 22.2, p. 442; 22.4, p. 460.

Yu. P. Drozhov
*Nonlinear Susceptibility of Impurity Centres in Semiconductors
(b) Vol. **116**, No. 1, p. 111–118 (1983).
See also 13.4, p. 176; 22, p. 404.

H. Stramska
*Nonlinear Magneto-Optical Phenomena of the Third Harmonic in Semiconductors
(b) Vol. **116**, No. 1, p. 129–135 (1983).
See also *18*, p. *273*; 22, p. 404.

S. Ghosh and S. Khan
*Stimulated Brillouin Scattering of Transversely Polarized Electromagnetic Waves in Magnetoactive Moderately Piezoelectric Semiconductor Plasmas
(b) Vol. **116**, No. 1, p. 323–334 (1983).
See also *13*, p. *147*; 22, p. 404.

K. P. Maheshwari and Ganesh Tarey
*Excitation of Magnetoacoustic Waves by Two Microwave Beams in InSb
(b) Vol. **116**, No. 1, p. 339–346 (1983).
See also *7*, p. *82*; 13, p. 147; 18, p. 274; 22, p. 405.

F. Comas and C. Trallero Giner
*Low-Frequency Secondary Radiation in Polar Semiconductors at Finite Temperatures
(b) Vol. **116**, No. 2, p. 445–453 (1983).
See also 6, p. 72; 22, p. 405.

Ch. Müller, W. Blau, and P. Ziesche
Local Partial DOS and Experimental Al K Spectra of Transition Metal Aluminides. Charge Transfer Trend
(b) Vol. **116**, No. 2, p. 561–573 (1983).
See also *13.1*, p. *157*; 21.1, p. 367; 21.1.1, p. 381.

K. H. J. Buschow and P. G. van Engen
*Note on the Magnetic and Magneto-Optical Properties of Ni_2In Type 3d Transition Metal Compounds
(a) Vol. **76**, No. 2, p. 615–620 (1983).
See also 1.1, p. 11; *18.2*, p. *282*; 21, p. 353; 21.1, p. 367; 21.1.1, p. 381.

Nguyen The Khoi, J. Ginter, and A. Twardowski
*Magnetooptical Study of the Fundamental Absorption Edge in $Cd_{1-x}Mn_xTe$
(b) Vol. **117**, No. 1, p. 67–74 (1983).
See also 13.5.1, p. 196; *18*, p. *274*; 22.4.4, p. 470.

V. N. Myasnikov, S. V. Marisova, and A. N. Lipovchenko
Optical Properties of Polaritons in Thin Crystal Films
(b) Vol. **117**, No. 1, p. 109–116 (1983).
See also *13.5*, p. *192*; 20.3, p. 338; 22.9, p. 527.

L. Wendler
*Effects of a Transition Layer on Slab Polaritons
(b) Vol. **117**, No. 1, p. 241–246 (1983).
See also *6*, p. *73*; 13.5, p. 192.

A. L. Ageev, Yu. A. Babanov, V. V. Vasin, N. V. Ershov, and A. V. Serikov
*Amorphous Problem in EXAFS Data Analysis
(b) Vol. **117**, No. 1, p. 345–350 (1983).
See also *2*, p. *48*; 21.1.1, p. 382.

N. T. Bagraev
*Electron Localization on Defects and Optical Nuclear Polarization in Disordered and Semimagnetic Semiconductors
(b) Vol. **117**, No. 1, p. 367–379 (1983).
See also *13.4*, p. *178*; 22.1.2, p. 429.

K. Biedrzycki, J. Chrzanowski, and B. Sujak
Some Electrical and Optical Effects Observed in Polycrystalline Ice at 200 K
(a) Vol. **77**, No. 1, p. K95–K98 (1983).
See also *1.2*, p. *16*; 14, p. 204; 22.6, p. 493.

A. M. Kanloev
*Kumakhov Radiation Energy Losses of Ultrarelativistic Electrons and Positrons
(b) Vol. **117**, No. 2, p. 507–511 (1983).
See also *11*, p. *128*; 22.1.2, p. 430.

H. Stramska
*The Magneto-Optical Effect under a Strong Electric Field in Semiconductors
(b) Vol. **117**, No. 2, p. 541 – 547 (1983).
See also 14.3.4, p. 245; *18*, p. *274*; 22, p. 407.

I. V. Stasyuk and S. S. Kotsur
*The Microscopic Theory of the Gyration and Electrogyration in Dielectric Crystals
(b) Vol. **117**, No. 2, p. 557 – 568 (1983).
See also *14.4*, p. *251*; 22, p. 407.

N. V. Ershov, Yu. A. Babanov, and V. R. Galakhov
*EXAFS Study of Crystalline Cu, Ni, and Fe
(b) Vol. **117**, No. 2, p. 749 – 753 (1983).
See also 21.1, p. 370; 21.1.1, p. 382.

A. S. Agabekyan
Resonant Coherent Energy Transfer. Role of Initial Conditions
(b) Vol. **118**, No. 1, p. 51 – 56 (1983).
See also *13.4*, p. *181*; 22, p. 408.

A. P. Deshpande, V. B. Sapre, and C. Mande
*A Study of the Chemical Shifts in X-Ray Absorption Spectra and the Sublattice Distortion in Some Ternary Chalcopyrite Compounds of the Type $CuMS_2$
(b) Vol. **118**, No. 1, p. K39 – K42 (1983).
See also *4*, p. *62*; 22.8, p. 507.

T. Kaczor, S. Nowak, and J. Zmija
*Description of Coloration and Discoloration Processes in Inorganic Photochromic Materials
(a) Vol. **78**, No. 1, p. 187 – 192 (1983).
See also *10.2*, p. *114*.

B. Jagannatha Rao and A. R. Chetal
*On the Chemical Shifts of the K-Absorption Discontinuity of Copper in Some of Its Systems
(b) Vol. **118**, No. 2, p. 551 – 555 (1983).
See also *13.1*, p. *160*; 21.1, p. 372; 22, p. 409.

K. Regiński
Phonon Polaritons in the Presence of a Magnetic Field
(b) Vol. **118**, No. 2, p. 587 – 593 (1983).
See also *6*, p. *76*; *13.5*, p. *193*; 18, p. 275; 22.5.2, p. 481.

Debendranath Sahoo
*Heterodyne Correlation of Scattered Light from a Sr^{++}-Doped NaCl Crystal. A Feasibility Study
(b) Vol. **118**, No. 2, p. 623 – 630 (1983).
See also *10.2*, p. *115*; 22.5.2, p. 482.

G. Dräger, R. Wippermann, O. Brümmer, and A. Šimůnek
*Polarized ZnK- and OK-Emission Bands of ZnO
(b) Vol. **118**, No. 2, p. K113 – K116 (1983).
See also *13.1*, p. *161*; 22.4, p. 460.

A. L. Aleksandrovskii and A. I. Nagaev
*Electrooptic Diffraction in Polydomain Barium-Sodium Niobate Crystals
(a) Vol. **78**, No. 2, p. 431–438 (1983).
See also *14.4.1*, p. *258*; 22.8.1, p. 516.

G. Kreysch, H. Kerkow, R. I. Bogdanov, V. P. Petukhov, and E. A. Romanovskii
Proton-Induced L-Shell X-Ray Emission Cross-Sections of Elements with $24 \leqq Z_2 \leqq 50$ for Projectile Energies between 30 and 350 keV
(a) Vol. **78**, No. 2, p. 507–525 (1983).
See also *11*, p. *130*; 21, p. 357.

W. Thulke, R. Haensel, and P. Rabe
Anisotropic Absorption of Unpolarized X-Radiation in Copper-Intercalated Niobium Disulfide
(a) Vol. **78**, No. 2, p. 539–545 (1983).
See also *1.3*, p. *25*; *4*, p. *63*; 22.8, p. 508.

P. Bhattacharya and A. R. Chetal
*Study of Cu K Edge in Some Copper (II) Octahedral Complexes
(b) Vol. **119**, No. 1, p. 179–183 (1983).
See also *13.1*, p. *162*; 22.9, p. 530.

J. Shanker and G. D. Jain
Analysis of the Dielectric Properties of Mixed Alkali Halide Crystals
(b) Vol. **119**, No. 1, p. 185–191 (1983).
See also *14.4*, p. *253*; 22.5.4, p. 489.

E. Ojala
*K and L Absorption Spectra of Aluminium from APW Calculation
(b) Vol. **119**, No. 1, p. 269–277 (1983).
See also *13.1*, p. *162*; 21, p. 358.

A. A. Kiselev
Circular Phonons and Optical Magnetization of Crystals
(b) Vol. **119**, No. 2, p. 525–530 (1983).
See also *6*, p. *78*; *18*, p. *275*; 20, p. 310; 22.5.2, p. 484.

N. S. Orlova
Variation of Phonon Dispersion Curves with Temperature in Indium Arsenide Measured by X-Ray Thermal Diffuse Scattering
(b) Vol. **119**, No. 2, p. 541–546 (1983).
See also *6*, p. *79*; 22.2.1, p. 446.

P. Bhattacharya and A. R. Chetal
Chemical Shift and EXAFS Studies of Some Copper Complexes
(b) Vol. **119**, No. 2, p. 735–740 (1983).
See also *13.1*, p. *163*; 22.8.1, p. 517.

A. V. Soldatov, A. N. Gusatinskii, M. G. Karin, and O. A. Sadovskaya
Energy Position and Participation of the Rare-Earth Element 5p-States in Chemical Interaction
(b) Vol. **119**, No. 2, p. K123–K126 (1983).
See also *13*, p. *151*; 21.4, p. 393.

A. V. Zamkov, I. T. Kokov, and A. T. Anistratov
*The Acousto-Optical Properties and Photoelasticity of $PbBr_2$ Single Crystals
(a) Vol. **79**, No. 2, p. K177–K180 (1983).
See also *7*, p. *83*; 12.2, p. 144; 22, p. 413.

Yu. P. Khapachev
*The Theory of Dynamical X-Ray Diffraction on a Superlattice
(b) Vol. **120**, No. 1, p. 155–163 (1983).
See also *1*, p. *10*.

P. Rennert, G. Carl, and W. Hergert
*Spin-Polarized Compton Profile of Nickel
(b) Vol. **120**, No. 1, p. 273–281 (1983).
See also *13.1*, p. *164*; 13.2, p. 166; 18.2, p. 287; 21.1, p. 376.

M. N. Ghatikar
*Chemical Transferability of Phase Shifts in Extended X-Ray Absorption Fine Structure
(b) Vol. **120**, No. 1, p. 445–450 (1983).
See also 21.1, p. 376; 21.4, p. 393; 22.5, p. 475; 22.6, p. 497; 22.7, p. 499.

N. G. Alexandropoulos, G. Boulakis, D. Papadimitriou, and I. Theodoridou
*A Note on X-Ray Raman Scattering from Atomic Hydrogen in $TiH_{1.9}$
(b) Vol. **120**, No. 1, p. K19–K22 (1983).
See also *10.1*, p. *106*; 21.1, p. 376.

L. Merten and R. Kotzott
*Excitation of Longitudinal Optical Phonons by Stimulated Raman Effect in Piezoelectric Cubic Crystals
(b) Vol. **120**, No. 2, p. 481–490 (1983).
See also *6*, p. *80*; 13.5, p. 193; 22, p. 414.

20.1 Spectra. Optical Constants (X-Ray Spectra See 20)

Review Articles

R. E. Hummel
*Differential Reflectometry and Its Application to the Study of Alloys, Ordering, Corrosion, and Surface Properties
(a) Vol. **76**, No. 1, p. 11–44 (1983).
See also *1.3*, p. *21*; 10.1, p. 99; 13.1, p. 155; 13.3, p. 167; 21, p. 349; 21,1, p. 363; 22.6, p. 490.

K. N. Shrivastava
Theory of Spin–Lattice Relaxation
(b) Vol. **117**, No. 2, p. 437–458 (1983).
See also *6*, p. *69*; *13*, p. *146*; 22.5, p. 473; 22.8.1, p. 512.

M. Cardona
Vibrational Spectra of Hydrogen in Silicon and Germanium
(b) Vol. **118**, No. 2, p. 463–481 (1983).
See also *6*, p. *69*; 22.1.1, p. 418; 22.1.2, p. 423

Original Papers and Short Notes

E. A. Radzhabov
A Model of Titanium Luminescence Centers in Lithium and Sodium Fluorides
(b) Vol. **115**, No. 1, p. 83–88 (1983).
See also *13.4*, p. *172*; *20.3*, p. *334*; 22.5.2, p. 477.

J. Prakash
*Effect of Ultraviolet Irradiation on Se^{2-}–Anion Vacancy Dipoles in Potassium Iodide
(b) Vol. **115**, No. 1, p. 131–139 (1983).
See also *10.2*, p. *108*; 22.5.2, p. 477.

M. P. Lisitsa, S. F. Terekhova, and N. A. Onishchenko
*Two-Beam Interferometry in the Exciton Spectroscopy of CdS
(b) Vol. **115**, No. 1, p. 187–194 (1983).
See also *13.5.1*, p. *194*; 22.4.1, p. 462.

E. Gurauskas, J. Kavaliauskas, G. Krivaitė, and A. Šileika
*Infrared Absorption Spectra of $Cd_{0.2}Hg_{0.8}Te$ Crystals
(b) Vol. **115**, No. 1, p. 195–202 (1983).
See also *6*, p. *69*; 22.4.4, p. 470.

C. W. M. Timmermans, S. O. Cholakh, R. L. van der Woude, and G. Blasse
*Some Optical and Electrical Measurements on $Cs_3Bi_2Br_9$ Single Crystals
(b) Vol. **115**, No. 1, p. 267–271 (1983).
See also 13.1, p. 155; *16*, p. *264*; 22.8, p. 502.

FAT DUEN HO
*Approximate Wave Functions for the F^+-Centers in $LiNbO_3$ and $LiTaO_3$
(b) Vol. **115**, No. 1, p. 299–303 (1983).
See also *10.2*, p. *108*; 13.4, p. 172; 22.8.1, p. 512.

Y. Y. ABOLINSH, S. V. KARPOV, and A. A. SHULTIN
Raman Studies of Ammonium Ion Motion in NH_4NO_3–V Crystals
(b) Vol. **115**, No. 1, p. K1–K5 (1983).
See also *1.2*, p. *14*; *6*, p. *70*; 22.8.1, p. 512.

G. HOEBEECK, J. PROVOOST, and G. JACOBS
*The F' Centre in RbBr
(b) Vol. **115**, No. 1, p. K33–K36 (1983).
See also *10.2*, p. *108*; 22.5.2, p. 477.

S. KETOLANEN
*On Mollwo-Ivey Relations of Mixed Alkali Halides
(b) Vol. **115**, No. 1, p. K41–K44 (1983).
See also *10.2*, p. *108*; 22.5.4, p. 488.

P. HARI, K. PURANDAR, J. LAKSHMANA RAO, and S. V. J. LAKSHMAN
*Optical Absorption Spectrum of Co^{2+} in Lithium Ammonium Sulphate Single Crystals
(b) Vol. **115**, No. 1, p. K45–K50 (1983).
See also *13.4*, p. *173*; 22.8.1, p. 512.

H. SOBOTTA, H. NEUMANN, N. N. SYRBU, and V. RIEDE
*Infrared Lattice Vibration Spectra of Monoclinic ZnP_2
(b) Vol. **115**, No. 1, p. K55–K58 (1983).
See also *6*, p. *70*; 22.3, p. 456.

K. TAKIYAMA, T. FUJITA, H. NOJIMA, and M. NISHI
*Vacuum Ultraviolet Absorption in γ-Irradiated LiCl
(b) Vol. **115**, No. 1, p. K59–K61 (1983).
See also 10.2, p. 108; 22.5.2, p. 477.

V. E. BORISENKO, V. V. GRIBKOVSKII, V. A. LABUNOV, V. A. SAMUILOV, and K. D. YASHIN
Incoherent Light Annealing of Phosphorus-Doped Polycrystalline Silicon
(a) Vol. **75**, No. 1, p. 117–120 (1983).
See also *14.3.1*, p. *229*; 22.1.2, p. 424.

H. POHLACK
*Thin-Film Photodetectors with High Quantum Yields Achieved by Induced Resonance Total Absorption
(a) Vol. **75**, No. 1, p. 179–185 (1983).
See also *14.3.1*, p. *229*; 14.3.3, p. 236; 22.1.2, p. 424.

W. KUCHARCZYK and P. GÓRSKI
*Measurement of Nonlinear Electrooptic Effects in KDP Crystals
(a) Vol. **75**, No. 1, p. K87–K90 (1983).
See also *14.4.1*, p. *256*; 22.8.1, p. 512.

C. S. SUNANDA and R. SINGH
*An Infrared Study of the Glass System $CaO \cdot B_2O_3 \cdot Al_2O_3 \cdot xFe_2O_3$ ($0 \leqq x \leqq 1$)
(a) Vol. **75**, No. 1, p. K91 – K94 (1983).
See also *2*, p. *44*; 22.6.1, p. 498.

K. TAKEDA, Y. MAEDA, E. OHTA, M. SAKATA, G. KIDO, and N. MIURA
Study of Energy Band Parameters in p-Type $Ge_{0.9}Si_{0.1}$Alloys
(b) Vol. **115**, No. 2, p. 369 – 379 (1983).
See also *13.1*, p. *156*; 14.3, p. 216; 22.1.1, p. 418; 22.1.2, p. 425.

YU. A. EKMANIS and I. S. RADCHENKO
*Optical Properties of Inhomogeneous Colloidal Centres in Alkali Halides
(b) Vol. **115**, No. 2, p. 389 – 397 (1983).
See also 22.5.2, p. 478.

A. N. NIKITINA, I. S. OSADKO, I. A. VASILIEVA, and L. A. YANOVSKAYA
Vibronic and Electron–Phonon Coupling in Spectra of Substituted Polyenes in Rigid Matrices
(b) Vol. **115**, No. 2, p. 493 – 498 (1983).
See also *6*, p. *70*; *13.4*, p. *174*; 20.3, p. 335; 22.9, p. 525.

S. V. MARISOVA, E. N. MYASNIKOV, and A. N. LIPOVCHENKO
*Phonon-Assisted Decay of Excitons in Anthracene Crystals
(b) Vol. **115**, No. 2, p. 649 – 654 (1983).
See also *13.5.1*, p. *195*; 22.9, p. 525.

B. M. NITSOVICH and G. M. PESTRYAKOV
Temperature Behaviour of the Integral Intensity of Exciton Absorption
(b) Vol. **115**, No. 2, p. K87 – K90 (1983).
See also *13.5.1*, p. *195*.

R. GRÜNDLER
*Superconvergent Sum Rules for Metals and Insulators
(b) Vol. **115**, No. 2, p. K147 – K150 (1983).

L. BOČÁNEK and J. MUSILOVÁ
*Modulation Piezoreflectance of Cubic Polycrystals
(a) Vol. **75**, No. 2, p. 555 – 558 (1983).
See also *12.2*, p. *141*; 13.1, p. 156; 22.1.1, p. 419.

V. D. AKHMETOV, V. V. BOLOTOV, L. S. SMIRNOV, and V. A. KHARCHENKO
*The Defect Accumulation and Irradiation-Induced Lattice Strain in Si at High-Dose Neutron Irradiation
(a) Vol. **75**, No. 2, p. 601 – 606 (1983).
See also 11, p. 124; *13.4*, p. *175*; 22.1.2, p. 426.

K. H. J. BUSCHOW, J. H. N. VAN VUCHT, P. G. VAN ENGEN, D. B. DE MOOIJ, and A. M. VAN DER KRAAN
Structure, Magnetic and Magneto-Optical Properties, and ^{57}Fe Mössbauer Effect in a New Variety of Heusler Alloy:PtFeSb
(a) Vol. **75**, No. 2, p. 617 – 623 (1983).
See also *1.3*, p. *22*; 5, p. 65; *18.2*, p. *281*; 21.1.1, p. 379.

M. I. MITKOWA and W. FALLMANN
Some Properties of Silver Photodiffused Chalcogenide Glasses
(a) Vol. **75**, No. 2, p. K145 – K148 (1983).
See also *2*, p. *45*; *14.3.3*, p. *237*; 22.8, p. 503.

R. A. ABREU, J. STANKIEWICZ, and W. GIRIAT
Temperature Dependence of the Absorption Edge in $Cd_{1-y}Mn_ySe$
(a) Vol. **75**, No. 2, p. K153 – K156 (1983).
See also 22.4.4, p. 470.

Z. BŁASZCZAK, M. LUDWICZAK, and F. KACZMAREK
*Bleaching of F_2 Centers in $LiF:F_2$ Color Center Lasers
(a) Vol. **75**, No. 2, p. K195 – K198 (1983).
See also *10.2*, p. *110*; 20.2, p. 332; 22.5.2, p. 478.

P. PAROLI, L. RYABOVA, V. SALUN, and A. TUCCIARONE
*Optical Filters Based on Pyrolytic Films of Metal Oxides
(a) Vol. **75**, No. 2, p. K209 – K210 (1983).
See also 22.6, p. 491.

V. B. SHARAN
*Absorption, Fluorescence, and Excitation Spectra of Thallium-Activated Rubidium Bromide
(b) Vol. **116**, No. 1, p. 41 – 47 (1983).
See also 13.4, p. 175; *20.3*, p. *336*; 22.5, p. 473.

M. ISHIGAME and S. YAMASHITA
*Raman Scattering in K_2SO_4 at High Temperatures
(b) Vol. **116**, No. 1, p. 49 – 56 (1983).
See also 1.2, p. 14; *6*, p. *71*; 22.8.1, p. 513.

K. SAKATA
*Study of Electronic Structure of Reduced TiO_2 and $V_xTi_{1-x}O_2$ Crystals by ESCA and Optical Absorption
(b) Vol. **116**, No. 1, p. 145 – 153 (1983).
See also *13.4*, p. *176*; 17, p. 269; 22.6.1, p. 498.

I. T. GODMANIS, A. N. TRUKHIN, and K. HÜBNER
*Exciton–Phonon Interaction in Crystalline and Vitreous SiO_2
(b) Vol. **116**, No. 1, p. 279 – 287 (1983).
See also *6*, p. *72*; *13.5.1*, p. *195*; 22.1.1, p. 419; 22.6, p. 492.

C. K. ONG and J. ADETUNJI
*The Optical Absorption of F and F_H Centres in Cesium Halides
(b) Vol. **116**, No. 1, p. K19 – K21 (1983).
See also *10.2*, p. *110*; 22.5.2, p. 478.

G. F. DOBRZHANSKII, O. V. FIALKOVSKAYA, L. I. KHIRUNENKO, and S. P. SMIRNOV
*Optical Absorption of CdS:Nd Crystals
(b) Vol. **116**, No. 1, p. K31 – K34 (1983).
See also 22.4.1, p. 463.

R. B. PODE, D. G. WAKDE, and B. T. DESHMUKH
*Effect of Z_1 Centres on the Isothermal Stability of Colour Centres in Microcrystalline Powder of KCl
(b) Vol. **116**, No. 1, p. K39–K42 (1983).
See also *10.2*, p. *110*; 22.5.2, p. 478.

V. I. IVANOV-OMSKII, V. K. OGORODNIKOV, and V. D. ROZUMNYI
Surface Defects of Closed-System-Grown $Cd_xHg_{1-x}Te$ Epitaxial Layers
(a) Vol. **76**, No. 1, p. 71–74 (1983).
See also *1.5*, p. *36*; 22.4.4, p. 470.

V. G. FEDOTOV, A. G. BYCHKOV, D. N. KARLIKOV, L. M. GORYNYA, A. U. SHELEG, E. M. SMOLYARENKO, and Z. K. ORLIK
*Temperature Studies of Optical Properties of Cadmium Diphosphide Crystals
(a) Vol. **76**, No. 1, p. 121–126 (1983).
See also 22.3, p. 456.

T. NANBA and T. P. MARTIN
*Raman Scattering from Metal Smokes
(a) Vol. **76**, No. 1, p. 235–239 (1983).
See also 1.6, p. 39; *6*, p. *72*; 21.1, p. 366; 21.6, p. 397.

J. M. NORAS
*Fluctuations in Light Transmission of Zinc Sulphide
(a) Vol. **76**, No. 1, p. K53–K56 (1983).
See also 22.4.1, p. 463.

A. F. RAKOV
*Neutron Induced Colour and Luminescence Centres in Yttrium–Aluminium Garnet Single Crystals
(a) Vol. **76**, No. 1, p. K57–K60 (1983).
See also *10.2*, p. *111*; *11*, p. *126*; 20.3, p. 337; 22.8.2, p. 520.

P. LOŠŤÁK, J. HORÁK, and L. KOUDELKA
*Some Physical Properties and Point Defects in $Bi_{2-x}In_xTe_3$ Mixed Crystals
(a) Vol. **76**, No. 1, p. K71–K75 (1983).
See also *1.3*, p. *22*; 14.3, p. 218; 22.7.1, p. 501.

N. M. GASANLY, A. F. GONCHAROV, N. N. MELNIK, A. S. RAGIMOV, and V. I. TAGIROV
*Optical Phonons and Structure of $TlGaS_2$, $TlGaSe_2$, and $TlInS_2$ Layer Single Crystals
(b) Vol. **116**, No. 2, p. 427–443 (1983).
See also *6*, p. *72*; 22.8, p. 504.

V. VORLÍČEK, I. GREGORA, and D. CHVOSTOVÁ
*Raman Scattering in GeS Single Crystals at Low Temperatures
(b) Vol. **116**, No. 2, p. 639–643 (1983).
See also *6*, p. *73*; 22, p. 405.

S. PAČESOVÁ, B. BŘEZINA, and L. JASTRABÍK
*Optical Properties of K_2SeO_4 in the Vicinity of Phase Transitions
(b) Vol. **116**, No. 2, p. 645–652 (1983).
See also *6*, p. *73*; 14.4.1, p. 257; 22.8.1, p. 513.

K. Takarabe and K. Wakamura
*Polarized Raman Spectra of InS Single Crystal
(b) Vol. **116**, No. 2, p. K79 – K83 (1983).
See also *6*, p. *73*; 22.3, p. 456.

F. R. Kessler, K. Dettmer, and E.-W. Ritters
*Ge–Al Thin Films for Selective Solar Absorbers
(a) Vol. **76**, No. 2, p. 533 – 541 (1983).
See also *14.3.1*, p. *231*; 22.1.1, p. 420.

K. Kambas
*Preparation and Optical Investigation of the $CdInGaS_4$ Compound
(a) Vol. **76**, No. 2, p. 735 – 741 (1983).
See also 1.3, p. 23; *6*, p. *73*; 22.8, p. 505.

J. Bonnafé, M. de Murcia, M. Castagné, and J. P. Fillard
*Infrared Photomodulation of the Band Edge Transmission in Semi-Insulating GaAs
(a) Vol. **76**, No. 2, p. K147 – K150 (1983).
See also 22.2.1, p. 444.

K. N. Reddy, Md. Iqbal Ahmed, and V. H. Babu
On Production of Z_1 Centres in $NaCl:Sm^{2+}$ Crystals
(a) Vol. **76**, No. 2, p. K167 – K170 (1983).
See also *10.2*, p. *112*; 20.3, p. 338; 22.5.2, p. 479.

H. Beez, D. Fasold, and H. Karge
*Refractive Index Changes of Ion-Implanted Quartz
(a) Vol. **76**, No. 2, p. K171 – K174 (1983).
See also *11*, p. *127*; 22.6, p. 493.

E. Wilk, W. Wesch, and K. Hehl
Defect Diffusion during Ion Implantation into GaAs
(a) Vol. **76**, No. 2, p. K197 – K201 (1983).
See also *11*, p. *127*; 22.1.2, p. 429.

G. Valuikonis, F. M. Gashimzade, D. A. Guseinova, G. Krivaité, A. M. Kulibekov, G. S. Orudzhev, and A. Šileika
*Reflectance and Thermoreflectance Spectra and Energy Band Structure of GeSe Crystals
(b) Vol. **117**, No. 1, p. 81 – 92 (1983).
See also *13.1*, p. *158*; 22, p. 406.

M. Burd, R. Stearns, and R. Braunstein
*De-Correlation Technique for Separation of Drude Parameters from Wavelength Modulation Spectroscopy Data
(b) Vol. **117**, No. 1, p. 101 – 107 (1983).
See also *13.1*, p. *158*; 21.1, p. 368; 21.6, p. 397.

J. Prakash
*Photo-Chemical Reaction and Photo-Induced Polarization Studies in Sulphur-Doped Potassium Bromide
(b) Vol. **117**, No. 1, p. 161 – 169 (1983).
See also *10.2*, p. *112*; 22.5.2, p. 479.

S. B. S. Sastry and S. Nagarajan
*Thermoluminescence and Optical Absorption Studies on Rubidium Bromide Doped with Barium
(b) Vol. **117**, No. 1, p. 171 – 175 (1983).
See also 10.2, p. 112; *20.3*, p. *338*; 22.5.2, p. 479.

H. Bergner, V. Brückner, D. Dietrich, F. Kerstan, and W. Nowick
*Picosecond Reflectivity Measurements on Glow Discharge a-Si:H
(b) Vol. **117**, No. 1, p. 197 – 202 (1983).
See also 2, p. 48; 22.1.2, p. 429.

K. M. Häussler, J. Brandmüller, L. Merten, F. Finsterhölzl, and A. Lehmeyer
*Observation of Dispersion on Magnetic Polaritons
(b) Vol. **117**, No. 1, p. 225 – 228 (1983).
See also *13.5.4*, p. *203*; 18, p. 274; 22.5, p. 474.

H. Zimmermann, R. Boyn, and N. Nagel
*Photoinduced Absorption Spectra of ZnS:Fe Crystals
(b) Vol. **117**, No. 1, p. 229 – 239 (1983).
See also 13.4, p. 178; 22.4.1, p. 463.

J. C. Parlebas
*Model for Optical Absorption in Transition Metals Containing d Impurities
(b) Vol. **117**, No. 1, p. 255 – 259 (1983).
See also 21.1, p. 368.

J. Gryko
Optical Conductivity in the Impurity Band of Doped Semiconductors
(b) Vol. **117**, No. 1, p. 313 – 322 (1983).
See also *13.4*, p. *178*; *14.3*, p. *219*; 22.1.1, p. 420.

J. Brandmüller, A. Lehmeyer, K. M. Häussler, and L. Merten
*Electromagnetic Wave Propagation in the Uniaxial Antiferromagnetic Crystals FeF_2, CoF_2, and MnF_2
(b) Vol. **117**, No. 1, p. 323 – 328 (1983).
See also *18.4*, p. *296*; 22.5, p. 474.

I. S. Edelman, G. Dustmuradov, and V. P. Kononov
*Polarized Light Absorption of the Magnetic Semiconductor $CdCr_2Se_4$
(b) Vol. **117**, No. 1, p. 351 – 354 (1983).
See also 13.1, p. 158; 22.8, p. 505.

P. Wiśniewski and M. Nawrocki
Composition Dependence of the Energy Gap in $Cd_{1-x}Mn_xSe$
(b) Vol. **117**, No. 1, p. K43 – K45 (1983).
See also *13.5.1*, p. *196*; 22.4.4, p. 470.

A. F. Andreeva and V. A. Ogorodnik
*Optical Absorption of Terbium Oxide
(b) Vol. **117**, No. 1, p. K57 – K59 (1983).
See also 22.6, p. 493.

A. Subrahmanyam
*Effect of DC Field and X-Irradiation on the Optical Absorption and Thermoluminescence of NaCl Single Crystals
(a) Vol. **77**, No. 1, p. 45–50 (1983).
See also *20.3*, p. *338*; 22.5.2, p. 479.

Mukesh Jain, H. K. Sehgal, and A. V. R. Warrier
*The Variation of Optical Energy Gaps with Composition in $Pb_{1-x}Hg_xS$ Alloy Films
(a) Vol. **77**, No. 1, p. 67–70 (1983).
See also 22.4.4, p. 471.

A. A. Kaminskii, V. A. Timofeeva, N. R. Agamalyan, A. B. Bykov, S. E. Sarkisov, and Yu. M. Gerasimov
* Luminescence Properties of Nd^{3+} Ions in Disordered $Na_5YSi_4O_{12}$ Crystals
(a) Vol. **77**, No. 1, p. 107–114 (1983).
See also *20.3*, p. *339*; 22.8, p. 505.

S. Jetschke, H. Karge, and K. Hehl
*Anisotropic Effects in N^+-Implanted $LiNbO_3$
(a) Vol. **77**, No. 1, p. 207–214 (1983).
See also *11*, p. *128*; 22.8.1, p. 514.

A. Remón, J. Piqueras, J. Llopis, and C. Ballesteros
*Cathodoluminescence and Optical Absorption in 20 kV Electron Irradiated MgO Single Crystals
(a) Vol. **77**, No. 1, p. K29–K32 (1983).
See also 11, p. 128; *20.3*, p. *339*; 22.6, p. 493.

D. A. Aronov and V. V. Rubinov
*On the Minimum Coefficient of Light Absorption by Free Carriers under the Conditions of Exclusion in "Pure" Semiconductors with Non-Injecting Contacts
(a) Vol. **77**, No. 1, p. K41–K44 (1983).
See also *14.3.4*, p. *245*; 22.1.1, p. 420.

W. Kowalsky, H.-H. Wehmann, F. Fiedler, and A. Schlachetzki
*Optical Absorption and Refractive Index near the Bandgap for InGaAsP
(a) Vol. **77**, No. 1, p. K75–K80 (1983).
See also 22.2.4, p. 454.

B. Vengalis and K. Valatska
Phase Segregation in Tellurium-Saturated GeTe
(b) Vol. **117**, No. 2, p. 471–476 (1983).
See also *1.2*, p. *16*; 8, p. 88; 13.4, p. 179; 14.3, p. 220; 22.3, p. 457.

G. Babonas, N. S. Korets, and S. Marcinkevičius
*Dispersion of Optical Constants in CdP_2
(b) Vol. **117**, No. 2, p. 477–481 (1983).
See also *4*, p. *61*; 22.3, p. 457.

H. Zimmermann and R. Boyn
On the Configuration Coordinate Diagram of ZnS:Fe
(b) Vol. **117**, No. 2, p. 611–618 (1983).
See also **6**, p. **74**; *13.4*, p. *179*; **20.3**, p. **339**; **22.4.1**, p. **463**.

A. Lehmann, L. Schumann, and K. Hübner
*Optical Phonons in Amorphous Silicon Oxides (I): Calculation of the Density of States and Interpretation of LO–TO Splittings of Amorphous SiO_2
(b) Vol. **117**, No. 2, p. 689–698 (1983).
See also **2**, p. **48**; *6*, p. *74*; **22.6**, p. **494**.

Y. Itoh and C. Hamaguchi
*Resonant Brillouin Scattering in HgI_2 Investigated by Acoustic Domain Injection Method
(b) Vol. **117**, No. 2, p. 725–732 (1983).
See also *7*, p. *83*; **22.5**, p. **474**.

R. Borromei, E. Cavalli, and G. Ingletto
*Electronic Spectrum of the Co(II) Ion in the Columbite $CoNb_2O_6$ Host Lattice at 4.2 K
(b) Vol. **117**, No. 2, p. 733–742 (1983).
See also *13.4*, p. *180*; **22.8**, p. **506**.

P. Kleinert
*Cluster–Bethe–Lattice Treatment of the Silicon Pair Defect in Gallium Arsenide
(b) Vol. **117**, No. 2, p. K91–K94 (1983).
See also *6*, p. *74*; **22.2.1**, p. **444**.

M. S. Brodin, I. V. Blonskii, and B. M. Nitsovich
Anisotropic Exciton Scattering by Bending Waves in Layer Crystals
(b) Vol. **117**, No. 2, p. K99–K103 (1983).
See also *13.5.1*, p. *196*; **22.5**, p. **474**.

K. R. Allakhverdiev, T. G. Mamedov, E. Yu. Salaev, and I. K. Efendieva
*The Fundamental Absorption Spectra of $TlInSe_2$ Crystals under Pressure
(b) Vol. **117**, No. 2, p. K109–K111 (1983).
See also *13.1*, p. *159*; **22.8**, p. **506**.

S. V. J. Lakshman and A. Sundar Jacob
*Absorption Spectrum of VO^{2+} in $LiKSO_4$ Single Crystals
(b) Vol. **117**, No. 2, p. K119–K122 (1983).
See also *13.4*, p. *180*; **19**, p. **301**; **22.8.1**, p. **514**.

C. Rincón, L. González, and G. Sánchez Pérez
*Temperature Dependence of the Energy Gap in $CuInSe_2$
(b) Vol. **117**, No. 2, p. K123–K126 (1983).
See also *13.1*, p. *159*; **22.8**, p. **506**.

T. L. Tansley, D. F. Neely, and C. P. Foley
*Optical Dispersion in Zinc Oxide
(a) Vol. **77**, No. 2, p. 491–496 (1983).
See also **13.1**, p. **159**; **22.4**, p. **460**.

A. J. COHEN and G. G. JANEZIC
*Relationships among Trapped Hole and Trapped Electron Centers in Oxidized Soda–Silica Glasses of High Purity
(a) Vol. **77**, No. 2, p. 619–624 (1983).
See also 2, p. 48; *13.4*, p. *180*; 22.6.1, p. 498.

K. N. REDDY, MD. IQBAL AHMED, U. V. SUBBA RAO, and V. HARI BABU
*Thermoluminescence and Optical Absorption Studies of Z_1-Centres in NaCl Crystals Doped with Terbium
(a) Vol. **77**, No. 2, p. 643–648 (1983).
See also 10.2, p. 113; *20.3*, p. *340*; 22.5.2, p. 480.

R. N. MUKHERJEE, D. N. S. SRIVASTWA, and S. C. SEN
*Ultraviolet Absorption of Tl^+ Ion in K_2O–KBr–B_2O_3 and Na_2O–NaBr–B_2O_3 Glasses
(a) Vol. **77**, No. 2, p. 799–802 (1983).
See also 2, p. 49; 22.8.1, p. 514.

A. A. KAMINSKII, A. G. PETROSYAN, K. L. OVANESYAN, and M. I. CHERTANOV
Stimulated Emission of Pr^{3+} Ions in $YAlO_3$ Crystals
(a) Vol. **77**, No. 2, p. K173–K178 (1983).
See also *20.2*, p. *332*; 20.3, p. 341; 22.8.1, p. 514.

I. SILDOS, O. SILD, and I. DOLINDO
Manifestation of Spin–Rotation Coupling in Intersystem Crossing in a Rotating Impurity Molecule
(b) Vol. **118**, No. 1, p. 25–29 (1983).
See also *13.4*, p. *181*; 22.5.2, p. 480.

NGUYEN VAN HUONG
*Theory of Free-Carrier Absorption in n-Type Lead Chalcogenides Due to Electron–Electron Scattering
(b) Vol. **118**, No. 1, p. 135–146 (1983).
See also 22, p. 408.

G. JEANDEL, A. WARIN et G. MORLOT
*Analyse du spectre d'absorption infrarouge de $BaF_2:Y^{3+}$
(Analysis of the Infrared Absorption Spectrum of $BaF_2:Y^{3+}$)
(b) Vol. **118**, No. 1, p. 167–172 (1983).
See also *6*, p. *75*; 22.5.3, p. 486.

K. HÜBNER, A. V. SHENDRIK, and A. M. PRAULINSH
Decomposition of SiO_x Films Due to Internal Stress
(b) Vol. **118**, No. 1, p. 239–244 (1983).
See also *1.5*, p. *37*; *4*, p. *61*; 19, p. 302; 22.6, p. 494.

R. KRÜSSMANN, H. VOLLMER, and R. LABUSCH
The Activation Energy of Phosphorus Donors in Silicon-Rich Silicon–Germanium Alloys
(b) Vol. **118**, No. 1, p. 275–282 (1983).
See also *13.4*, p. *182*; 22.1.1, p. 421; 22.1.2, p. 432.

P. Kleinert
Vibrational Behaviour of Mixed Diatomic Crystals $AB_{1-c}C_c$ within the Bethe Lattice
(b) Vol. **118**, No. 1, p. 283–291 (1983).
See also *6*, p. *75*; 22.2.4, p. 455; 22.4.4, p. 471.

A. V. Malakhovskii and G. G. Vasilev
*Covalency as Source of Intensity for d–d Transitions in Octahedral Complexes
(b) Vol. **118**, No. 1, p. 337–344 (1983).
See also 4, p. 61; 22.5, p. 474.

A. P. Fedotov and V. F. Shabanov
*Deuteron Dynamics in Crystals of Sodium Trideuterium Selenite
(b) Vol. **118**, No. 1, p. 345–351 (1983).
See also *6*, p. *75*; 22.8.1, p. 515.

B. Rambabu, C. Ramasastry, and B. V. R. Chowdari
*Radiation Damage Studies of Co^{2+} and Ni^{2+} Doped NH_4Cl Crystals
(b) Vol. **118**, No. 1, p. 381–387 (1983).
See also *10.2*, p. *114*; 19, p. 302; 22.5, p. 474.

L. M. Kanskaya, S. I. Kokhanovskii, R. P. Seysyan, and Al. L. Efros
Direct Determination of Indium Antimonide Energy Band Parameters from Diamagnetic Exciton Spectra
(b) Vol. **118**, No. 1, p. 447–452 (1983).
See also *13.1*, p. *160*; *13.5.1*, p. *197*; 22.2.3, p. 453.

C. Rincón and J. González
*Residual Optical Absorption below the Band Gap in $CuInSe_2$
(b) Vol. **118**, No. 1, p. K21–K24 (1983).
See also *13.1*, p. *160*; 22.8, p. 507.

O. A. C. Nunes
*On the Multiphoton Laser Absorption in Free-Carrier Semiconductors
(b) Vol. **118**, No. 1, p. K25–K29 (1983).
See also 6, p. 75; *13*, p. *150*; 22, p. 409.

K. Kamarás, R. Swietlik, and A. Graja
*Polarized IR Reflectance Spectra of Pure and Neutron Irradiated $Qn(TCNQ)_2$ Mosaics
(b) Vol. **118**, No. 1, p. K43–K46 (1983).
See also 11, p. 129; 22.9, p. 528.

M. El Sherif, F. Brehat, and B. Wyncke
*A Simple Best Fit Method for the Far Infrared Reflectivity Spectra by Using the Four-Parameter Semi-Quantum Model: Application on the Sodium Nitrite Spectrum
(b) Vol. **118**, No. 1, p. K47–K50 (1983).
See also *6*, p. *76*; 22.8.1, p. 515.

H. Neumann, R. D. Tomlinson, W. Kissinger, and N. Avgerinos
*Polarization-Dependent Infrared Reflectivity Spectra of $CuInSe_2$
(b) Vol. **118**, No. 1, p. K51–K56 (1983).
See also *6*, p. *76*; 22.8, p. 507.

Hj. Bernhardt
*Vacuum-UV Absorption and Thermoluminescence of X-Rayed Quartz Crystals
(a) Vol. **78**, No. 1, p. 61–67 (1983).
See also 10.2, p. 114; *20.3*, p. *342*; 22.6, p. 494.

J. García Solé, C. Zaldo, and F. Agulló-López
*Optical Spectra of Pb^{2+} to Monitor the Precipitation Behavior of NaCl:Cd
(a) Vol. **78**, No. 1, p. 85–91 (1983).
See also *10.2*, p. *114*; 20.3, p. 342; 22.5.2, p. 481.

J. Szczyrbowski, A. Dietrich, and H. Hoffmann
*Optical and Electrical Properties of RF-Sputtered Indium–Tin Oxide Films
(a) Vol. **78**, No. 1, p. 243–252 (1983).
See also *14.3.1*, p. *231*; 22.6.1, p. 498.

Yu. B. Bolkhovityanov, B. A. Bobylev, V. I. Korzhov, V. I. Yudaev, and E. H. Hairi
Growth Peculiarities of Graded Band Gap GaAsSb/GaAs Heterostructures and Effective Light Modulation in the 1.06 μm Wavelength Region
(a) Vol. **78**, No. 1, p. 343–347 (1983).
See also *3*, p. *58*; 22.2.4, p. 455.

B. S. Acharya and A. K. Panda
*Optical Absorption Spectra of Mercury Centres in KCl and NaCl
(a) Vol. **78**, No. 1, p. K81–K85 (1983).
See also *10.2*, p. *115*; 22.5.2, p. 481.

S. W. Biernacki
*Splitting of the 3d Levels of Iron Group Impurities in Crystals with Zincblende Structure
(b) Vol. **118**, No. 2, p. 525–533 (1983).
See also *13.4*, p. *183*; 22.4.2, p. 467.

B. Rambabu, C. Ramasastry, and B. V. R. Chowdari
*Cobalt Centres in Ammonium Chloride: Optical Absorption Studies
(b) Vol. **118**, No. 2, p. 541–549 (1983).
See also *10.2*, p. *115*; 13.4, p. 183; 22.5, p. 475.

A. K. Ahuja, S. Auluck, and S. K. Joshi
*Application of Special-Directions Method to the Optical Properties of Copper and Silver
(b) Vol. **118**, No. 2, p. 575–586 (1983).
See also *13.1*, p. *160*; 21.1, p. 372; 21.6, p. 398.

Nguyen van Huong
*Free Carrier-Absorption in n-PbSe at Low Temperatures Due to Electron–Electron Scattering
(b) Vol. **118**, No. 2, p. 631–640 (1983).
See also *13.1*, p. *161*; 22, p. 409.

W. Burckhardt and A. Feltz
*The Influence of the Structure of Different Arsenic Chalcogenide Glasses on Refractive Index and Dispersion
(b) Vol. **118**, No. 2, p. 653–660 (1983).
See also *2*, p. *50*; 22, p. 410.

A. LEHMEYER, L. MERTEN, and K. M. HÄUSSLER
*On the Reflectivity for Oblique Incidence of the Uniaxial Antiferromagnetic Crystals FeF_2, CoF_2, and MnF_2
(b) Vol. **118**, No. 2, p. 675–679 (1983).
See also *18.4*, p. *296*; 22.5, p. 475.

C. RAZZETTI, P. P. LOTTICI, L. ZANOTTI, and M. CURTI
*A Raman Study of Compositional Disorder in Defect Chalcopyrites
(b) Vol. **118**, No. 2, p. 743–749 (1983).
See also *6*, p. *76*; 22.8, p. 508.

S. ZAZUBOVICH, A. USAROV, and ZH. EGEMBERDIEV
Photothermal Creation of $Cu^{\circ}v_a^+$ Centres and Mobility of Anion Vacancies in X-Irradiated KCl:Cu Crystals
(b) Vol. **118**, No. 2, p. 789–798 (1983).
See also *10.2*, p. *116*; *20.3*, p. *342*; 22.5.2, p. 482.

K. SIERAŃSKI and J. SZATKOWSKI
*Intraband Absorption Due to Alloy Scattering in Lead Chalcogenide Semiconductors
(b) Vol. **118**, No. 2, p. 871–879 (1983).
See also *13.1*, p. *161*; 22, p. 410.

R. TOVAR BARRADAS, J. GONZÁLEZ, and C. RINCÓN
*Impurity States near the Fundamental Absorption Edge in p-$CuInS_2$
(b) Vol. **118**, No. 2, p. K103–K106 (1983).
See also 13.1, p. 161; 13.4, p. 184; 22.8, p. 508.

I. I. GEGUSIN and V. N. DATSYUK
*APW Approach to the Momentum Matrix Element Calculation
(b) Vol. **118**, No. 2, p. K121–K124 (1983).
See also *13.1*, p. *162*.

M. M. HAFIZ, H. M. SHAFEY, A. M. HAFEZ, and M. DONGOL
*Irradiation-Induced Photodarkening in As–Se–Cu Amorphous Films
(a) Vol. **78**, No. 2, p. 449–455 (1983).
See also *11*, p. *130*; 22, p. 410.

H. EZBAKHE et A. DONNADIEU
*Rugosité de surface et sélectivité de couches minces dioxyde molybdène
(Surface Roughness and Selectivity of Molybdenum Dioxide Thin Films)
(a) Vol. **78**, No. 2, p. 607–614 (1983).
See also *1.5*, p. *38*; 22.6, p. 495.

A. SUBRAHMANYAM
*Optical Absorption and Thermoluminescence of KCl Single Crystals Coloured under DC Fields
(a) Vol. **78**, No. 2, p. 665–670 (1983).
See also *20.3*, p. *343*; 22.5.2, p. 482.

A. A. Kaminskii, B. V. Mill, and A. V. Butashin
Growth and Stimulated Emission Spectroscopy of $Ca_3Ga_2Ge_3O_{12}$-Nd^{3+} Garnet Crystals
(a) Vol. **78**, No. 2, p. 723–732 (1983).
See also *20.2*, p. *333*; *20.3*, p. *343*; 22.8.2, p. 522.

E. Yu. Brailovskii, G. N. Eritsyan, and N. E. Grigoryan
*Optical Absorption in InP Crystals with Point Radiation Defects
(a) Vol. **78**, No. 2, p. K113–K115 (1983).
See also *11*, p. *131*; 22.2.2, p. 450.

V. Ziraps
Ionic and Electronic Processes during Frenkel Defect Annealing in KBr Crystals above LHeT
(b) Vol. **119**, No. 1, p. 49–59 (1983).
See also *10.2*, p. *117*; 11, p. 131; 13.4, p. 185; 22.5.2, p. 482.

P. K. Katana
*Two-Photon Interband Absorption of Light in Semiconductors in External Parallel Electric and Magnetic Fields
(b) Vol. **119**, No. 1, p. 89–98 (1983).
See also *13.1*, p. *162*; 14.3.4, p. 247; 18, p. 275; 22, p. 411.

G. D. Shibarshina, S. A. Moskalenko, and P. I. Khadzhi
*The Polariton Soliton Formation Due to the Effect of the Dipole Momentum Saturation
(b) Vol. **119**, No. 1, p. 153–158 (1983).
See also *13.5*, p. *193*; 13.5.1, p. 197; 22.4.1, p. 464.

V. I. Ivanov-Omskii, A. Sh. Mekhtiev, S. A. Rustambekova, and E. N. Ukraintsev
*Plasma Reflection and Hole Optical Effective Mass in p-HgTe
(b) Vol. **119**, No. 1, p. 159–169 (1983).
See also *13.1*, p. *162*; 22.4.3, p. 469.

L. Stauffer, B. Stébé, and G. Munschy
Optical Absorption by an Exciton Bound to an Ionized Donor Impurity in Anisotropic Semiconductors
(b) Vol. **119**, No. 1, p. 193–198 (1983).
See also *13.4*, p. *185*; *13.5.1*, p. *197*; 22.4.1, p. 464.

K.-P. Oboth and K. Schmitt
*Effective Dynamical Coupling Constants of the A-Band of Tl^+ Centers in NaCl, KCl, KBr, and KI
(b) Vol. **119**, No. 1, p. 215–219 (1983).
See also *13.4*, p. *185*; 22.5.2, p. 482.

B. H. Bairamov, A. Heinrich, G. Irmer, V. V. Toporov, and E. Ziegler
*Raman Study of the Phonon Halfwidths and the Phonon-Plasmon Coupling in ZnO
(b) Vol. **119**, No. 1, p. 227–234 (1983).
See also *6*, p. *77*; 13.5.3, p. 202; 22.4, p. 460.

E. Fiorino and R. Del Sole
*Some Remarks on the Theory of the Dielectric Response of Crystals and on the Lorentz-Lorenz Formula
(b) Vol. **119**, No. 1, p. 315–326 (1983).
See also 13.5.1, p. 198.

A. V. Malakhovskii, T. P. Morozova, and V. I. Yuzvak
*Magneto-Optical Properties of $RbMnF_3$ in the Paramagnetic State
(b) Vol. **119**, No. 1, p. 411–417 (1983).
See also 13.1, p. 163; *18*, p. *275*; 22.5, p. 475.

I. J. Lalov, K. T. Stoychev, and S. P. Penchev
Non-Linear Optical Susceptibilities of Molecular Crystals in the Region of Vibronic Excitations (I): The Three-Dimensional Case
(b) Vol. **119**, No. 1, p. 419–427 (1983).
See also 6, p. 77; *13.5.1*, p. *198*; 22, p. 411.

P. Chandrasekharam, D. Raja Reddy, and B. K. Reddy
*Optical Properties of $(ZnSe)_x(CdTe)_{1-x}$Single Crystals
(b) Vol. **119**, No. 1, p. K5–K9 (1983).
See also *13.1*, p. *163*; 22.4.4, p. 471.

J. Horák, P. Lošťák, and A. Montaner
Point Defects in Ga-Doped Bi_2Se_3 Single Crystals
(b) Vol. **119**, No. 1, p. K17–K21 (1983).
S3e also *10.2*, p. *117*; 14.3, p. 223; 22.7, p. 499.

V. V. Mazurenko, V. V. Petrov, and V. D. Tkachev
Investigation of the Single-Phonon Absorption Band at 230 cm^{-1} in Silicon Irradiated by Neutrons
(b) Vol. **119**, No. 1, p. K69–K71 (1983).
See also *6*, p. *78*; *11*, p. *131*; 22.1.2, p. 434.

L. V. Nomerovannaya and V. A. Rassokhin
Anomalies in the Optical and Electrical Properties of Cr–Al Alloys
(a) Vol. **79**, No. 1, p. 87–93 (1983).
See also *14.1*, p. *209*; 21, p. 359; 21.1, p. 374.

K. Somaiah and V. Hari Babu
Thermoluminescence of Thallium-Doped BaFCl
(a) Vol. **79**, No. 1, p. 237–242 (1983).
See also 13.4, p. 186; *20.3*, p. *344*; 22.5, p. 475.

K. Biskupska, J. Tatarkiewicz, and K. Wieteska
Comparison of Fast Electron and Thermal Annealing of Proton Bombarded Silicon – X-Ray and IR Spectroscopy Study
(a) Vol. **79**, No. 1, p. K13–K15 (1983).
See also 10.2, p. 118; *11*, p. *132*; 22.1.2, p. 435.

H. Bergner and V. Brückner
Dependence of the Ambipolar Diffusion in Silicon on the Carrier Density
(a) Vol. **79**, No. 1, p. K85–K88 (1983).
See also *14.3*, p. *224*; 22.1.2, p. 435.

V. M. Fomin and E. P. Pokatilov
*Nonlinear Kinetics and Optical Properties of Band Charge Carriers
(b) Vol. **119**, No. 2, p. 483–492 (1983).
See also 6, p. 78; 13, p. 151; 14.3.4, p. 247; 22, p. 412.

Z. G. Akhvlediani, I. G. Akhvlediani, and T. L. Kalabegishvili
Effect of Combined Mechanical Load and Irradiation on Hydrogen Stabilization in LiF:OH
(b) Vol. **119**, No. 2, p. 503–506 (1983).
See also *10.2*, p. *119*; 19, p. 304; 22.5.2, p. 483.

V. D. Iskra
On the Problem of an Intermediate Order in Amorphous Semiconductors
(b) Vol. **119**, No. 2, p. 515–524 (1983).
See also *2*, p. *52*; 22, p. 412.

G. Mützenich, R. Faymonville, and P. Grosse
One- and Multiphonon Absorption in EuSe and $Eu_{1-x}Sr_xS$ $(0 < x < 1)$
(b) Vol. **119**, No. 2, p. 531–539 (1983).
See also *6*, p. *78*; 22, p. 412.

G. Irmer, V. V. Toporov, B. H. Bairamov, and J. Monecke
*Determination of the Charge Carrier Concentration and Mobility in n-GaP by Raman Spectroscopy
(b) Vol. **119**, No. 2, p. 595–603 (1983).
See also 6, p. 79; 13.5.2, p. 200; *14.3* p. *224*; 22.2.2, p. 450.

K. Köhler, H. P. Geserich, Th. Wolf, and G. Müller-Vogt
*Optical Properties of TiN Single Crystals
(b) Vol. **119**, No. 2, p. 605–610 (1983).
See also *13.1*, p. *163*; 22, p. 412.

K. H. Goetz and W. Richter
Interaction of Free Hole Intraband Excitations with the A_1-Phonon in Highly Doped (p-Type) Tellurium
(b) Vol. **119**, No. 2, p. 631–638 (1983).
See also *6*, p. *79*; *13.1*, p. *163*; 22.1.3, p. 440.

A. Morimoto, S. Oozora, M. Kumeda, and T. Shimizu
Annealing Behaviour of Hydrogenated Amorphous Silicon–Nitrogen Alloy Films Prepared by Sputtering
(b) Vol. **119**, No. 2, p. 715–720 (1983).
See also *2*, p. *52*; *10.2*, p. *119*; 19, p. 304; 22, p. 412.

H. Tichá, L. Tichý, and E. Černošková
Temperature Dependence of the Optical Gap in Glassy GeS_2
(b) Vol. **119**, No. 2, p. K135–K137 (1983).
See also *2*, p. *52*; 6, p. 79; 22, p. 412.

B. G. Tagiev, G. M. Niftiev, S. A. Abushov, and A. O. Khalilov
*Optical Properties of $InSe_{1-x}S_x$ Single Crystals in the Region of the Fundamental Absorption Edge
(b) Vol. **119**, No. 2, p. K143–K148 (1983).
See also 13.5.1, p. 198; 14.3.4, p. 247; *20.3*, p. *345*; 22.3.1, p. 459.

S. A. Abo-Namous, Y. Zaka, and R. W. Fane
*Dependence of the Electronic and Optical Properties of Unhydrogenated a-Si on Preparation Conditions
(a) Vol. **79**, No. 2, p. 477–482 (1983).
See also *14.3*, p. *225*; 16, p. 267; 22.1.2, p. 436.

I. Gregora, V. Vorlíček, and R. Uecker
*Determination of the Fresnel Indicatrix of the Monoclinic $NaVO_3$ Crystal in the Visible and Near Infrared Spectral Range
(a) Vol. **79**, No. 2, p. 497–501 (1983).
See also 22.8.1, p. 517.

L. Komitov and A. G. Petrov
Quasihomeotropic, Reversely Pretilted Nematic Liquid Crystal Layers. Electrooptical Behaviour
(a) Vol. **79**, No. 2, p. 623–629 (1983).
See also *2*, p. *53*; 14.4, p. 253; 22.9, p. 531.

P. Fedorko and F. Masin
*DC Conductivity and Optical Absorption Spectra of Polyacetylene Doped with HSO_3F
(a) Vol. **79**, No. 2, p. K155–K158 (1983).
See also *14.4*, p. *254*; 22.9, p. 531.

P. Petelenz
Optical Detrapping in Molecular Crystals (1): Vibrational Structure of Detrapping Transitions
(b) Vol. **120**, No. 1, p. 31–40 (1983).
See also *13.4*, p. *188*; 22.9, p. 531.

V. S. Gorelik, B. S. Umarov, and M. Umarov
*On the Connection between Isofrequency Temperature Dependences of Inelastic Light Scattering and Dielectric Anomalies in Lithium Niobate Crystals
(b) Vol. **120**, No. 1, p. 131–136 (1983).
See also *6*, p. *79*; 14.4.1, p. 259; 22.8.1, p. 517.

N. M. Gasanly, A. F. Goncharov, N. N. Melnik, and A. S. Ragimov
*Optical Phonons in $GaS_{1-x}Se_x$ Layer Mixed Crystals
(b) Vol. **120**, No. 1, p. 137–147 (1983).
See also *6*, p. *79*; 22.3.1, p. 459.

L. Wendler and E. Jäger
*Phonon-Polaritons in Bilayer Systems. Surface Phonon-Polaritons
(b) Vol. **120**, No. 1, p. 235–247 (1983).
See also *6*, p. *80*; 13.5, p. 193; 22.2.1, p. 446; 22.2.2, p. 451; 22.5.3, p. 487.

J. Szatkowski and K. Sierański
*Theory of Free Carrier-Magnetoabsorption in Mixed Zincblende Narrow-Gap Semiconductors
(b) Vol. **120**, No. 1, p. 263–272 (1983).
See also 13.1, p. 164; 22.4.4, p. 472.

M. N. Sapozhnikov and V. I. Alekseev
Study of Optical Relaxation and Inhomogeneous Distribution Function of Impurity Organic Molecules by Site-Selective Laser Spectroscopy
(b) Vol. **120**, No. 1, p. 435–444 (1983).
See also *6*, p. *80*; 13.4 p. 189; *20.3*, p. *346*; 22.9, p. 531.

L. E. Kholodenkov and A. G. Makhanek
*Laser-Induced Reversible Deformation of Europium Stark Levels in Fluorite
(b) Vol. **120**, No. 1, p. K5–K8 (1983).
See also 13.4, p. 189; *20.3*, p. *346*; 22.5.3, p. 487.

V. Riede, H. Neumann, F. Lévy, and H. Sobotta
*Two-Phonon Absorption Spectra and Lattice Vibration Anisotropy in HfS_2
(b) Vol. **120**, No. 1, p. K9 –K14 (1983).
See also *6*, p. *80*; 22, p. 413.

A. A. Kaminskii, B. V. Mill, G. G. Khodzhabagyan, A. F. Konstantinova, A. I. Okorochkov, and I. M. Silvestrova
*Investigation of Trigonal $(La_{1-x}Nd_x)_3Ga_5SiO_{14}$ Crystals (I): Growth and Optical Properties
(a) Vol. **80**, No. 1, p. 387–398 (1983).
See also *3*, p. *59*; 22.8.1, p. 518.

A. N. Georgobiani, A. N. Gruzintsev, S. I. Radautsan, and I. M. Tiginyanu
The Contribution of Iodine Impurity to the Formation of Centers in $CdGa_2S_4$ Single Crystals
(a) Vol. **80**, No. 1, p. K45–K48 (1983).
See also *13.4*, p. *190*; 16, p. 268; 22.8, p. 510.

J. Tatarkiewicz
*Fast Electron Annealing of IR Absorption Bands of Proton Bombarded Silicon
(a) Vol. **80**, No. 1, p. K49–K51 (1983).
See also *11*, p. *134*; 22.1.2, p. 438.

A. N. Botvich, V. G. Podoprigora, V. F. Shabanov, N. P. Shestakov, and A. N. Vtyurin
*Optical Properties of Molecular Crystals. Phenomenological Approach
(b) Vol. **120**, No. 2, p. 491–501 (1983).
See also *6*, p. *80*; 22.9, p. 532.

I. Savatinova and E. Anachkova
*Correlations between OH(OD)Stretching Vibrations and Structural Characteristics of Hydrogen Bonds in $K_4Fe(CN_6) \cdot 3\ H_2O$ Crystal
(b) Vol. **120**, No. 2, p. 539–545 (1983).
See also 4, p. 64; *6*, p. *80*; 22.8, p. 510.

D. Suisky
*Multiphonon Structure of Optical Transitions to Impurity Resonance States. The Influence of Anharmonicity
(b) Vol. **120**, No. 2, p. 645 – 654 (1983).
See also 6, p. 81; *13.4*, p. *190*; 22.4, p. 461.

H. Bergner, V. Brückner, D. Dietrich, and W. Nowick
*Picosecond Recombination of Photoexcited Carriers in Glow Discharge a-Si:H Films of Different Deposition Temperature
(b) Vol. **120**, No. 2, p. 655 – 658 (1983).
See also 2, p. 55; 22.1.2, p. 439.

I. J. Lalov, K. T. Stoychev, and S. P. Penchev
Non-Linear Optical Susceptibilites of Molecular Crystals in the Region of Vibronic Excitations (II): Totally Symmetric Vibrations
(b) Vol. **120**, No. 2, p. 671 – 678 (1983).
See also 6, p. 81; *13.5.1*, p. *199*; 22, p. 414.

C. Pedrini, H. Chermette, A. B. Goldberg, D. S. McClure, and B. Moine
*Optical Absorption and Electronic Structure of NaCl:Ag^+. A Multiple Scattering Xα Study of the $(AgCl_6)^{5-}$ Cluster
(b) Vol. **120**, No. 2, p. 753 – 766 (1983).
See also *13.4*, p. *190*; 22.5.2, p. 485.

P. D. Alekseev, G. I. Baranov, E. P. Kurakina, and K. A. Maltsev
*Formation of Hydrogen Bonding in Doped OH-Group Alkali Halide Crystals by the Action of γ-Radiation
(b) Vol. **120**, No. 2, p. K119 – K121 (1983).
See also 10.2, p. 121; *11*, p. *134*; 22.5.2, p. 485.

B. Susła, R. Czajka, and J. Raułuszkiewicz
Electronic Excitation of Ni^{2+} and Co^{2+} by Tunneling Electrons
(b) Vol. **120**, No. 2, p. K129 – K134 (1983).
See also *13.4*, p. *190*; *14.3.3*, p. *242*; 19, p. 305; 22.6, p. 497.

V. Koubele, M. Zvára, and P. Hlídek
*Compositional Variations of the Nonresonant Interband Faraday Rotation in $Cd_xHg_{1-x}Te$ Single Crystals
(b) Vol. **120**, No. 2, p. K173 – K177 (1983).
See also 13, p. 153; *18*, p. *276*; 22.4.4, p. 472.

K. Asami and M. Ishiguro
Pressure Effects on the Absorption and Emission of F_A Centers in KCl
(b) Vol. **120**, No. 2, p. K205 – K209 (1983).
See also *10.2*, p. *121*; *13.4*, p. *191*; 20.3, p. 347; 22.5.2, p. 485.

W. Burckhardt and A. Feltz
*The Influence of the Structure of Different Germanium Chalcogenide Glasses on the Molar Refraction and Its Wavelength Dependence
(a) Vol. **80**, No. 2, p. 463 – 470 (1983).
See also *2*, p. *55*; 22, p. 415.

A. A. Kaminskii, I. M. Silvestrova, S. E. Sarkisov, and G. A. Denisenko
Investigation of Trigonal $(La_{1-x}Nd_x)_3Ga_5SiO_{14}$ Crystals (II): Spectral Laser and Electromechanical Properties
(a) Vol. **80**, No. 2, p. 607 – 620 (1983).
See also 12.2, p. 144; *14.4*, p. *254*; *20.2*, p. *333*; 20.3, p. 347; 22.8.1, p. 518.

20.2 Lasers

Review Article

V. Kh. Brikenshtein, V. A. Benderskii, and P. G. Filippov
Fluorescence of Molecular Crystals at High Levels of Optical Pumping
(b) Vol. **117**, No. 1, p. 9–39 (1983).
See also 6, p. 69; *13.5.1*, p. *194*; *20.3*, p. *334*; 22.9, p. 525.

Original Papers and Short Notes

V. M. Garmash, A. A. Kaminskii, M. I. Polyakov, S. E. Sarkisov, and A. A. Filimonov
*Luminescence and Stimulated Emission of Nd^{3+} Ions in $LaMgAl_{11}O_{19}$ Crystals in the $^4F_{3/2} \rightarrow {}^4I_{11/2}$ and $^4F_{3/2} \rightarrow {}^4I_{13/2}$ Transitions
(a) Vol. **75**, No. 2, p. K111–K116 (1983).
See also 13.4, p. 175; *20.3*, p. *336*; 22.8, p. 503.

Z. Błaszczak, M. Ludwiczak, and F. Kaczmarek
Bleaching of F_2 Centers in LiF:F_2 Color Center Lasers
(a) Vol. **75**, No. 2, p. K195–K198 (1983).
See also *10.2*, p. *110*; *20.1*, p. *315*; 22.5.2, p. 478.

D. Dădârlat, Rodica M. Cândea, and M. Bârlea
*Majority Carrier Concentration and Quasi-Fermi Level for $Pb_{1-x}Sn_xTe$ DH Diode Lasers
(a) Vol. **76**, No. 1, p. K61–K65 (1983).
See also *14.3.3*, p. *238*; 22.8, p. 504.

S. K. Avetissian, E. M. Kazarian, A. O. Melikian, and H. R. Minasian
*Parametric Generation of Submillimeter Radiation in Semiconductors in the Field of a Bichromatic Pulse
(b) Vol. **116**, No. 2, p. 455–463 (1983).
See also 22, p. 405.

A. A. Kaminskii, A. G. Petrosyan, K. L. Ovanesyan, and M. I. Chertanov
*Stimulated Emission of Pr^{3+} Ions in $YAlO_3$ Crystals
(a) Vol. **77**, No. 2, p. K173–K178 (1983).
See also 20.1, p. 321; 20.3, p. 341; 22.8.1, p. 514.

M. S. Brodin, N. I. Vitrikhovskii, A. A. Kypen, S. G. Shevel, and N. I. Yanushevskii
*Spatial and Spectral Characteristics and a New Model of Laser Generation for CdS-Type Single Crystals under One-Photon Excitation
(a) Vol. **78**, No. 1, p. 349–363 (1983).
See also 22.4.1, p. 464; 22.4.4, p. 471.

A. A. Kaminskii, B. P. Sobolev, and T. V. Uvarova
*Stimulated Emission of Ho^{3+} in $BaTm_2F_8$ and Tm^{3+} Ions in $BaYb_2F_8$:Er^{3+} Crystals
(a) Vol. **78**, No. 1, p. K13–K17 (1983).
See also 13.4, p. 183; *20.3*, p. *342*; 22.8, p. 507.

A. A. Kaminskii, B. V. Mill, and A. V. Butashin
*Growth and Stimulated Emission Spectroscopy of $Ca_3Ga_2Ge_3O_{12}$-Nd^{3+} Garnet Crystals
(a) Vol. **78**, No. 2, p. 723–732 (1983).
See also 20.1, p. 325; *20.3*, p. *343*; 22.8.2, p. 522.

A. A. Kaminski, I. M. Silvestrova, S. E. Sarkisov, and G. A. Denisenko
*Investigation of Trigonal $(La_{1-x}Nd_x)_3Ga_5SiO_{14}$ Crystals (II): Spectral Laser and Electromechanical Properties
(a) Vol. **80**, No. 2, p. 607–620 (1983).
See also 12.2, p. 144; *14.4*, p. *254*; 20.1, p. 331; 20.3, p. 347; 22.8.1, p. 518.

K. Herrmann, J. W. Tomm, A. Jalyschko, and W. Bremser
*On the Behaviour of the Threshold Current Densities in Dual-Wavelength $Pb_{1-x}Sn_xTe$ Injection Lasers
(a) Vol. **80**, No. 2, p. K113–K116 (1983).
See also *14.3.4*, p. *249*; 22, p. 415.

20.3 Luminescence (X-Ray Spectra See 20)

Review Article

V. KH. BRIKENSHTEIN, V. A. BENDERSKII, and P. G. FILIPPOV
*Fluorescence of Molecular Crystals at High Levels of Optical Pumping
(b) Vol. **117**, No. 1, p. 9–39 (1983).
See also 6, p. 69; *13.5.1*, p. *194*; 20.2, p. 332; 22.9, p. 525.

Original Papers and Short Notes

A. A. MANENKOV, G. N. MIKHAILOVA, A. M. PROKHOROV, A. V. SIDORIN, S. YU. SOKOLOV, and S. G. TIKHODEEV
*Interaction of High Power CO_2 Laser Radiation with a Nonequilibrium Carrier System in Ge at Low Temperatures
(b) Vol. **115**, No. 1, p. 75–81 (1983).
Erratum
(b) Vol. **117**, No. 2, p. 756 (1983).
See also *13.5.1*, p. *194*; 22.1.1, p. 418.

E. A. RADZHABOV
*A Model of Titanium Luminescence Centers in Lithium and Sodium Fluorides
(b) Vol. **115**, No. 1, p. 83–88 (1983).
See also *13.4*, p. *172*; 20.1, p. 312; 22.5.2, p. 477.

H. YOSHIDA and S. SHINOYA
*Luminescence of High Density Electron–Hole Plasma in CdS and CdSe in a Wide Temperature Range
(b) Vol. **115**, No. 1, p. 203–208 (1983).
See also 13.5, p. 192; 22.4.1, p. 462; 22.4.2, p. 466.

T. TAKASE, E. MORIKAWA, and M. KOTANI
*Triplet Exciton Trapping in p-Terphenyl Crystals Studied by Delayed Fluorescence and Phosphorescence
(b) Vol. **115**, No. 1, p. 293–298 (1983).
See also *13.5.1*, p. *194*; 22.9, p. 525.

E. A. RADZHABOV
*Luminescence Centers in $LiF:Li_2O$
(b) Vol. **115**, No. 1, p. K25–K28 (1983).
See also *13.4*, p. *173*; 22.5.2, p. 477.

P. L. KUKK and A. J. ERM
*Luminescent Centre Formation Processes in Activated CdS Phosphors
(a) Vol. **75**, No. 1, p. 91–100 (1983).
See also *10.2*, p. *108*; 13.4, p. 173; 22.4.1, p. 462.

R. Mach, G. O. Müller, W. Gericke, and J. von Kalben
*The Efficiency of Electron Impact Excited Luminescence in ZnS Thin Film Devices
(a) Vol. **75**, No. 1, p. 187–194 (1983).
See also 14.3.1, p. 229; *14.3.3*, p. *236*; 22.4.1, p. 462; 22.4.2, p. 466.

A. R. Lakshmanan, Bhuwan Chandra, and R. C. Bhatt
*Thermoluminescence of Z Centres in LiF (TLD-100) at Elevated Irradiation Temperatures
(a) Vol. **75**, No. 1, p. 263–272 (1983).
See also *10.2*, p. *109*; 11, p. 123; 22.5.2, p. 477.

A. A. Kaminskii, N. R. Agamalyan, L. P. Kozeeva, V. F. Nesterenko, and A. A. Pavlyuk
*New Data on Stimulated Emission of Nd^{3+} Ions in Disordered Crystals with Scheelite Structure
(a) Vol. **75**, No. 1, p. K1–K4 (1983).
See also 13.4, p. 173; 22.8, p. 502.

G. Blasse
*Energy Transfer from Ce^{3+} to Eu^{3+} in $(Y, Cd)F_3$
(a) Vol. **75**, No. 1, p. K41–K43 (1983).
See also *13.4*, p. *173*; 22.8, p. 502.

A. V. Malyavkin
*Optically Detected Magnetic Resonance in $Zn_{1-x}Mn_xTe$
(b) Vol. **115**, No. 2, p. 353–358 (1983).
See also 13.5.1, p. 194; *19*, p. *299*; 22.4.4, p. 470.

M. Elango, J. Pruulmann, and A. P. Zhurakovskii
*Recombination Luminescence and Energy Transfer in Ionic Crystals at XUV Excitation by Synchrotron Radiation
(b) Vol. **115**, No. 2, p. 399–407 (1983).
See also 22.5.2, p. 478; 22.6, p. 491.

M. Mocker and O. Ziep
Intrinsic Recombination in Dependence on Doping Concentration and Excitation Level. Application to Lead Chalcogenides
(b) Vol. **115**, No. 2, p. 415–425 (1983).
See also *13.1*, p. *156*; *14.3*, p. *216*; 22.8, p. 503.

A. N. Nikitina, I. S. Osadko, I. A. Vasilieva, and L. A. Yanovskaya
Vibronic and Electron–Phonon Coupling in Spectra of Substituted Polyenes in Rigid Matrices
(b) Vol. **115**, No. 2, p. 493–498 (1983).
See also *6*, p. *70*; *13.4*, p. *174*; 20.1, p. 314; 22.9, p. 525.

D. Verity, J. Nahum, and F. J. Bryant
*Donor-Related Edge Luminescence in Ion-Implanted ZnTe
(b) Vol. **115**, No. 2, p. 499–508 (1983).
See also *13.4*, p. *174*; 22.4.3, p. 468.

M. S. Brodin, N. A. Davydova, and I. Yu. Shablii
Relation of Recombination Radiation Spectra to the Concentration of Structure Defects in CdS Single Crystals
(b) Vol. **115**, No. 2, p. 641–647 (1983).
See also *13.5.1*, p. *195*; 22.4.1, p. 462.

A. P. Litvinchuk and N. I. Vitrikhovskii
*The Variation of the Energy Gap with Composition in the Quaternary Alloy System $ZnTe_{1-2x}S_xSe_x$
(b) Vol. **115**, No. 2, p. K151–K154 (1983).
See also *13.1*, p. *156*; 13.5.1, p. 195; 22.4.4, p. 470.

P. L. Kukk and A. J. Erm
*Isochronal Annealing of Quenched CdS Phosphors
(a) Vol. **75**, No. 2, p. 393–399 (1983).
See also *10.2*, p. *109*; 13.4, p. 175; 22.4.1, p. 462.

R. Mach, J. van Kalben, W. Gericke, G. O. Müller, and G. U. Reinsperger
*ZnSe:Mn in AC Thin Film Electroluminescence Devices
(a) Vol. **75**, No. 2, p. 489–494 (1983).
See also 14.3.1, p. 230; *14.3.3*, p. *237*; 22.4.2, p. 466.

V. D. Rodriguez, V. M. Orera, Y. Chen, and R. Alcala
*Thermoluminescence of Low-Temperature X-Irradiated MgO and MgO:Li Single Crystals
(a) Vol. **75**, No. 2, p. 577–582 (1983).
See also 22.6, p. 491.

V. M. Garmash, A. A. Kminskii, M. I. Polyakov, S. E. Sarkisov, and A. A. Filimonov
*Luminescence and Stimulated Emission of Nd^{3+} Ions in $LaMgAl_{11}O_{19}$ Crystals in the $^4F_{3/2} \rightarrow {}^4I_{11/1}$ and $^4F_{3/2} \rightarrow {}^4I_{13/2}$ Transitions
(a) Vol. **75**, No. 2, p. K111–K116 (1983).
See also 13.4, p. 175; *20.2*, p. *322*; 22.8, p. 503.

V. B. Sharan
*Absorption, Fluorescence, and Excitation Spectra of Thallium-Activated Rubidium Bromide
(b) Vol. **116**, No. 1, p. 41–47 (1983).
See also 13.4, p. 175; *20.1*, p. *315*; 22.5, p. 473.

A. Haufe, K. Henneberger, J. Röseler, and H.-J. Wünsche
Parameters of Bound Multiexciton Complexes in Direct Polar Semiconductors
(b) Vol. **116**, No. 1, p. 191–203 (1983).
See also *13.5.1*, p. *195*; 22.4.1, p. 463; 22.4.2, p. 466.

M. Mocker and M. Beiler
Theory of Intrinsic Recombination at Zero Temperature in Small Gap Semiconductors. Estimation for PbSSe
(b) Vol. **116**, No. 1, p. 205–215 (1983).
See also *13.4*, p. *176*; *14.3*, p. *217*; 22.8, p. 504.

R. BAUMERT, I. BROSER, J. GUTOWSKI, and A. HOFFMANN
*Forbidden Luminescence and Resonance Raman Scattering of Bound Exciton States in CdS
(b) Vol. **116**, No. 1, p. 261–267 (1983).
See also *13.5.1*, p. *195*; 22.4.1, p. 463.

V. AGEKYAN, R. BINDEMANN, R. SCHWABE, YU. STEPANOV, and I. STREIT
*Luminescence of Excitons Bound by Disorder-Potential Fluctuations in $CdS_{1-x}Se_x$: Spatial Diffusion, Excitation Level Dependence, and Kinetics
(b) Vol. **116**, No. 1, p. K43–K46 (1983).
See also *13.5.1*, p. *196*; 22.4.4, p. 470.

A. N. GEORGOBIANI, M. V. GLUSHKOV, E. S. LOGOZINSKAYA, ZH. A. PUKHLII, I. M. TIGINYANU, and I. A. SHCHERBAKOV
*Radiative Recombination in γ-La_2S_3 Single Crystals
(a) Vol. **76**, No. 1, p. 311–317 (1983).
See also *13.4*, p. *176*; 22.3, p. 456.

A. F. RAKOV
Neutron Induced Colour and Luminescence Centres in Yttrium–Aluminium Garnet Single Crystals
(a) Vol. **76**, No. 1, p. K57–K60 (1983).
See also *10.2*, p. *111*; *11*, p. *126*; 20.1, p. 316; 22.8.2, p. 520.

S. PAKEVA and R. DAFINOVA
*Electroluminophors on the Basis of ZnS:Sn
(a) Vol. **76**, No. 1, p. K77–K80 (1983).
See also 22.4.1, p. 463.

H. WEINERT
*Decay Behaviour of High-Excitation Luminescence in GaP:N (I): Model for the Deacy of Two Coexisting Phases
(b) Vol. **116**, No. 2, p. 525–532 (1983).
See also *13.5*, p. *192*; 22.2.2, p. 449.

M. YA. VALAKH, A. P. LITVINCHUK, and G. G. TARASOV
*Effect of the Pumping Intensity on the Resonant Secondary Emission Spectra of Optically Aligned Hot Excitons in ZnTe Crystals
(b) Vol. **116**, No. 2, p. K61–K65 (1983).
Erratum
(b) Vol. **118**, No. 2, p. K167 (1983).
See also *6*, p. *73*; 13.5.1, p. 196; 22.4.3, p. 468.

A. I. BASYK, V. F. KOVALENKO, and G. P. PEKA
*Transfer Mechanisms of Nonequilibrium Charge Carriers in Graded Band-Gap Semiconductors
(a) Vol. **76**, No. 2, p. 429–435 (1983).
See also *13.1*, p. *157*; 22.2.4, p. 454.

W. Busse, H.-E. Gumlich, R. O. Törnqvist, and V.-P. Tanninen
*Zero-Phonon Lines in Electroluminescence and Photoluminescence of ZnS:Mn Thin Films Grown by Atomic Layer Epitaxy
(a) Vol. **76**, No. 2, p. 553–558 (1983).
See also 22.4.1, p. 463.

F. Ekpo Eno, R. Grasser, and A. Scharmann
*Radiation Effects in $CaWO_4$ Crystals
(a) Vol. **76**, No. 2, p. K113–K116 (1983).
See also *11*, p. *126*; 22.8.1, p. 513.

K. N. Reddy, M. D. Iqbal Ahmed, and V. H. Babu
On Production of Z_1 Centres in $NaCl:Sm^{2+}$ Crystals
(a) Vol. **76**, No. 2, p. K167–K170 (1983).
See also *10.2*, p. *112*; 20.1, p. 317; 22.5.2, p. 479.

V. N. Myasnikov, S. V. Marisova, and A. N. Lipovchenko
Optical Properties of Polaritons in Thin Crystal Films
(b) Vol. **117**, No. 1, p. 109–116 (1983).
See also *13.5*, p. *192*; 20, p. 308; 22.9, p. 527.

S. B. S. Sastry and S. Nagarajan
*Thermoluminescence and Optical Absorption Studies on Rubidium Bromide Doped with Barium
(b) Vol. **117**, No. 1, p. 171–175 (1983).
Se also 10.2, p. 112; *20.1*, p. *318*; 22.5.2, p. 479.

H. Weinert, R. Bindemann, and K. Kreher
*Decay Behaviour of the High-Excitation Luminescence in GaP:N (II): Experimental Results on the Densities of the Coexisting Phases and the Relevant Recombination Processes
(b) Vol. **117**, No. 1, p. 247–253 (1983).
See also *13.5*, p. *192*; 22.2.2, p. 449.

M. Heming, G. Lehmann, K. Recker, and F. Wallrafen
*EPR and Luminescence of Mn^{2+} in Distorted Tetrahedral Symmetry in $CaZnF_4$ Crystals
(b) Vol. **117**, No. 1, p. 271–276 (1983).
See also 3, p. 58; *19*, p. *301*; 22.5, p. 474.

Yu. P. Gnatenko, Z. D. Kovalyuk, P. A. Skubenko, and Yu. I. Zhirko
*Emission of Free and Bound Excitons in Layered GaSe Crystals
(b) Vol. **117**, No. 1, p. 283–287 (1983).
See also *13.5.1*, p. *196*; 22.3, p. 457.

A. Subrahmanyam
*Effect of DC Field and X-Irradiation on the Optical Absorption and Thermoluminescence of NaCl Single Crystals
(a) Vol. **77**, No. 1, p. 45–50 (1983).
See also *20.1*, p. *319*; 22.5.2, p. 479.

A. A. KAMINSKII, V. A. TIMOFEEVA, N. R. AGAMALYAN, A. B. BYKOV, S. E. SARKISOV, and YU. M. GERASIMOV
*Luminescence Properties of Nd^{3+} Ions in Disordered $Na_5YSi_4O_{12}$ Crystals
(a) Vol. **77**, No. 2, p. 107–114 (1983).
See also *20.1*, p. *319*; 22.8, p. 505.

J. W. TOMM, K. HERRMANN, A. JALYSCHKO, A. E. YUNOVICH, and V. TEN
*Direct Comparison of Photo- and Electroluminescence in $Pb_{1-x}Sn_xTe$ Diode Lasers
(a) Vol. **77**, No. 1, p. 175–179 (1983).
See also 22, p. 407.

S. I. GOLOVKOVA, A. M. GURVICH, A. I. KRAVCHENKO, V. V. MIKHAILIN, and A. N. VASILEV
*Enhancing and Quenching of Intrinsic Luminescence and Characteristic Features of Calcium Tungstate Phosphorescence in the Presence of Lanthanides
(a) Vol. **77**, No. 1, p. 375–380 (1983).
See also *13.4*, p. *179*; 22.8.1, p. 514.

A. NIKLAS and W. JELEŃSKI
*X-Ray Luminescence of YAG:Nd^{3+}
(a) Vol. **77**, No. 1, p. 393–398 (1983).
See also *13.4*, p. *179*; 22.8.2, p. 521.

S. V. MOHARIL
*A Phenomenological Model for Supralinearity and Sensitization in LiF-TLD 100
(a) Vol. **77**, No. 1, p. 413–420 (1983).
See also 13.4, p. 179; 22.5.2, p. 479.

A. REMOŃ, J. PÍQUERAS, J. LLOPIS, and C. BALLESTEROS
*Cathodoluminescence and Optical Absorption in 20 kV Electron Irradiated MgO Single Crystals
(a) Vol. **77**, No. 1, p. K29–K32 (1983).
See also 11, p. 128; *20.1*, p. *319*; 22.6, p. 493.

M. CASALBONI, R. FRANCINI, U. M. GRASSANO, and R. PIZZOFERRATO
*Nonlinear Excitation of New Emission Bands in KCl Crystals Containing Color Centers
(b) Vol. **117**, No. 2, p. 493–498 (1983).
See also *10.2*, p. *113*; 22.5.2, p. 479.

H. BERGNER, V. BRÜCKNER, F. KERSTAN, and W. NOWICK
*Comparative Studies of Fast Recombination Processes in Amorphous Silicon Films
(b) Vol. **117**, No. 2, p. 603–610 (1983).
See also *2*, p. *48*; 22.1.2, p. 430.

H. ZIMMERMANN and R. BOYN
On the Configuration Coordinate Diagram of ZnS:Fe
(b) Vol. **117**, No. 2, p. 611–618 (1983).
See also 6, p. 74; *13.4*, p. *179*; 20.1, p. 320; 22.4.1, p. 463.

R. SCHWABE, I. STREIT, F. THUSELT, and K. UNGER
*Experimental Investigations of Electron–Hole Plasma in Tellurium-Doped GaP
(b) Vol. **117**, No. 2, p. 625–634 (1983).
See also *13.5*, p. *192*; 22.2.2, p. 449.

V. ŽELEZNÝ, L. JASTRABÍK, and L. DOBIÁŠOVÁ
*The Temperature Dependence of Emission Spectra in α-9,10-Dichloroanthracene
(b) Vol. **117**, No. 2, p. K95–K97 (1983).
See also *13.5.1*, p. *196*; 22.9, p. 528.

P. PIPINYS, A. KIVERIS, and A. RIMEIKA
* Temperature Dependence of Electroluminescence in CdF_2 Thin Films Doped with Rare Earth
(a) Vol. 77, No. 2, p. 471–476 (1983).
See also 22.5, p. 474.

S. P. KUILA
*Analysis of the 215 °C Glow Peak in Li-Doped KCl Single Crystals
(a) Vol. 77, No. 2, p. 477–482 (1983).
See also *13.4*, p. *180*; 22.5.2, p. 480.

B. RHEINLÄNDER, G. OELGART, R. REULKE, H. HAEFNER, and R. STEGMANN
*Degradation of Yellow Light-Emitting $GaAs_{0.1}P_{0.9}$:N Diodes (I): Implanted Diodes
(a) Vol. 77, No. 2, p. 527–534 (1983).
See also *14.3.3*, p. *239*; 14.3.4, p. 245; 22.2.4, p. 454.

V. M. ORERA, V. D. RODRIGUEZ, R. ALCALÁ, and Y. CHEN
*Thermoluminescence of X-Irradiated CaO and CaO:Li Single Crystals
(a) Vol. 77, No. 2, p. 625–631 (1983).
See also 22.6, p. 494.

K. N. REDDY, MD. IQBAL AHMED, U. V. SUBBA RAO, and V. HARI BABU
*Thermoluminescence and Optical Absorption Studies of Z_1-Centres in NaCl Crystals Doped with Terbium
(a) Vol. 77, No. 2, p. 643–648 (1983).
See also 10.2, p. 113; *20.1*, p. *321*; 22.5.2, p. 480.

J. C. LOU, M. S. LIN, J. I. CHYI, M. D. LIN, and S. SCHE
*Electrical and Optical Characteristics of $GaAs_{0.6}P_{0.4}$ LEDs Fabricated by Zn Semi-Closed Diffusion Method
(a) Vol. 77, No. 2, p. 741–747 (1983).
See also *14.3.3*, p. *239*; 14.3.4, p. 245; 22.2.4, p. 454.

V. P. GRIBKOVSKII, A. A. GLADYSHCHUK, V. V. ZUBRITSKII, V. V. PARASHCHUK, and G. P. YABLONSKII
*Streamer Discharges in Semiconductors
(a) Vol. 77, No. 2, p. 765–774 (1983).
See also *14.3.4*, p. *245*; 22.4.1, p. 463.

B. RHEINLÄNDER
*Diffusion-Limited Carrier Lifetime in GaP:N LED
(a) Vol. 77, No. 2, p. K111–K114 (1983).
See also *14.3.3*, p. *239*; 22.2.2, p. 450.

T. TAGUCHI
*Crystal Growth and Neutral-Acceptor Bound-Exciton Emission of ZnCdTe by THM with Te Solvent
(a) Vol. **77**, No. 2, p. K115–K119 (1983).
See also *3*, p. *58*; 13.4, p. 181; 13.5.1, p. 196; 22.4.4, p. 471.

A. A. KAMINSKII, A. G. PETROSYAN, K. L. OVANESYAN, and M. I. CHERTANOV
Stimulated Emission of Pr^{3+} Ions in $YAlO_3$ Crystals
(a) Vol. 77, No. 2, p. K173–K178 (1983).
See also 20.1, p. 321; *20.2*, p. *332*; 22.8.1, p. 514.

G. O. MÜLLER and R. MACH
*On the Mechanism of Electron Impact Excited Luminescence Devices
(a) Vol. **77**, No. 2, p. K179–K183 (1983).
See also *14.3.4*, p. *246*; 22.4.1, p. 464.

A. V. EFANOV and M. V. ENTIN
*Optical Orientation and Polarized Luminescence in Silicon
(b) Vol. **118**, No. 1, p. 63–72 (1983).
See also 13.5.1, p. 197; 22.1.2, p. 432.

S. ASANO, N. YAMASHITA et Y. OGAWA
*Luminescence et structure vibrationnelle des luminophores $CaS:Ce^{3+}$ et $CaSe:Ce^{3+}$
(Luminescence and Vibrational Structure of the Luminophors $CaS:Cr^{3+}$ and $CaSe:Cr^{3+}$)
(b) Vol. **118**, No. 1, p. 89–100 (1983).
See also *13.4*, p. *181*; 22.4.1, p. 464; 22.4.2, p. 467.

T. AMAND, A. CORNET, M. PUGNET, J. COLLET, M. BROUSSEAU, and B. S. RAZBIRIN
Picosecond Study of the Exciton Gas Cooling
(b) Vol. **118**, No. 1, p. 113–119 (1983).
See also *13.5.1*, p. *197*; 22.4.2, p. 467; 22.5, p. 474.

J. M. HVAM, G. BLATTNER, M. REUSCHER, and C. KLINGSHIRN
*The Biexciton Levels and Nonlinear Optical Transitions in ZnO
(b) Vol. **118**, No. 1, p. 179–189 (1983).
See also *13.5.1*, p. *197*; 22.4, p. 460.

C. W. M. TIMMERMANS and G. BLASSE
*The Luminescence of $K_4Bi_2Br_{10} \cdot 4\,H_2O$ and Rb_3BiBr_6 Single Crystals
(b) Vol. **118**, No. 1, p. 353–361 (1983).
See also 22.8, p. 507.

B. G. TAGIEV, G. M. NIFTIEV, and S. A. ABUSHOV
*Intrashell Yb^{3+} Transitions in Gallium Sulphide Single Crystals
(b) Vol. **118**, No. 1, p. K13–K16 (1983).
See also *13.4*, p. *182*; 22.3, p. 457.

M. SUHRKE
*On the Time Dependence of the Luminescence Spectrum in Chalcogenide Glasses
(b) Vol. **118**, No. 1, p. K79–K83 (1983).
See also *2*, p. *49*; 22, p. 409.

R. Capelletti, O. Carranza Torres, J. García Solé, and M. Manfredi
*On the 430 nm Emission Band in the NaCl:Pb^{++} System
(a) Vol. 78, No. 1, p. 35–39 (1983).
See also *10.2*, p. *114*; 13.4, p. 182; 22.5.2, p. 481.

Hj. Bernhardt
*Vacuum-UV Absorption and Thermoluminescence of X-Rayed Quartz Crystals
(a) Vol. 78, No. 1, p. 61–67 (1983).
See also 10.2, p. 114; *20.1*, p. *323*; 22.6, p. 494.

J. García Solé, C. Zaldo, and F. Agulló-López
Optical Spectra of Pb^{2+} to Monitor the Precipitation Behaviour of NaCl:Cd
(a) Vol. 78, No. 1, p. 85–91 (1983).
See also *10.2*, p. *114*; *20.1*, p. *323*; 22.5.2, p. 481.

G. Oelgart, B. Rheinländer, R. Reulke, H. Haefner, and R. Stegmann
*Degradation of Yellow Light-Emitting $GaAs_{0.1}P_{0.9}$:N Diodes (II): Diffused Diodes
(a) Vol. 78, No. 1, p. 169–176 (1983).
See also *14.3.3*, p. *240*; 14.3.4, p. 246; 22.2.4, p. 455.

A. A. Kaminskii, B. P. Sobolev, and T. V. Uvarova
*Stimulated Emission of Ho^{3+} in $BaTm_2F_8$ and Tm^{3+} Ions in $BaYb_2F_8$:Er^{3+} Crystals
(a) Vol. 78, No. 1, p. K13–K17 (1983).
See also 13.4, p. 183; *20.2*, p. *332*; 22.8, p. 507.

C. De Blasi, C. Manfredotti, G. Micocci, and A. Tepore
*Thermoluminescence of n-GaS
(a) Vol. 78, No. 1, p. K31–K34 (1983).
See also 13.4, p. 183; 22.3, p. 457.

N. Takeuchi
*On the General-Order Kinetics in Thermoluminescence
(a) Vol. 78, No. 1, p. K93–K96 (1983).
See also 22.5.2, p. 481.

M. Saakes and G. Blasse
*The Luminescence of Bi^{3+} and Mn^{2+} in Lanthanum Magnesium Borate ($LaMgB_5O_{10}$)
(a) Vol. 78, No. 1, p. K97–K99 (1983).
See also 22.8.1, p. 515.

N. R. Nurtdinov, R. Stegmann, and M. Griepentrog
*The Influence of the Variation of the Chemical Composition and of the Nitrogen Doping Level on the Energy Spectra of the Bound Excitons in $GaAs_{1-x}P_x$:N
(b) Vol. 118, No. 2, p. 613–621 (1983).
See also 10.2, p. 115; *13.5.1*, p. *197*; 22.2.4, p. 455.

S. Zazubovich, A. Usarov, and Zh. Egemberdiev
*Photothermal Creation of $Cu^\circ v_a^+$ Centres and Mobility of Anion Vacancies in X-Irradiated KCl:Cu Crystals
(b) Vol. 118, No. 2, p. 789–798 (1983).
See also *10.2*, p. *116*; 20.1, p. 324; 22.5.2, p. 482.

J. Kalinowski and R. Signerski
Exciton-Enhanced Double Injection Currents in Tetracene Crystals
(b) Vol. **118**, No. 2, p. K147 – K150 (1983).
See also *13.5.1*, p. *197*; *14.3.4*, p. *247*; 22.9, p. 529.

B. Rauscher, H. Pfleiderer, and B. Bullemer
*Tail–Tail Luminescence in Amorphous Silicon
(a) Vol. **78**, No. 2, p. 623 – 632 (1983).
See also *2*, p. *50*; 22.1.2, p. 433.

M. Suezawa and K. Sumino
*The Nature of Photoluminescence from Plastically Deformed Silicon
(a) Vol. **78**, No. 2, p. 639 – 645 (1983).
See also *10.2*, p. *116*; 22.1.2, p. 433.

A. Subrahmanyam
*Optical Absorption and Thermoluminescence of KCl Single Crystals Coloured under DC Fields
(a) Vol. **78**, No. 2, p. 665 – 670 (1983).
See also *20.1*, p. *324*; 22.5.2, p. 482.

J. Llopis, C. Ballesteros, J. Piqueras, A. Remon, and R. González
*Thermally Induced Changes in the Cathodòluminescence Image of Deformed MgO
(a) Vol. **78**, No. 2, p. 679 – 684 (1983).
See also 22.6, p. 495.

J. Manam, R. K. Gartia, and V. V. Ratnam
*Thermoluminescence of Z_3 Centres in X-Irradiated Ca-Doped KCl Crystals
(a) Vol. **78**, No. 2, p. 685 – 688 (1983).
See also *10.2*, p. *116*; 22.5.2, p. 482.

A. A. Kaminskii, B. V. Mill, and A. V. Butashin
*Growth and Stimulated Emission Spectroscopy of $Ca_3Ga_2Ge_3O_{12}$–Nd^{3+} Garnet Crystals
(a) Vol. **78**, No. 2, p. 723 – 732 (1983).
See also 20.1, p. 325; *20.2*, p. *333*; 22.8.2, p. 522.

Vu Xuan Quang, Nguyen Quang Liem, Nguyen Cong Thanh, Thuong Van Chuong, and Le Thi Le Thanh
*Photoluminescence of Rare-Earth Centers in ZnO Phosphors
(a) Vol. **78**, No. 2, p. K161 – K165 (1983).
See also 22.4, p. 460.

M. S. Brodin, V. M. Bandura, and M. G. Matsko
*Direct Luminescence Evidence for the Degenerate Electron–Hole Plasma Formation in $A^{II}B^{VI}$ Semiconductors
(b) Vol. **119**, No. 1, p. 83 – 87 (1983).
See also *13*, p. *151*; 13.5.1, p. 197; 22.4.2, p. 467; 22.4.3, p. 468.

J. Puls, F. Henneberger, and J. Voigt
*Excited States of Bound Excitons in Wurtzite-Type Semiconductors
(b) Vol. **119**, No. 1, p. 291 – 298 (1983).
See also *13.5.1*, p. *198*; 22.4, p. 461.

S. Benci, F. Fermi, and G. Lenzi
*Dimer Center Luminescence in Low-Doped KI:Tl^+ Crystals
(b) Vol. **119**, No. 1, p. 337–342 (1983).
See also 13.4, p. 185; 22.5.2, p. 482.

B. P. Chandra, M. Elyas, and B. Majumdar
*Pressure and Temperature Dependence of the Mechanoluminescence of γ-Irradiated Alkali Halide Crystals
(a) Vol. **79**, No. 1, p. 95–102 (1983).
See also 12.2, p. 143; 22.5.2, p. 483.

G. Koschek and E. Kubalek
*Micron-Scaled Spectral-Resolved Cathodoluminescence of Grains in Bariumtitanate Ceramics
(a) Vol. **79**, No. 1, p. 131–139 (1983).
See also 22.8.1, p. 516.

B. Rauscher and B. Bullemer
*Photoluminescence Decay in Amorphous Silicon Alloys
(a) Vol. **79**, No. 1, p. 161–166 (1983).
See also *2*, p. *51*; 22.1.2, p. 434.

J. Manam, R. K. Gartia, and V. V. Ratnam
*Effect of Post-Irradition Deformation on Thermoluminescence Spectra of KCl:Ca and NaCl:Ca Crystals
(a) Vol. **79**, No. 1, p. 167–171 (1983).
See also *10.2*, p. *118*; 22.5.2, p. 483.

M. Suezawa, Y. Sasaki, and K. Sumino
*Dependence of Photoluminescence on Temperature in Dislocated Silicon Crystals
(a) Vol. **79**, No. 1, p. 173–181 (1983).
See also *13.4*, p. *186*; 22.1.2, p. 435.

K. Somaiah and V. Hari Babu
*Thermoluminescence of Thallium-Doped BaFCl
(a) Vol. **79**, No. 1, p. 237–242 (1983).
See also 13.4, p. 186; 20.1, p. 326 ; 22.5, p. 475.

R. Chen, D. J. Huntley, and G. W. Berger
Analysis of Thermoluminescence Date Dominated by Second-Order Kinetics
(a) Vol. **79**, No. 1, p. 251–261 (1983).
See also *13.4*, p. *186*.

O. K. Gorodnichenko, E. A. Glushkov, V. F. Kovalenko, and N. D. Vasilenko
Strengthening of GaAs by Impurity Atom–Vacancy Complexes
(a) Vol. **79**, No. 1, p. 331–335 (1983).
See also *10.2*, p. *118*; 22.2.1, p. 446.

R. Pickenhain, B. Rheinländer, and S. Brehme
P_{Ga} Antisite Defect in Zn-Implanted $GaAs_{0.1}P_{0.9}$ LED
(a) Vol. **79**, No. 1, p. K61–K63 (1983).
See also *13.4*, p. *186*; 22.2.4, p. 455.

S. P. Kathuria and C. M. Sunta
*Fractional Order of Kinetics and Its Variation with Trap Occupancy in LiF TLD-100
(a) Vol. **79**, No. 1, p. K77 – K80 (1983).
See also *13.4*, p. *186*; 22.5.2, p. 483.

S. V. Moharil
*Reply to "Fractional Order of Kinetics and Its Variation with Trap Occupancy in LiF TLD-100" by S. P. Kathuria and C. S. Sunta
(a) Vol. **79**, No. 1, p. K81 – K84 (1983).
See also *13.4*, p. *187*; 22.5.2, p. 483.

B. T. Kolomiets, T. N. Mamontova, A. V. Chernyshov, G. Z. Vinogradova, and N. V. Demokritova
*The Effect of Low Oxygen Concentrations on the Photoluminescence of Vitreous Selenium
(a) Vol. **79**, No. 1, p. K89 – K92 (1983).
See also 2, p. 52; *10.2*, p. *118*; 22.3, p. 458.

R. Baltramiejūnas, E. Kuokštis, and A. Žukauskas
Self-Heating of Dense Exciton Gas in CdTe Single Crystals
(b) Vol. **119**, No. 2, p. 453 – 457 (1983).
See also *13.5.1*, p. *198*; 22.4.3, p. 469.

V. G. Plekhanov, V. V. Shepelev, and A. U. Grinfelds
Spectroscopic Manifestation of Self-Trapping Excitons in Ionic Insulators
(b) Vol. **119**, No. 2, p. 493 – 502 (1983).
See also 6, p. 78; *13.5.1*, p. *198*; 22.5.2, p. 483.

B. G. Tagiev, G. M. Niftiev, S. A. Abushov, and A. O. Khalilov
*Optical Properties of $InSe_{1-x}S_x$ Single Crystals in the Region of the Fundamental Absorption Edge
(b) Vol. **119**, No. 2, p. K143 – K148 (1983).
See also 13.5.1, p. 198; 14.3.4, p. 247; *20.1*, p. *328*; 22.3.1, p. 459.

I. D. Henning and H. Thomas
Process-Induced Defect States in GaAsP Light Emitting Diodes
(a) Vol. **79**, No. 2, p. 567 – 574 (1983).
See also *13.4*, p. *187*; 14.3.3, p. 241; 22.2.4, p. 455.

S. V. Moharil
*Track Interaction Model for Supralinearity and Sensitization in LiF-TLD 100
(a) Vol. **79**, No. 2, p. 599 – 605 (1983).
See also 10.2, p. 119; 13.4, p. 187; 22.5.2, p. 484.

G. P. Kothiyal, B. Ghosh, and B. M. Arora
*Photoluminescence Studies on Synthetic Crystals of As_2Se_3
(a) Vol. **79**, No. 2, p. K171 – K175 (1983).
See also 22, p. 413.

V. X. Quang, T. Kim Anh, N. Q. Liem, K. Maass, and B. Selle
*A New Luminescence Band Related to Radiation Defects in Ion-Implanted Zinc Sulphide
(a) Vol. **79**, No. 2, p. K181 – K185 (1983).
See also *11*, p. *133*; 22.4.1, p. 464.

L. Bosi, P. Gagliardelli, and M. Nimis
*More on the F-Center Decay in Additively Colored NaF:LiF
(b) Vol. **120**, No. 1, p. 41–48 (1983).
See also *10.2*, p. *119*; 22.5.2, p. 484.

D. V. Korbutyak and V. G. Litovchenko
*Expansion of the Electron–Hole Liquid Plasma at GaAs Surface
(b) Vol. **120**, No. 1, p. 87–97 (1983).
See also *13.3*, p. *170*; 13.5.1, p. 198; 22.2.1, p. 446.

V. Hizhnyakov, G. Liidya, V. Nagirnyi, T. Soovik, and S. Zazubovich
*Effect of Magnetic Field on the Decay Kinetics and Polarization of the A_T Emission of KCl:Ga at 0.38 K
(b) Vol. **120**, No. 1, p. 105–115 (1983).
See also *13.4*, p. *188*; 22.5.2, p. 484.

V. Nagirnyi, T. Soovik, S. Zazubovich, and N. Janson
*Polarized A_T Emission of In^+ Centres in KI:In
(b) Vol. **120**, No. 1, p. 117–122 (1983).
See also *13.4*, p. *188*; 22.5.2, p. 484.

M. N. Sapozhnikov and V. I. Alekseev
*Study of Optical Relaxation and Inhomogeneous Distribution Function of Impurity Organic Molecules by Site-Selective Laser Spectroscopy
(b) Vol. **120**, No. 1, p. 435–444 (1983).
See also *6*, p. *80*; 13.4, p. 189; 20.1, p. 329; 22.9, p. 531.

S. Bhan
Erratum to: Thermoluminescence of NaF:Ca^{2+} Single Crystals
(b) Vol. **120**, No. 1, p. 451 (1983).
See also *10.2*, p. *119*; 22.5.2, p. 484.

L. E. Kholodenkov and A. G. Makhanek
*Laser-Induced Reversible Deformation of Europium Stark Levels in Fluorite
(b) Vol. **120**, No. 1, p. K5–K8 (1983).
See also 13.4, p. 189; *20.1*, p. *329*; 22.5.3, p. 487.

G. O. Müller, S. L. Karpenko, R. P. Seisyan, and M. A. Yakobson
*Influence of Doping on Excitonic Luminescence Spectra in CdS
(b) Vol. **120**, No. 1, p. K93–K95 (1983).
See also 13.5.1, p. 198; 22.4.1, p 465.

A. N. Georgobiani, A. V. Mikulyonok, I. G. Stoyanova, and I. M. Tiginyanu
*Non-Equilibrium Carrier Radiative Recombination in Indium Phosphide Single Crystals
(a) Vol. **80**, No. 1, p. 109–118 (1983).
See also *13.4*, p. *189*; 22.2.2, p. 451.

K. Löhnert and E. Kubalek
*On the Degradation of Electroluminescence Efficiency in Gallium Phosphide Green Light Emitting Diodes
(a) Vol. **80**, No. 1, p. 173–183 (1983).
See also *14.3.3*, p. *242*; 22.2.2, p. 451.

D. LAPRAZ and A. BAUMER
*Thermoluminescent Properties of Synthetic and Natural Fluorapatite, $Ca_5(PO_4)_3F$
(a) Vol. **80**, No. 1, p. 353–366 (1983).
See also *10.2*, p. *120*; 22.8, p. 509.

L. BRAGINA, V. HIZHNYAKOV, G. LIIDYA, V. NAGIRNYI, T. SOOVIK, S. ZAZUBOVICH, and N. JANSON
*Effect of Magnetic Field on the Decay Kinetics and Polarization of the A_T Emission of KCl:Tl and KI:In Crystals at 0.4 to 0.7 K
(b) Vol. **120**, No. 2, p. 463–470 (1983).
See also *13.4*, p. *190*; 22.5.2, p. 485.

J. L. JANSONS, P. A. KULIS, Z. A. RACHKO, M. J. SPRINGIS, I. A. TALE, and J. A. VALBIS
*Luminescence of Ga-Doped α-Al_2O_3 Crystals
(b) Vol. **120**, No. 2, p. 511–518 (1983).
See also 22.6, p. 497.

P. K. CHAKRABORTY and K. V. RAO
*F-Band Absorption and Thermoluminescence of KBr Single Crystals X-Ray Irradiated under High AC Fields
(b) Vol. **120**, No. 2, p. K107–K110 (1983).
See also *10.2*, p. *120*; 11, p. 134; 22.5.2, p. 485.

DANG DINH THONG and O. GOEDE
*Optical Study of Highly Mn-Doped ZnS Crystals
(b) Vol. **120**, No. 2, p. K145–K150 (1983).
See also 22.4.1, p. 465.

H. ZIMMERMANN und R. BOYN
*Excitation Behaviour of Blue ($^1G_4 \rightarrow {}^3H_6$) and Infrared ($^3F_4 \rightarrow {}^3H_6$) Photoluminescence of ZnS:Tm, Ag Crystals
(b) Vol. **120**, No. 2, p. K189–K193 (1983).
See also *13.4*, p. *191*; 22.4.1, p. 465.

K. ASAMI and M. ISHIGURO
Pressure Effects on the Absorption and Emission of F_A Centers in KCl
(b) Vol. **120**, No. 2, p. K205–K209 (1983).
See also *10.2*, p. *121*; *13.4*, p. *191*; 20.1, p. 330; 22.5.2, p. 485.

A. A. KAMINSKII, I. M. SILVESTROVA, S. E. SARKISOV, and G. A. DENISENKO
Investigation of Trigonal $(La_{1-x}Nd_x)_3Ga_5SiO_{14}$ Crystals (II): Spectral Laser and Electromechanical Properties
(a) Vol. **80**, No. 2, p. 607–620 (1983).
See also 12.2, p. 144; *14.4*, p. *254*; 20.1, p. 331; *20.2*, p. *333*; 22.8.1, p. 518.

V. A. BYKOVSKII, A. V. MUDRYI, V. P. POSKREBYSHEV, and V. D. TKACHEV
*Radiative Recombination Centers in Irradiated Germanium
(a) Vol. **80**, No. 2, p. K133–K136 (1983).
See also 11, p. 135; *13.4*, p. *191*; 22.1.1, p. 422.

SH. A. VAKHIDOV and A. F. RAKOV
*Radiation Stimulated Recombination Processes in $Y_3Al_5O_{12}$ Crystals
(a) Vol. **80**, No. 2, p. K175–K178 (1983).
See also *11*, p. *135*; 13.4, p. 191; 13.5.1, p. 199; 22.8.2, p. 523.

M. PARDAVI-HORVÁTH and M. OSVAY
*Thermoluminescent Properties of Gadolinium Gallium Garnet Crystals Containing Ca^{2+} Impurity
(a) Vol. **80**, No. 2, p. K183–K185 (1983).
See also 22.8.2, p. 524.

W. UCHIDA
*Photoluminescence in Self-Activated $Zn_xCd_{1-x}S$ Powders
(a) Vol. **80**, No. 2, p. K199–K202 (1983).
See also 22.4.4, p. 472.

B. RHEINLÄNDER, R. PICKENHAIN, H. HAEFNER, and G. OELGART
*Degradation of LPE–GaP:N LED
(a) Vol. **80**, No. 2, p. K203–K207 (1983).
See also *14.3.4*, p. *249*; 22.2.2, p. 451.

21. Metals. Alloys

Review Articles

N. C. Halder and K. C. Phillips
Electronic Density of States of Liquid Metals Studied by Perturbation and Green Function Methods
(b) Vol. **115**, No. 1, p. 9–35 (1983).
See also *2*, p. *43*; *13*, p. *146*.

R. E. Hummel
Differential Reflectrometry and Its Application to the Study of Alloys, Ordering, Corrosion, and Surface Properties
(a) Vol. **76**, No. 1, p. 11–44 (1983).
See also *1.3*, p. *21*; 10.1, p. 99; 13.1, p. 155; 13.3, p. 167; *20.1*, p. *312*; 21.1, p. 363; 22.6, p. 490.

Original Papers and Short Notes

B. I. Craig
The Electronic Structure of a Hydrogen Impurity in Aluminium (III): The Energy and the Metal–Impurity Interaction
(b) Vol. **115**, No. 1, p. 53–61 (1983).
See also *13.4*, p. *172*.

R. H. Rautioaho
A Pair Potential Calculation of Vacancy Formation and Migration Energies for Aluminium and Copper
(b) Vol. **115**, No. 1, p. 95–103 (1983).
See also 9, p. 93; *10.1*, p. *99*; 21.1, p. 363.

E. Krüger
Exchange Interaction in Superconducting Metals
(b) Vol. **115**, No. 1, p. 181–185 (1983).
See also 13.1, p. 155; *14.2*, p. *212*.

R. Khanna
Huang Diffuse Scattering from Self-Interstitials in a Trigonal Lattice
(b) Vol. **115**, No. 1, p. 305–310 (1983).
See also *10.1*, p. *99*.

E. A. Pamyatnykh and A. V. Poltavets
Short-Wavelength Cyclotron Waves in Metals in Quantizing Magnetic Fields
(b) Vol. **115**, No. 1, p. K51–K54 (1983).
See also *13*, p. *146*; *19*, p. *299*.

T. A. Jenkins and S. Arajs
Electrical Resistivity of Some Directionally-Solidified Aluminium-Based Eutectic Alloys
(a) Vol. **75**, No. 1, p. 73–82 (1983).
See also *14.1*, p. *206*; 21.1, p. 363; 21.1.1, p. 379.

B. RAUSCHENBACH and K. HOHMUTH
A Simple Approach to the Analysis of Ion Collision Cascade in Solids Based on the Shock Wave Model
(a) Vol. **75**, No. 1, p. 159 – 168 (1983).
See also *11*, p. *122*.

R. GERLACH and H. LÖFFLER
On the Structure of the Precipitation-Free Zone in Two Al–Zn–Mg Alloys after Various Heat Treatments
(a) Vol. **75**, No. 1, p. 169 – 178 (1983).
See also *10.1*, p. *99*; 21.1, p. 364; 21.3, p. 389.

R. H. J. FASTENAU and M. I. BASKES
A Combined Atomistic and Monte Carlo Simulation of Point Defect–Dislocation Interactions
(a) Vol. **75**, No. 1, p. 323 – 334 (1983).
See also *10.1*, p. *99*.

V. ATZRODT, W. TITEL, TH. WIRTH, and H. LANGE
AES Studies on Thin Film $MoSi_2$
(a) Vol. **75**, No. 1, p. K15 – K20 (1983).
See also *1.6*, p. *39*; *11*, p. *123*.

P. FELTHAM
Low-Temperature Anomalies in the Plasticity of Crystalline Materials
(a) Vol. **75**, No. 1, p. K95 – K98 (1983).
See also *10.1*, p. *99*.

C. MALINOWSKA-ADAMSKA, M. OWSIEJCZYK, and L. WOJTCZAK
Thermodynamics of Strongly Anharmonic Crystals
(b) Vol. **115**, No. 2, p. 335 – 345 (1983).
See also *6*, p. *70*; 8, p. 85; 21.2, p. 387.

K. H. J. BUSCHOW, H.-J. EIFERT, and B. ELSCHNER
Electronic Structure and Stability of Amorphous Pd–Si Alloys
(b) Vol. **115**, No. 2, p. 455 – 462 (1983).
See also *2*, p. *44*; *13*, p. *147*; 19, p. 299; 22.1.2, p. 425.

H. MÜLLER, E. HEGENBARTH, and S. MAHMOUD
Thermal Conductivity of the Intermetallic Rare Earth Compound $PrAl_2$ in the Temperature Range between 4.2, and 100 K
(b) Vol. **115**, No. 2, p. K69 – K73 (1983).
See also *8*, p. *85*; 21.4, p. 390.

I. GYÉMÁNT and B. VASVÁRI
Cluster Phase Shifts and Virtual Bond States for V, Mn, and Fe Impurities in Aluminium
(b) Vol. **115**, No. 2, p. K83 – K86 (1983).
See also *13.4*, p. *174*.

R. H. Rautioaho
Energies of Self-Interstitials for Aluminium and Copper
(b) Vol. **115**, No. 2, p. K121 – K125 (1983).
See also *10.1*, p. *100*; 21.1, p. 364.

J. S. Brown
Some Exchange-Correlation Approximations Applied to Stopping Powers of Slow Ions in Metals
(b) Vol. **115**, No. 2, p. K141 – K146 (1983).
See also *13.2*, p. *166*; *17*, p. *269*; 21.6, p. 396.

F. Abd El-Salam, A. I. Eatah, and A. Tawfik
Effect of Li Loss on Some Physical Properties of Al–Li Alloy
(a) Vol. **75**, No. 2, p. 379 – 384 (1983).
See also *1.3*, p. *21*; 10.1, p. 100; 12.1, p. 137; 14.1, p. 206; 21.2, p. 387.

J. M. Lang, U. Dahmen, and K. H. Westmacott
The Origin of Mo_2C Precipitate Morphology in Molybdenum
(a) Vol. **75**, No. 2, p. 409 – 420 (1983).
See also *1.3*, p. *21*; 1.4, p. 29.

A. Baldan
Dislocation Structures in Aluminium-Rich Ni_3(Al, Ta) and Ni_3(Al, Hf)
(a) Vol. **75**, No. 2, p. 441 – 449 (1983).
See also *1.3*, p. *21*; *10.1*, p. *100*; 21.1, p. 364.

A. Korner and H. P. Karnthaler
Glide Dislocations on Cube Planes in a Low Stacking-Fault Energy Alloy
(a) Vol. **75**, No. 2, p. 525 – 535 (1983).
See also *10.1*, p. *100*; 12.1, p. 137; 21.1, p. 365.

V. Navrátil, M. Hamerský, P. Lukáč, V. P. Soldatov, and V. I. Startsev
Solid Solution Hardening of Cd – Zn Single Crystals at 1.5 K
(a) Vol. **75**, No. 2, p. K133 – K135 (1983).
See also *1.3*, p. *22*; *10.1*, p. *100*; 21.1, p. 365.

I. M. Chapnik
Penetration of Electromagnetic Fields through Thin Foils of Niobium and Indium
(a) Vol. **75**, No. 2, p. K157 – K160 (1983).
See also *14.2*, p. *212*.

I. Kanazawa, H. Murakami, T. Kurihara, Y. Sakurai, and M. Doyama
Positron Annihilation Study of Quenched-in Defects in Dilute Al–Ge and Al–Si Alloys
(a) Vol. **75**, No. 2, p. K175 – K177 (1983).
See also *10.1*, p. *100*.

V. V. Zoninashvili, I. A. Naskidashvili, U. Messerschmidt, V. A. Melik-Shakhnazarov, N. M. Yastrebova, and L. M. Kolesnikova
The Dependence of the Elastic Relaxation on a Bias Stress in Pure Nb in the Temperature Region of the α-Maximum
(a) Vol. **75**, No. 2, p. K205 – K208 (1983).
See also *7*, p. *82*; *10.1*, p. *100*; 12.1, p. 137.

I. Kanazawa, H. Murakami, and M. Tanaka
Internal Friction by a Second-Phase Particle in Al–Ge Alloy
(a) Vol. **75**, No. 2, p. K211 – K213 (1983).
See also *10.1*, p. *101*; *12.1*, p. *137*.

P. Streitenberger
Generalized Pseudoatom Model of Inhomogeneous Systems and Its Application to a Simple-Metal Surface
(b) Vol. **116**, No. 1, p. 179 – 190 (1983).
See also *1.6*, p. *39*; *13*, p. *147*.

S. V. Yalyshko and R. Wedell
Stopping Power for N^+ Ions with Energies from 160 to 440 keV in Aluminium
(b) Vol. **116**, No. 1, p. K47 – K50 (1983).
See also *11*, p. *124*.

V. A. Klimenko
Modification of the Pair Approximation of the Quasi-Chemical Ordering Theory of Binary Systems
(a) Vol. **76**, No. 1, p. 45 – 55 (1983).
See also 1.2, p. 14; *1.3*, p. *22*.

S. V. Sudareva, V. A. Rassokhin, and A. F. Prekul
The Structure of Cr–Al Alloys Exhibiting Anomalous Physical Properties
(a) Vol. **76**, No. 1, p. 101 – 106 (1983).
See also *1.2*, p. *15*; *1.4*, p. *29*; 21.1, p. 366.

H. Mizubayashi, T. Arai, and S. Okuda
Low-Temperature Dislocation Pinning and Breakaway Observed in Ta Containing H
(a) Vol. **76**, No. 1, p. 165 – 177 (1983).
See also *10.1*, p. *101*; *12.1*, p. *138*.

W. Zag and K. Urban
Temperature Dependence of the Threshold Energy for Atom Displacement in Irradiated Molybdenum
(a) Vol. **76**, No. 1, p. 285 – 295 (1983).
See also *10.1*, p. *101*; 11, p. 125.

H. Misiorek, T. Zakrzewski, and J. Rafałowicz
The Lorenz Function for Aluminium of Different Chemical Purity, Neutron Irradiated or Plastically Deformed
(a) Vol. **76**, No. 1, p. K25 – K26 (1983).
See also *8*, p. *87*; *11*, p. *125*; 12.1, p. 138.

J. Richter and G. Schubert
The Electrical Resistivity of Short-Range Ordered Alloys and Its Temperature Dependence
(b) Vol. **116**, No. 2, p. 597 – 605 (1983).
See also *1.3*, p. *23*; *14.1*, p. *207*.

J. Zieliński and J. Kowalska
On the Two-Site Spinless Falicov-Kimball Model with Hybridization
(b) Vol. **116**, No. 2, p. 677 – 683 (1983).
See also *13*, p. *148*.

B. I. Craig
Structure of Muons Trapped in Aluminium
(b) Vol. **116**, No. 2, p. K91 – K94 (1983).
See also *10.1*, p. *101*; *13.4*, p. *177*.

D. Schryvers, J. van Landuyt, G. van Tendeloo, and S. Amelinckx
Electron Microscopy Study of Two New Long Period Structures in Pt_xTi Alloys ($3 \leq x \leq 8$)
(a) Vol. **76**, No. 2, p. 575 – 593 (1983).
Erratum
(a) Vol. **78**, No. 2, p. 761 (1983).
See also *1.3*, p. *23*; *1.4*, p. *30*; 21.6, p. 397.

K. H. J. Buschow and P. G. van Engen
Note on the Magnetic and Magneto-Optical Properties of Ni_2In Type 3d Transition Metal Compounds
(a) Vol. **76**, No. 2, p. 615 – 620 (1983).
See also 1.1, p. 11; *18.2*, p. *282*; *20*, p. *308*; 21.1, p. 367; 21.1.1, p. 381.

J. M. López and J. A. Alonso
Semiempirical Theory of Solid Solubility in Metallic Alloys
(a) Vol. **76**, No. 2, p. 675 – 682 (1983).
See also *1.3*, p. *23*; 8, p. 88; 21.1, p. 368.

G. Cocco, L. Schiffini, M. Sampoli, A. Lucci, and R. Riontino
Compositional Effects in Pd–B Glassy Alloys from X-Ray Determined Radial Distribution Functions
(a) Vol. **76**, No. 2, p. 753 – 762 (1983).
See also *2*, p. *47*.

D. Sundaraman, A. L. E. Terrance, G. van Tendeloo, and J. van Landuyt
High-Resolution Electron Microscopy Study of a Ti–1.6 at% N Alloy
(a) Vol. **76**, No. 2, p. K109 – K111 (1983).
See also *1.4*, p. *31*.

H.-J. Fitting and R. Technow
Electron Backscattering at Various Angles of Incidence
(a) Vol. **76**, No. 2, p. K151 – K154 (1983).
See also *11*, p. *127*; 17, p. 269; 22.1.2, p. 428.

H. Hermann
Influence of Directed Chemical Bonding on the Structure of Metallic Glasses. Two-Dimensional Model
(b) Vol. **117**, No. 1, p. 185 – 195 (1983).
See also *2*, p. *47*; 4, p. 61.

O. HARDOUIN DUPARC and V. R. VELASCO
Influence of a Thin Overlayer on the Surface Brillouin Scattering. Cubic Crystals
(b) Vol. **117**, No. 1, p. K37 – K41 (1983).
See also *13.3*, p. *168*.

A. K. ZHETBAEV and A. N. OZERNOI
Mössbauer Study of Radiation Effect of the ^{181}Ta(p, n)^{181}W Reaction
(a) Vol. **77**, No. 1, p. 63 – 66 (1983).
See also *5*, p. *66*; 11, p. 127.

V. E. ANTONOV, T. E. ANTONOVA, I. T. BELASH, E. G. PONYATOVSKII, V. I. RASHUPKIN, and V. G. THIESSEN
The Pd–Ni–H System at High Pressure
(a) Vol. **77**, No. 1, p. 71 – 79 (1983).
See also *1.3*, p. *23*; 14.2, p. 212; 18.2, p. 283; 21.1, p. 369.

C. W. LUNG and L. Y. XIONG
The Dislocation Distribution Function in the Plastic Zone at a Crack Tip
(a) Vol. **77**, No. 1, p. 81 – 86 (1983).
See also *10.1*, p. *102*; 12.1, p. 138.

G. I. GUDZENKO, V. B. MOLODKIN, and V. G. NOVITSKII
Dynamical Diffraction in Alloys. Influence of Long-Range Order
(a) Vol. **77**, No. 1, p. 115 – 120 (1983).
See also *1.3*, p. *23*; *10.1*, p. *102*; 21.1, p. 369.

M. ANGLADA and F. GUIU
Temperature and Strain Rate Dependence of the Cyclic Flow Stress of Molybdenum Single Crystals
(a) Vol. **77**, No. 1, p. 283 – 289 (1983).
See also *10.1*, p. *102*; *12.1*, p. *138*.

K. T-Y. KUNG, B. X. LIU, and M-A. NICOLET
Study of Ni–Nb System by Ion Mixing
(a) Vol. **77**, No. 1, p. 355 – 359 (1983).
See also 1.5, p. 37; *2*, p. *48*; 14.1, p. 207; 21.1, p. 369.

Y. MOMOSE, M. OHKUBO, and J. GUNJI
Interaction of Mechanically Rubbed Aluminium with Polar Organic Vapors Found by Measurements of Optically Stimulated Exoelectron Emission under the Influence of Applied Electric Potential
(a) Vol. **77**, No. 1, p. K45 – K48 (1983).
See also *17*, p. *270*.

M. OHTA, T. KANADANI, A. SAKAKIBARA, H. YAMADA, and M. YAMADA
Effects of Grain Boundary on Age Hardening and Reversion in Al–Zn Alloys
(a) Vol. **77**, No. 1, p. K49 – K51 (1983).
See also *1.3*, p. *24*; 10.1, p. 102; 12.1, p. 138; 21.1, p. 369.

A. I. OSETSKII
Low-Temperature Fluctuational Motion of Dislocations in Crystals (II): Manifestations of Dissipative and Quantum Effects in the Low-Temperature Plasticity
(b) Vol. **117**, No. 2, p. 589 – 602 (1983).
See also *10.1*, p. *103*; 21.1, p. 370.

E. Dębowska
Investigation of the Electronic Structure of TiH_x and $Ti_{1-y}V_yH_x$ by Positron Annihilation
(b) Vol. **117**, No. 2, p. 699–706 (1983).
See also *13*, p. *149*.

T. Soma, Y. Konno, and H.-M. Kagaya
Local Heine-Abarenkov Potential with Volume-Dependent Parameters of Al
(b) Vol. **117**, No. 2, p. 743–748 (1983).
See also *4*, p. *61*; 12.1, p. 138.

K. Elk
Susceptibility of the Periodic Anderson Model: Hartree-Fock Approximation
(b) Vol. **117**, No. 2, p. K87–K90 (1983).
See also 13, p. 149; *18.1*, p. *278*.

B. A. Men, M. L. Levitan, and A. N. Men
Influence of the Ground State Symmetry on Qualitative Form of Ordering Phase Diagram
(a) Vol. **77**, No. 2, p. 455–461 (1983).
See also *1.2*, p. *17*; 18, p. 274.

C. M. van Baal
The Influence of Texture on the X-Ray Determination of Residual Strains in Ground or Worn Surfaces
(a) Vol. **77**, No. 2, p. 521–526 (1983).
See also *1.6*, p. *40*; 10.1, p. 103; 12.1, p. 138.

N. Kuramochi, H. Mizubayashi, and S. Okuda
Dislocation Breakaway at Low Temperatures in Nb
(a) Vol. **77**, No. 2, p. 633–642 (1983).
See also *10.1*, p. *103*; *12.1*, p. *138*.

Z. Gierak, J. W. Moroń, and A. Lehr
Recovery of Structural Defects in Molybdenum
(a) Vol. **77**, No. 2, p. 775–783 (1983).
See also *10.1*, p. *103*; 14.1, p. 208.

P. K. Acharyya, A. K. Tewari, S. K. Roy, and D. L. Bhattacharya
Anharmonic Study on the Debye-Waller Factor for a Vacancy-Trapped Mössbauer Impurity in Solid Solution of Aluminium
(b) Vol. **118**, No. 1, p. 41–50 (1983).
See also *6*, p. *74*; 10.1, p. 103.

V. S. Patil, R. G. Pillay, P. N. Tandon, and H. G. Devare
Hyperfine Field Investigations in Some Ru and Rh Based Heusler Alloys
(b) Vol. **118**, No. 1, p. 57–61 (1983).
See also *5*, p. *67*; 18.2, p. 284; 21.1.1, p. 382.

A. Jezierski
Application of CPA to the Magnon Hamiltonian
(b) Vol. **118**, No. 1, p. 73–80 (1983).
See also *13.5.4*, p. *203*; 18.2, p. 284.

L. Schwartz and A. Paja
On the Electrical Conductivity of Disordered Alloys in Muffin-Tin Model
(b) Vol. **118**, No. 1, p. 101 – 104 (1983).
See also *14.1*, p. *208*.

V. A. Muralev
Channeling of Ultrarelativistic Electrons in Crystals
(b) Vol. **118**, No. 1, p. 363 – 372 (1983).
See also *11*, p. *129*; 22.1.2, p. 432.

M. S. Wartak
Cyclotron Resonance in Metals with g-Factor Anisotropy
(b) Vol. **118**, No. 1, p. K1 – K2 (1983).
See also 13.2, p. 166; *19*, p. *302*.

S. Maciejewski
Cooper Pair Condensation Energy Dependence on Electron Structure
(b) Vol. **118**, No. 1, p. K3 – K5 (1983).
See also *13.1*, p. *160*; *14.2*, p. *213*.

H. Englisch
*There is No Energy Interval Lying in the Spectra of All One-Dimensional Periodic Alloys
(b) Vol. **118**, No. 1, p. K17 – K19 (1983).
See also *13.1*, p. *160*.

U. Messerschmidt, F. Appel, M. Bartsch, and R. Gerlach
Dislocation Motion in a Dispersion Hardened Al–Zn–Mg Alloy and Its Plastic Properties
(a) Vol. **78**, No. 1, p. 93 – 102 (1983).
See also *10.1*, p. *103*; 21.1, p. 371; 21.3, p. 389.

V. E. Antonov, T. E. Antonova, I. T. Belash, E. G. Ponyatovskii, and V. I. Rashupkin
The Pd–Pt–H System: Phase Transformations at High Pressure and Superconductivity
(a) Vol. **78**, No. 1, p. 137 – 146 (1983).
See also *1.2*, p. *17*; *14.2*, p. *213*; 21.1, p. 371.

G. Dlubek, O. Brümmer, R. Krause, A. Baranowski, and B. Rozenfeld
A Positron Study of Guinier-Preston Zones and Precipitates in Al–Zn Alloys Having Mg or Ge Additions
(a) Vol. **78**, No. 1, p. 217 – 228 (1983).
See also *10.1*, p. *104*; 21.1, p. 371.

E. K. Polychroniadis and P. Delavignette
Dislocation Structures and Subgrain Boundaries in Al_3Zr
(a) Vol. **78**, No. 1, p. 287 – 301 (1983).
See also *1.4*, p. *32*; *10.1*, p. *104*.

A. Le Lann and A. Dubertret
A Structural Description for the F.C.C. → B.C.C. Martensitic Transformation in Terms of $\{011\}\langle 011\rangle_{\text{f.c.c.}}$ Simultaneous Double Shear. Relation with the Atomic Configuration in $\{3, 15, 10\}_{\text{f.c.c.}}$ Habit Plane
(a) Vol. **78**, No. 1, p. 309 – 321 (1983).
See also *1.2*, p. *17*; 1.4, p. 32.

M. Ohta, T. Kanadani, A. Sakakibara, H. Yamada, and M. Yamada
Effects of Surface and Thickness of Specimens on the Aging in Al–Zn Alloys
(a) Vol. **78**, No. 1, p. K23–K26 (1983).
See also *1.3*, p. *25*; 10.1, p. 104; 12.1, p. 138; 21.1, p. 372.

B. Babić, M. Stojić, and J. Konstantinović
Electrical Properties of the Metastable $Mn_{1.11}Al_{0.89}$ Alloy
(a) Vol. **78**, No. 1, p. K77–K80 (1983).
See also *13*, p. *150*; *14.1*, p. *208*; 21.1, p. 372.

W. Puff, P. Mascher, P. Kindl, and H. Sormann
Precise Positron Lifetime Measurements in Indium
(b) Vol. **118**, No. 2, p. 799–803 (1983).
See also *10.1*, p. *104*.

L. Bottyán, D. L. Beke, and K. Tompa
Diffusion-Induced Quadrupole Relaxation of ^{27}Al Nuclei in Dilute Al–Ti, Al–Cr, Al–Mn, and Al–Cu Alloys at High Temperatures
(b) Vol. **118**, No. 2, p. 835–842 (1983).
See also *9*, p. *95*; 10.1, p. 104; *19*, p. *303*; 21.1, p. 372.

T. Soma, Y. Konno, and H.-M. Kagaya
Compression Effect on the Bulk Modulus of Al
(b) Vol. **118**, No. 2, p. K155–K157 (1983).
See also *4*, p. *62*; *12.1*, p. *139*.

O. V. Presnyakova, V. I. Zaitsev, and V. A. Fomchenko
Creep of Aluminium Single Crystals under High Pressure
(a) Vol. **78**, No. 2, p. 377–383 (1983).
See also *10.1*, p. *104*; 12.1, p. 139.

G. Wendrock
Study of the Behaviour of Al–Zn Alloys by SAXS in the Temperature Range of the Instable Solid Solution
(a) Vol. **78**, No. 2, p. 497–506 (1983).
See also *1.2*, p. *18*; 1.3, p. 25; 21.1, p. 373.

G. Kreysch, H. Kerkow, R. I. Bogdanov, V. P. Petukhov, and E. A. Romanovskii
Proton-Induced L-Shell X-Ray Emission Cross-Sections of Elements with $24 \leqq Z_2 \leqq 50$ for Projectile Energies between 30 and 350 keV
(a) Vol. **78**, No. 2, p. 507–525 (1983).
See also *11*, p. *130*; 20, p. 310.

A. R. Yavari
Hardening by Thermal Vacancies and Quenching Strains in Singly-Oriented Pb Single Crystals
(a) Vol. **78**, No. 2, p. 583–588 (1983).
See also *10.1*, p. *105*; 12.1, p. 139.

SHU ZHEN and G. J. DAVIES
Calculation of the Lennard-Jones $n-m$ Potential Energy Parameters for Metals
(a) Vol. **78**, No. 2, p. 595–605 (1983).
See also 1.3, p. 25; *4*, p. *63*; 12.1, p. 139.

V. NAVRÁTIL, M. HAMERSKÝ, P. LUKÁČ, V. P. SOLDATOV, and V. I. STARTSEV
Solid Solution Hardening of Cd–Zn Single Crystals at 1.5, 2.8, and 4.2 K
(a) Vol. **78**, No. 2, p. 633–638 (1983).
See also *10.1*, p. *105*; *12.1*, p. *139*; 21.1, p. 373.

A. A. MOHAMED, S. A. SALEH, E. A. ABOU-SAIF, M. ABD-RABO, and G. A. EL-GANAINY
On the Identification of New and Normal Phases in the System Cadmium–Antimony Alloys
(a) Vol. **78**, No. 2, p. 705–716 (1983).
See also *1.3*, p. *25*; 21.7, p. 401.

R. SCHUMANN and K. ELK
On Thermopower and Thermal Conductivity of the Periodic Anderson Model
(b) Vol. **119**, No. 1, p. 221–225 (1983).
See also *8*, p. *90*; *15*, p. *262*.

E. OJALA
K and L Absorption Spectra of Aluminium from APW Calculation
(b) Vol. **119**, No. 1, p. 269–277 (1983).
See also *13.1*, p. *162*; *20*, p. *310*.

S. MAHAJAN and S. PRAKASH
Electronic Structure of Impurities in Dilute Alloys of Al
(b) Vol. **119**, No. 1, p. 381–389 (1983).
See also *13.4*, p. *185*; 14.1, p. 209.

B. RAYKOV
Density Functional Theory of Metal Surface Properties
(b) Vol. **119**, No. 1, p. 437–444 (1983).
See also *13.3*, p. *170*.

E. CANESSA and J. VIGNOLO
The Resistivity of Liquid Mercury at 150 °C
(b) Vol. **119**, No. 1, p. K83–K85 (1983).
See also *2*, p. *51*; *14.1*, p. *209*.

P. M. BOERRIGTER, A. LODDER, and J. MOLENAAR
Calculated Residual Resistivities in Al and Cu Alloys
(b) Vol. **119**, No. 1, p. K91–K96 (1983).
See also *14.1*, p. *209*; 21.1, p. 374.

N. V. MOSEEV, V. G. CHUDINOV, B. N. GOSHCHITSKII, and V. I. PROTASOV
Atomic Configurations of Interstitials in Intermetallic Compound Nb_3Sn
(a) Vol. **79**, No. 1, p. 67–70 (1983).
See also *10.1*, p. *105*; 21.7, p. 401.

L. V. Nomerovannaya and V. A. Rassokhin
Anomalies in the Optical and Electrical Properties of Cr–Al Alloys
(a) Vol. **79**, No. 1, p. 87 – 93 (1983).
See also *14.1*, p. *209*; 20.1, p. 326; 21.1, p. 374.

K. Iwasaki
High-Temperature Internal Friction Peaks of Pure Aluminium
(a) Vol. **79**, No. 1, p. 115 – 124 (1983).
See also *10.1*, p. *105*; *12.1*, p. *139*.

R. Ramlau and H. Löffler
On the Size–Shape Relation of Guinier-Preston Zones and α'_R-Precipitates in an Al–6.8 at% Zn Alloy
(a) Vol. **79**, No. 1, p. 141 – 148 (1983).
See also *1.3*, p. *26*; *10.1*, p. *105*; 21.1, p. 375.

J. Sosnowski
On the Flux Jumps in Hard Superconductors
(a) Vol. **79**, No. 1, p. 207 – 212 (1983).
See also *14.2*, p. *213*.

G. K. Baranova
On the Role of Surface Oxides in the Processes of Metal Etching
(a) Vol. **79**, No. 1, p. K9 – K12 (1983).
See also *1.6*, p. *41*; 10.1, p. 106.

V. G. Chudinov, V. P. Gogolin, B. N. Goshchitskii, and V. I. Protasov
Simulation of Collision Cascades in Intermetallic V_2Hf Compounds
(a) Vol. **79**, No. 1, p. K57 – K60 (1983).
See also *11*, p. *132*; 14.2, p. 213; 21.1, p. 375.

D. J. González, M. L. Zorita, J. A. Alonso, and M. P. Iñiguez
Density Functional Pseudopotential Calculation of the Cohesive Properties of Disordered Solid Alloys of Nontransition Metals. Application to Zn – Cd
(b) Vol. **119**, No. 2, p. 589 – 594 (1983).
See also *1.3*, p. *26*; *13*, p. *151*; 21.1, p. 375.

S. M. Rafique, N. R. Mitra, and P. L. Srivastava
On Choosing a Set of Eigenvalues and Wave Functions in Harrison's First Principle Method
(b) Vol. **119**, No. 2, p. K113 – K116 (1983).
See also *13.1*, p. *164*; *14.1*, p. *210*.

J. Richter
Disorder and Competing Interactions in Spin Systems: Finite-Cluster Calculations
(b) Vol. **119**, No. 2, p. K159 – K163 (1983).
See also *2*, p. *52*; *18.4*, p. *297*.

B. Rauschenbach and K. Hohmuth
Analysis of Heavy-Ion Sputtering Based on the Shock Wave Model
(a) Vol. **79**, No. 2, p. 395 – 401 (1983).
See also 1.6, p. 41; *11*, p. *132*.

V. G. Chudinov and V. I. Protasov
Radiation Diffusion–Dislocation Creep in the Presence of Weertman Mechanism of Dislocation Multiplication
(a) Vol. **79**, No. 2, p. 433 – 443 (1983).
See also *9*, p. *95*; 10.1, p. 106.

V. Atzrodt and H. Lange
Principal Component Analysis as a Method for Silicide Investigation with Auger Electron Spectroscopy
(a) Vol. **79**, No. 2, p. 489 – 496 (1983).
See also *1.6*, p. *42*; 17, p. 270; 22.1.2, p. 436.

M. Gasgnier, L. Nevot, P. Baillif, and J. Bardolle
Characterization and Crystalline Structures of Tungsten Thin Films
(a) Vol. **79**, No. 2, p. 531 – 542 (1983).
See also 1.3, p. 26; *1.5*, p. *38*.

S. E. Donnelly, A. A. Lucas, Ph. Lambin, and J. P. Vigneron
A Possible Mechanism for Electron-Bombardment-Induced Loop Punching in Helium-Implanted Materials
(a) Vol. **79**, No. 2, p. 543 – 548 (1983).
See also *11*, p. *133*.

H. Hahn, G. Frohberg, and H. Wever
Self-Diffusion in the Intermetallic B2 Electron Compound PdIn
(a) Vol. **79**, No. 2, p. 559 – 565 (1983).
See also *9*, p. *95*.

G. J. Chashoo and G. Dharmadurai
Role of Thermal Transport in the Non-Equilibrium Behaviour of a Cylindrical Ferromagnetic Superconductor
(b) Vol. **120**, No. 1, p. 397 – 400 (1983).
See also *14.2*, p. *214*.

D. E. G. Williams and A. Jezierski
The Spin Wave Stiffness Constant in $Pt_c(Mn_{0.5}Cr_{0.5})_{1-c}$ Alloys
(b) Vol. **120**, No. 1, p. K87 – K91 (1983).
See also *18.2*, p. *288*.

C. B. Carter
Multiple Ribbons in Face-Centered Cubic Materials (IV): The Interaction between Double Ribbons and Inclined Dislocations
(a) Vol. **80**, No. 1, p. 51 – 64 (1983).
See also *10.1*, p. *106*.

R. Tognato
Why Metals Do Not Show Overheating
(a) Vol. **80**, No. 1, p. 147 – 157 (1983).
See also 1.2, p. 19; *10.1*, p. *107*.

U. Dahmen and K. H. Westmacott
Ledge Structure and the Mechanism of θ′ Precipitate Growth in Al–Cu
(a) Vol. **80**, No. 1, p. **249**–**262** (1983).
See also *3*, p. *59*; 10.1, p. 107; 21.1, p. 377.

D. J. Farnum, O. T. Inal, and R. J. Walko
A Field-Ion Microscopy Study of Ion Damage to Tungsten
(a) Vol. **80**, No. 1, p. 287–303 (1983).
See also *11*, p. *134*; 17.1, p. 272.

S. K. Pabi
On the Variation of Interface Composition around a Dissolving Precipitate
(a) Vol. **80**, No. 1, p. 305–316 (1983).
See also *1.3*, p. *27*; 9, p. 95; 21.6, p. 399.

A. Charai and C. Boulesteix
Visualization and Study of the Shape and Size of Very Small Coherent Precipitates by a Phase Contrast Method
(a) Vol. **80**, No. 1, p. 333–341 (1983).
See also *1.4*, p. *34*; *10.1*, p. *107*.

V. G. Chudinov, V. P. Gogolin, B. N. Goshchitskii, and V. I. Protasov
Evolution of the Cascade Region in Nb_3Sn during the Thermal Spike under Fast-Neutron Irradiation
(a) Vol. **80**, No. 1, p. 349–352 (1983).
See also *10.1*, p. *107*; 11, p. 134; 21.7, p. 402.

P. Meisterle and W. Pfeiler
Short-Range Ordering in the Matrix of a Two-Phase AgAl Alloy
(a) Vol. **80**, No. 1, p. K37–K40 (1983).
See also *1.3*, p. *27*; *14.1*, p. *210*.

B. Glowacki and J. Chojcan
Studies on the Condition of Obtaining a Single Phase. Superconducting Nb_3Sn Layer by Mössbauer Spectroscopy
(a) Vol. **80**, No. 1, p. K93–K96 (1983).
See also *5*, p. *68*; 9, p. 96; 14.2, p. 214.

F. Goedsche, R. Richter, H. Wegener, and G. Röpke
Correlation Function Method for the Thermoelectric Power of Metals
(b) Vol. **120**, No. 2, p. **631**–**644** (1983).
See also 13, p. 153; *14.1*, p. *210*; *15*, p. *263*.

J. M. Khalifeh
Asymptotic Elastic Energy in Simple Metals
(b) Vol. **120**, No. 2, p. K161–K166 (1983).
See also *10.1*, p. *107*; 13.4, p. 190.

R. Schumann
Temperature Dependence of the Thermopower of the Periodic Anderson Model
(b) Vol. **120**, No. 2, p. K183–K188 (1983).
See also 8, p. 92; 14.1, p. 210; *15*, p. *263*.

B. I. Craig

Low-Temperature Trapping of Positive Muons in Metals
(b) Vol. **120**, No. 2, p. K211 – K214 (1983).
See also *10.1*, p. *107*; *13.4*, p. *191*.

S. V. Tsivinskii

The Theory of Electric Resistance of Pure Metals and Metals with a Given Concentration of Impurities, Vacancies, and Dislocations
(a) Vol. **80**, No. 2, p. 439 – 446 (1983).
See also 10.1, p. 107; *14.1*, p. *210*; 21.1, p. 377; 21.1.1, p. 385; 21.2, p. 388; 21.6, p. 399.

V. I. Motorin

Vitrification Kinetics of Pure Metals
(a) Vol. **80**, No. 2, p. 447 – 455 (1983).
See also *2*, p. *55*; 21.1, p. 362; 21.6, p. 399.

B. Rauschenbach, A. Kolitsch, and K. Hohmuth

Iron Nitride Phases Formed by Nitrogen Ion Implantation and Thermal Treatment
(a) Vol. **80**, No. 2, p. 471 – 482 (1983).
See also *1.3*, p. *27*; 1.5, p. 38; *11*, p. *135*; 21.1.1, p. 385.

C. J. Rossouw

A γ- to α-Phase Transition Induced by Heavy Ion Bombardment of a CuAl Alloy
(a) Vol. **80**, No. 2, p. 631 – 636 (1983).
See also *1.2* p. *20*; 11, p. 135; 21.1, p. 378.

M. Z. Butt, P. Feltham, and I. M. Ghauri

Stress-Equivalence of Solid Solution Hardening in Concentrated Ag–Al and Ag–Zn Alloys
(a) Vol. **80**, No. 2, p. K125 – K128 (1983).
See also *10.1*, p. *107*; 21.1, p. 378; 21.6, p. 400.

21.1 Transition Metals (Group Sc to Zn) and Their Alloys

Review Articles

R. E. Hummel
Differential Reflectometry and Its Application to the Study of Alloys, Ordering, Corrosion, and Surface Properties
(a) Vol. **76**, No. 1, p. 11 – 44 (1983).
See also *1.3*, p. *21*; 10.1, p. 99; 13.1, p. 155; 13.3, p. 167; *20.1*, p. *312*; 21, p. 349; 22.6, p. 490.

K. H. Fischer
Spin Glasses (I)
(b) Vol. **116**, No. 2, p. 357 – 414 (1983).
See also 2, p. 43; 8, p. 85; 13, p. 146; 14.1, p. 206; *18.2*, p. *280*; *18.4*, p. *295*; 21.1.1, p. 379.

Original Papers and Short Notes

R. H. Rautioaho
A Pair Potential Calculation of Vacancy Formation and Migration Energies for Aluminium and Copper
(b) Vol. **115**, No. 1, p. 95 – 103 (1983).
See also 9, p. 93; *10.1*, p. *99*; 21, p. 349.

T. Nautiyal and S. Auluck
Fermi Surface of Noble Metals Using the Eight-Cone Model
(b) Vol. **115**, No. 1, p. 127 – 130 (1983).
See also *13.2*, p. *166*; 21.6, p. 396.

F. Goedsche, A. Richter, and G. Vojta
Dynamical Susceptibility and Electrical Resistivity of Disordered Spin Systems
(b) Vol. **115**, No. 1, p. 209 – 218 (1983).
See also 13, p. 146; *14.1*, p. *206*; *18.2*, p. *280*.

V. A. Maslov, I. I. Naumov, and V. E. Panin
On the Problem of Electron Phase Stability
(b) Vol. **115**, No. 1, p. 317 – 324 (1983).
See also *1.3*, p. *21*; *13.2*, p. *166*.

Sh. Sh. Ibragimov, O. P. Maksimkin, and Sh. B. Shiganakov
Grain-Boundary Deformation in Nickel Irradiated by α-Particles
(a) Vol. **75**, No. 1, p. 55 – 58 (1983).
See also *1.4*, p. *29*; *11*, p. *122*.

T. A. Jenkins and S. Arajs
Electrical Resistivity of Some Directionally-Solidified Aluminium-Based Eutectic Alloys
(a) Vol. **75**, No. 1, p. 73 – 82 (1983).
See also *14.1*, p. *206*; 21, p. 349; 21.1.1, p. 379.

R. Gerlach and H. Löffler
On the Structure of the Precipitation-Free Zone in Two Al–Zn–Mg Alloys after Various Heat Treatments
(a) Vol. **75**, No. 1, p. 169–178 (1983).
See also *10.1*, p. *99*; 21, p. 350; 21.3, p. 389.

F. Walz, H. J. Blythe, and F. Dworschak
Annealing Behaviour of Dilute FeCu, FeMn, and FeTi Alloys in the Temperature Range up to Stage III, Following Low-Temperature Electron Irradiation
(a) Vol. **75**, No. 1, p. 235–242 (1983).
See also *10.1*, p. *99*; 11, p. 122; *18.2*, p. *280*; 21.1.1, p. 379.

F. T. Parker and H. Oesterreicher
Temperature Dependence of Intrinsic Magnetic Hardness in Some Rare-Earth Pseudobinary Compounds and Amorphous Materials
(a) Vol. **75**, No. 1, p. 273–282 (1983).
See also 2, p. 43; *18.2*, p. *280*; 21.4, p. 390.

A. S. Lileev and V. P. Menushenkov
The Crystal Structure and Coercive Force of $SmCo_5$ Sintered Permanent Magnets
(a) Vol. **75**, No. 1, p. 341–346 (1983).
See also *1.3*, p. *21*; *18.2*, p. *280*; 21.4, p. 390.

Bai-Xin Liu
A Metastable Electron Compound Formed by Ion Irradiation
(a) Vol. **75**, No. 1, p. K77–K81 (1983).
See also *1.5*, p. *35*; *11*, p. *123*; 21.6, p. 396.

L. Kornblit
Divacancy Binding Energy in F.C.C. Metals
(b) Vol. **115**, No. 2, p. 485–491 (1983).
See also *10.1*, p. *99*; 21.6, p. 396.

R. H. Routioaho
Energies of Self-Interstitials for Aluminium and Copper
(b) Vol. **115**, No. 2, p. K121–K125 (1983).
See also *10.1*, p. *100*; 21, p. 351.

W. Zarek, A. Kasprzyk, and E. Kozubek
Magnetic and ESR Investigations of $Sm_6(Mn_{1-x}Fe_x)_{23}$ Compounds for $0 < x < 0.55$ in the Paramagnetic Range
(a) Vol. **75**, No. 2, p. 359–366 (1983).
See also *18.1*, p. *278*; *19*, p. *300*; 21.1.1, p. 379.

A. I. Lotkov, V. N. Grishkov, A. V. Kuznetsov, and S. N. Kulkov
TiNi Aging and Its Effect on the Start Temperature of the Martensitic Transformation
(a) Vol. **75**, No. 2, p. 373–377 (1983).
See also *1.2*, p. *14*.

A. Baldan
Dislocation Structures in Aluminium-Rich Ni_3(Al, Ta) and Ni_3(Al, Hf)
(a) Vol. **75**, No. 2, p. 441–449 (1983).
See also *1.3*, p. *21*; *10.1*, p. *100*; 21, p. 351.

K. K. Ray and A. K. Mallik
An Investigation on the Activation Volume for Plastic Flow in Nickel Determined by Micro and Macro Tests
(a) Vol. **75**, No. 2, p. 451 – 458 (1983).
See also *10.1*, p. *100*; 12.1, p. 137.

A. Korner and H. P. Karnthaler
Glide Dislocations on Cube Planes in a Low Stacking-Fault Energy Alloy
(a) Vol. **75**, No. 2, p. 525 – 535 (1983).
See also *10.1*, p. *100*; 12.1, p. 137; 21, p. 351.

D. Schryvers, G. van Tendeloo, J. van Landuyt, and S. Amelinckx
High-Resolution Study of the One- and Two-Dimensional Long-Period Antiphase Boundary Structures in Au_4Zn
(a) Vol. **75**, No. 2, p. 607 – 616 (1983).
See also *1.3*, p. *22*; *1.4*, p. *29*; 21.6, p. 396.

P. Delobelle and C. Oytana
Creep at Low and Average Stress of Several Highly Concentrated Solid Solutions of the Cubic Centered Type
(a) Vol. **75**, No. 2, p. 625 – 634 (1983).
See also *10.1*, p. *100*; 12.1, p. 137; 21.1.1, p. 380.

K. K. Misra and A. N. Nigam
Development of Stress in Nickel during Hydrogen Diffusion
(a) Vol. **75**, No. 2, p. 635 – 639 (1983).
See also *9*, p. *93*; *12.1*, p. *137*.

V. Navrátil, M. Hamerský, P. Lukáč, V. P. Soldatov, and V. I. Startsev
Solid Solution Hardening of Cd–Zn Single Crystals at 1.5 K
(a) Vol. **75**, No. 2, p. K133 – K135 (1983).
See also *1.3*, p. *22*; *10.1*, p. *100*; 21, p. 351.

K. Masuda
On the Interaction between a Vacancy and an Interstitial Impurity Atom in B.C.C. Transition Metals
(b) Vol. **116**, No. 1, p. 9 – 16 (1983).
See also *10.1*, p. *101*; 21.1.1, p. 380.

V. Drchal and J. Kudrnovský
Second-Order Perturbation Treatment of Correlations in Transition Metal Alloys
(b) Vol. **116**, No. 1, p. 119 – 128 (1983).
See also *13*, p. *147*.

T. G. Abdel-Malik, A. A. Aly, A. M. Abdeen, and H. A. Motaweh
The Influence of the Surface Barrier on the Trajectory of Sputtered Particles
(b) Vol. **116**, No. 1, p. 169 – 172 (1983).
See also 11, p. 124; *13.3*, p. *168*.

A. V. Zaborov and M. V. Medvedev
A Binary Ferromagnetic Alloy with Magnetic Moments Dependent on Local Environment (I): The Jaccarino-Walker Model in the Effective Field Approximation
(b) Vol. **116**, No. 1, p. 227 – 238 (1983).
See also *18.2*, p. *281*; 21.1.1, p. 380.

R. N. Joarder and R. V. Gopala Rao
Temperature Dependence of the Structure Factor of Liquid Metals
(b) Vol. **116**, No. 1, p. 299 – 306 (1983).
See also *2*, p. *45*; 4, p. 61; 9, p. 93; 21.2, p. 387; 21.6, p. 396.

J. Chojcan, M. Szuszkiewicz, S. Szuszkiewicz, and B. Rozenfeld
Localization of Electron in Fe–Co Alloys
(b) Vol. **116**, No. 1, p. K57 – K60 (1983).
See also *13*, p. *148*; 21.1.1, p. 380.

S. V. Sudareva, V. A. Rassokhin, and A. F. Prekul
The Structure of Cr–Al Alloys Exhibiting Anomalous Physical Properties
(a) Vol. **76**, No. 1, p. 101 – 106 (1983).
See also *1.2*, p. *15*; *1.4*, p. *29*; 21, p. 352.

T. Nanba and T. P. Martin
Raman Scattering from Metal Smokes
(a) Vol. **76**, No. 1, p. 235 – 239 (1983).
See also 1.6, p. 39; *6*, p. *72*; *20.1*, p. *316*; 21.6, p. 397.

G. L. Whittle
Magnetic Order in Isoelectronic Co(GaVFe) Alloys
(a) Vol. **76**, No. 1, p. 351 – 356 (1983).
See also *5*, p. *66*; *18*, p. *274*.

P. Čapková, R. Kužel, and J. Šedivý
Thermal Vibrations and Static Displacements of Atoms in Nonstoichiometric Titanium Carbide
(a) Vol. **76**, No. 1, p. 383 – 390 (1983).
See also *6*, p. *72*; 8, p. 87.

P. Czarnecki and G. Fritzkowski
Transverse Magnetoresistance of Amorphous $Fe_{40-x}Ni_{40}Cr_xB_{20}$ Alloys in Magnetic Fields up to 20 T
(a) Vol. **76**, No. 1, p. K93 – K96 (1983).
See also *2*, p. *46*; *14.1*, p. *207*; 18.2, p. 282; 21.1.1, p. 380.

S. Takahashi and K. Ikeda
Magnetic Interactions in Plastically Deformed Ni_3Mn Alloy
(a) Vol. **76**, No. 1, p. K97 – K100 (1983).
See also 10.1, p. 101; *18.2*, p. *282*; *18.4*, p. *295*.

A. V. Zaborov and M. V. Medvedev
A Binary Ferromagnetic Alloy with Magnetic Moments Dependent on Local Environment (II): Neutron Magnetic Elastic Scattering and Hyperfine Field Distribution
(b) Vol. **116**, No. 2, p. 511 – 523 (1983).
See also *18.2*, p. *282*; 21.1.1, p. 381.

Ch. Müller, W. Blau, and P. Ziesche
Local Partial DOS and Experimental Al K Spectra of Transition Metal Aluminides. Charge Transfer Trend
(b) Vol. **116**, No. 2, p. 561 – 573 (1983).
See also *13.1*, p. *157*; 20, p. 308; 21.1.1, p. 381.

K.-H. Steinmetz, H.-P. Fischer, M. Martin, and E. Nembach
Positron-Annihilation Measurements of the Fermi Surface of Copper-Base Solid Solutions with Gallium, Germanium, and Nickel
(b) Vol. **116**, No. 2, p. 575 – 583 (1983).
See also *1.3*, p. *22*; *13.2*, p. *166*.

S. N. Kaul
Two-Current Conduction Model and Anisotropic Magnetoresistance in Transition Metal–Metalloid Glasses
(b) Vol. **116**, No. 2, p. K99 – K104 (1983).
See also *2*, p. *46*; *14.1*, p. *207*; 21.1.1, p. 381.

M. E. Gurevich, L. N. Larikov, V. G. Novitskii, A. E. Pogorelov, and A. F. Zhuravlev
Laser-Stimulated Mass Transfer in Metals
(a) Vol. **76**, No. 2, p. 479 – 484 (1983).
See also *9*, p. *94*; *11*, p. *126*; 21.1.1, p. 381.

H. Börner and R. Höhne
Study of Magnetic After-Effects in the Amorphous Alloy $Fe_{40}Ni_{40}P_{14}B_6$
(a) Vol. **76**, No. 2, p. 499 – 503 (1983).
See also *2*, p. *47*; *18.2*, p. *282*; 21.1.1, p. 381.

G. H. van Haastert and A. van den Beukel
Static and Dynamic Strain Aging in a Ni(C) Solid Solution
(a) Vol. **76**, No. 2, p. 609 – 614 (1983).
See also *10.1*, p. *101*.

K. H. J. Buschow and P. G. van Engen
Note on the Magnetic and Magneto-Optical Properties of Ni_2In Type 3d Transition Metal Compounds
(a) Vol. **76**, No. 2, p. 615 – 620 (1983).
See also 1.1, p. 11; *18.2*, p. *282*; *20*, p. *308*; 21, p. 353; 21.1.1, p. 381.

E. Balanzat and M. Halbwachs
A Comparative Analysis of the Ordering Kinetic Laws for the Derivation of the Vacancy Mobility Parameters in Concentrated Alloys
(a) Vol. **76**, No. 2, p. 667 – 673 (1983).
See also 9, p. 94; *10.1*, p. *102*; 14.1, p. 207; 21.6, p. 397.

J. M. LÓPEZ and J. A. ALONSO
Semiempirical Theory of Solid Solubility in Metallic Alloys
(a) Vol. **76**, No. 2, p. 675–682 (1983).
See also *1.3*, p. *23*; 8, p. 88; 21, p. 353.

W. BAŻELA-WRČBEL and A. SZYTUŁA
Neutron Diffraction Study of the $Ni_2Mn_{1-x}Cr_xSn$ System
(a) Vol. **76**, No. 2, p. K117–K120 (1983).
See also *1.3*, p. *23*; 18.2, p. 283.

M. BURD, R. STEARNS, and R. BRAUNSTEIN
De-Correlation Technique for Separation of Drude Parameters from Wavelength Modulation Spectroscopy Data
(b) Vol. **117**, No. 1, p. 101–107 (1983).
See also *13.1*, p. *158*; *20.1*, p. *317*; 21.6, p. 397.

D. DĂDÂRLAT, T. PETRIŞOR, A. GIURGIU, and I. POP
Localization Effect in CrRh Alloys
(b) Vol. **117**, No. 1, p. 155–159 (1983).
See also *18.4*, p. *296*; 21.4, p. 391.

J. C. PARLEBAS
Model for Optical Absorption in Transition Metals Containing d Impurities
(b) Vol. **117**, No. 1, p. 255–259 (1983).
See also *20.1*, p. *318*.

I. MERTIG, E. MROSAN, R. ZELLER, P. H. DEDERICHS, and P. ZIESCHE
Electronic Properties of Dilute Copper Alloys (I): Theory
(b) Vol. **117**, No. 1, p. 335–344 (1983).
See also *13.4*, p. *178*.

M. SOROHAN
Magnetoresistance of Polycrystalline Ni Wire under Hydrostatic Pressure
(b) Vol. **117**, No. 1, p. K1–K4 (1983).
See also *14.1*, p. *207*; 18.2, p. 283.

N. N. DELYAGIN, V. I. NESTEROV, and S. I. SEMENOV
The Oscillations of the Mössbauer Line Shift near T_c in the Ferromagnetic Ni–Rh Alloy
(b) Vol. **117**, No. 1, p. K27–K29 (1983).
See also *5*, p. *66*; *18.2*, p. *283*.

N. V. VOLKENSHTEIN, E. V. GALOSHINA, M. E. KOST, and T. S. SHUBINA
Magnetic Properties of the Sc–H System
(b) Vol. **117**, No. 1, p. K47–K52 (1983).
See also *13.1*, p. *158*; *18.1*, p. *278*.

A. A. ALEKHIN
The Bloch Lines Dynamics in Permalloy Films
(a) Vol. **77**, No. 1, p. 51–61 (1983).
See also *18.2.1*, p. *290*; 21.1.1, p. 382.

V. E. ANTONOV, T. E. ANTONOVA, I. T. BELASH, E. G. PONYATOVSKII, V. I. RASHUPKIN, and V. G. THIESSEN
The Pd–Ni–H System at High Pressure
(a) Vol. 77, No. 1, p. 71–79 (1983).
See also *1.3*, p. *23*; 14.2, p. 212; 18.2, p. 283; 21, p. 354.

G. I. GUDZENKO, V. B. MOLODKIN, and V. G. NOVITSKII
Dynamical Diffraction in Alloys. Influence of Long-Range Order
(a) Vol. 77, No. 1, p. 115–120 (1983).
See also *1.3*, p. *23*; *10.1*, p. *102*; 21, p. 354.

E. BURZO, I. LUPSA, and I. COSMA
On the Magnetic Behaviour of NiCuSi Solid Solutions
(a) Vol. 77, No. 1, p. 127–132 (1983).
See also *18*, p. *274*; 18.2, p. 283.

H.-H. POTTHOFF
Velocities of Dislocation Groups in Very Thin Neutron-Irradiated Copper Single Crystals Measured by Slip Line Cinematography
(a) Vol. 77, No. 1, p. 215–224 (1983).
See also *10.1*, p. *102*; 11, p. 128.

J. TEUHO and K. MÄKINEN
A Study of Non-Periodic Antiphase Boundaries in Long-Period Ordered $Au_3Zn(H)$ (I): Antiphase Boundary Planes
(a) Vol. 77, No. 1, p. 263–268 (1983).
See also *1.3*, p. *24*; *1.4*, p. *31*; 21.6, p. 397.

B. COCHRANE and P. J. GOODHEW
TEM Images of Faceted Bubbles and Voids
(a) Vol. 77, No. 1, p. 269–282 (1983).
See also 1.4, p. 31; 10.1, p. 102; *11*, p. *128*.

K. T.-Y. KUNG, B. X. LIU, and M.-A. NICOLET
Study of Ni–Nb System by Ion Mixing
(a) Vol. 77, No. 1, p. 355–359 (1983).
See also 1.5, p. 37; *2*, p. *48*; 14.1, p. 207; 21, p. 354.

G. VAN TENDELOO and S. AMELINCKX
Real Space Evidence for the Occurrence of Microdomains in "Disordered" Cu_3Pd
(a) Vol. 77, No. 1, p. K9–K11 (1983).
See also *1.4*, p. *31*.

M. OHTA, T. KANADANI, A. SAKAKIBARA, H. YAMADA, and M. YAMADA
Effects of Grain Boundary on Age Hardening and Reversion in Al–Zn Alloys
(a) Vol. 77, No. 1, p. K49–K51 (1983).
See also *1.3*, p. *24*; 10.1, p. 102; 12.1, p. 138; 21, p. 354.

S. L. Drechsler and G. M. Vujičić
On the Influence of Tunneling States on the Transition Temperature of Metallic Glassy Superconductors
(b) Vol. **117**, No. 2, p. 569–574 (1983).
See also 2, p. 48; *14.2*, p. *213*.

A. I. Oestskii
Low-Temperature Fluctuational Motion of Dislocations in Crystals (II): Manifestations of Dissipative and Quantum Effects in the Low-Temperature Plasticity
(b) Vol. **117**, No. 2, p. 589–602 (1983).
See also *10.1*, p. *103*; 21, p. 354.

I. Mertig, E. Mrosan, R. Zeller, and P. H. Dederichs
Electronic Properties of Dilute Copper Alloys (II): Residual Resistivity
(b) Vol. **117**, No. 2, p. 619–623 (1983).
See also *13.4*, p. *180*.

M. Szuszkiewicz, J. Chojcan, and A. Ostrasz
Investigation of the Electron Structure of V–Fe–H Alloys by Positron Annihilation
(b) Vol. **117**, No. 2, p. 645–651 (1983).
See also *13*, p. *149*; 21.1.1, p. 382.

H. Grimm and H. Kornmüller
Investigation of Structural Defects in the Amorphous Ferromagnetic Alloy $Fe_{40}Ni_{40}P_{14}B_6$
(b) Vol. **117**, No. 2, p. 663–674 (1983).
See also 2, p. 48; *10.1*, p. *103*; *18.2*, p. *284*; 21.1.1, p. 382.

S. Mróz, A. Mróz, A. Grudniewski, and E. Bąk
Effective Debye Temperature of the (110) Ni Face for Different Azimuths
(b) Vol. **117**, No. 2, p. 683–688 (1983).
See also *6*, p. *74*.

N. V. Ershov, Yu. A. Babanov, and V. R. Galakhov
EXAFS Study of Crystalline Cu, Ni, and Fe
(b) Vol. **117**, No. 2, p. 749–753 (1983).
See also *20*, p. *309*; 21.1.1, p. 382.

M. Čerňanský
Erratum to: Broadening and Shift of Diffraction Lines Due to Partial Coherence of X-Rays
(b) Vol. **117**, No. 2, p. 755 (1983).
See also *1.4*, p. *32*; *10.1*, p. *103*.

D. W. Field
A Reexamination of Diffraction Data on Zinc to Determine Anharmonic Components in the One-Particle-Potential Model
(a) Vol. **77**, No. 2, p. 733–740 (1983).
See also *6*, p. *74*.

H. Kido, T. Hoshikawa, M. Shimada, and M. Koizumi
Synthesis and Magnetic Properties of RCuSi (R = Y, Ce, Nd, Sm, Gd, Ho)
(a) Vol. **77**, No. 2, p. K121 – K123 (1983).
See also *18.2*, p. *284*; *18.4*, p. *296*; 21.4, p. 391.

H. K. Phan and T. H. Nguyen
Measurement of the Electrothermal Nonlinearity and Thermal Relaxation Times of Thin Metal Films
(a) Vol. **77**, No. 2, p. K135 – K138 (1983).
See also *8*, p. *88*; *14.1*, p. *208*.

A. K. Ahuja, S. Auluck, and S. K. Joshi
Density of States and Optical Conductivity of Ferromagnetic Nickel
(b) Vol. **118**, No. 1, p. 105 – 111 (1983).
See also *13.1*, p. *159*; 14.1, p. 208.

A. K. Ghosh, S. Chakraborty, and A. Manna
Density Functional Approach for Metals. The Matheiss Prescription
(b) Vol. **118**, No. 1, p. 373 – 379 (1983).
See also 4, p. 62; *13*, p. *150*; 21.1.1, p. 382; 21.6, p. 398.

V. V. Kelarev, V. V. Chuev, A. N. Pirogov, S. K. Sidorov, and V. N. Dvinyaninov
Magnetization of the Rare-Earth Sublattice in Intermetallic Compounds of the Type $R(Co_{1-x}Ni_x)_5$
(b) Vol. **118**, No. 1, p. K75 – K78 (1983).
See also *18*, p. *274*; 21.4, p. 392.

U. Messerschmidt, F. Appel, M. Bartsch, and R. Gerlach
Dislocation Motion in a Dispersion Hardened Al–Zn–Mg Alloy and Its Plastic Properties
(a) Vol. **78**, No. 1, p. 93 – 102 (1983).
See also *10.1*, p. *103*; 21, p. 356; 21.3, p. 389.

V. E. Antonov, T. E. Antonova, I. T. Belash, E. G. Ponyatovskii, amd V. I. Rashupkin
The Pd–Pt–H System: Phase Transformations at High Pressure and Superconductivity
(a) Vol. **78**, No. 1, p. 137 – 146 (1983).
See also *1.2*, p. *17*; *14.2*, p. *213*; 21, p. 356.

C. M. Mô and P. Moser
Hydrogen in $Fe_{40}Ni_{40}P_{14}B_6$ Amorphous Alloy Studied by Magnetic After Effect, Internal Friction, and Positron Annihilation
(a) Vol. **78**, No. 1, p. 201 – 210 (1983).
See also *2*, p. *49*; 10.1, p. 103; *18.2*, p. *285*; 21.1.1, p. 383.

R. Höhne, H. Börner, and M. Krötzsch
On the Annealing Behaviour and Crystallization of the Amorphous Alloy $Fe_{40}Ni_{40}P_{14}B_6$
(a) Vol. **78**, No. 1, p. 211 – 216 (1983).
See also *2*, p. *50*; *18.2*, p. *285*; 21.1.1, p. 383.

G. Dlubek, O. Brümmer, R. Krause, A. Baranowski, and B. Rozenfeld
A Positron Study of Guinier-Preston Zones and Precipitates in Al–Zn Alloys Having Mg or Ge Additions
(a) Vol. **78**, No. 1, p. 217 – 228 (1983).
See also *10.1*, p. *104*; 21, p. 356.

L. Z. MEZEY, O. NISHIKAWA, and J. GIBER
Monotonic and Alternating Surface Segregation Profiles in Binary Alloys
(a) Vol. 78, No. 1, p. 323–331 (1983).
See also *1.3*, p. *24*; *1.6*, p. *40*; 21.6, p. 398.

G. D. SUKHOMLIN and A. V. ANDREEVA
Particular Properties of $\Sigma = 3^n$ Boundaries in F.C.C. Polycrystals (I): Crystallographical Parameters and Grain Boundary Faceting during Annealing
(a) Vol. 78, No. 1, p. 333–341 (1983).
See also *1.4*, p. *32*; 21.1.1, p. 383.

M. OHTA, T. KANADANI, A. SAKAKIBARA, H. YAMADA, and M. YAMADA
Effects of Surface and Thickness of Specimens on the Aging in Al–Zn Alloys
(a) Vol. 78, No. 1, p. K23–K26 (1983).
See also *1.3*, p. *25*; 10.1, p. 104; 12.1, p. 138; 21, p. 357.

B. BABIĆ, M. STOJIĆ, and J. KONSTANTINOVIĆ
Electrical Properties of the Metastable $Mn_{1.11}Al_{0.89}$ Alloy
(a) Vol. 78, No. 1, p. K77–K80 (1983).
See also *13*, p. *150*; *14.1*, p. *208*; 21, p. 357.

B. JAGANNATHA RAO and A. R. CHETAL
On the Chemical Shifts of the K-Absorption Discontinuity of Copper in Some of Its Systems
(b) Vol. 118, No. 2, p. 551–555 (1983).
See also *13.1*, p. *160*; *20*, p. *309*; 22, p. 409.

A. K. AHUJA, S. AULUCK, and S. K. JOSHI
Application of Special-Directions Method to the Optical Properties of Copper and Silver
(b) Vol. 118, No. 2, p. 575–586 (1983).
See also *13.1*, p. *160*; 20.1, p. 323; 21.6, p. 398.

K. HRICOVÍNI, P. MIKUŠÍK, and J. KREMPASKÝ
Photoemission Studies of $Pd_{80-x}Si_{20}Co_x$ Amorphous Alloys
(b) Vol. 118, No. 2, p. 823–827 (1983).
See also *2*, p. *50*; 13.1, p. 161; *17*, p. *270*.

L. BOTTYÁN, D. L. BEKE, and K. TOMPA
Diffusion-Induced Quadrupole Relaxation of ^{27}Al Nuclei in Dilute Al–Ti, Al–Cr, Al–Mn, and Al–Cu Alloys at High Temperatures
(b) Vol. 118, No. 2, p. 835–842 (1983).
See also *9*, p. *95*; 10.1, p. 104; *19*, p. *303*; 21, p. 357.

S. LIPIŃSKI
Effects of Order on the Electronic Structure of Ni_3Fe
(b) Vol. 118, No. 2, p. 861–869 (1983).
See also *13*, p. *150*; 18.2, p. 285; 21.1.1, p. 383.

CH. OPITZ, H. MÜLLER, and H. MIESSNER
EHT Calculations for Interstitial Hydrogen Bonding in Nickel
(b) Vol. 118, No. 2, p. K85–K87 (1983).
See also *4*, p. *62*; *10.1*, p. *104*; 13.4, p. 184.

P. Muth
On the Resistivity Behaviour of Some Weak-Ferromagnetic Nickel-Based Alloys
(b) Vol. **118**, No. 2, p. K117 – K120 (1983).
See also *13.1*, p. *161*; *14.1*, p. *208*.

P. Muth
A Critical Concentration of Ni–Mn Alloys at about 6 at% Mn
(b) Vol. **118**, No. 2, p. K137 – K139 (1983).
See also *13.1*, p. *162*; *14.1*, p. *208*.

R. Z. Valiev, O. A. Kaibyshev, V. V. Astanin, and A. K. Emaletdinov
The Nature of Grain Boundary Sliding and the Superplastic Flow
(a) Vol. **78**, No. 2, p. 439 – 448 (1983).
See also *1.4*, p. *33*; *10.1*, p. *104*.

A. R. Ali and G. Said
On the Recovery and Annealing Kinetics of Cold-Worked Ni–4.5 at% Cu Alloy
(a) Vol. **78**, No. 2, p. 463 – 467 (1983).
See also *10.1*, p. *104*; 14.1, p. 209.

G. Wendrock
Study of the Behaviour of Al–Zn Alloys by SAXS in the Temperature Range of the Instable Solid Solution
(a) Vol. **78**, No. 2, p. 497 – 506 (1983).
See also *1.2*, p. *18*; 1.3, p. 25; 21, p. 357.

P. J. Ratcliffe and R. A. Collins
The Influence of Ion Implantation on the Thermal Oxidation of Copper
(a) Vol. **78**, No. 2, p. 547 – 553 (1983).
See also *1.6*, p. *40*; *11*, p. *130*.

E. Nembach
Precipitation Hardening Caused by a Difference in Shear Modulus between Particle and Matrix
(a) Vol. **78**, No. 2, p. 571 – 581 (1983).
See also *10.1*, p. *105*; 12.1, p. 139.

E. Bonetti, E. Evangelista, A. Cavallini, and C. Patuelli
Internal Friction Peaks and Structural Relaxation on Some Electroless Deposited Amorphous Alloys
(a) Vol. **78**, No. 2, p. 615 – 621 (1983).
See also *2*, p. *50*; *12.1*, p. *139*.

V. Navrátil, M. Hamerský, P. Lukáč, V. P. Soldatov, and V. I. Startsev
Solid Solution Hardening of Cd–Zn Single Crystals at 1.5, 2.8, and 4.2 K
(a) Vol. **78**, No. 2, p. 633 – 638 (1983).
See also *10.1*, p. *105*; *12.1*, p. *139*; 21, p. 358.

I. Lupsa and E. Burzo
On the Resistive Anomaly in NiCuSi and NiCuZn Alloys near the Curie Temperature
(a) Vol. **78**, No. 2, p. K155 – K158 (1983).
See also *14.1*, p. *209*; 18.2, p. 285.

E. König
Variation of the Magnetic Properties of $SmCo_5$ Powders by Water Milling
(a) Vol. **78**, No. 2, p. K159 – K160 (1983).
See also *18.2*, p. *286*; 21.4, p. 392.

S. F. Kolodeznaya and E. V. Kozlov
Calculation of the Properties and the Structure of Ordering Alloys of the Cu–Au System
(b) Vol. **119**, No. 1, p. 131 – 139 (1993).
See also *1.3*, p. *25*; 13.1, p. 162; 21.6, p. 398.

Yu. S. Stark, A. S. Steinberg, V. M. Vasilev, and V. I. Dimitrov
Monte Carlo Determination of Long-Range Pair Interaction Energies from Diffuse Scattering of X-Rays by Alloys
(b) Vol. **119**, No. 1, p. 147 – 151 (1983).
See also *1.3*, p. *25*; *4*, p. *63*; 21.6, p. 398.

H.-J. Rockosch and Ch. Herzig
Isotope Effect of Manganese and Simultaneous Impurity Diffusion of Chromium and Zinc in Copper
(b) Vol. **119**, No. 1, p. 199 – 207 (1983).
See also *9*, p. *95*; 10.1, p. 105.

I. Mertig, E. Mrosan, R. Zeller, and P. H. Dederichs
Electronic Properties of Dilute Copper Alloys (III): Influence on the de Haas-van Alphen Effect
(b) Vol. **119**, No. 1, p. 251 – 260 (1983).
See also *13.4*, p. *185*.

S. P. Singh and R. P. S. Rathorne
Van der Waals' Forces in Face Centred Cubic Cobalt
(b) Vol. **119**, No. 1, p. 343 – 348 (1983).
See also *4*, p. *63*; 6, p. 77; 12.1, p. 139.

P. M. Boerrigter, A. Lodder, and J. Molenaar
Calculated Residual Resistivities in Al and Cu Alloys
(b) Vol. **119**, No. 1, p. K91 – K96 (1983).
See also *14.1*, p. *209*; 21, p. 358.

V. V. Kelarev, V. V. Chuev, A. N. Pirogov, and S. K. Sidorov
Anisotropy and Exchange Effects in Heavy Rare-Earth–Cobalt Compounds of the RCo_5 Type
(a) Vol. **79**, No. 1, p. 57 – 66 (1983).
See also *1.3*, p. *25*; *18.3*, p. *293*; 21.4, p. 392.

L. V. Nomerovannaya and V. A. Rassokhin
Anomalies in the Optical and Electrical Properties of Cr–Al Alloys
(a) Vol. **79**, No. 1, p. 87 – 93 (1983).
See also *14.1*, p. *209*; 20.1, p. 326; 21, p. 359.

W. BECK and H. KORNMÜLLER
On the Coercive Field of the Nearly Non-Magnetostrictive Amorphous Ferromagnetic Alloy $Co_{58}Ni_{10}Fe_5Si_{11}B_{16}$
(a) Vol. **79**, No. 1, p. 109 – 114 (1983).
See also 2, p. 51; 10.1, p. 105; *18.2*, p. *286*; 21.1.1, p. 383.

R. RAMLAU and H. LÖFFLER
On the Size-Shape Relation of Guinier-Preston Zones and α'_R-Precipitates in an Al–6.8 at% Zn Alloy
(a) Vol. **79**, No. 1, p. 141 – 148 (1983).
See also *1.3*, p. *26*; *10.1*, p. *105*; 21, p. 359.

C. F. NILSEN and K. N. SUBRAMANIAN
Deformation of α–β Brass with Oriented and Equiaxed Duplex Boundaries
(a) Vol. **79**, No. 1, p. 281 – 292 (1983).
See also 1.4, p. 33; *12.1*, p. *140*.

Y. D. YAO, S. ARAJS, and E. E. ANDERSON
Deviations from Matthiessen's Rule in the Nickel–Chromium System
(a) Vol. **79**, No. 1, p. 327 – 330 (1983).
See also *13.4*, p. *186*; *14.1*, p. *209*; 18.2, p. 286.

V. G. CHUDINOV, V. P. GOGOLIN, B. N. GOSHCHITSKII, and V. I. PROTASOV
Simulation of Collision Cascades in Intermetallic V_2Hf Compounds
(a) Vol. **79**, No. 1, p. K57 – K60 (1983).
See also *11*, p. *132*; 14.2, p. 213; 21, p. 359.

G. POMPE, M. GAAFAR, P. BÜTTNER, and TH. FRANCKE
Low Temperature Thermal Conductivity of Amorphous (Fe, Ni, Co) (P, B, Si) Alloys and Their Change by Heat Treatment
(b) Vol. **119**, No. 2, p. 579 – 587 (1983).
See also *2*, p. *52*; *8*, p. *90*; 21.1.1, p. 384.

D. J. GONZÁLEZ, M. L. ZORITA, J. A. ALONSO, and M. P. IÑIGUEZ
Density Functional Pseudopotential Calculation of the Cohesive Properties of Disordered Solid Alloys of Nontransition Metals. Application to Zn–Cd
(b) Vol. **119**, No. 2, p. 589 – 594 (1983).
See also *1.3*, p. *26*; *13*, p. *151*; 21, p. 359.

S. ENGELS and W. MÖRKE
Silica Supported Nickel Clusters: The Particle Size Dependence of FMR Intensity and Magnetization
(b) Vol. **119**, No. 2, p. K109 – K112 (1983).
See also *18.2*, p. *287*; *19*, p. *304*.

H.-G. FRITSCHE
The Electronic Charge Distribution of Hydrogen Absorbed in Niobium – a Comparison of SW-Xα Cluster Calculations and Neutron Diffraction Experiments
(b) Vol. **119**, No. 2, p. K127 – K130 (1983).
See also *13*, p. *151*.

E. M. Gololobov, E. L. Mager, Z. V. Mezhevich, and L. K. Pan
Elastic Properties and Root-Mean-Square Dynamic Displacement of Niobium Atoms at High Pressures
(b) Vol. **119**, No. 2, p. K139–K142 (1983).
See also *7*, p. *83*; *12.1*, p. *140*.

W. Beck and H. Kronmüller
The Influence of Surface Roughness on the Coercive Field of the Nearly Non-Magnetostrictive Amorphous Ferromagnetic Alloy $Co_{58}Ni_{10}Fe_5Si_{11}B_{16}$
(a) Vol. **79**, No. 2, p. 389–394 (1983).
See also 1.6, p. 41; 2, p. 53; *18.2*, p. *287*; 21.1.1, p. 384.

V. I. Stashenko, O. A. Troitskii, and V. I. Spitsyn
Action of Current Pulses on Zinc Single Crystals during Creep
(a) Vol. **79**, No. 2, p. 549–557 (1983).
See also *10.1*, p. *106*.

J. Filipecki, B. Całusiński, and Z. Mandecki
Investigation of Electron Phenomena in the Metallic Glasses $Fe_xNi_{78-x}B_{10}Si_{12}$ with the Use of Positron Annihilation Methods before and after Their Crystallization
(a) Vol. **79**, No. 2, p. 585–591 (1983).
See also *2*, p. *53*; 13, p. 152; 21.1.1, p. 384.

P. Rennert, G. Carl, and W. Hergert
Spin-Polarized Compton Profile of Nickel
(b) Vol. **120**, No. 1, p. 273–281 (1983).
See also *13.1*, p. *164*; 13.2, p. 166; 18.2, p. 287; *20*, p. *311*.

C. Regnaut, E. Fusco, and J. P. Badiali
Structure and Thermodynamics of the Liquid Noble Metals from the Generalized OPW and Resonant Model Approaches
(b) Vol. **120**, No. 1, p. 373–382 (1983).
See also *2*, p. *54*; 8, p. 91; 13.1, p. 164; 21.6, p. 399.

M. N. Ghatikar
Chemical Transferability of Phase Shifts in Extended X-Ray Absorption Fine Structure
(b) Vol. **120**, No. 1, p. 445–450 (1983).
See also *20*, p. *311*; 21.4, p. 393; 22.5, p. 475; 22.6, p. 497; 22.7, p. 499.

N. G. Alexandropoulos, G. Boulakis, D. Papadimitriou, and I. Theodoridou
A Note on X-Ray Raman Scattering from Atomic Hydrogen in $TiH_{1.9}$
(b) Vol. **120**, No. 1, p. K19–K22 (1983).
See also *10.1*, p. *106*; *20*, p. *311*.

K. H. Georgy and R. Kamel
The Transient Creep of Cobalt near the Temperature of Transformation
(a) Vol. **80**, No. 1, p. 95–100 (1983).
See also *10.1*, p. *106*.

G. L. Whittle and A. M. Stewart
Mössbauer Effect Study of $Fe_{85-x}Ni_xB_{15}$ Metallic Glasses
(a) Vol. **80**, No. 1, p. 101 – 107 (1983).
See also 2, p. 54; *5*, p. *68*; 18.2, p. 288; 21.1.1, p. 385.

M. Vázquez, W. Fernengel, and H. Kronmüller
The Effect of Tensile Stresses on the Magnetic Properties of $Co_{58}Fe_5Ni_{10}Si_{11}B_{16}$ Amorphous Alloys
(a) Vol. **80**, No. 1, p. 195 – 204 (1983).
See also 2, p. 54; 12.1, p. 140; *18.2*, p. *288*; 21.1.1, p. 385.

U. Dahmen and K. H. Westmacott
Ledge Structure and the Mechanism of θ′ Precipitate Growth in Al–Cu
(a) Vol. **80**, No. 1, p. 249 – 262 (1983).
See also *3*, p. *59*; 10.1, p. 107; 21, p. 361.

C. H. Ling
Resistivity Change in Ti/Au Bimetal Films Annealed in the Temperature Range 300 to 400 °C
(a) Vol. **80**, No. 1, p. K59 – K62 (1983).
See also *9*, p. *96*; *14.3.2*, p. *235*; 21.6, p. 399.

C. Colinet and A. Pasturel
A Thermodynamic Study of Cerium Behaviour in Hexagonal $CeNi_5$ Compound
(a) Vol. **80**, No. 1, p. K75 – K79 (1983).
See also *8*, p. *91*; 21.4, p. 393.

Y. Shimada
Fabrication and Some Magnetization Characteristics of Amorphous Cores with Multilayered Structure
(a) Vol. **80**, No. 1, p. K87 – K91 (1983).
See also *18.2*, p. *288*.

H. Fehske
Spin Fluctuations in Alloys with Random Transfer
(b) Vol. **120**, No. 2, p. 611 – 620 (1983).
See also *13*, p. *153*; 14, p. 204; 18.1, p. 279; 18.2, p. 288.

S. V. Tsivinskii
The Theory of Electric Resistance of Pure Metals and Metals with a Given Concentration of Impurities, Vacancies, and Dislocations
(a) Vol. **80**, No. 2, p. 439 – 446 (1983).
See also 10.1, p. 107; *14.1*, p. *210*; 21, p. 362; 21.1.1, p. 385; 21.2, p. 388; 21.6, p. 399.

V. I. Motorin
Vitrification Kinetics of Pure Metals
(a) Vol. **80**, No. 2, p. 447 – 455 (1983).
See also *2*, p. *55*; 21, p. 362; 21.6, p. 399.

M. Vázquez, W. Fernengel, and H. Kronmüller
Influence of Applied Stresses on the Magnetic Properties of Annealed $Co_{58}Fe_5Ni_{10}Si_{11}B_{16}$ Amorphous Alloy
(a) Vol. **80**, No. 2, p. 513–518 (1983).
See also 2, p. 55; 12.1, p. 140; *18.2*, p. *289*; 21.1.1, p. 385.

M. Ellner, M. El-Boragy, and B. Predel
On the Formation of Palladium-Rich Metallic Glasses in Binary and Ternary Systems with Copper, Silver, and Gold as One Component and Phosphorus and Arsenic as Further Component
(a) Vol. **80**, No. 2, p. 519–533 (1983).
See also *2*, p. *55*; 13.1, p. 165; 21.6, p. 399.

J. Teuho
A Study of Non-Periodic Antiphase Boundaries in Long-Period Ordered $Au_3Zn(H)$ (II): Dissociated Antiphase Boundaries
(a) Vol. **80**, No. 2, p. 589–600 (1983).
See also *1.3*, p. *27*; *1.4*, p. *34*; 21.6, p. 400.

H. Kido, T. Hoshikawa, M. Shimada, and M. Koizumi
Preparation and Magnetic and Electrical Properties of RZnSi
(a) Vol. **80**, No. 2, p. 601–605 (1983).
See also *1.3*, p. *27*; 14.1, p. 211; *18.1*, p. *279*; 21.4, p. 394.

C. J. Rossouw
A γ- to α-Phase Transition Induced by Heavy Ion Bombardment of a CuAl Alloy
(a) Vol. **80**, No. 2, p. 631–636 (1983).
See also *1.2*, p. *20*; 11, p. 135; 21, p. 362.

M. Jurczyk
Magnetic Properties of Substituted $Pr_2(Co_{1-x}M_x)_{17}$ Compounds (M = Fe, Mn, and Cr)
(a) Vol. **80**, No. 2, p. 657–662 (1983).
See also *1.3*, p. *28*; *18.2*, p. *289*; 21.4, p. 394.

L. Amaral, F. P. Livi, and A. A. Gomes
Remarks on Alloying-Induced Lattice Parameter Changes in Intermetallic Compounds
(a) Vol. **80**, No. 2, p. 669–677 (1983).
See also *1.3*, p. *28*; 21.1.1, p. 385; 21.4, p. 394.

M. Z. Butt, P. Feltham, and I. M. Ghauri
Stress-Equivalence of Solid Solution Hardening in Concentrated Ag–Al and Ag–Zn Alloys
(a) Vol. **80**, No. 2, p. K125–K128 (1983).
See also *10.1*, p. *107*; 21, p. 362; 21.6, p. 400.

A. San Martin, K. Kleinstück, N. H. Quyen, and P. Paufler
Influence of a Direct Current and a Temperature Gradient on the Creep Rate in V_3Si
(a) Vol. **80**, No. 2, p. K171–K174 (1983).
See also *12.1*, p. *140*; *14.1*, p. *211*.

21.1.1 Fe and Fe Alloys

Review Article

K. H. Fischer
Spin Glasses (I)
(b) Vol. **116**, No. 2, p. 357–414 (1983).
See also 2, p. 43; 8, p. 85; 13, p. 146; 14.1, p. 206; *18.2*, p. *280*; *18.4*, p. *295*; 21.1, p. 363.

Original Papers and Short Notes

T. A. Jenkins and S. Arajs
Electrical Resistivity of Some Directionally-Solidified Aluminium-Based Eutectic Alloys
(a) Vol. **75**, No. 1, p. 73–82 (1983).
See also *14.1*, p. *206*; 21, p. 349; 21.1, p. 363.

F. Walz, H. J. Blythe, and F. Dworschak
Annealing Behaviour of Dilute FeCu, FeMn, and FeTi Alloys in the Temperature Range up to Stage III, Following Low-Temperature Electron Irradiation
(a) Vol. **75**, No. 1, p. 235–242 (1983).
See also *10.1*, p. *99*; 11, p. 122; *18.2*, p. *280*; 21.1, p. 364.

T. M. Robinson
Xenon Ion Irradiation of α-Fe
(a) Vol. **75**, No. 1, p. 243–249 (1983).
See also *11*, p. *122*.

S. Tyagi
Ferromagnetic Resonance in Single Particles of Amorphous $Fe_{81.5}B_{14.5}Si_4$
(a) Vol. **75**, No. 1, p. K29–K32 (1983).
See also *2*, p. *43*; *18.2*, p. *281*; 19, p. 299.

W. Zarek, A. Kasprzyk, and E. Kozubek
Magnetic and ESR Investigations of $Sm_6(Mn_{1-x}Fe_x)_{23}$ Compounds for $0 < x < 0.55$ in the Paramagnetic Range
(a) Vol. **75**, No. 2, p. 359–366 (1983).
See also *18.1*, p. *278*; *19*, p. *300*; 21.1, p. 364.

H. O. Martikainen, M. A. Korhonen, and V. K. Lindroos
Determination of Grain Boundary Rotation Parameters from Kikuchi Patterns (I): A Simplified Method for Determining the Accurate Orientation Relationship of Two Grains
(a) Vol. **75**, No. 2, p. 559–566 (1983).
See also *1.4*, p. *29*.

K. H. J. Buschow. J. H. N. van Vucht, P. G. van Engen, D. B. de Mooij, and A. M. van der Kraan
Structure, Magnetic and Magneto-Optical Properties, and ^{57}Fe Mössbauer Effect in a New Variety of Heusler Alloy: PtFeSb
(a) Vol. **75**, No. 2, p. 617–623 (1983).
See also *1.3*, p. *22*; 5, p. 65; *18.2*, p. *281*; 20.1, p. 314.

P. Delobelle and C. Oytana
Creep at Low and Average Stress of Several Highly Concentrated Solid Solutions of the Cubic Centered Type
(a) Vol. **75**, No. 2, p. 625–634 (1983).
See also *10.1*, p. *100*; 12.1, p. 137; 21.1, p. 365.

K. Masuda
On the Interaction between a Vacancy and an Interstitial Impurity Atom in B.C.C. Transition Metals
(b) Vol. **116**, No. 1, p. 9–16 (1983).
See also *10.1*, p. *101*; 21.1, p. 365.

A. V. Zaborov and M. V. Medvedev
A Binary Ferromagnetic Alloy with Magnetic Moments Dependent on Local Environment (I): The Jaccarino-Walker Model in the Effective Field Approximation
(b) Vol. **116**, No. 1, p. 227–238 (1983).
See also *18.2*, p. *281*; 21.1, p. 366.

J. Chojcan, M. Szuszkiewicz, S. Szuszkiewicz, and B. Rozenfeld
Localization of Electron in Fe–Co Alloys
(b) Vol. **116**, No. 1, p. K57–K60 (1983).
See also *13*, p. *148*; 21.1, p. 366.

V. M. Pan, I. Ya. Dekhtyar, M. E. Osinovskii, S. E. Litvin, B. G. Nikitin, M. M. Nishenko, and M. P. Voronko
Thermodynamical and Structural Properties of Amorphous $Fe_{80}B_{20}$ and $Fe_{80}P_{13}C_7$ Alloys
(a) Vol. **76**, No. 1, p. 197–206 (1983).
See also *2*, p. *45*; 5, p. 65; 8, p. 87; 14.1, p. 206; *18.2.1*, p. *290*.

F. Walz, H. J. Blythe, and F. Dworschak
Annealing Behaviour of Dilute $\underline{Fe}Ti$, $\underline{Fe}Cu$, and $\underline{Fe}Mn$ Alloys in the Temperature Range above Stage III Following Low-Temperature Electron Irradiation
(a) Vol. **76**, No. 1, p. 267–276 (1983).
See also *10.1*, p. *101*; 11, p. 125; *18.2*, p. *281*.

W. Matz, B. Kunsch, and E. Wieser
The Structure of Liquid $Fe_{75}B_{25}$ by Neutron Scattering
(a) Vol. **76**, No. 1, p. K17–K20 (1983).
See also *2*, p. *46*.

J. Weis and I. Červeň
Structural Changes in Amorphous $Fe_{83}B_{17}$ by Neutron Irradiation
(a) Vol. **76**, No. 1, p. K49–K51 (1983).
See also *2*, p. *46*; *11*, p. *126*.

P. Czarnecki and G. Fritzkowski
Transverse Magnetoresistance of Amorphous $Fe_{40-x}Ni_{40}Cr_xB_{20}$ Alloys in Magnetic Fields up to 20 T
(a) Vol. **76**, No. 1, p. K93–K96 (1983).
See also *2*, p. *46*; *14.1*, p. *207*; 18.2, p. 282; 21.1, p. 366.

A. V. Zaborov and M. V. Medvedev
A Binary Ferromagnetic Alloy with Magnetic Moments Dependent on Local Environment (II): Neutron Magnetic Elastic Scattering and Hyperfine Field Distribution
(b) Vol. **116**, No. 2, p. 511–523 (1983).
See also *18.2*, p. *282*; 21.1, p. 367.

Ch. Müller, W. Blau, and P. Ziesche
Local Partial DOS and Experimental Al K Spectra of Transition Metal Aluminides. Charge Transfer Trend
(b) Vol. **116**, No. 2, p. 561–573 (1983).
See also *13.1*, p. *157*; 20, p. 308; 21.1, p. 367.

S. N. Kaul
Two-Current Conduction Model and Anisotropic Magnetoresistance in Transition Metal-Metalloid Glasses
(b) Vol. **116**, No. 2, p. K99–K104 (1983).
See also *2*, p. *46*; *14.1*, p. *207*; 21.1, p. 367.

P. Duhaj and F. Hanic
Electron Microscopy and Electron Diffraction Study of Crystallization of Metastable Phases in Amorphous Fe–B Alloys
(a) Vol. **76**, No. 2, p. 467–477 (1983).
See also *1.2*, p. *15*; 1.3, p. 23; *2*, p. *47*.

M. E. Gurevich, L. N. Larikov, V. G. Novitskii, A. E. Pogorelov, and A. F. Zhuravlev
Laser-Stimulated Mass Transfer in Metals
(a) Vol. **76**, No. 2, p. 479–484 (1983).
See also *9*, p. *94*; *11*, p. *126*; 21.1, p. 367.

H. Börner and R. Höhne
Study of Magnetic After-Effects in the Amorphous Alloy $Fe_{40}Ni_{40}P_{14}B_6$
(a) Vol. **76**, No. 2, p. 499–503 (1983).
See also *2*, p. *47*; *18.2*, p. *282*; 21.1, p. 367.

K. H. J. Buschow and P. G. van Engen
Note on the Magnetic and Magneto-Optical Properties of Ni_2In Type 3d Transition Metal Compounds
(a) Vol. **76**, No. 2, p. 615–620 (1983).
See also 1.1, p. 11; *18.2*, p. *282*; *20*, p. *308*; 21, p. 353; 21.1, p. 367.

H. O. Martikainen, M. A. Korhonen, and V. K. Lindroos
Determination of Grain Boundary Rotation Parameters from Kikuchi Patterns (II): Experimental Optimization of the Analysis of Grain Boundary Orientation Relationships and Verification of the Accuracy of the Proposed Method
(a) Vol. **76**, No. 2, p. 709–719 (1983).
See also *1.4*, p. *30*.

K. Sumiyama, M. Shiga, and Y. Nakamura
Characteristic Martensitic Transformation of the Fe–Pt Disordered Alloys around the γ–α Phase Boundary
(a) Vol. **76**, No. 2, p. 747–751 (1983).
See also *1.2*, p. *16*; 18.2, p. 282.

G. Knuyt, L. M. Stals, L. de Schepper, and H. van Swygenhoven
The Inhomogeneity Effect in the Vacancy–Dislocation Interaction for α-Iron
(a) Vol. **76**, No. 2, p. K175 – K178 (1983).
See also *10.1*, p. *102*.

J. Schneider, J. Kamarád, Z. Arnold and M. Bitterlich
The Pressure Dependence of the Curie Temperature of Amorphous Fe–P–Ga and Fe–P–Ge Alloys
(a) Vol. **76**, No. 2, p. K183 – K188 (1983).
See also *2*, p. *47*; *18.2*, p. *283*.

A. L. Ageev, Yu. A. Babanov, V. V. Vasin, N. V. Ershov, and A. V. Serikov
Amorphous Problem in EXAFS Data Analysis
(b) Vol. **117**, No. 1, p. 345 – 350 (1983).
See also *2*, p. *48*; *20*, p. *308*.

A. A. Alekhin
The Bloch Lines Dynamics in Permalloy Films
(a) Vol. **77**, No. 1, p. 51 – 61 (1983).
See also *18.2.1*, p. *290*; 21.1, p. 368.

M. Szuszkiewicz, J. Chojcan, and A. Ostrasz
Investigation of the Electronic Structure of V–Fe–H Alloys by Positron Annihilation
(b) Vol. **117**, No. 2, p. 645 – 651 (1983).
See also *13*, p. *149*; 21.1, p. 370.

H. Grimm and H. Kronmüller
Investigation of Structural Defects in the Amorphous Ferromagnetic Alloy $Fe_{40}Ni_{40}P_{14}B_6$
(b) Vol. **117**, No. 2, p. 663 – 674 (1983).
See also 2, p. 48; *10.1*, p. *103*; *18.2*, p. *284*; 21.1, p. 370.

N. V. Ershov, Yu. A. Babanov, and V. R. Galakhov
EXAFS Study of Crystalline Cu, Ni, and Fe
(b) Vol. **117**, No. 2, p. 749 – 753 (1983).
See also *20*, p. *309*; 21.1, p. 370.

H. Ratajczak, P. Slančo, T. Tima, A. Zentko, E. Demkovicsová, and M. Konč
Electrical Resistivity and Hall Effect in Thin Amorphous UFe Films
(a) Vol. **77**, No. 2, p. 785 – 791 (1983).
See also *2*, p. *49*; *14.1*, p. *208*; 18.2, p. 284.

V. S. Patil, R. G. Pillay, P. N. Tandon, and H. G. Devare
Hyperfine Field Investigations in Some Ru and Rh Based Heusler Alloys
(b) Vol. **118**, No. 1, p. 57 – 61 (1983).
See also *5*, p. *67*; 18.2, p. 284; 21, p. 355.

A. K. Ghosh, S. Chakraborty, and A. Manna
Density Functional Approach for Metals. The Matheiss Prescription
(b) Vol. **118**, No. 1, p. 373 – 379 (1983).
See also 4, p. 62; *13*, p. *150*; 21.1, p. 371; 21.6, p. 398.

C. M. Mô and P. Moser
Hydrogen in $Fe_{40}Ni_{40}P_{14}B_6$ Amorphous Alloy Studied by Magnetic After Effect, Internal Friction, and Positron Annihilation
(a) Vol. **78**, No. 1, p. 201–210 (1983).
See also *2*, p. *49*; 10.1, p. 103; *18.2*, p. *285*; 21.1, p. 371.

R. Höhne, H. Börner, and M. Krötzsch
On the Annealing Behaviour and Crystallization of the Amorphous Alloy $Fe_{40}Ni_{40}P_{14}B_6$
(a) Vol. **78**, No. 1, p. 211–216 (1983).
See also *2*, p. *50*; *18.2*, p. *285*; 21.1, p. 371.

G. D. Sukhomlin and A. V. Andreeva
Particular Properties of $\Sigma = 3^n$ Boundaries in F.C.C. Polycrystals (I): Crystallographical Parameters and Grain Boundary Faceting During Annealing
(a) Vol. **78**, No. 1, p. 333–341 (1983).
See also *1.4*, p. *32*; 21.1, p. 372.

H. Kronmüller
The Rôle of Two-Level Systems in Amorphous Metallic Alloys (I): Induced Anisotropy in Amorphous Ferromagnets
(b) Vol. **118**, No. 2, p. 661–673 (1983).
See also *2*, p. *50*; *18.2*, p. *285*.

S. Lipiński
Effects of Order on the Electronic Structure of Ni_3Fe
(b) Vol. **118**, No. 2, p. 861–869 (1983).
See also *13*, p. *150*; 18.2, p. 285; 21.1, p. 372.

D. Fruchart, B. Malaman, G. Le Caer, and B. Roques
Magnetic Studies of Monoclinic FeGe
(a) Vol. **78**, No. 2, p. 555–569 (1983).
See also 5, p. 67; *18.4*, p. *296*.

Y. Muraoka, M. Shiga, and Y. Nakamura
Magnetic Properties and Thermal Expansion of $(Zr_{1-x}Ta_x)Fe_2$
(a) Vol. **78**, No. 2, p. 717–722 (1983).
See also *5*, p. *67*; 8, p. 89; *18.2*, p. *285*; 18.4, p. 297.

J. M. Szlanta
Temperature Shift of the Maximum of the Critically Scattered Neutrons
(b) Vol. **119**, No. 1, p. K23–K25 (1983).
See also *18.2*, p. *286*.

W. Beck and H. Kronmüller
On the Coercive Field of the Nearly Non-Magnetostrictive Amorphous Ferromagnetic Alloy $Co_{58}Ni_{10}Fe_5Si_{11}B_{16}$
(a) Vol. **79**, No. 1, p. 109–114 (1983).
See also 2, p. 51; 10.1, p. 105; *18.2*, p. *286*; 21.1, p. 375.

O. SCHNEEWEISS, T. ZEMČÍK, T. ŽÁK, and S. MAGER
Atomic Structure and Magnetic Properties of the Pseudobinary Alloys Fe_3(Al, Si)
(a) Vol. **79**, No. 1, p. 125–129 (1983).
See also *1.3*, p. *26*; *5*, p. *67*; 14.1, p. 209; 18.2, p. 286.

G. POMPE, M. GAAFAR, P. BÜTTNER, and TH. FRANCKE
Low Temperature Thermal Conductivity of Amorphous (Fe, Ni, Co) (P, B, Si) Alloys and Their Change by Heat Treatment
(b) Vol. **119**, No. 2, p. 579–587 (1983).
See also *2*, p. *52*; *8*, p. *90*; 21.1, p. 375.

Z. Z. AKSELROD, M. BUDZYNSKI, B. A. KOMISSAROVA, L. N. KRYUKOVA, G. K. RYASNY, and A. A. SOROKIN
Hyperfine Magnetic Field at ^{181}Ta in YFe_2
(b) Vol. **119**, No. 2, p. 667–673 (1983).
See also *18.2*, p. *287*; 21.4, p. 392.

W. BECK and H. KRONMÜLLER
The Influence of Surface Roughness on the Coercive Field of the Nearly Non-Magnetostrictive Amorphous Ferromagnetic Alloy $Co_{58}Ni_{10}Fe_5Si_{11}B_{16}$
(a) Vol. **79**, No. 2, p. 389–394 (1983).
See also 1.6, p. 41; 2, p. 53; *18.2*, p. *287*; 21.1, p. 376.

M. KATO
Activation Energy for Dislocation Motion in a One-Dimensional Periodic Field
(a) Vol. **79**, No. 2, p. 517–522 (1983).
See also *10.1*, p. *106*; 21.2, p. 388.

J. ŠTERNBERK, E. KRATOCHVÍLOVÁ, J. HŘEBÍK, and A. GEMPERLE
Coercivity and Microstructure of Low-Alloy Cr–Mo Steel
(a) Vol. **79**, No. 2, p. 523–529 (1983).
See also *1.4*, p. *34*; *18.2*, p. *287*.

J. FILIPECKI, B. CAŁUSIŃSKI, and Z. MANDECKI
Investigation of Electron Phenomena in the Metallic Glasses $Fe_xNi_{78-x}B_{10}Si_{12}$ with the Use of Positron Annihilation Methods before and after Their Crystallization
(a) Vol. **79**, No. 2, p. 585–591 (1983).
See also *2*, p. *53*; 13, p. 152; 21.1, p. 376.

J. WEIS, D. BARANČOK, and I. ČERVEŇ
RDF Analysis of Structural Evolution during the Primary Crystallization of the $Fe_{83}B_{17}$ Glassy Alloy
(a) Vol. **79**, No. 2, p. K117–K122 (1983).
See also *2*, p. *53*.

R. PFRANGER, D. PŁUSA, and B. WYSŁOCKI
Magnetic Domain Structure in Polycrystalline $DyFe_2$ Compounds
(a) Vol. **80**, No. 1, p. 73–78 (1983).
See also *18.3*, p. *294*; 21.4, p. 393.

G. L. WHITTLE and A. M. STEWART
Mössbauer Effect Study of $Fe_{85-x}Ni_xB_{15}$ Metallic Glasses
(a) Vol. **80**, No. 1, p. 101–107 (1983).
See also 2, p. 54; *5*, p. *68*; 18.2, p. 288; 21.1, p. 377.

M. VÁZQUEZ, W. FERNENGEL, and H. KRONMÜLLER
The Effect of Tensile Stresses on the Magnetic Properties of $Co_{58}Fe_5Ni_{10}Si_{11}B_{16}$ Amorphous Alloys
(a) Vol. **80**, No. 1, p. 195–204 (1983).
See also 2, p. 54; 12.1, p. 140; *18.2*, p. *288*; 21.1, p. 377.

B. RAUSCHENBACH and A. KOLITSCH
Formation of Compounds by Nitrogen Ion Implantation in Iron
(a) Vol. **80**, No. 1, p. 211–222 (1983).
See also *1.3*, p. *26*; *11*, p. *133*.

P. DUHAJ and P. ŠVEC
Transmission Electron Microscopy Study of Crystallization in Amorphous $Fe_{10}Ni_{40}B_{20}$ Alloy
(a) Vol. **80**, No. 1, p. 231–236 (1983).
See also *2*, p. *55*; *3*, p. *59*.

E. A. VASILEV and V. A. VIRCHENKO
Hyperfine Interactions of ^{57}Fe and ^{119}Sn in Iron Antimonide Base Alloys
(a) Vol. **80**, No. 1, p. K41–K44 (1983).
See also *5*, p. *68*.

S. V. TSIVINSKII
The Theory of Electric Resistance of Pure Metals and Metals with a Given Concentration of Impurities ,Vacancies, and Dislocations
(a) Vol. **80**, No. 2, p. 439–446 (1983).
See also 10.1, p. 107; *14.1*, p. *210*; 21, p. 362; 21.1, p. 377; 21.2, p. 388; 21.6, p. 399.

B. RAUSCHENBACH, A. KOLITSCH, and K. HOHMUTH
Iron Nitride Phases Formed by Nitrogen Ion Implantation and Thermal Treatment
(a) Vol. **80**, No. 2, p. 471–482 (1983).
See also *1.3*, p. *27*; 1.5, p. 38; *11*, p. *135*; 21, p. 362.

M. VÁZQUEZ, W. FERNENGEL, and H. KRONMÜLLER
Influence of Applied Stresses on the Magnetic Properties of Annealed $Co_{58}Fe_5Ni_{10}Si_{11}B_{16}$ Amorphous Alloy
(a) Vol. **80**, No. 2, p. 513–518 (1983).
See also 2, p. 55; 12.1, p. 140; *18.2*, p. *289*; 21.1, p. 378.

L. AMARAL, F. P. LIVI, and A. A. GOMES
Remarks on Alloying-Induced Lattice Parameter Changes in Intermetallic Compounds
(a) Vol. **80**, No. 2, p. 669–677 (1983).
See also *1.3*, p. *28*; 21.1, p. 378; 21.4, p. 394.

C. HIDALGO and N. DE DIEGO
Positron Trapping in the Alloy Fe–1.25 wt% C
(a) Vol. **80**, No. 2, p. K145–K148 (1983).
See also *1.4*, p. *34*; *10.1*, p. *107*.

21.2 Alkali Metals

Original Papers and Short Notes

T. Soma, H.-Matsuo Kagaya, and M. Nishigaki
Mean-Square Displacement and Lindeman's Criterion for Melting of Alkali Metals
(b) Vol. **115**, No. 1, p. 273–276 (1983).
See also *6*, p. *69*; 8, p. 85.

C. Malinowska-Adamska, M. Owsiejczyk, and L. Wojtczak
Thermodynamics of Strongly Anharmonic Crystals
(b) Vol. **115**, No. 2, p. 335–345 (1983).
See also *6*, p. *70*; 8, p. 85; 21, p. 350.

D. Sen, S. K. Sarkar, S. Sengupta, and D. Roy
Consistency Condition in the Energy Expression of the Pseudopotential Theory of Metals and Unified Study of Alkali Metals
(b) Vol. **115**, No. 2, p. 593–602 (1983).
See also *6*, p. *71*; *12.1*, p. *137*.

F. Abd El-Salam, A. I. Eatah, and A. Tawfik
Effect of Li Loss on Some Physical Properties of Al–Li Alloy
(a) Vol. **75**, No. 2, p. 379–384 (1983).
See also *1.3*, p. *21*; 10.1, p. 100; 12.1, p. 137; 14.1, p. 206; 21, p. 351.

T. Soma, H.-M. Kagaya, and Y. Kimura
Thermal Expansion Coefficient of Alkali Metals
(b) Vol. **116**, No. 1, p. 57–61 (1983).
See also *6*, p. *71*; 8, p. 86.

R. N. Joarder and R. V. Gopala Rao
Temperature Dependence of the Structure Factor of Liquid Metals
(b) Vol. **116**, No. 1, p. 299–306 (1983).
See also *2*, p. *45*; 4, p. 61; 9, p. 93; 21.1, p. 366; 21.6, p. 396.

T. Soma, H.-Matsuo Kagaya, and M. Nishigaki
Volume Dependence of the Melting Temperature for Alkali Metals with Debye's Model
(b) Vol. **116**, No. 2, p. 673–676 (1983).
See also *6*, p. *73*; 8, p. 87.

P. Beauchamp and J. Prevost
Influence of Interaction Range on Computed Peierls Stress and Dislocation Displacement in B.C.C. Lithium
(b) Vol. **117**, No. 2, p. 499–506 (1983).
See also *10.1*, p. *102*.

C. Malinowska-Adamska
Dynamics and Thermodynamics of Anharmonic Crystals near the Instability Point
(b) Vol. **118**, No. 2, p. 535–539 (1983).
See also *6*, p. *76*; 8, p. 89; 12.1, p. 139.

E. Canessa and J. Vignolo
Electrical Resistivities of the Liquid Alkali Metals
(b) Vol. **119**, No. 1, p. 429–436 (1983).
See also 2, p. 51; *14.1*, p. *209*.

G. P. Das and P. Chaddah
Correlation Energy from Compton Scattering and Positron Annihilation Data
(b) Vol. **119**, No. 2, p. 467–472 (1983).
See also *13*, p. *151*.

M. Kato
Activation Energy for Dislocation Motion in a One-Dimensional Periodic Field
(a) Vol. **79**, No. 2, p. 517–522 (1983).
See also *10.1*, p. *106*; 21.1.1, p. 384.

S. V. Tsivinskii
The Theory of Electric Resistance of Pure Metals and Metals with a Given Concentration of Impurities, Vacancies, and Dislocations
(a) Vol. **80**, No. 2, p. 439–446 (1983).
See also 10.1, p. 107; *14.1*, p. *210*; 21, p. 362; 21.1, p. 377; 21.1.1, p. 385; 21.6, p. 399.

21.3 Alkali-Earth Metals

Original Papers and Short Notes

P. RENNERT, G. CARL, and U.-H. GLÄSER
Theoretical Compton Profiles of Magnesium
(b) Vol. **115**, No. 1, p. 219–224 (1983).
See also *13.1*, p. *155*; 13.2, p. 166; *20*, p. *307*.

R. GERLACH and H. LÖFFLER
On the Structure of the Precipitation-Free Zone in Two Al–Zn–Mg Alloys after Various Heat Treatments
(a) Vol. **75**, No. 1, p. 169–178 (1983).
See also *10.1*, p. *99*; 21, p. 350; 21.1, p. 364.

U. MESSERSCHMIDT, F. APPEL, M. BARTSCH, and R. GERLACH
Dislocation Motion in a Dispersion Hardened Al–Zn–Mg Alloy and Its Plastic Properties
(a) Vol. **78**, No. 1, p. 93–102 (1983).
See also *10.1*, p. *103*; 21, p. 356; 21.1, p. 371.

21.4 Rare-Earth Metals

Review Article

V. S. Shpinel
Exchange Interaction and Hyperfine Fields at Nonmagnetic Impurities in Rare-Earth Metals
(b) Vol. **118**, No. 1, p. 11–24 (1983).
See also 5, p. 65; *18.2*, p. *280*; *18.4*, p. *295*.

Original Papers and Short Notes

F. T. Parker and H. Oesterreicher
Temperature Dependence of Intrinsic Magnetic Hardness in Some Rare-Earth Pseudobinary Compounds and Amorphous Materials
(a) Vol. **75**, No. 1, p. 273–282 (1983).
See also 2, p. 43; *18.2*, p. *280*; 21.1, p. 364.

A. S. Lileev and V. P. Menushenkov
The Crystal Structure and Coercive Force of $SmCo_5$ Sintered Permanent Magnets
(a) Vol. **75**, No. 1, p. 341–346 (1983).
See also *1.3*, p. *21*; *18.2*, p. *280*; 21.1, p. 364.

H. Müller, E. Hegenbarth, and S. Mahmoud
Thermal Conductivity of the Intermetallic Rare Earth Compound $PrAl_2$ in the Temperature Range between 4.2 and 100 K
(b) Vol. **115**, No. 2, p. K69–K73 (1983).
See also *8*, p. *85*; 21, p. 350.

A. Apostolov, H. Hristov, M. Mikhov, and V. Skumriev
Magnetic Properties of Bulk Amorphous Alloys Gd_4Co_3, Er_4Co_3, and Sm_4Co_3
(a) Vol. **75**, No. 2, p. 401–404 (1983).
See also *2*, p. *44*; *18.1*, p. *278*; 18.2, p. 281.

T. J. McKenna, S. J. Campbell, D. H. Chaplin, and G. V. H. Wilson
Temperature Modulation and Temperature Hysteresis Studies of Dysprosium
(a) Vol. **75**, No. 2, p. 421–432 (1983).
See also *18.2*, p. *281*; *18.4*, p. *295*.

R. Nagarajan, E. V. Sampathkumaran, R. Vijayaraghavan, and Bhaktdarshan
^{151}Eu Mössbauer Effect Measurements in Eu_4As_3
(a) Vol. **75**, No. 2, p. K149–K151 (1983).
See also *5*, p. *65*.

N. Gopi Krishna and D. B. Sirdeshmukh
X-Ray Determination of the Mean-Square Atomic Displacement and Associated Debye Temperature of Erbium
(b) Vol. **116**, No. 2, p. K105–K108 (1983).
See also *6*, p. *73*; *8*, p. *87*.

Z. Drzazga
Magnetic Anisotropy of $ErCo_{4.5}Cu$ Single Crystals
(a) Vol. **76**, No. 2, p. 647–649 (1983).
See also *18.3*, p. *291*.

S. I. Tyukhtenko, D. A. Stakhov, and G. I. Batalin
High-Temperature Paramagnetic Susceptibility of the Erbium–Germanium Alloys
(a) Vol. **76**, No. 2, p. K179–K181 (1983).
See also *18.1*, p. *278*.

D. Dădârlat, T. Petrişor, A. Giurgiu, and I. Pop
Localization Effect in CrRh Alloys
(b) Vol. **117**, No. 1, p. 155–159 (1983).
See also *18.4*, p. *296*; 21.1, p. 368.

P. C. M. Gubbens, W. Ras, A. M. van der Kraan, and K. H. J. Buschow
Magnetization and ^{161}Dy Mössbauer Effect Study of $DyMn_2$ and Dy_6Mn_{23} and Their Ternary Hydrides
(b) Vol. **117**, No. 1, p. 277–282 (1983).
See also *5*, p. *66*; *18.3*, p. *292*.

L. G. Khvostantsev and N. A. Nikolaev
Electroresistance of Cerium at Phase Transitions under Hydrostatic Pressure up to 8.0 GPa. *P–T* Diagram of the Cerium Phase Transitions
(a) Vol. **77**, No. 1, p. 161–166 (1983).
See also *1.2*, p. *16*; 12.1, p. 138; *14.1*, p. *207*.

V. Brédimas, C. Papatriantafyllou, and H. Gamari-Seale
On the Magnetic Structure of the Intermetallic Compounds Gd_2Al and Gd_2Au
(b) Vol. **117**, No. 2, p. 513–519 (1983).
See also *18.2*, p. *284*; 18.4, p. 296.

O. J. Żogał
Knight Shifts of Protons in Some Rare-Earth Dihydrides
(b) Vol. **117**, No. 2, p. 717–723 (1983).
See also 18.1, p. 278; *19*, p. *301*.

G. Theodorou
Critical Behaviour at the Incommensurate–Commensurate Transition of Ho
(b) Vol. **117**, No. 2, p. K67–K69 (1983).
See also *1.2*, p. *17*.

M. A. Angadi and P. V. Ashrit
The Effect of Deposition Parameters on the Electrical Properties of Thin Ytterbium Films
(a) Vol. **77**, No. 2, p. 685–692 (1983).
See also *14.1*, p. *207*.

H. Kido, T. Hoshikawa, M. Shimada, and M. Koizumi
Synthesis and Magnetic Properties of RCuSi (R = Y, Ce, Nd, Sm, Gd, Ho)
(a) Vol. **77**, No. 2, p. K121–K123 (1983).
See also *18.2*, p. *284*; *18.4*, p. *296*; 21.1, p. 371.

J. C. Upadhyaya and M. L. Sharma
Model Potential Calculation of Phonon Dispersion in Tb and Ho
(b) Vol. **118**, No. 1, p. 121–128 (1983).
See also *6*, p. *74*.

V. V. Kelarev, V. V. Chuev, A. N. Pirogov, S. K. Sidorov, and V. N. Dvinyaninov
Magnetization of the Rare-Earth Sublattice in Intermetallic Compounds of the Type $R(Co_{1-x}Ni_x)_5$
(b) Vol. **118**, No. 1, p. K75–K78 (1983).
See also *18*, p. *274*; 21.1, p. 371.

H. V. Sharma and I. Singh
Valence Fluctuations and Magnetic-Nonmagnetic Transitions in a Model for Intermediate-Valence Systems
(b) Vol. **118**, No. 2, p. 511–520 (1983).
See also 8, p. 89; *13*, p. *150*; 18.1, p. 278; *18.2*, p. *285*.

N. Shohoji
Statistical Thermodynamic Approach to Some Hyperstoichiometric Dihydrides of Lanthanides (La, Ce, Pr, Nd, Sm) and Actinides (Np, Pu, Am)
(b) Vol. **118**, No. 2, p. 811–822 (1983).
See also *4*, p. *62*; *8*, p. *89*; 10.1, p. 104; 21.5, p. 395.

E. König
Variation of the Magnetic Properties of $SmCo_5$ Powders by Water Milling
(a) Vol. **78**, No. 2, p. K159–K160 (1983).
See also *18.2*, p. *286*; 21.1, p. 374.

N. Shohoji
Statistical Thermodynamics of Hydrogen Solution in Some Lanthanides
(b) Vol. **119**, No. 1, p. K87–K90 (1983).
See also *1.3*, p. *25*; 4, p. 63.

V. V. Kelarev, V. V. Chuev, A. N. Pirogov, and S. K. Sidorov
Anisotropy and Exchange Effects in Heavy Rare-Earth–Cobalt Compounds of the RCo_5 Type
(a) Vol. **79**, No. 1, p. 57–66 (1983).
See also *1.3*, p. *25*; *18.3*, p. *293*; 21.1, p. 374.

P. A. Alekseev, E. A. Goremychkin, B. Lippold, E. Mühle, and I. P. Sadikov
Crystal Field Splitting in $NdAl_3$ Studied by Inelastic Neutron Scattering
(b) Vol. **119**, No. 2, p. 651–658 (1983).
See also *13.4*, p. *187*.

Z. Z. Akselrod M. Budzynski, B. A. Komissarova, L. N. Kryukova, G. K. Ryasny, and A. A. Sorokin
Hyperfine Magnetic Field at ^{181}Ta in YFe_2
(b) Vol. **119**, No. 2, p. 667–673 (1983).
See also *18.2*, p. *287*; 21.1.1, p. 384.

A. V. Soldatov, A. N. Gusatinskii, M. G. Karin, and O. A. Sadovskaya
Energy Position and Participation of the Rare-Earth Element 5p-States in Chemical Interaction
(b) Vol. **119**, No. 2, p. K123 – K126 (1983).
See also *13*, p. *151*; *20*, p. *311*.

K. Witański
Theoretical Calculations of the Curie Temperature for $Gd_{1-x}RE_xAl_2$ Alloys
(b) Vol. **120**, No. 1, p. 149 – 154 (1983).
See also *18.2*, p. *287*.

A. F. Sadreev, I. S. Sandalov, and A. N. Podmarkov
The Spin-Wave Theory of Electron and s–f-Exchange Magnetostriction in the Rare Earth Metals
(b) Vol. **120**, No. 1, p. 383 – 395 (1983).
See also *18.2*, p. *288*.

M. N. Ghatikar
Chemical Transferability of Phase Shifts in Extended X-Ray Absorption Fine Structure
(b) Vol. **120**, No. 1, p. 445 – 450 (1983).
See also *20*, p. *311*; 21.1, p. 376; 22.5, p. 475; 22.6, p. 497; 22.7, p. 499.

R. Pfranger, D. Płusa, and B. Wysłocki
Magnetic Domain Structure in Polycrystalline $DyFe_2$ Compounds
(a) Vol. **80**, No. 1, p. 73 – 78 (1983).
See also *18.3*, p. *294*; 21.1.1, p. 384.

S. Surma, M. Błaszkiewicz, R. Meclewski, and W. Gubernator
A Rearranged Structure on a Rhenium Surface
(a) Vol. **80**, No. 1, p. 269 – 276 (1983).
See also *1.6*, p. *42*; 17.1, p. 272.

Z. Drzazga and T. Mydlarz
Magnetic Properties of the $ErCo_{5.5-x}Cu_x$ Alloys
(a) Vol. **80**, No. 1, p. 403 – 409 (1983).
See also *18.3*, p. *294*.

M. Gasgnier
Complex Crystallization Phenomena in Rare-Earth Binary Alloys and Compounds as Thin Films
(a) Vol. **80**, No. 1, p. K1 – K6 (1983).
See also *1.5*, p. *38*; 2, p. 55.

C. Colinet and A. Pasturel
A Thermodynamic Study of Cerium Behaviour in Hexagonal $CeNi_5$ Compound
(a) Vol. **80**, No. 1, p. K75 – K79 (1983).
See also *8*, p. *91*; 21.1, p. 377.

O. L. T. de Menezes and A. Troper
A Model for Inhomogeneous Mixed-Valence f-Semiconductors
(b) Vol. **120**, No. 2, p. 767 – 774 (1983).
See also *13*, p. *153*; 14, p. 205.

H. Kido, T. Hoshikawa, M. Shimada, and M. Koizumi
Preparation and Magnetic and Electrical Properties of RZnSi
(a) Vol. **80**, No. 2, p. 601–605 (1983).
See also *1.3*, p. *27*; 14.1, p. 211; *18.1*, p. *279*; 21.1, p. 378.

M. Jurczyk
Magnetic Properties of Substituted $Pr_2(Co_{1-x}M_x)_{17}$ Compounds (M = Fe, Mn, and Cr)
(a) Vol. **80**, No. 2, p. 657–662 (1983).
See also *1.3*, p. *28*; *18.2*, p. *289*; 21.1, p. 378.

L. Amaral, F. P. Livi, and A. A. Gomes
Remarks on Alloying-Induced Lattice Parameter Changes in Intermetallic Compounds
(a) Vol. **80**, No. 2, p. 669–677 (1983).
See also *1.3*, p. *28*; 21.1, p. 378; 21.1.1, p. 385.

21.5 Actinides

Original Papers and Short Notes

N. Shohoji
Statistical Thermodynamic Approach to Some Hyperstoichiometric Dihydrides of Lanthanides (La, Ce, Pr, Nd, Sm) and Actinides (Np, Pu, Am)
(b) Vol. **118**, No. 2, p. 811–822 (1983).
See also *4*, p. *62*; *8*, p. *89*; 10.1, p. 104; 21.4, p. 392.

V. Heera, G. Seifert, and P. Ziesche
Semirelativistic SCF-SW-Xα Calculations of Uranium Compounds. Electronic Structure and Chemical Bond
(b) Vol. **118**, No. 2, p. K107–K112 (1983).
Erratum
(b) Vol. **120**, No. 2, p. K227 (1983).
See also **4**, p. 62; *13.1*, p. *161*.

V. Heera, G. Seifert, and P. Ziesche
Semirelativistic SCF-SW-X_α Calculations of Uranium Compounds. Charge Density Differences at the Uranium Nucleus
(b) Vol. **119**, No. 1, p. K1–K4 (1983).
See also *13.1*, p. *163*.

M. Tamaki, H. Matsui, G. Matsumoto, and T. Kirihara
Critical Composition of Antiferromagnetic Ordering in UN_xC_{1-x}
(a) Vol. **79**, No. 1, p. K93–K98 (1983).
See also *14.1*, p. *209*; *18.4*, p. *297*.

21.6 Noble Metals

Original Papers and Short Notes

T. Nautiyal and S. Auluck
Fermi Surface of Noble Metals Using the Eight-Cone Model
(b) Vol. **115**, No. 1, p. 127–130 (1983).
See also *13.2*, p. *166*; 21.1, p. 363.

W. Kleinn
Erratum to: Determination of the Residual Resistivity from Current–Voltage Characteristics of Short Support-Free Samples of Epitaxial Gold Films
(a) Vol. **75**, No. 1, p. 347 (1983).
See also *1.5*, p. *35*; *14.1*, p. *206*.

G. Cocco, L. Schiffini, A. Lucci, L. Battezzati, and M. Sampoli
Structural Features of $Pd_{76}B_{24}$ Amorphous Alloy
(a) Vol. **75**, No. 1, p. K53–K56 (1983).
See also *2*, p. *44*.

Bai-Xin Liu
A Metastable Electron Compound Formed by Ion Irradiation
(a) Vol. **75**, No. 1, p. K77–K81 (1983).
See also *1.5*, p. *35*; *11*, p. *123*; 21.1, p. 364.

L. Kornblit
Divacancy Binding Energy in F.C.C. Metals
(b) Vol. **115**, No. 2, p. 485–491 (1983).
See also *10.1*, p. *99*; 21.1, p. 364.

J. S. Brown
Some Exchange-Correlation Approximations Applied to Stopping Powers of Slow Ions in Metals
(b) Vol. **115**, No. 2, p. K141–K146 (1983).
See also *13.2*, p. *166*; *17*, p. *269*; 21, p. 351.

D. Schryvers, G. van Tendeloo, J. van Landuyt, and S. Amelinckx
High-Resolution Study of the One- and Two-Dimensional Long-Period Antiphase Boundary Structures in Au_4Zn
(a) Vol. **75**, No. 2, p. 607–616 (1983).
See also *1.3*, p. *22*; *14*, p. *29*; 21.1, p. 365.

R. N. Joarder and R. V. Gopala Rao
Temperature Dependence of the Structure Factor of Liquid Metals
(b) Vol. **116**, No. 1, p. 299–306 (1983).
See also *2*, p. *45*; 4, p. 61; 9, p. 93; 21.1, p. 366; 21.2, p. 387.

T. NANBA and T. P. MARTIN
Raman Scattering from Metal Smokes
(a) Vol. **76**, No. 1, p. 235–239 (1983).
See also 1.6, p. 39; *6*, p. *72*; *20.1*, p. *316*; 21.1, p. 366.

D. SCHRYVERS, J. VAN LANDUYT, G. VAN TENDELOO, and S. AMELINCKX
Electron Microscopy Study of Two New Long Period Structures in Pt_xTi Alloys ($3 \leqq x \leqq 8$)
(a) Vol. **76**, No. 2, p. 575–593 (1983).
Erratum
(a) Vol. **78**, No. 2, p. 761 (1983).
See also *1.3*, p. *23*; *1.4*, p. *30*; 21, p. 353.

A. H. KING
Evidence of the Formation of Twins by Deformation and "Growth-Accidents" in Evaporated Thin Films of Gold
(a) Vol. **76**, No. 2, p. 629–636 (1983).
See also *1.5*, p. *36*; 10.1, p. 101.

E. BALANZAT and M. HALBWACHS
A Comparative Analysis of the Ordering Kinetic Laws for the Derivation of the Vacancy Mobility Parameters in Concentrated Alloys
(a) Vol. **76**, No. 2, p. 667–673 (1983).
See also 9, p. 94; *10.1*, p. *102*; 14.1, p. 207; 21.1, p. 367.

M. BURD, R. STEARNS, and R. BRAUNSTEIN
De-Correlation Technique for Separation of Drude Parameters from Wavelength Modulation Spectroscopy Data
(b) Vol. **117**, No. 1, p. 101–107 (1983).
See also *13.1*, p. *158*; *20.1*, p. *317*; 21.1, p. 368.

J. TEUHO and K. MÄKINEN
A Study of Non-Periodic Antiphase Boundaries in Long-Period Ordered $Au_3Zn(H)$ (I): Antiphase Boundary Planes
(a) Vol. **77**, No. 1, p. 263–268 (1983).
See also *1.3*, p. *24*; *1.4*, p. *31*; 21.1, p. 369.

K. ABDELMOULA, J. J. MÉTOIS, and G. NIHOUL
Reconstructed Surface Structure of (111) Gold Platelets, Observed by High Resolution Techniques
(a) Vol. **77**, No. 1, p. K5–K7 (1983).
See also *1.6*, p. *40*.

V. E. ANTONOV, T. E. ANTONOVA, I. T. BELASH, E. G. PONYATOVSKII, and V. I. RASHUPKIN
Superconductivity of Pd–Au–H Solid Solution
(a) Vol. **77**, No. 1, p. K23–K27 (1983).
See also 1.3, p. 24; *14.2*, p. *213*.

M. DĘBOWSKA and A. BARANOWSKI
Positron Diffusion Constant for Gold
(b) Vol. **117**, No. 2, p. K127–K129 (1983).
See also *13*, p. *149*.

M. M. Jakas, G. H. Lantschner, J. C. Eckardt, and V. H. Ponce
Angular Dependence of Stopping Power
(b) Vol. **117**, No. 2, p. K131 – K134 (1983).
Erratum
(b) Vol. **120**, No. 2, p. K225 (1983).
See also *11*, p. *128*.

W. Kleinn
Determination of Residual Resistivity and Observation of Resistance Recovery from Current–Voltage Curves of Short Epitaxial Gold Thin Films
(a) Vol. **77**, No. 2, p. K149 – K153 (1983).
See also *1.5*, p. *37*; *14.1*, p. *208*.

F. Bechstedt, R. Enderlein, and D. Reichardt
Electronic Relaxation Effects in Core Level Spectra of Surfaces and Interfaces
(b) Vol. **118**, No. 1, p. 327 – 336 (1983).
See also *13.3*, p. *169*; 14.4, p. 252; 22.1.2, p. 432; 22.2.1, p. 445; 22.6, p. 494.

A. K. Ghosh, S. Chakraborty, and A. Manna
Density Functional Approach for Metals. The Matheiss Prescription
(b) Vol. **118**, No. 1, p. 373 – 379 (1983).
See also 4, p. 62; *13*, p. *150*; 21.1, p. 371; 21.1.1, p. 382.

L. Z. Mezey, O. Nishikawa, and J. Giber
Monotonic and Alternating Surface Segregation Profiles in Binary Alloys
(a) Vol. **78**, No. 1, p. 323 – 331 (1983).
See also *1.3*, p. *24*; *1.6*, p. *40*; 21.1, p. 372.

A. K. Ahuja, S. Auluck, and S. K. Joshi
Application of Special-Directions Method to the Optical Properties of Copper and Silver
(b) Vol. **118**, No. 2, p. 575 – 586 (1983).
See also *13.1*, p. *160*; 20.1, p. 323; 21.1, p. 372.

J. Woltersdorf and E. Pippel
Influence of Substrate Deformations on Interface Structures (I): Theory
(a) Vol. **78**, No. 2, p. 475 – 483 (1983).
See also *1.5*, p. *37*; *10.1*, p. *105*.

S. F. Kolodeznaya and E. V. Kozlov
Calculation of the Properties and the Structure of Ordering Alloys of the Cu–Au System
(b) Vol. **119**, No. 1, p. 131 – 139 (1983).
See also *1.3*, p. *25*; 13.1, p. 162; 21.1, p. 374.

Yu. S. Stark, A. S. Steinberg, V. M. Vasilev, and V. I. Dimitrov
Monte Carlo Determination of Long-Range Pair Interaction Energies from Diffuse Scattering of X-Rays by Alloys
(b) Vol. **119**, No. 1, p. 147 – 151 (1983).
See also *1.3*, p. *25*; *4*, p. *63*; 21.1, p. 374.

E. Pippel and J. Woltersdorf
Influence of Substrate Deformations on Interface Structures (II): Experimental Investigations for Epitaxial Ag/Pd and Au/Pd Bicrystals
(a) Vol. **79**, No. 1, p. 189 – 198 (1983).
See also *1.5*, p. *38*; *10.1*, p. *106*.

V. V. Nemoshkalenko, A. E. Krasovskii, V. N. Antonov, Vl. N. Antonov, U. Fleck, H. Wonn, and P. Ziesche
The Relativistic Linear Muffin-Tin Orbital Method. Application to Au
(b) Vol. **120**, No. 1, p. 283–296 (1983).
See also *13.1*, p. *164*.

C. Regnaut, E. Fusco, and J. P. Badiali
Structure and Thermodynamics of the Liquid Noble Metals from the Generalized OPW and Resonant Model Approaches
(b) Vol. **120**, No. 1, p. 373–382 (1983).
See also *2*, p. *54*; 8, p. 91; 13.1, p. 164; 21.1, p. 376.

K. S. Bhadauria and S. B. Shrivastava
Positron Annihilation Rate at Metal Surfaces
(b) Vol. **120**, No. 1, p. K27–K31 (1983).
See also *1.6*, p. *42*.

S. K. Pabi
On the Variation of Interface Composition around a Dissolving Precipitate
(a) Vol. **80**, No. 1, p. 305–316 (1983).
See also *1.3*, p. *27*; 9, p. 95; 21, p. 361.

C. H. Ling
Resistivity Change in Ti/Au Bimetal Films Annealed in the Temperature Range 300 to 400 °C
(a) Vol. **80**, No. 1, p. K59–K62 (1983).
See also *9*, p. *96*; *14.3.2*, p. *235*; 21.1, p. 377.

C. Malinowska-Adamska
Self-Consistent Calculations of the Lattice Dynamics and Thermodynamics of Anharmonic Noble Metals
(b) Vol. **120**, No. 2, p. 601–610 (1983).
See also *6*, p. *81*; 8, p. 92; 12.1, p. 140.

S. V. Tsivinskii
The Theory of Electric Resistance of Pure Metals and Metals with a Given Concentration of Impurities, Vacancies, and Dislocations
(a) Vol. **80**, No. 2, p. 439–446 (1983).
See also 10.1, p. 107; *14.1*, p. *210*; 21, p. 362; 21.1, p. 377; 21.1.1, p. 385; 21.2, p. 388.

V. I. Motorin
Vitrification Kinetics of Pure Metals
(a) Vol. **80**, No. 2, p. 447–455 (1983).
See also *2*, p. *55*; 21, p. 362; 21.1, p. 377.

M. Ellner, M. El-Boragy, and B. Predel
On the Formation of Palladium-Rich Metallic Glasses in Binary and Ternary Systems with Copper, Silver, and Gold as One Component and Phosphorus and Arsenic as Further Component
(a) Vol. **80**, No. 2, p. 519–533 (1983).
See also *2*, p. *55*; 13.1, p. 165; 21.1, p. 378.

J. TEUHO
A Study of Non-Periodic Antiphase Boundaries in Long-Period Ordered $Au_3Zn(H)$ (II): Dissociated Antiphase Boundaries
(a) Vol. **80**, No. 2, p. 589–600 (1983).
See also *1.3*, p. *27*; *1.4*, p. *34*; 21.1, p. 378.

M. Z. BUTT, P. FELTHAM and I. M. GHAURI
Stress-Equivalence of Solid Solution Hardening in Concentrated Ag–Al and Ag–Zn Alloys
(a) Vol. **80**, No. 2, p. K125–K128 (1983).
See also *10.1*, p. *107*; 21, p. 362; 21.1, p. 378.

21.7 Semi-Metals (Semiconducting Alloys See 22.7)

Original Papers and Short Notes

U. Dillner
Analysis of Resistivity, Hall Effect, and Magnetoresistance Measurements of Thin Bismuth Films by Means of an Anisotropic Two-Carrier Model
(a) Vol. **75**, No. 1, p. 227–233 (1983).
See also *14.1*, p. *206*; *14.3.1*, p. *229*.

G. V. Dedkov and A. M. Gorshokov
Study of Various Approximations for the Continuum Planar Potentials at Channeling (I) Monoatomic Crystals
(b) Vol. **117**, No. 1, p. 57–66 (1983).
See also *11*, p. *127*; 22.1, p. 417; 22.1.2, p. 429.

J. Rose and R. Schuchardt
A Theoretical Investigation of the Fermi Surfaces of Bismuth and Antimony
(b) Vol. **117**, No. 1, p. 213–224 (1983).
See also *13.1*, p. *158*; *13.2*, p. *166*.

V. V. Kechin and A. I. Likhter
Hall Effect and Magnetoresistance of Antimony up to 5 GPa
(b) Vol. **117**, No. 1, p. K13–K17 (1983).
See also *14.1*, p. *207*.

P. Vašek and P. Svoboda
Thermopower of Bismuth in Low Longitudinal Magnetic Fields
(b) Vol. **118**, No. 1, p. 405–408 (1983).
See also *15*, p. *262*.

A. A. Mohamed, S. A. Saleh, E. A. Abou-Saif, M. Abd-Rabo, and G. A. El-Ganainy
On the Identification of New and Normal Phases in the System Cadmium–Antimony Alloys
(a) Vol. **78**, No. 2, p. 705–716 (1983).
See also *1.3*, p. *25*; 21, p. 358.

N. V. Moseev, V. G. Chudinov, B. N. Goshchitskii, and V. I. Protasov
Atomic Configurations of Interstitials in Intermetallic Compound Nb_3Sn
(a) Vol. **79**, No. 1, p. 67–70 (1983).
See also *10.1*, p. *105*; 21, p. 358.

N. B. Mustafaev and M. G. Shakhtakhtinskii
Density of Electron States in Thin Bi Films
(b) Vol. **120**, No. 1, p. K23–K26 (1983).
See also *13.1*, p. *164*.

D. V. Gitsu and F. M. Muntyanu
Commutation Effect of Transport Phenomena in Antimony–Bismuth Alloys in Quantizing Magnetic Fields
(b) Vol. **120**, No. 1, p. K45 – K49 (1983).
See also *14.1*, p. *210*; *15*, p. *263*.

B. G. S. Doman
Electron Energy Levels in a Thin Wire in a Magnetic Field
(b) Vol. **120**, No. 1, p. K79 – K82 (1983).
See also *13.1*, p. *164*.

V. G. Chudinov, V. P. Gogolin, B. N. Goshchitskii, and V. I. Protasov
Evolution of the Cascade Region in Nb_3Sn during the Thermal Spike under Fast-Neutron Irradiation
(a) Vol. **80**, No. 1, p. 349 – 352 (1983).
See also *10.1*, p. *107*; 11, p. 134; 21, p. 361.

P. Svoboda and P. Vašek
The Generalized Thomson-Voigt Relations for Transport Coefficients of Bismuth
(a) Vol. **80**, No. 2, p. 567 – 571 (1983).
See also *14.1*, p. *210*; *15*, p. *263*.

22. Semiconductors and Ionic Crystals

Review Articles

R. A. Evarestov and V. P. Smirnov
Special Points of the Brillouin Zone and Their Use in the Solid State Theory
(b) Vol. **119**, No. 1, p. 9–40 (1983).
See also *13.1*, p. *155*; 13.3, p. 167; 13.4, p. 172.

T. D. Dzhafarov
Photostimulated Diffusion in Semiconductors
(a) Vol. **79**, No. 1, p. 11–45 (1983).
See also *9*, p. *93*; 11, p. 122; 14.3.3, p. 236; 14.3.4, p. 243; 16, p. 264.

B. Pałósz
Reasons for Polytypism of Crystals of the Type MX_2 (II): Classification of Faults and Structural Series of Polytypes; Conditions of Polytypic Growth of CdI_2, PbI_2, $CdBr_2$, SnS_2, $SnSe_2$ and $Ti_{1.2}S_2$
(a) Vol. **80**, No. 1, p. 11–41 (1983).
See also *1*, p. *9*; *1.1*, p. *11*; 10.2, p. 108; 22.5, p. 473.

Original Papers and Short Notes

H. Kahnt and F. Schirrmeister
Electric Field Dependent Conductivity in Amorphous Semiconductors
(b) Vol. **115**, No. 1, p. 171–179 (1983).
See also *2*, p. *43*; 13.4, p. 172; *14.3.4*, p. *243*.

C. Gunkel, D. Schalch, and A. Scharmann
Electrical Conduction in Silicon Nitride Thin Films under 15 keV He^+ Irradiation
(a) Vol. **75** No. 1, p. K57–K62 (1983).
See also 2, p. 44; *11*, p. *123*; *14.3.1*, p. *229*.

I. Bartoš, M. Kolář, and G. Paasch
Modified Thomas-Fermi Approximation for Systems with Finite Surface Barriers
(b) Vol. **115**, No. 2, p. 437–442 (1983).
See also *13.3*, p. *167*.

G. Paasch
Current Density near the Semiconductor–Insulator Interface
(b) Vol. **115**, No. 2, p. 471–475 (1983).
See also *13.3*, p. *167*; *14.3.2*, p. *233*.

S. Eränen and J. Sinkkonen
Conduction Electron Self-Energy Effects in Ferromagnetic Semiconductors
(b) Vol. **115**, No. 2, p. 519–527 (1983).
See also *13,1*, p. *156*; 14.3, p. 216.

F. M. Peeters and J. T. Devreese
The $\frac{3}{2}kT$ Problem in the Low-Temperature Polaron Mobility Theories
(b) Vol. **115**, No. 2, p. 539–543 (1983).
See also *13.5.3*, p. *201*; *14,3*, p. *216*.

M. BANCEWICZ and A. STASCH
Theoretical Investigation of the Resonant Magnon–Phonon Interaction in FeF_2 in High Magnetic Field
(b) Vol. **115**, No. 2, p. 563–574 (1983).
See also *6*, p. *71*; *13.5.4*, p. *203*; 18, p. 273.

D. P. MAHAPATRA and H. C. PADHI
The Compton Profiles of TiB_2 and VB_2
(b) Vol. **115**, No. 2, p. K101–K106 (1983).
See also *20*, p. *307*.

V. K. ARORA
Nonohmic Conductivity for Acoustic-Phonon Scattering
(b) Vol. **115**, No. 2, p. K107–K112 (1983).
See also *6*, p. *71*; *14.3*, p. *216*.

L. TICHÝ, H. TICHÁ, M. FRUMAR, and A. TŘÍSKA
Compositional Dependence of the Average Gap in the Ge-Rich Region of the $Ge_{1-x}S_x$ Non-Crystalline System
(b) Vol. **115**, No. 2, p. K127–K131 (1983).
See also *2*, p. *44*; *4*, p. *60*.

F. SCHAUER, O. ZMEŠKAL, and S. NEŠPŮREK
A New Approach to the Analysis of Steady-State Space-Charge-Limited Currents Using Their Activation Energies. An Application to a-As_2Se_3
(a) Vol. **75**, No. 2, p. 591–594 (1983).
See also 2, p. 45; *13.4*, p. *175*; *14.3.4*, p. *244*.

G. RUDLOF and H. GLAEFEKE
Analysis of First-Order Glow Curves by the Balarin-Zetzsche Method in Differential Form
(a) Vol. **75**, No. 2, p. K169–K173 (1983).
See also *13.4*, p. *175*.

YU. P. DROZHOV
Nonlinear Susceptibility of Impurity Centres in Semiconductors
(b) Vol. **116**, No. 1, p. 111–118 (1983).
See also 13.4, p. 176; *20*, p. *307*.

H. STRAMSKA
Nonlinear Magneto-Optical Phenomena of the Third Harmonic in Semiconductors
(b) Vol. **116**, No. 1, p. 129–135 (1983).
See also *18*, p. *273*; *20*, p. *307*.

S. GHOSH and S. KHAN
Stimulated Brillouin Scattering of Transversely Polarized Electromagnetic Waves in Magnetoactive Moderately Piezoelectric Semiconductor Plasmas
(b) Vol. **116**, No. 1, p. 323–334 (1983).
See also *13*, p. *147*; *20*, p. *307*.

K. P. Maheshwari and Ganesh Tarey
Excitation of Magnetoacoustic Waves by Two Microwave Beams in InSb
(b) Vol. **116**, No. 1, p. 339–346 (1983).
See also *7*, p. *82*; 13, p. 147; 18, p. 274; *20*, p. *307*.

V. N. Pavlovich
Photostimulated Diffusion of Light Substitutional Impurities in Semiconductors and Insulators
(b) Vol. **116**, No. 1, p. K9–K14 (1983).
See also *9*, p. *93*; 22.1.2, p. 426.

P. Enders
Universal Tight-Binding Parameters for Cubic IV–VI Compounds
(b) Vol. **116**, No. 1, p. K15–K18 (1983).
See also *13.1*, p. *157*.

L. N. Christophorov and V. N. Kharkyanen
Theory of Interimpurity Transitions in Condensed Medium
(b) Vol. **116**, No. 2, p. 415–425 (1983).
See also *13.4*, p. *176*.

F. Comas and C. Trallero Giner
Low-Frequency Secondary Radiation in Polar Semiconductors at Finite Temperatures
(b) Vol. **116**, No. 2, p. 445–453 (1983).
See also 6, p. 72; *20*, p. *308*.

S. K. Avetissian, E. M. Kazarian, A. O. Melikian, and H. R. Minasian
Parametric Generation of Submillimeter Radiation in Semiconductors in the Field of a Bichromatic Pulse
(b) Vol. **116**, No. 2, p. 455–463 (1983).
See also *20.2*, p. *332*.

D. Ihle and B. Lorenz
Theory of the Verwey Transition in the Mixed Valence Compound Eu_3S_4
(b) Vol. **116**, No. 2, p. 539–546 (1983).
See also *6*, p. *72*; *13.5.3*, p. *201*; 18.2, p. 282.

M. Satarić and R. Žakula
On the Role of the Soliton Mechanism in the Electron Transfer along the Quasi-One-Dimensional Molecular Structures
(b) Vol. **116**, No. 2, p. 617–632 (1983).
See also *13.4*, p. *177*.

V. Vorlíček, I. Gregora, and D. Chvostová
Raman Scattering in GeS Single Crystals at Low Temperatures
(b) Vol. **116**, No. 2, p. 639–643 (1983).
See also *6*, p. *73*; *20.1*, p. *316*.

J. Sinkkonen and P. Kuivalainen
Many Electron Magnetic Polaron in Magnetic Semiconductors
(b) Vol. **116**, No. 2, p. K73–K76 (1983).
See also *13.5.3*, p. *201*; *18.2*, p. *282*.

M. MÜHLBERG and D. HESSE
TEM Precipitation Studies in Te-Rich As-Grown PbTe Single Crystals
(a) Vol. **76**, No. 2, p. 513–524 (1983).
See also *1.4*, p. *30*; 10.2, p. 112.

O. BOSTANJOGLO
Multiple Phase Transitions in GeTe Films. Dynamics and Film Structure
(a) Vol. **76**, No. 2, p. 525–531 (1983).
See also *1.2*, p. *16*; *1.5*, p. *36*; 2, p. 47; 3, p. 58.

T. K. CHATTOPADHYAY, E. GMELIN, and H. G. VON SCHNERING
Heat Capacity Study of the Phase Transitions in As_4S_3 and As_4S_4
(a) Vol. **76**, No. 2, p. 543–551 (1983).
See also 1.2, p. 16; *8*, p. *87*.

A. M. BELYANTSEV and M. N. DROZDOV
On *I–U* Characteristics of Semiconductor Structures with Inhomogeneous Mobility Profile
(a) Vol. **76**, No. 2, p. 721–728 (1983).
See also *14.3.4*, p. *244*.

A. JAKUBOWICZ
On EBIC versus Time Measurements in Thin Semiconductor Layers with the Scanning Electron Microscope
(a) Vol. **76**, No. 2, p. 729–734 (1983).
See also *14.3.1*, p. *231*; *14.3.3*, p. *238*.

A. BENNOUNA, P. Y. TESSIER, M. PRIOL, Q. DANG TRAN, and S. ROBIN
Far Ultraviolet Photoelectric Study of Thin SnSe Evaporated Films
(b) Vol. **117**, No. 1, p. 51–56 (1983).
See also *13.1*, p. *158*; *17*, p. *270*.

G. VALIUKONIS, F. M. GASHIMZADE, D. A. GUSEINOVA, G. KRIVAITÉ, A. M. KULIBEKOV, G. S. ORUDZHEV, and A. ŠILEIKA
Reflectance and Thermoreflectance Spectra and Energy Band Structure of GeSe Crystals
(b) Vol. **117**, No. 1, p. 81–92 (1983).
See also *13.1*, p. *158*; *20.1*, p. *317*.

N. A. POKLONSKI and V. F. STELMAKH
Screening of Electrostatic Fields in Crystalline Semiconductors by Electrons Hopping over Defects
(b) Vol. **117**, No. 1, p. 93–99 (1983).
See also *13.4*, p. *178*; *14.3*, p. *218*.

F. BECHSTEDT, R. ENDERLEIN, and D. REICHARDT
Inverse Dielectric Function for a Semi-Infinite Solid
(b) Vol. **117**, No. 1, p. 261–270 (1983).
See also *13.3*, p. *168*; *14.4*, p. *251*.

M. STORDEUR and H. T. LANGHAMMER
Anisotropy of Transport Properties Due to Direction-Dependent Band Nonparabolicity
(b) Vol. **117**, No. 1, p. 329–333 (1983).
See also *13.1* p. *158*; 14.3, p. 219; 15, p. 261.

R. A. Evarestov, V. A. Lovchikov, and I. I. Tupitsin
Hartree-Fock Exchange and LCAO Approximation in the Band Structure Calculations of Solids
(b) Vol. **117**, No. 1, p. 417–427 (1983).
See also *13.1*, p. *158*.

L. V. Vassilev and G. Ch. Christakudis
On the Calculation of Transport Integrals in Lead Chalcogenides
(b) Vol. **117**, No. 1, p. K53–K56 (1983).
See also *14.3*, p. *219*; **15**, p. **261**.

J. W. Tomm, K. Herrmann, A. Jalyschko, A. E. Yunovich, and V. Ten
Direct Comparison of Photo- and Electroluminescence in $Pb_{1-x}Sn_xTe$ Diode Lasers
(a) Vol. **77**, No. 1, p. 175–179 (1983).
See also *20.3*, p. *339*.

M. Balarin
On the Influence of a Temperature Dependence of the Frequency Factor on the Shope and Width of Glow Peaks
(a) Vol. **77**, No. 1, p. 181–194 (1983).
See also *13.4*, p. *178*.

R. Deblieck, G. A. Wiegers, K. D. Bronsema, D. van Dyck, G. van Tendeloo, J. van Landuyt, and S. Amelinckx
Clustering and "Phase Ordering" Transitions in Molybdenum (III) Sulphide, $Mo_{2.065}S_3$
(a) Vol. **77**, No. 1, p. 249–261 (1983).
See also *1.3*, p. *23*.

J. Mahy, J. van Landuyt, and S. Amelinckx
Electron Diffraction Evidence for Superstructures in $TaTe_4$ and $NbTe_4$
(a) Vol. **77**, No. 1, p. K1–K4 (1983).
See also *1.3*, p. *24*.

H. Stramska
The Magneto-Optical Effect under a Strong Electric Field in Semiconductors
(b) Vol. **117**, No. 2, p. 541–547 (1983).
See also 14.3.4, p. 245; *18*, p. *274*; *20*, p. *309*.

I. V. Stasyuk and S. S. Kotsur
The Microscopic Theory of the Gyration and Electrogyration in Dielectric Crystals
(b) Vol. **117**, No. 2, p. 557–568 (1983).
See also *14.4*, p. *251*; *20*, p. *309*.

A. M. Berezhkovskii and A. A. Ovchinnikov
The Interband Interaction and the Tunnel Level Width of the Wannier-Stark Ladder in a Semiconductor Subject to a Strong Electric Field
(b) Vol. **117**, No. 2, p. 585–588 (1983).
See also *13*, p. *149*.

R. Zimmermann
Thermodiffusion in Electron–Hole Plasmas Revised
(b) Vol. **117**, No. 2, p. K135 – K140 (1983).
See also *13.5*, p. *193*.

A. M. Belyantsev and M. N. Drozdov
The Effect of Spatial Variation of a High Frequency-Field on the Instability Occurrence in Inhomogeneous Semiconductors
(a) Vol. 77, No. 2, p. 437 – 442 (1983).
See also *14.3.4*, p. *245*.

V. S. Lysenko, R. N. Litovskii, M. M. Lokshin, and M. F. Sherbakova
Double Injection Current and Field Effect in SOS Diodes
(a) Vol. 77, No. 2, p. 443 – 448 (1983).
See also *14.3.3*, p. *239*; *14.3.4*, p. *245*.

O. A. Golikova and S. Samatov
Conduction Mechanism of Some "Quasi-Amorphous" Semiconductors on the Basis of Boron
(a) Vol. 77, No. 2, p. 449 – 453 (1983).
See also *13.4*, p. *180*; *14.3*, p. *220*; 15, p. 262.

K. Ram Kumar and M. Satyam
Minority Carrier Lifetime in Polycrystalline Semiconductors
(a) Vol. 77, No. 2, p. 467 – 470 (1983).
See also *14.3*, p. *220*.

Z. Korczak and M. Subotowicz
Structural and Galvonomagnetic Properties of $Pb_{1-x}Mn_xTe$ Single Crystals Grown by the Bridgman-Stockbarger Method
(a) Vol. 77, No. 2, p. 497 – 503 (1983).
See also *1.3*, p. *24*; 3, p. 58; *14.3*, p. *220*.

N. P. Kalmykova, B. T. Kolomiets, V. Kh. Shpunt, and E. A. Smorgonskaya
On the Influence of the Non-Homogeneous Structure on the Current–Voltage Behaviour of Chalcogenide Vitreous Semiconductors
(a) Vol. 77, No. 2, p. 803 – 808 (1983).
See also *2*, p. *49*; *14.3.4*, p. *246*.

A. S. Agabekyan
Resonant Coherent Energy Transfer. Role of Initial Conditions
(b) Vol. **118**, No. 1, p. 51 – 56 (1983).
See also *13.4*, p. *181*; 20, p. 309.

Nguyen van Huong
Theory of Free-Carrier Absorption in n-Type Lead Chalcogenides Due to Electron–Electron Scattering
(b) Vol. **118**, No. 1, p. 135 – 146 (1983).
See also *20.1*, p. *321*.

K. Regiński
Some Extensions of the Shell Model of Ionic Crystals
(b) Vol. **118**, No. 1, p. 147–157 (1983).
See also *6*, p. *75*.

O. A. C. Nunes
On the Multiphoton Laser Absorption in Free-Carrier Semiconductors
(b) Vol. **118**, No. 1, p. K25–K29 (1983).
See also 6, p. 75; *13*, p. *150*; *20.1*, p. *322*.

M. Suhrke
On the Time Dependence of the Luminescence Spectrum in Chalcogenide Glasses
(b) Vol. **118**, No. 1, p. K79–K83 (1983).
See also *2*, p. *49*; *20.3*, p. *341*.

J. Vetter and H. Bethge
Thermal Self-Decoration of GeS Crystals
(a) Vol. **78**, No. 1, p. 103–109 (1983).
See also 1.3, p. 24; *1.6*, p. *40*.

G. D. Bagratishvili, R. B. Dzhanelidze, and D. A. Jishiashvili
Nonstoichiometric Germanium Nitride. A New Insulating Material for MIS Microelectronics (I): The Choice of the Insulator and Deposition
(a) Vol. **78**, No. 1, p. 115–123 (1983).
See also *13.3* p. *169*; *14.3.2*, p. *234*; 14.3.3, p. 239.

T. D. Sokolovskii
Phonon Spectrum of Boron Nitride with Wurtzite Structure
(b) Vol. **118**, No. 2, p. 493–498 (1983).
See also *6*, p. *76*; 8, p. 88.

K. O. Kechechyan and A. A. Kirakosyan
Magneto-Resistance of a Semiconductor with the Dislocation Wall in the Quantum Limit
(b) Vol. **118**, No. 2, p. 521–524 (1983).
See also *10.2*, p. *115*; *14.3*, p. *222*.

B. Jagannatha Rao and A. R. Chetal
On the Chemical Shifts of the K-Absorption Discontinuity of Copper in Some of Its Systems
(b) Vol. **118**, No. 2, p. 551–555 (1983).
See also *13.1*, p. *160*; *20*, p. *309*; 21.1, p. 372.

Nguyen van Huong
Free-Carrier Absorption in n-PbSe at Low Temperatures Due to Electron–Electron Scattering
(b) Vol. **118**, No. 2, p. 631–640 (1983).
See also *13.1*, p. *161*; *20.1*, p. *323*.

G. Paasch, T. Fiedler, M. Kolář, and I. Bartoš
Quasi-Classical Quantization for Potentials with an Infinite Surface Barrier. Subbands in Inversion Layers
(b) Vol. **118**, No. 2, p. 641–652 (1983).
See also *13.3*, p. *169*.

W. Burckhardt and A. Feltz
The Influence of the Structure of Different Arsenic Chalcogenide Glasses on Refractive Index and Dispersion
(b) Vol. **118**, No. 2, p. 653–660 (1983).
See also *2*, p. *50*; *20.1*, p. *323*.

K. Sierański and J. Szatkowski
Intraband Absorption Due to Alloy Scattering in Lead Chalcogenide Semiconductors
(b) Vol. **118**, No. 2, p. 871–879 (1983).
See also *13.1*, p. *161*; *20.1*, p. *324*.

S. Anzai, S. Ohta, A. Yoshino, J. Nishio, and M. Hatori
Charge-Density-Wave-Like Transition in V_3Te_4
(b) Vol. **118**, No. 2, p. K99–K102 (1983).
See also *14.1*, p. *208*; 15, p. 262; *18.1*, p. *279*.

D. Genzow
Transverse Dember Effect in Lead Chalcogenides under Linearly Polarized Light Excitation
(b) Vol. **118**, No. 2, p. K159–K162 (1983).
See also *14.3*, p. *222*; 16, p. 266.

B. V. Kobrin, R. M. Kupriyanova, V. S. Minaev, E. P. Prokopiev, and V. P. Shantarovich
Microstructure and Imperfections of Glassy and Glass-Ceramic Telluride Semiconductors on the Basis of Positron Annihilation
(a) Vol. **78**, No. 2, p. 385–389 (1983).
See also *2*, p. *50*; *10.2*, p. *116*.

G. D. Bagratishvili, R. B. Dzhanelidze, and D. A. Jishiashvili
Nonstoichiometric Germanium Nitride – A New Insulating Material for MIS Microelectronics (II): Composition, Properties, and the Interface with a Semiconductor
(a) Vol. **78**, No. 2, p. 391–400 (1983).
See also *13.3*, p. *169*; *14.3.2*, p. *234*; 14.3.3, p. 240.

M. M. Hafiz, H. M. Shafey, A. M. Hafez, and M. Dongol
Irradiation-Induced Photodarkening in As–Se–Cu Amorphous Films
(a) Vol. **78**, No. 2, p. 449–455 (1983).
See also *11*, p. *130*; *20.1*, p. *324*.

S. Katsura and A. Matsuno
Phase Diagram of a Spin Glass of $Eu_pSr_{1-p}S$
(b) Vol. **119**, No. 1, p. 73–77 (1983).
See also *2*, p. *51*; *18*, p. *275*; 18.1, p. 279; 18.2, p. 286.

P. K. KATANA
Two-Photon Interband Absorption of Light in Semiconductors in External Parallel Electric and Magnetic Fields
(b) Vol. **119**, No. 1, p. 89–98 (1983).
See also *13.1*, p. *162*; 14.3.4, p. 247; 18, p. 275; *20.1*, p. *325*.

B. V. EGOROV and B. B. TIMOFEEV
The Theory of Magnetic Polarons in Antiferromagnetic Semiconductors
(b) Vol. **119**, No. 1, p. 99–112 (1983).
See also *13.5.3*, p. *201*; *18.4*, p. *297*.

A. I. BOBRYSHEVA and S. A. MOSKALENKO
Biexcitons in Crystals with Large Ortho–Para-Exciton Splitting
(b) Vol. **119**, No. 1, p. 141–146 (1983).
See also *13.5.1*, p. *197*.

WAN-SUN TSE and CHIA-NAN CHANG
Lattice Vibrations of Solid Hydrofluoric Acid
(b) Vol. **119**, No. 1, p. 349–354 (1983).
See also *6*, p. *77*.

T. M. GASYMOV, A. A. KATANOV, and M. M. BABAEV
Thermopower and Thermomagnetic Effects in Kane Semiconductors under Conditions of Electron and Phonon Heating and Their Mutual Drag
(b) Vol. **119**, No. 1, p. 391–399 (1983).
See also 6, p. 77; *14.3.4*, p. *247*; *15*, p. *262*.

A. V. SKRIPOV, A. P. STEPANOV, A. D. SHEVCHENKO, and Z. D. KOVALYUK
NMR Study of the Charge-Density-Wave State in VSe_2
(b) Vol. **119**, No. 1, p. 401–410 (1983).
See also *13*, p. *151*; *19*, p. *303*.

I. J. LALOV, K. T. STOYCHEV, and S. P. PENCHEV
Non-Linear Optical Susceptibilities of Molecular Crystals in the Region of Vibronic Excitations (I): The Three-Dimensional Case
(b) Vol. **119**, No. 1, p. 419–427 (1983).
See also 6, p. 77; *13.5.1*, p. *198*; 20.1, p. 326.

S. K. PLACHKOVA, G. CH. CHRISTAKUDIS, and L. V. VASSILEV
Density-of-States Effective Mass in the System $(GeTe)_{1-x}(AgSbTe_2)_x$
(b) Vol. **119**, No. 1, p. K73–K76 (1983).
See also *13.1*, p. *163*; *14.3*, p. *223*; 15, p. 262.

D. CHATTOPADHYAY
Transient Behaviour of Hot Electrons in Quantum-Well Structures of Polar Semiconductors
(b) Vol. **119**, No. 1, p. K77–K81 (1983).
See also 6, p. 78; *13*, p. *151*; *14.3*, p. *223*.

M. MERTIG, G. POMPE, and E. HEGENBARTH
Low-Temperature Specific Heat of Vapour-Deposited Ge_xSe_{1-x} Glass ($x = 0.3$)
(a) Vol. **79**, No. 1, p. K49 – K52 (1983).
See also *2*, p. *52*; *8*, p. *90*.

A. V. EFANOV and M. V. ENTIN
Photogalvanic Effect in Crystal with Dislocations
(b) Vol. **119**, No. 2, p. 473 – 481 (1983).
See also *16*, p. *267*.

V. M. FOMIN and E. P. POKATILOV
Nonlinear Kinetics and Optical Properties of Band Charge Carriers
(b) Vol. **119**, No. 2, p. 483 – 492 (1983).
See also 6, p. 78; 13, p. 151; 14.3.4, p. 247; *20.1*, p. *327*.

V. D. ISKRA
On the Problem of an Intermediate Order in Amorphous Semiconductors
(b) Vol. **119**, No. 2, p. 515 – 524 (1983).
See also *2*, p. *52*; 20.1, p. 327.

G. MÜTZENICH, R. FAYMONVILLE, and P. GROSSE
One-Phonon and Multiphonon Absorption in EuSe and $Eu_{1-x}Sr_xS$ ($0 < x < 1$)
(b) Vol. **119**, No. 2, p. 531 – 539 (1983).
See also *6*, p. *78*; *20.1*, p. *327*.

K. KÖHLER, H. P. GESERICH, TH. WOLF, and G. MÜLLER-VOGT
Optical Properties of TiN Single Crystals
(b) Vol. **119**, No. 2, p. 605 – 610 (1983).
See also *13.1*, p. *163*; *20.1*, p. *327*.

A. MORIMOTO, S. OOZORA, M. KUMEDA, and T. SHIMIZU
Annealing Behaviour of Hydrogenated Amorphous Silicon-Nitrogen Alloy Films Prepared by Sputtering
(b) Vol. **119**, No. 2, p. 715 – 720 (1983).
See also *2*, p. *52*; *10.2*, p. *119*; 19, p. 304; 20.1, p. 327.

H. TICHÁ, L. TICHÝ, and E. ČERNOŠKOVÁ
Temperature Dependence of the Optical Gap in Glassy GeS_2
(b) Vol. **119**, No. 2, p. K135 – K137 (1983).
See also *2*, p. *52*; 6, p. 79; *20.1*, p. *327*.

G. A. SCHOLZ and R. F. FRINDT
Electrical Resistivity of Silver Intercalated $2H\text{-}TaS_2$
(a) Vol. **79**, No. 2, p. 483 – 488 (1983).
See also 13, p. 152; *14.1*, p. *210*; *14.3*, p. *22*5.

B. M. WILAMOWSKI
The Punch-Through Transistor with MOS Controled Gate
(a) Vol. **79**, No. 2, p. 631 – 637 (1983).
See also *14.3.3*, p. *241*; 14.3.4, p. 248.

E. Fritzsch and C. Pietzsch
Lamb Mössbauer Fraction and Mean Square Displacement of Iron in Pyrite (FeS_2)
(a) Vol. **79**, No. 2, p. K113–K115 (1983).
See also *5*, p. *68*; 6, p. 79.

P. G. Rustamov, Ch. I. Abilov, and S. G. Agdamskaya
Thermoelectric Properties of $(PTe)_{1-x}[(Cr, Fe, Ni)_2Te_3]_x$
(a) Vol. **79**, No. 2, p. K135–K137 (1983).
See also 8, p. 91; 14.3, p. 226; *15*, p. *263*.

G. P. Kothiyal, B. Ghosh, and B. M. Arora
Photoluminescence Studies on Synthetic Crystals of As_2Se_3
(a) Vol. **79**, No. 2, p. K171–K175 (1983).
See also *20.3*, p. *345*.

A. V. Zamkov, I. T. Kokov, and A. T. Anistratov
The Acousto-Optical Properties and Photoelasticity of $PbBr_2$ Single Crystals
(a) Vol. **79**, No. 2, p. K177–K180 (1983).
See also *7*, p. *83*; 12.2, p. 144; *20*, p. *311*.

N. M. Khamidullina
Carrier Mobility in Doped Nondegenerate Semiconductors
(b) Vol. **120**, No. 1, p. 55–63 (1983).
See also *14.3*, p. *226*.

Ch. Schnittler, J. Kelber, G. Paasch, I. Bartoš and H. Übensee
A Modified Local Density Approximation for Potentials with a Finite Step
(b) Vol. **120**, No. 1, p. 297–309 (1983).
See also *13.3*, p. *170*.

H.-J. Unger
Localization of Electronic States in Disordered Tight-Binding Systems (I): Derivation and Study of the Model
(b) Vol. **120**, No. 1, p. 401–413 (1983).
See also *2*, p. *54*; *13.4*, p. *189*.

V. Riede, H. Neumann, F. Lévy, and H. Sobotta
Two-Phonon Absorption Spectra and Lattice Vibration Anisotropy in HfS_2
(b) Vol. **120**, No. 1, p. K9–K14 (1983).
See also *6*, p. *80*; *20.1*, p. *329*.

A. Antonini, V. Fano, and I. Ortalli
Mössbauer Study of the $Pb_xSn^{119}_{1-x}Te$ System
(b) Vol. **120**, No. 1, p. K15–K18 (1983).
See also *5*, p. *68*.

L. N. Vasilev, V. M. Grabov, A. V. Golubkov, A. G. Gorobets, V. S. Oskotskii, I. A. Smirnov, and V. V. Tikhonov
Physical Properties and Phase Transitions of the Rare Earth Monosulphides in the Homogeneity Range
(a) Vol. **80**, No. 1, p. 237–244 (1983).
See also *1.3*, p. *26*; *8*, p. *91*; 13, p. 152; 14.3, p. 226; 18.4, p. 297.

B.-P. Kietis, R. Reksnys, and K. Tumkevičius
Hall Effect and Acoustoelectric Phenomena in PbTe Film
(a) Vol. **80**, No. 1, p. K57 – K58 (1983).
See also *7*, p. *84*; *14.3.1*, p. *232*.

K. H. Herrmann and K.-P. Möllmann
Curie Temperature as a Critical Temperature for Dielectric, Galvanomagnetic, and Photoelectric Phenomena in Strongly Doped $Pb_{1-x}Sn_xTe$
(a) Vol. **80**, No. 1, p. K101 – K104 (1983).
See also *16*, p. *268*.

N. N. Koren, V. V. Kindyak, and E. E. Matyas
Phase Diagram of the System GeS–GeSe
(a) Vol. **80**, No. 1, p. K105 – K108 (1983).
See also *1.2*, p. *20*.

V. A. Sablikov and V. B. Sandomirskii
The Photoacoustic Effect in Semiconductors
(b) Vol. **120**, No. 2, p. 471 – 480 (1983).
See also 8, p. 92; *13.3*, p. *171*; *14.3*, p. *227*.

L. Merten and R. Kotzott
Excitation of Longitudinal Optical Phonons by Stimulated Raman Effect in Piezoelectric Cubic Crystals
(b) Vol. **120**, No. 2, p. 481 – 490 (1983).
See also *6*, p. *80*; 13.5, p. 193; *20*, p. *311*.

V. A. Kochelap and V. N. Sokolov
Deformational Phase Transitions and Periodic Structures in a Magnetic Field
(b) Vol. **120**, No. 2, p. 565 – 575 (1983).
See also *6*, p. *81*; *13*, p. *153*.

I. J. Lalov, K. T. Stoychev, and S. P. Penchev
Non-Linear Optical Susceptibilities of Molecular Crystals in the Region of Vibronic Excitations (II): Totally Symmetric Vibrations
(b) Vol. **120**, No. 2, p. 671 – 678 (1983).
See also 6, p. 81; *13.5.1*, p. *199*; 20.1, p. 330.

S. Ichinose and I. Tamura
Acoustic Phonon Anomaly in Mixed-Valent TmSe
(b) Vol. **120**, No. 2, p. 703 – 710 (1983).
See also *6*, p. *81*.

P. Enders
Structure and Bonding in Cubic IV–VI Crystals (I): Evidence for Covalent Bonding from LCAO Parameters
(b) Vol. **120**, No. 2, p. 735 – 744 (1983).
See also *4*, p. *64*; *13*, p. *153*.

H. Stramska
Higher Harmonics Generated by Mobile Carriers in Semiconductors
(b) Vol. **120**, No. 2, p. 783–788 (1983).
See also *13*, p. *153*; 14, p. 205; 18, p. 276.

A. H. Nayyar and P. Shukla
Spin Glass Behaviour in $Eu_xSr_{1-x}S$: Calculation of the Phase Diagram in Mean Field Approximation
(b) Vol. **120**, No. 2, p. K141–K144 (1983).
See also *18*, p. *276*; 18.2, p. 289.

E. I. Goldman, Yu. V. Markin, P. S. Sulzhenko, and A. G. Zhdan
Grain Boundary Tunneling Conductivity of Semiconductor Poly- and Bicrystals
(a) Vol. **80**, No. 2, p. 423–429 (1983).
See also 1.4, p. 34; *14.3.2*, p. *235*.

W. Burckhardt and A. Feltz
The Influence of the Structure of Different Germanium Chalcogenide Glasses on the Molar Refraction and Its Wavelength Dependence
(a) Vol. **80**, No. 2, p. 463–470 (1983).
See also *2*, p. *55*; *20.1*, p. *330*.

K. H. Herrmann, K.-P. Möllmann, and M. Wendt
Photoeffects in Strongly Gallium-Doped Lead Telluride above Critical Temperature
(a) Vol. **80**, No. 2, p. 541–546 (1983).
See also *16*, p. *268*.

K. Herrmann, J. W. Tomm, A. Jalyschko, and W. Bremser
On the Behaviour of the Threshold Current Densities in Dual-Wavelength $Pb_{1-x}Sn_xTe$ Injection Lasers
(a) Vol. **80**, No. 2, p. K113–K116 (1983).
See also *14.3.4*, p. *249*; *20.2*, p. *333*.

I. Manika and J. Teteris
Compositional Dependence of Photo-Induced Microhardness Changes in As–Se Films
(a) Vol. **80**, No. 2, p. K121–K123 (1983).
See also *10.2*, p. *121*; *12.2*, p. *145*.

Yu. Yu. Sidorin and Yu. A. Zakharov
The Light Stimulated Charge Carrier Transfer in Silver Azide
(a) Vol. **80**, No. 2, p. K157–K160 (1983).
See also *14.4*, p. *255*; *16*, p. *268*.

L. I. Ryabinkina and G. V. Loseva
Influence of Non-Stoichiometry and Cation Substitution on the Electrical Properties of α-MnS
(a) Vol. **80**, No. 2, p. K179–K182 (1983).
See also *14.3*, p. *228*.

P. N. Dashuk, A. V. Kolobov, B. T. Kolomiets, and V. M. Lyubin
Plasma-Induced Processes in Chalcogenide Vitreous Semiconductors
(a) Vol. **80**, No. 2, p. K187 – K192 (1983).
See also *2*, p. *56*; *11*, p. *135*.

W. Hoerstel, W. Kraak, and A. F. Rudolph
The Influence of Hydrostatic Pressure on the *I–U* Characteristic of $Pb_{1-x}Sn_xTe$ Diodes
(a) Vol. **80**, No. 2, p. K193 – K197 (1983).
See also *14.3.3*, p. *242*.

22.1 Elements

Original Papers and Short Notes

P. V. Smith
A Localized Orbital Description of Ge and α-Sn
(b) Vol. **116**, No. 1, p. 101 – 109 (1983).
See also *13.1*, p. *157*; 22.1.1, p. 419.

G. V. Dedkov and A. M. Gorshokov
Study of Various Approximations for the Continuum Planar Potentials at Channeling (I): Monoatomic Crystals
(b) Vol. **117**, No. 1, p. 57 – 66 (1983).
See also *11*, p. *127*; 21.7, p. 401; 22.1.2, p. 429.

O. Bílek, L. Skála, and L. Künne
Tight-Binding Cluster Model for C, Si, and Ge (I): Analytic Solution for s-States in Diamond Lattice
(b) Vol. **117**, No. 2, p. 675 – 681 (1983).
See also *13*, p. *149*; 22.1, p. 417; 22.1.2, p. 431.

L. Künne, L. Skála, and O. Bílek
Tight-Binding Cluster Model for C, Si, and Ge (II): Size and Shape Dependence of Some Cluster Properties
(b) Vol. **118**, No. 1, p. 173 – 178 (1983).
See also *13*, p. *149*; 22.1.1, p. 420; 22.1.2, p. 432.

M. Ortuño
Impurity Potentials in Diamond
(b) Vol. **119**, No. 2, p. 709 – 713 (1983).
See also *13.4*, p. *187*.

A. N. Rodionov, R. I. Kalendarev, A. V. Shendrik, and Yu. R. Zakis
Photoinduced Paramagnetic States of Yellow Arsenic
(a) Vol. **79**, No. 2, p. K151 – K154 (1983).
See also *2*, p. *53*; *19*, p. *305*.

22.1.1 Germanium

Review Articles

L. N. Aleksandrov and F. L. Edelman
Shock Crystallization of Films
(a) Vol. **76**, No. 2, p. 409–427 (1983).
See also 1.3, p. 21; *1.5*, p. *35*; *3*, p. *57*; 22.1.2, p. 423; 22.1.3, p. 440; 22.6, p. 490.

M. Cardona
Vibrational Spectra of Hydrogen in Silicon and Germanium
(b) Vol. **118**, No. 2, p. 463–481 (1983).
See also *6*, p. *69*; *20.1*, p. *312*; 22.1.2, p. 423.

Original Papers and Short Notes

A. A. Manenkov, G. N. Mikhailova, A. M. Prokhorov, A. V. Sidorin, S. Yu. Sokolov, and S. G. Tikhodeev
Interaction of High Power CO_2 Laser Radiation with a Nonequilibrium Carrier System in Ge at Low Temperatures
(b) Vol. **115**, No. 1, p. 75–81 (1983).
Erratum
(b) Vol. **117**, No. 2, p. 756 (1983).
See also *13.5.1*, p. *194*; *20.3*, p. *334*.

M. Stutzmann, J. Stuke, and H. Dersch
Electron Spin Resonance of Doped Glow-Discharge Amorphous Germanium
(b) Vol. **115**, No. 1, p. 141–151 (1983).
See also 2, p. 43; *13.4*, p. *172*; 14.3, p. 215; *19*, p. *299*; 22.1.2, p. 423.

V. M. Ivashchenko, V. V. Mitin, and N. A. Zakhlenjuk
Galvanomagnetic Effects in Semiconductors of p-Ge Type
(b) Vol. **115**, No. 1, p. 245–254 (1983).
See also *14.3*, p. *215*.

J. L. Demenet, P. Grosbras, and J. C. Desoyer
Creept of Germanium under Different Orientations
(a) Vol. **75**, No. 1, p. K33–K36 (1983).
See also *10.2*, p. *109*.

K. Takeda, Y. Maeda, E. Ohta, M. Sakata, G. Kido, and N. Miura
Study of Energy Band Parameters in p-Type $Ge_{0.9}Si_{0.1}$Alloys
(b) Vol. **115**, No. 2, p. 369–379 (1983).
See also *13.1*, p. *156*; 14.3, p. 216; 20.1, p. 314; 22.1.2, p. 425.

O. G. Alaverdova, M. Ya. Fuks, L. S. Khazan, L. P. Koval, L. A. Matveeva, I. F. Mikhailov, N. N. Soldatenko, and Yu. A. Tkhorik
Coherency of the Interphase Boundary and Elastic Strain in the Epitaxial System Ge/GaAs
(a) Vol. **75**, No. 2, p. 367 – 371 (1983).
See also *1.5*, p. *35*; *10.2*, p. *109*; 12.2, p. 141; 22.2.1, p. 443.

L. Bočánek and J. Musilová
Modulation Piezoreflectance of Cubic Polycrystals
(a) Vol. **75**, No. 2, p. 555 – 558 (1983).
See also *12.2*, p. *141*; 13.1, p. 156; *20.1*, p. *314*.

P. V. Smith
A Localized Orbital Description of Ge and α-Sn
(b) Vol. **116**, No. 1, p. 101 – 109 (1983).
See also *13.1*, p. *157*; 22.1, p. 417.

M. A. Ansari, Anil Kumar, and B. K. Srivastava
Thermal Conductivity of Germanium Containing Dislocations at Low Temperatures
(b) Vol. **116**, No. 1, p. 163 – 167 (1983).
See also 6, p. 71; *8*, p. *86*.

I. T. Godmanis, A. N. Trukhin, and K. Hübner
Exciton–Phonon Interaction in Crystalline and Vitreous SiO_2
(b) Vol. **116**, No. 1, p. 279 – 287 (1983).
See also *6*, p. *72*; *13.5.1*, p. *195*; 20.1, p. 315; 22.6, p. 492.

L. N. Aleksandrov
Formation of Semiconductor Epitaxial Films by Pulse Heating Crystallization or Regrowth
(a) Vol. **76**, No. 1, p. 179 – 190 (1983).
See also *1.5*, p. *36*; *3*, p. *57*; 8, p. 86; 22.1.2, p. 427.

N. E. Belova, F. Eichhorn, V. A. Somenkov, K. Utemisov, and S. Sh. Shilshtein
Analysis of the Inclination Method for Investigating Pendellösung Interferences of Neutrons and X-Rays
(a) Vol. **76**, No. 1, p. 257 – 265 (1983).
See also *1.1*, p. *11*; 22.1.2, p. 427.

A. Chovet, S. Cristoloveanu, Y. Daher, and V. K. Malyutenko
Useful Relations for Highly Sensitive Sensors Based on Carrier Concentration
(a) Vol. **76**, No. 1, p. K43 – K48 (1983).
See also *14.3.3*, p. *238*; 22.2.3, p. 452.

V. N. Ermakov, V. V. Kolomoets, and V. S. Timoshuk
The Deformation Potential Ξ_d^X, Ξ_u^X of the X-Valleys of the Germanium Conduction Band
(b) Vol. **116**, No. 2, p. K77 – K78 (1983).
See also *13.1*, p. *157*.

E. M. Trukhanov
Accommodation of Misfit in Heterostructures with Continuous and Island Films
(a) Vol. **76**, No. 2, p. 437–445 (1983).
See also *1.5*, p. *36*; *10.2*, p. *111*.

F. R. Kessler, K. Dettmer, and E.-W. Ritters
Ge–Al Thin Films for Selective Solar Absorbers
(a) Vol. **76**, No. 2, p. 533–541 (1983).
See also *14.3.1*, p. *231*; *20.1*, p. *317*.

E. G. Vyatkin, Yu. M. Filimonov, A. M. Taratin, and S. A. Vorobiev
Molecular-Type Channeling of Relativistic Electrons in Crystals
(b) Vol. **117**, No. 1, p. 41–50 (1983).
See also *11*, p. *127*; 13, p. 148.

J. Gryko
Optical Conductivity in the Impurity Band of Doped Semiconductors
(b) Vol. **117**, No. 1, p. 313–322 (1983).
See also *13.4*, p. *178*; *14.3*, p. *219*; 20.1, p. 318.

D. Ghose, D. Basu, and S. B. Karmohapatro
Sputtering of Germanium by 15 to 35 keV Noble Gas Ions
(a) Vol. **77**, No. 1, p. 121–125 (1983).
See also *11*, p. *127*.

D. A. Aronov and V. V. Rubinov
On the Minimum Coefficient of Light Absorption by Free Carriers under the Conditions of Exclusion in "Pure" Semiconductors with Non-Injecting Contacts
(a) Vol. **77**, No. 1, p. K41–K44 (1983).
See also *14.3.4*, p. *245*; *20.1*, p. *319*.

O. Bílek, L. Skála, and L. Künne
Tight-Binding Cluster Model for C, Si, and Ge (I): Analytic Solution for s-States in Diamond Lattice
(b) Vol. **117**, No. 2, p. 675–681 (1983).
See also *13*, p. *149*; 22.1, p. 417; 22.1.2, p. 431.

L. Künne, L. Skála, and O. Bílek
Tight-Binding Cluster Model for C, Si, and Ge (II): Size and Shape Dependence of Some Cluster Properties
(b) Vol. **118**, No. 1, p. 173–178 (1983).
See also *13*, p. *149*; 22.1, p. 417; 22.1.2, p. 432.

G. Paasch and H. Übensee
Carrier Density near the Semiconductor–Insulator Interface. Local Density Approximation for Non-Isotropic Effective Mass
(b) Vol. **118**, No. 1, p. 255–266 (1983).
See also 13.1, p. 160; *13.3*, p. *169*; *14.3.2*, p. *234*; 22.1.2, p. 432.

R. Krüssmann, H. Vollmer, and R. Labusch
The Activation Energy of Phosphorus Donors in Silicon-Rich Silicon–Germanium Alloys
(b) Vol. **118**, No. 1, p. 275–282 (1983).
See also *13.4*, p. *182*; 20.1, p. 321; 22.1.2, p. 432.

N. Ishii, M. Kumeda, and T. Shimizu
Defect States in Amorphous Germanium
(b) Vol. **118**, No. 1, p. K65–K68 (1983).
See also 2, p. 49; *13.4*, p. *182*; 19, p. 302.

S. J. Pearton and J. M. Kahn
Dislocations in Germanium: Effects of Plasma Hydrogenation
(a) Vol. **78**, No. 1, p. K65–K69 (1983).
See also *10.2*, p. *115*; *13.4*, p. *183*.

A. A. El-Sharkawy, M. S. Zaghloul, and A. M. Sanad
Study of Thermal Properties, Electrical Conductivity, and the Crystallization Process of the Glassy Alloy $SeGe_{20}$
(a) Vol. 78, No. 2, p. 457–462 (1983).
See also *3*, p. *59*; 8, p. 89; 14.3, p. 222; 22.1.3, p. 440.

A. F. Akkerman, A. L. Gibrekhterman, and A. V. Dvurechenskii
Absorbed Energy Distribution in Heterogeneous Semiconductor Structures
(a) Vol. 78, No. 2, p. K131–K135 (1983).
See also *11*, p. *131*; 22.1.2, p. 434.

V. V. Aristov, Yu. A. Osipyan, R. Scholz, I. I. Snighireva, I. I. Khodos, and S. A. Shevchenko
The Effect of Annealing on the Defect Structure and Electrical Properties of Deformed Single Crystals of Ge
(a) Vol. **79**, No. 1, p. 47–55 (1983).
See also *10.2*, p. *117*; *14.3*, p. *224*.

V. P. Dobrego and A. K. Fedotov
Influence of Growth Dislocations on Low Temperature Conductivity in Irradiated p-Germanium
(a) Vol. **79**, No. 1, p. K45–K48 (1983).
See also *10.2*, p. *118*; 11, p. 132; *14.3*, p. *224*.

F. H. Baumann and W. Schröter
Deformation-Induced Defects in p-Type Germanium
(b) Vol. **79**, No. 2, p. K123–K126 (1983),
See also *10.2*, p. *119*.

M. Stutzmann and J. Stuke
Temperature Dependence of ESR Spectra of Doped Amorphous Germanium
(b) Vol. **120**, No. 1, p. 225–234 (1983).
See also 2, p. 54; *13.4*, p. *189*; *19*, p. *305*.

M. Pasemann, D. Hoehl, A. L. Aseev, and O. P. Pchelyakov
Analysis of Rod-Like Defects in Silicon and Germanium by Means of High-Resolution Electron Microscopy
(a) Vol. **80**, No. 1, p. 135–139 (1983).
See also *1.4*, p. *34*; *11*, p. *133*; 22.1.2, p. 438.

U. Pietsch and K. Unger
The Influence of Free Carriers on the Equilibrium Lattice Parameter of Semiconductor Materials
(a) Vol. **80**, No. 1, p. 165–172 (1983).
See also *1.1*, p. *12*; 4, p. 64; 22.1.2, p. 438; 22.2.1, p. 447; 22.2.2, p. 451; 22.2.4, p. 455.

V. A. Bykovskii, A. V. Mudryi, V. P. Poskrebyshev, and V. D. Tkachev
Radiative Recombination Centers in Irradiated Germanium
(a) Vol. **80**, No. 2, p. K133–K136 (1983).
See also 11, p. 135; *13.4*, p. *191*; *20.3*, p. *347*.

22.1.2 Silicon

Review Articles

N. D. Stojadinović and S. D. Ristić
Failure Physics of Integrated Circuits and Relationship to Reliability
(a) Vol. **75**, No. 1, p. 11–48 (1983).
See also *14.3.3*, p. *236*.

L. N. Aleksandrov and F. L. Edelman
Shock Crystallization of Films
(a) Vol. **76**, No. 2, p. 409–427 (1983).
See also 1.3, p. 21; *1.5*, p. *35*; *3*, p. *57*; 22.1.1, p. 418; 22.1.3, p. 440; 22.6, p. 490.

M. Cardona
Vibrational Spectra of Hydrogen in Silicon and Germanium
(b) Vol. **118**, No. 2, p. 463–481 (1983).
See also *6*, p. *69*; *20.1*, p. *312*; 22.1.1, p. 418.

E. G. Sieverts
Classification of Defects in Silicon after Their *g*-Values
(b) Vol. **120**, No. 1, p. 11–29 (1983).
See also *10.2*, p. *108*; 13.4, p. 172; 19, p. 299.

Original Papers and Short Notes

A. M. Taratin and S. A. Vorobev
Electronic Energy Loss of Channeled Negative Pions
(b) Vol. **115**, No. 1, p. 37–46 (1983).
See also *11*, p. *122*; 13, p. 146.

M. Stutzmann, J. Stuke, and H. Dersch
Electron Spin Resonance of Doped Glow-Discharge Amorphous Germanium
(b) Vol. **115**, No. 1, p. 141–151 (1983).
See also 2, p. 43; *13.4*, p. *172*; 14.3, p. 215; *19*, p. *299*; 22.1.2, p. 418.

M. Downey and P. V. Smith
A Localized Orbital Description of Si
(b) Vol. **115**, No. 1, p. 255–266 (1983).
See also *13.1*, p. *155*.

A. Ferreira da Silva, M. Fabbri, and I. C. da Cunha Lima
Electron Hopping Energy Influence on the Specific Heat of Phosphorus-Doped Silicon
(b) Vol. **115**, No. 1, p. 311–316 (1983).
See also *8*, p. *85*; 13.4, p. 172.

M. Sato
Development of Faulted Dislocation Dipoles in Silicon during High Temperature Deformation
(a) Vol. **75**, No. 1, p. 107–116 (1983).
See also *10.2*, p. *109*.

V. E. Borisenko, V. V. Gribkovskii, V. A. Labunov, V. A. Samuilov, and K. D. Yashin
Incoherent Light Annealing of Phosphorus-Doped Polycrystalline Silicon
(a) Vol. **75**, No. 1, p. 117–120 (1983).
See also *14.3.1*, p. *229*; 20.1, p. 313.

S. Unterricker and F. Schneider
TDPAC Investigation of ^{111}Cd in Silicon after Recoil Implantation of the Parent Nuclei ^{111}In
(a) Vol. **75**, No. 1, p. 155–158 (1983).
See also *11*, p. *122*.

H. Pohlack
Thin-Film Photodetectors with High Quantum Yields Achieved by Induced Resonance Total Absorption
(a) Vol. **75**, No. 1, p. 179–185 (1983).
See also *14.3.1*, p. *229*; 14.3.3, p. 236; *20.1* p. *313*.

R. Mattheis
The Influence of Substrate Temperature on the Oriented Growth of Ni and Ni_2Si on Si (111)
(a) Vol. **75**, No. 1, p. 195–205 (1983).
See also 1.5, p. 35; *3*, p. *57*.

H. Kerkow, G. Kreysch, B. Lukasch, and G. Lück
Calorimetric Determination of the Formation Energy for an Amorphous Layer on a Crystalline Silicon Substrate
(a) Vol. **75**, No. 1, p. 251–254 (1983).
See also 2, p. 43; *11*, p. *122*.

D. Mergel
Recombination at Dislocations in Silicon
(a) Vol. **75**, No. 1, p. 255–262 (1983).
See also *13.4*, p. *173*; *16*, p. *264*.

M. K. Han, W. A. Anderson, P. Sung, and R. Lahri
Photogenerated Carrier Conduction Mechanisms for Schottky a-Si:H Solar Cells
(a) Vol. **75**, No. 1, p. 283–288 (1983).
See also *14.3.4* p. *243*; *16*, p. *264*.

W. Bergholz, S. Damgaard, J. W. Petersen, and G. Weyer
A Mössbauer-Spectroscopy Study of the Annealing of Supersaturated Solutions of ^{57}Co in Silicon
(a) Vol. **75**, No. 1, p. 289–300 (1983).
See also *5*, p. *65*.

D. P. Joshi and K. Sen
Effective Lifetime for Polycrystalline Solar Cells
(a) Vol. **75**, No. 1, p. 311–315 (1983).
See also *14.3.3*, p. *236*; *16*, p. *264*.

K. Irmscher, H. Klose, and K. Maass
Electric Field Enhanced Electron Emission from Gold Acceptor Level and A-Centre in Silicon
(a) Vol. **75**, No. 1, p. K25 – K28 (1983).
See also *13.4*, p. *173*; 14.3.4, p. 243.

H. Lemke
Ion Pairs between Chronium and Shallow Acceptors in Silicon
(a) Vol. **75**, No. 1, p. K49 – K51 (1983).
See also *13.4*, p. *174*; 14.3, p. 215.

K. B. Kadyrakunov, E. V. Nidaev, I. E. Tyshchenko, and L. S. Smirnov
Decomposition of Supersaturated Antimony Sólutions in Silicon Created by Pulsed Annealing of Ion-Doped Layers
(a) Vol. **75**, No. 1, p. K107 – K109 (1983).
See also *11*, p. *123*; *14.3*, p. *215*.

N. M. Ravindra, C. Ance, J. P. Ferraton, A. Donnadieu, and S. Robin
Penn-Like Model Calculations. Application to Hydrogenated Amorphous Silicon Films Prepared by Glow Discharge
(b) Vol. **115**, No. 2, p. 347 – 352 (1983).
See also *2*, p. *44*; 13.1, p. 156; 14.4, p. 250.

K. Takeda, Y. Maeda, E. Ohta, M. Sakata, G. Kido, and N. Miura
Study of Energy Band Parameters in p-Type $Ge_{0.9}Si_{0.1}$ Alloys
(b) Vol. **115**, No. 2, p. 369 – 379 (1983).
See also *13.1*, p. *156*; 14.3, p. 216; 20.1, p. 314; 22.1.1, p. 418.

W. Gehlhoff and K. H. Segsa
EPR of Iron–Boron Centres in Silicon
(b) Vol. **115**, No. 2, p. 443 – 453 (1983).
See also *13.4*, p. *174*; *19*, p. *299*.

K. H. J. Buschow, H.-J. Eifert, and B. Elschner
Electronic Structure and Stability of Amorphous Pd–Si Alloys
(b) Vol. **115**, No. 2, p. 455 – 462 (1983).
See also *2*, p. *44*; *13*, p. *147*; 19, p. 299; 21, p. 350.

J. Reichel
The Distinction of Different Kinds of Thermally Induced Defects in Silicon
(a) Vol. **75**, No. 2, p. 459 – 464 (1983).
See also *10.2*, p. *109*; 14.3, p. 216.

H. Lemke
Properties of Some Impurity Complexes of Gold in Silicon
(a) Vol. **75**, No. 2, p. 473 – 482 (1983).
See also *13.4*, p. *175*; 14.3, p. 216.

K. B. Kadyrakunov, E. V. Nidaev, A. E. Plotnikov, L. S. Smirnov, I. G. Melnik, and M. V. Makeev
Flash Lamp Annealing of Ion-Implanted Polycrystalline Silicon
(a) Vol. **75**, No. 2, p. 483–488 (1983).
See also 11, p. 123; *14.3.1*, p. *229*.

P. Türkes
An Ion-Implanted Resistor as Thermal Transient Sensor for the Determination of the Thermal Diffusivity in Silicon
(a) Vol. **75**, No. 2, p. 519–523 (1983).
See also *8*, p. *86*.

H. Pfleiderer and B. Rauscher
Capacitance of Amorphous Silicon pin Solar Cells
(a) Vol. **75**, No. 2, p. 537–545 (1983).
See also 2, p. 44; 13.4, p. 175; *14.3.1*, p. *230*; *14.3.3*, p. *237*; 16, p. 264.

G. Weber, B. Gillot, and P. Barret
Interface Structure in Relation with the Mechanisms in the Reaction Copper–Silicon
(a) Vol. **75**, No. 2, p. 567–576 (1983).
See also *1.3*, p. *22*; *1.6*, p. *39*.

V. D. Akhmetov, V. V. Bolotov, L. S. Smirnov, and V. A. Kharchenko
The Defect Accumulation and Irradiation-Induced Lattice Strain in Si at High-Dose Neutron Irradiation
(a) Vol. **75**, No. 2, p. 601–606 (1983).
See also 11, p. 124; *13.4*, p. *175*; *20.1*, p. *314*.

K. Sen, R. S. Srivastava, D. P. Joshi, and V. K. Goyal
Mobility and Its Effect on the Performance of Polycrystalline Solar Cells
(a) Vol. **75**, No. 2, p. 657–661 (1983).
See also 13.3, p. 167; *14.3.3*, p. *237*; *16*, p. *264*.

F. Fenske
Simple Improvement of the Lindner High-Frequency MOS Capacitance Approximation
(a) Vol. **75**, No. 2, p. K141–K144 (1983).
See also *13.3*, p. *167*; *14.3.3*, p. *237*.

N. Ishii, M. Kumeda, and T. Shimizu
Defect States in Amorphous Silicon
(b) Vol. **116**, No. 1, p. 91–100 (1983).
See also *2*, p. *45*; 10.2, p. 110; *13.4*, p. *176*.

V. N. Pavlovich
Photostimulated Diffusion of Light Substitutional Impurities in Semiconductors and Insulators
(b) Vol. **116**, No. 1, p. K9–K14 (1983).
See also *9*, p. *93*; 22, p. 405.

Yu.,V. Bulgakov, K. Lenkeit, and R. Stolle
Energy Distribution Measurements of 300 keV Transmitted Protons at the Axial-to-Planar Channeling Transition in Silicon
(b) Vol. **116**, No. 1, p. K23 – K26 (1983).
See also *11*, p. *124*.

Yu. V. Bulgakov and K. Lenkeit
The Energy Dependence of the Angles of Resonance Dechanneling of Ions in the ⟨0110⟩ → → (110) Axial-to-Planar Channeling Transition in Silicon
(b) Vol. **116**, No. 1, p. K27 – K30 (1983).
See also *11*, p. *124*.

V. V. Kalinin and N. N. Gerasimenko
The Interaction between the Point Defects Introduced by Implantation and Dislocations in Silicon
(a) Vol. **76**, No. 1, p. 65 – 70 (1983).
See also *10.2*, p. *110*; *11*, p. *124*.

L. N. Aleksandrov
Formation of Semiconductor Epitaxial Films by Pulse Heating Crystallization or Regrowth
(a) Vol. **76**, No. 1, p. 179 – 190 (1983).
See also *1.5*, p. *36*; *3*, p. *57*; 8, p. 86; 22.1.1, p. 419.

H. Lemke
Energy Levels and Binding Energies of Ion Pairs in Silicon
(a) Vol. **76**, No. 1, p. 223 – 234 (1983).
See also *13.4*, p. *176*; 14.3, p. 217.

N. E. Belova, F. Eichhorn, V. A. Somenkov, K. Utemisov, and S. Sh. Shilshtein
Analysis of the Inclination Method for Investigating Pendellösung Interferences of Neutrons and X-Rays
(a) Vol. **76**, No. 1, p. 257 – 265 (1983).
See also *1.1*, p. *11*; 22.1.1, p. 419.

A. I. Pekarev, G. F. Krasnova, and G. Z. Nemtsev
Orientation Dependence of Oxidation Stacking Fault Density in Silicon
(a) Vol. **76**, No. 1, p. 327 – 330 (1983).
See also *1.6*, p. *39*; *10.2*, p. *111*.

A. G. Kazanskii, V. F. Kiselev, E. A. Silaev, and V. S. Vavilov
On the Role of Surface Phenomena in the Staebler-Wronski Effect
(a) Vol. **76**, No. 1, p. 337 – 343 (1983).
See also 2, p. 46; *13.3*, p. *168*; *14.3.2*, p. *233*; 16, p. 265.

H. J. Goldsmid, M. M. Kaila, and G. L. Paul
Thermal Conductivity of Amorphous Silicon
(a) Vol. **76**, No. 1, p. K31 – K33 (1983).
See also *2*, p. *46*; *8*, p. *87*.

J. Belz, K. F. Heidemann, H. F. Kappert, and E. te Kaat
Anomalous Defect Interaction and Amorphization during Self Irradiation of Si Crystals at 450 K
(a) Vol. **76**, No. 1, p. K81 – K84 (1983).
See also *11*, p. *126*.

Dao Khac An, Le Hoang Mai, and Pham Hoi
Concentration Dependence of the Boron Diffusion Coefficient in Silicon
(a) Vol. **76**, No. 1, p. K85 – K88 (1983).
See also *9*, p. *93*.

E. Kührt and R. Wedell
Energy Losses in Crystalline and Amorphous Solids in the Intermediate Energy Region
(b) Vol. **116**, No. 2, p. 585 – 595 (1983).
See also *11*, p. *126*; 13.5.2, p. 200.

P. I. Baranskii, V. V. Kolomoets, and S. S. Korolyuk
The Non-Parabolicity of the n-Si Conduction Band, Caused by Elastic Deformation along the [111] Direction
(b) Vol. **116**, No. 2, p. K109 – K112 (1983).
See also *13.1*, p. *157*; *14.3*, p. *218*.

G. N. Gaidukov, S. K. Maksimov, and A. A. Podrezov
The Effect of the Crystal Anisotropy on the Energy of the Dislocation–Impurity Atmosphere Elastic Interaction in Silicon
(a) Vol. **76**, No. 2, p. 453 – 459 (1983).
See also *10.2*, p. *111*; *12.2*, p. *142*.

V. V. Aristov, M. N. Zolotukhin, V. V. Kveder, Yu. A. Osipyan, I. I. Snighireva, and I. I. Khodos
The Effect of Annealing on the Dislocation Dissociation in Plastically Deformed Silicon
(a) Vol. **76**, No. 2, p. 485 – 491 (1983).
See also *1.4*, p. *30*; *10.2*, p. *112*.

V. Kumar and S. B. Iyer
Characterization of Surface States in MOS Capacitors by a Modified DLTS Technique
(a) Vol. **76**, No. 2, p. 637 – 640 (1983).
See also *13.3*, p. *168*; 14.3.3, p. 238.

J. Dziesaty, F.-G. Kirscht, M. Mai, and K-R. Sprung
Influence of Backside Argon Implantation on the Si–SiO_2 Interface
(a) Vol. **76**, No. 2, p. K105 – K108 (1983).
See also *11*, p. *126*; *14.3.2*, p. *233*.

H.-J. Fitting and R. Technow
Electron Backscattering at Various Angles of Incidence
(a) Vol. **76**, No. 2, p. K151 – K154 (1983).
See also *11*, p. *127*; 17, p. 269; 21, p. 353.

H. Lemke
On the Chemical Nature of Recombination Centres in Si Power Devices
(a) Vol. **76**, No. 2, p. K193–K196 (1983).
See also *13.4*, p. *177*; *14.3.3*, p. *238*.

E. Wilk, W. Wesch, and K. Hehl
Defect Diffusion during Ion Implantation into GaAs
(a) Vol. **76**, No. 2, p. K197–K201 (1983).
See also *11*, p. *127*; *20.1*, p. *317*.

H. Kerkow, B. Lukasch, and G. Kreysch
Kinetics of Furnace Annealing of As-Implanted Silicon
(a) Vol. **76**, No. 2, p. K203–K207 (1983).
See also *11*, p. *127*.

I. V. Antonova, K. B. Kadyrakunov, E. V. Nidaev, and L. S. Smirnov
Diffusion of Iron and Gold in Silicon Annealed with Millisecond Pulses
(a) Vol. **76**, No. 2, p. K213–K215 (1983).
See also *9*, p. *94*.

J. Kočka, O. Štika, and J. Stuchlík
Comments on the Interpretation of DLTS Spectra of a-Si:H
(a) Vol. **76**, No. 2, p. K217–K220 (1983).
See also 2, p. 47; *13.4*, p. *178*; *14.3.2*, p. *233*.

G. V. Dedkov and A. M. Gorshokov
Study of Various Approximations for the Continuum Planar Potentials at Channeling (I): Monoatomic Crystals
(b) Vol. **117**, No. 1, p. 57–66 (1983).
See also *11*, p. *127*; 21.7, p. 401; 22.1, p. 417.

H. Bergner, V. Brückner, D. Dietrich, F. Kerstan, and W. Nowick
Picosecond Reflectivity Measurements on Glow Discharge a-Si:H
(b) Vol. **117**, No. 1, p. 197–202 (1983).
See also 2, p. 48; *20.1*, p. *318*.

N. T. Bagraev
Electron Localization on Defects and Optical Nuclear Polarization in Disordered and Semimagnetic Semiconductors
(b) Vol. **117**, No. 1, p. 367–379 (1983).
See also *13.4*, p. *178*; *20*, p. *308*.

H. van Cong, S. Brunet, S. Charar, S. Benet, C. Delseny, and J. C. Martin
Effective Electric Field in the n-Type Moderately Doped Region of Silicon Devices at 300 K
(b) Vol. **117**, No. 1, p. K23–K26 (1983).
See also *14.3* p. *219*.

T. Gessner, M. Pasemann, and B. Schmidt
Calculation of Neutron-Induced Defect Clusters in Silicon and Comparison with TEM Investigations
(a) Vol. **77**, No. 1, p. 133–138 (1983).
See also 1.4, p. 31; *11*, p. *127*.

M. KITTLER and K.-W. SCHRÖDER
Determination of Semiconductor Parameters and of the Vertical Structure of Devices by Numerical Analysis of Energy-Dependent EBIC Measurements
(a) Vol. **77**, No. 1, p. 139–151 (1983).
See also 1.5, p. 36; 14.3.2, p. 233; *14.3.3*, p. *238*.

H. G. SCHNEIDER, H. KARGE, and B. WEBER
Characterization of Heteroepitaxial Silicon Layers on Sapphire by Rutherford-Backscattering (RBS) Analyses
(a) Vol. **77**, No. 1, p. 201–205 (1983).
See also *1.5*, p. *37*.

H. KLOSE, A. MERTENS, R. REETZ, and NG. TANG
Ion Beam Contacting of Silicon (I): Aluminium in p-Silicon
(a) Vol. **77**, No. 1, p. 233–239 (1983).
See also *11*, p. *128*; *14.3.4*, p. *244*.

C. ANCE and N. M. RAVINDRA
Departure of Hydrogen from a-Si:H. A Spectroscopic Study
(a) Vol. **77**, No. 1, p. 241–248 (1983).
See also *2*, p. *48*; *10.2*, p. *112*.

A. HÜBLER and D. MARTON
Energy Dependence of the Si–SiO_2 Interface State Cross-Section Measured by the TSC Method
(a) Vol. **77**, No. 1, p. K17–K21 (1983).
See also *13.3*, p. *168*; 22.6, p. 493.

F. GĂISEANU
On the Diffusion Profile of Boron in Silicon at High Concentrations
(a) Vol. **77**, No. 1, p. K59–K61 (1983).
See also *9*, p. *94*.

A. L. GOLOVIN and R. M. IMAMOV
Investigation of the X-Ray Scattering Intensity for the Laue-Case Diffraction under Total-External-Reflection Conditions
(a) Vol. **77**, No. 1, p. K91–K94 (1983).
See also *1.6*, p. *40*.

A. M. KANLOEV
Kumakhov Radiation Energy Losses of Ultrarelativistic Electrons and Positrons
(b) Vol. **117**, No. 2, p. 507–511 (1983).
See also *11*, p. *128*; *20*, p. *308*.

H. VAN CONG
Band-Gap Narrowing in n-Type Moderately Doped Silicon at 300 K
(b) Vol. **117**, No. 2, p. 575–584 (1983).
See also *13.1*, p. *159*; *13.4*, p. *179*.

H. BERGNER, V. BRÜCKNER, F. KERSTAN, and W. NOWICK
Comparative Studies of Fast Recombination Processes in Amorphous Silicon Films
(b) Vol. **117**, No. 2, p. 603–610 (1983).
See also *2*, p. *48*; *20.3*, p. *339*.

O. Bílek, L. Skála, and L. Künne
Tight-Binding Cluster Model for C, Si, and Ge (I): Analytic Solution for s-States in Diamond Lattice
(b) Vol. **117**, No. 2, p. 675–681 (1983).
See also *13*, p. *149*; 22.1, p. 417; 22.1.1, p. 420.

H. van Cong, S. Charar, S. Brunet, and J. C. Martin
Transport Parameters in the n-Type Moderately Doped Region of Silicon Devices at 300 K
(b) Vol. **117**, No. 2, p. K61–K65 (1983).
See also *14.3*, p. *220*.

D. Panknin, E. Wieser, R. Klabes and H. Syhre
Dose Dependence of the Flash Lamp Annealing of Arsenic-Implanted Silicon
(a) Vol. **77**, No. 2, p. 553–559 (1983).
See also *11*, p. *128*; 14.3, p. 220.

P. Gaworzewski and K. Schmalz
Oxygen-Related Donors Formed at 600 °C in Silicon in Dependence on Oxygen and Carbon Content
(a) Vol. 77, No. 2, p. 571–582 (1983).
See also *13.4*, p. *180*; *14.3*, p. *220*.

R. N. Litovskii, V. S. Lysenko, A. N. Nazarov, T. E. Rudenko, S. B. Kaschieva, and N. R. Nedev
The Influence of Defect Surface Layers on the Capacitive Properties of MOS Structures
(a) Vol. **77**, No. 2, p. 699–707 (1983).
See also 13.3, p. 169; *14.3.3* p. *239*.

M. M. Chandra and M. Prasad
Schottky Barrier Characteristics at Low Temperatures
(a) Vol. **77**, No. 2, p. 715–719 (1983).
See also *14.3.3*, p. *239*.

A. Szekeres and S. Alexandrova
RF Annealing of Defects Induced in SiO_2 by Oxygen Plasma
(a) Vol. **77**, No. 2, p. 721–724 (1983).
See also *10.2*, p. *113*.

I. L. Baginskii, E. G. Kostsov, and E. E. Meerson
Pulse Technique for Flat-Band Voltage Measurements in MIS Structures
(a) Vol. **77**, No. 2, p. K99–K102 (1983).
See also *14.3.3*, p. *239*.

P. Zaumseil and U. Winter
Detection of Surface Imperfections at Polished Silicon Wafers by TCD Measurements
(a) Vol. **77**, No. 2, p. K139–K141 (1983).
See also *1.6*, p. *40*; *10.2*, p. *113*.

A. V. Efanov and M. V. Entin
Optical Orientation and Polarized Luminescence in Silicon
(b) Vol. **118**, No. 1, p. 63–72 (1983).
See also 13.5.1, p. 197; *20.3*, p. *341*.

L. Künne, L. Skála, and O. Bílek
Tight-Binding Cluster Model for C, Si, and Ge (II): Size and Shape Dependence of Some Cluster Properties
(b) Vol. **118**, No. 1, p. 173–178 (1983).
See also *13*, p. *149*; 22.1, p. 417; 22.1.1, p. 420.

G. Paasch and H. Übensee
Carrier Density near the Semiconductor–Insulator Interface. Local Density Approximation for Non-Isotropic Effective Mass
(b) Vol. **118**, No. 1, p. 255–266 (1983).
See also 13.1, p. 160; *13.3*, p. *169*; *14.3.2*, p. *234*; 22.1.1, p. 420.

R. Krüssmann, H. Vollmer, and R. Labusch
The Activation Energy of Phosphorus Donors in Silicon-Rich Silicon–Germanium Alloys
(b) Vol. **118**, No. 1, p. 275–282 (1983).
See also *13.4*, p. *182*; 20.1, p. 321; 22.1.1, p. 421.

F. Bechstedt, R. Enderlein, and D. Reichardt
Electronic Relaxation Effects in Core Level Spectra of Surfaces and Interfaces
(b) Vol. **118**, No. 1, p. 327–336 (1983).
See also *13.3*, p. *169*; 14.4, p. 252; 21.6, p. 398; 22.2.1, p. 445; 22.6, p. 494.

V. A. Muralev
Channeling of Ultrarelativistic Electrons in Crystals
(b) Vol. **118**, No. 1, p. 363–372 (1983).
See also *11*, p. *129*; 21, p. 356.

P. H. Bezirganyan and V. G. Aslanyan
About a Feature of X-Ray Diffraction in Imperfect Crystals
(a) Vol. **78**, No. 1, p. 157–161 (1983).
See also *1.1*, p. *12*; 22.6, p. 494.

H.-D. Geiler, K. Hehl, and D. Stock
Deposition of Laser Energy into Inhomogeneous Layer Systems
(a) Vol. **78**, No. 1, p. 193–199 (1983).
See also *8*, p. *88*; *11*, p. *129*.

H. Hedler and G. Götz
Incorporation of Sb in Si during CW-Laser Stimulated Solid Phase Epitaxy
(a) Vol. **78**, No. 1, p. 237–241 (1983).
See also *1.2* p. *17*; 11, p. 129; 14.3, p. 221.

M. Katayama, H. Morimoto, and K. Sugawara
Electron Irradiation Effects on Amorphous Silicon Solar Cells
(a) Vol. **78**, No. 1, p. K5–K8 (1983).
See also 11, p. 129; *14.3.3*, p. *240*; *16*, p. *266*.

B. N. Mukashev, Zh. A. Kusainov, K. Kh. Nusupov, and S. Zh. Tokmoldin
Electrical Properties of Shallow Implanted Layers in Silicon
(a) Vol. 78, No. 1, p. K19–K22 (1983).
See also *11*, p. *129*; *14.3*, p. *221*.

W. Gehlhoff and K. H. Segsa
Determination of Stresses in Si–SiO_2 and Mechanically Affected Si by EPR
(b) Vol. **118**, No. 2, p. 703–712 (1983).
See also *12.2*, p. *143*; *19*, p. *303*; 22.6, p. 495.

K. Hübner, H. Köster, Jr., B. Derlich, and W. Ecke
DLTS Investigations of Si–SiO_2 Interface States of Electron-Beam Irradiated MOS Structures
(b) Vol. **118**, No. 2, p. K133–K136 (1983).
See also *11*, p. *130*; *13.3*, p. *169*; 14.3.2, p. 234; 22.6, p. 495.

R. Köhler and W. Möhling
Sensitivity of Plane Wave Topography to Microdefects
(a) Vol. **78**, No. 2, p. 489–496 (1983).
See also *1.4*, p. *33*; 10.2, p. 116.

R. Gleichmann, H. Blumtritt, and J. Heydenreich
New Morphological Types of CuSi Precipitates in Silicon and Their Electrical Effects
(a) Vol. **78**, No. 2, p. 527–538 (1983).
See also *10.2*, p. *116*; 14.3.3, p. 240.

B. Rauscher, H. Pfleiderer, and B. Bullemer
Tail–Tail Luminescence in Amorphous Silicon
(a) Vol. **78**, No. 2, p. 623–632 (1983).
See also *2*, p. *50*; *20.3*, p. *343*.

M. Suezawa and K. Sumino
The Nature of Photoluminescence from Plastically Deformed Silicon
(a) Vol. **78**, No. 2, p. 639–645 (1983).
See also *10.2*, p. *116*; *20.3*, p. *343*.

Y. Nishino and T. Imura
Dislocation Configurations Induced by Cyclic Deformation in Silicon Crystals
(a) Vol. **78**, No. 2, p. 655–663 (1983).
See also *10.2*, p. *116*.

P. I. Baranskii, V. M. Babich, N. P. Baran, Yu. P. Dotsenko, V. B. Kovalchuk, and V. A. Shershel
Investigation of Formation Conditions of Thermal Donors-I and -II in Oxygen-Containing n-Type Silicon within the Temperature Range 400 to 800 °C
(a) Vol. **78**, No. 2, p. 733–739 (1983).
See also *13.4*, p. *184*; *14.3*, p. *223*; 19, p. 303.

P. V. Petrashen and I. L. Shulpina
X-Ray Determination of the Surface Compression Force in the Ion-Doped Layer in Silicon
(a) Vol. 78, No. 2, p. K105–K108 (1983).
See also *10.2*, p. *117*; *11*, p. *131*.

N. I. Berezhnov, V. F. Stelmakh, and A. R. Chelyadinskii
The Interstitial Type Defects in Ion Implanted Silicon
(a) Vol. 78, No. 2, p. K121–K125 (1983).
See also *10.2*, p. *117*; *11*, p. *131*.

A. F. Akkerman, A. L. Gibrekhterman, and A. V. Dvurechenskii
Absorbed Energy Distribution in Heterogeneous Semiconductor Structures
(a) Vol. 78, No. 2, p. K131–K135 (1983).
See also *11*, p. *131*; 22.1.1, p. 421.

K. Schmalz and P. Gaworzewski
On the Effect of Ambients on the Formation of Oxygen-Related Donors in Cs-Si
(a) Vol. 78, No. 2, p. K141–K145 (1983).
See also *10.2*, p. *117*; 13.4, p. 184; 14.3, p. 223.

N. N. Gerasimenko, A. E. Gershinskii, A. Yu. Surtaev, S. A. Sokolov, E. I. Cherepov, B. I. Fomin, and G. V. Timofeeva
Silicon–Nickel Silicide Schottky Barriers Formed by Ion Mixing
(a) Vol. 78, No. 2, p. K151–K154 (1983).
See also 11, p. 131; *14.3.3*, p. *240*.

L. I. Datsenko, E. N. Kislovskii, V. I. Khrupa, and K. M. Shevtsov-Kazmirchuk
Effect of Twisting of the Diffraction Net Planes on the Integral Intensities at Laue Diffraction of X-Rays in Elastically Bent Silicon Crystals
(a) Vol. 78, No. 2, p. K175–K178 (1983).
See also *1.1*, p. *12*; *10.2*, p. *117*.

V. V. Mazurenko, V. V. Petrov, and V. D. Tkachev
Investigation of the Single-Phonon Absorption Band at 230 cm^{-1} in Silicon Irradiated by Neutrons
(b) Vol. 119, No. 1, p. K69–K71 (1983).
See also *6*, p. *78*; *11*, p. *131*; 20.1, p. 326.

A. V. Dvurechenskii, I. A. Ryazantsev, L. S. Smirnov, H. Klose, and M. Rieth
Amorphous Silicon–Phosphorus Layers
(a) Vol. 79, No. 1, p. 83–86 (1983).
See also 2, p. 51; 13.4, p. 186; *14.3.1*, p. *231*; 16, p. 267; 19, p. 304.

B. Rauscher and B. Bullemer
Photoluminescence Decay in Amorphous Silicon Alloys
(a) Vol. 79, No. 1, p. 161–166 (1983).
See also *2*, p. *51*; *20.3*, p. *344*.

M. Suezawa, Y. Sasaki, and K. Sumino
Dependence of Photoluminescence on Temperature in Dislocated Silicon Crystals
(a) Vol. **79**, No. 1, p. 173–181 (1983).
See also *13.4*, p. *186*; *20.3*, p. *344*.

H. Tsuya and F. Shimura
Transient Behaviour of Intrinsic Gettering in CZ Silicon Wafers
(a) Vol. **79**, No. 1, p. 199–206 (1983).
See also *1.6*, p. *41*; *10.2*, p. *118*; 14.3.3, p. 240.

A. Wolkenberg
A Mechanism for the Effect of Doping on the Silicon Native Oxide Thickness
(a) Vol. **79**, No. 1, p. 313–322 (1983).
See also *1.6*, p. *41*; 10.2, p. 118; 22.6, p. 496.

L. Nanu, Doina Bărbulescu, I. N. Mihăilescu, V. Teodorescu, and Leona C. Nistor
Annealing of Implanted Silicon under the Action of Microsecond Pulsed TEA-CO_2 Laser Radiation
(a) Vol. **79**, No. 1, p. K5–K8 (1983).
See also *1.5*, p. *38*; *11*, p. *132*.

K. Biskupska, J. Tatarkiewicz, and K. Wieteska
Comparison of Fast Electron and Thermal Annealing of Proton Bombarded Silicon – X-Ray and IR Spectroscopy Study
(a) Vol. **79**, No. 1, p. K13–K15 (1983).
See also 10.2, p. 118; *11*, p. *132*; 20.1, p. 326.

J. Koshy
Electrical Properties of Junctions between Sputter Deposited Silicon Films and Monocrystalline Silicon
(a) Vol. **79**, No. 1, p. K21–K24 (1983).
See also *14.3.2*, p. *234*; *14.3.3*, p. *240*.

H. K. Phan, L. H. Phu, and C. V. Chiem
Stress Effects on the Switching Voltage of Metal–SnO_2–Si(n)–Si(p^+) Devices
(a) Vol. **79**, No. 1, p. K25–K28 (1983).
See also *14.3.3*, p. *241*; *14.3.4*, p. *247*.

H. Bergner and V. Brückner
Dependence of the Ambipolar Diffusion in Silicon on the Carrier Density
(a) Vol. **79**, No. 1, p. K85–K88 (1983).
See also *14.3*, p. *224*; 20.1, p. 327.

V. E. Arkhincheev and S. K. Savvinikh
Semiempirical Theory of Shallow Short-Range Electron Traps in Silicon in a Magnetic Field
(b) Vol. **119**, No. 2, p. 459–466 (1983).
See also *13.4*, p. *187*.

M. Höhne
Gold-Related EPR Centres of Low Symmetry in Silicon
(b) Vol. **119**, No. 2, p. K117–K121 (1983).
See also 13.4, p. 187; *19*, p. *304*.

L. A. Kazakevich and P. F. Lugakov
The Effect of Dislocations on the Interaction Processes of Point and Group Radiation Defects in Silicon
(a) Vol. **79**, No. 2, p. 377–380 (1983).
See also *11*, p. *132*; 14.3, p. 225.

V. I. Kuznetsov and P. F. Lugakov
Parameters of Defect Clusters in Irradiated Silicon and a Method for Their Determining
(a) Vol. **79**, No. 2, p. 381–388 (1983).
See also *11*, p. *132*; 14.3, p. 225.

F. Fenske
Influence of Lateral Charge Inhomogeneities in MOS Structures on the Determination of the Interface State Density Distribution from HF CV Curves
(a) Vol. **79**, No. 2, p. 417–421 (1983).
See also *13.3*, p. *170*; *14.3.3*, p. *241*.

S. A. Abo-Namous, Y. Zaka, and R. W. Fane
Dependence of the Electronic and Optical Properties of Unhydrogenated a-Si on Preparation Conditions
(a) Vol. **79**, No. 2, p. 477–482 (1983).
See also *14.3*, p. *225*; 16, p. 267; *20.1*, p. *328*.

V. Atzrodt and H. Lange
Principal Component Analysis as a Method for Silicide Investigation with Auger Electron Spectroscopy
(a) Vol. **79**, No. 2, p. 489–496 (1983).
See also *1.6*, p. *42*; 17, p. 270; 21, p. 360.

A. A. Frantsuzov and A. P. Kovchavtsev
The Determination of Surface Potential Fluctuation Values and Spatial Distribution of Traps in Insulators from the Temperature Dependence of the $(G_p/\omega) - \omega$ Curves of MOS Structures
(a) Vol. **79**, No. 2, p. 503–511 (1983).
See also *13.3*, p. *170*; *14.3.3*, p. *241*.

H. Feichtinger and R. Czaputa
On the Question of Multiple Charge States of Interstitial Fe and Cr in Silicon
(a) Vol. **79**, No. 2, p. K143–K146 (1983).
See also *13.4*, p. *188*.

K. D. Glinchuk, N. M. Litovchenko, Z. A. Salnik, and S. I. Skryl
Effect of Heat Treatment on the Minority Carrier Lifetime in Oxygen-Containing Silicon
(a) Vol. **79**, No. 2, p. K159–K163 (1983).
See also *10.2*, p. *119*; *13.4*, p. *188*.

P. I. Baranskii, V. V. Savyak, and Y. V. Simonenko
Effect of Heat Treatment and Cooling Conditions on the Anisotropy Parameters of Drag TEMF in Neutron-Doped Silicon Crystals
(a) Vol. **79**, No. 2, p. K165–K170 (1983).
See also *14.3*, p. *226*; *15*, p. *263*.

G. Kaden
On the Influence of the Surface Potential on the Photoinjection from the Metal Electrode in MOS Structures
(a) Vol. **79**, No. 2, p. K187–K191 (1983).
See also *14.3.3*, p. *241*.

W. Frentrup, M. Griepentrog, and U. Müller-Jahreis
The Influence of Cesium Implanted in Silicon on the Secondary Ion Emission
(a) Vol. **79**, No. 2, p. K193–K196 (1983).
See also **11**, p. **133**; *17*, p. *270*.

J. Gong, C. M. van Vliet, D. Schoenfeld, and G. Bosman
Noise from Hole-Initiated Photoavalanche Processes
(a) Vol. **79**, No. 2, p. K201–K203 (1983).
See also *14.3.3*, p. *242*; *14.3.4*, p. *248*; **16**, p. **267**.

U. Pietsch
Hydrostatic Pressure and Uniaxial Stress Dependence of the Silicon (222) X-Ray Reflection Power
(b) Vol. **120**, No. 1, p. 183–188 (1983).
See also *4*, p. *63*; *13*, p. *152*.

W. Fuhs, H. M. Welsch, and D. C. Booth
Optical Quenching of Photoconductivity in a-Si:H Films
(b) Vol. **120**, No. 1, p. 197–205 (1983).
See also 2, p. 53; 14.3.1, p. 232; *16*, p. *267*.

D. Della Sala, F. Evangelisti, P. Fiorini, and C. Giovannella
Influence of Oxygen on the Photoconductivity of Hydrogenated Amorphous Silicon
(b) Vol. **120**, No. 1, p. K59–K62 (1983).
See also 2, p. 54; *16*, p. *267*.

M. Asche and H. Kostial
Experimental Indication of the Presence of Loops for Hot Electron Properties in Many-Valley Semiconductors
(b) Vol. **120**, No. 1, p. K83–K85 (1983).
See also *13*, p. *152*; *14.3.4*, p. *248*.

S. S. Simeonov and M. D. Ivanovitch
Determination of Quasi-Fermi Level at the Metal–Semiconductor Interface in a Forward-Biased Schottky Barrier Diode
(a) Vol. **80**, No. 1, p. 79–84 (1983).
See also *14.3.3*, p. *242*.

M. PASEMANN, D. HOEHL, A. L. ASEEV, and O. P. PCHELYAKOV
Analysis of Rod-Like Defects in Silicon and Germanium by Means of High-Resolution Electron Microscopy
(a) Vol. **80**, No. 1, p. 135–139 (1983).
See also *1.4*, p. *34*; *11*, p. *133*; 22.1.1, p. 422.

U. PIETSCH and K. UNGER
The Influence of Free Carriers on the Equilibrium Lattice Parameter of Semiconductor Materials
(a) Vol. **80**, No. 1, p. 165–172 (1983).
See also *1.1*, p. *12*; 4, p. 64; 22.1.1, p. 422; 22.2.1, p. 447; 22.2.2, p. 451; 22.2.4, p. 455.

M. SERVIDORI, C. DAL MONTE, and Q. ZINI
Double Crystal X-Ray Analysis of Phosphorus Precipitation in Supersaturated Si–P Solid Solutions
(a) Vol. **80**, No. 1, p. 277–285 (1983).
See also *10.2*, p. *120*; 11, p. 133.

L. HRIVNÁK
On the Determination of Carrier Mobilities and Densities in Semi-Insulating GaAs
(a) Vol. **80**, No. 1, p. 317–323 (1983).
See also *14.3*, p. *227*.

V. V. BORIMSKII, K. D. GLINCHUK, N. M. LITOVCHENKO, and Z. A. SALNIK
Annealing Behaviour of Oxygen-Induced Recombination Centres in Silicon
(a) Vol. **80**, No. 1, p. 343–348 (1983).
See also *13.4*, p. *189*.

A. K. RAI, R. S. BHATTACHARYA, and P. P. PRONKO
Annealing of Amorphous Layer in Si Implanted GaAs
(a) Vol. **80**, No. 1, p. 367–372 (1983).
See also *1.4*, p. *34*; 10.2, p. 120; 11, p. 134.

J. TATARKIEWICZ
Fast Electron Annealing of IR Absorption Bands of Proton Bombarded Silicon
(a) Vol. **80**, No. 1, p. K49–K51 (1983).
See also *11*, p. *134*; *20.1*, p. *329*.

A. L. GOLOVIN and R. M. IMANOV
Obtaining Quantitative Information on Amorphous Layer Thickness on Crystal Surface Using X-Ray Diffraction under Specular Reflection Conditions
(a) Vol. **80**, No. 1, p. K63–K65 (1983).
See also *1.6*, p. *42*.

V. S. ARAKELYAN and A. A. AVAKYAN
Diffusion of Interstitial Admixture in Solids Due to Standing Ultrasound Waves
(a) Vol. **80**, No. 1, p. K71–K74 (1983).
See also *7*, p. *84*; *9*, p. *96*; 22.2.1, p. 447.

P. B. Allen
A Tetrahedron Method for Doubly Constrained Brillouin Zone Integrals. Application to Silicon Optic Phonon Decay
(b) Vol. **120**, No. 2, p. 529–538 (1983).
See also *6*, p. *80*; *13.1*, p. *165*.

H. Bergner, V. Brückner, D. Dietrich, and W. Nowick
Picosecond Recombination of Photoexcited Carriers in Glow Discharge a-Si:H Films of Different Deposition Temperature
(b) Vol. **120**, No. 2, p. 655–658 (1983).
See also 2, p. 55; *20.1*, p. *330*.

K. Lenkeit, R. Wedell, and P. Gehrmann
Nonlinear Effect of Particle Motion under Axial-to-Planar Channeling Transition Conditions
(b) Vol. **120**, No. 2, p. K195–K199 (1983).
See also *11*, p. *134*.

Yu. V. Bulgakov, K. Lenkeit, and R. Wedell
Measurements of the Ionic Charge of Hydrogen and Helium Ions Moving through Silicon by Means of the Resonance Dechanneling Effect
(b) Vol. **120**, No. 2, p. K201–K204 (1983).
See also *11*, p. *134*.

F. Eichhorn, J. Kulda, and P. Mikula
Study of Asymmetric Neutron Diffraction Including the Transition from Bragg to Laue Case
(a) Vol. **80**, No. 2, p. 483–489 (1983).
See also *1.1*, p. *12*.

K. Lal and P. Thoma
Study of Microstructural Changes in Silicon Single Crystals Subject to High Electric Fields by Using a High Resolution X-Ray Diffraction Method
(a) Vol. **80**, No. 2, p. 491–502 (1983).
See also *1.1*, p. *13*; *14.4*, p. *254*.

B. P. Tyagi and K. Sen
On the Resistivity of Polycrystalline Silicon
(a) Vol. **80**, No. 2, p. 679–684 (1983).
See also *13.4*, p. *191*; *14.3.4*, p. *249*.

D. S. Gupta, M. M. Chandra, and V. Kumar
Direct Measurement of Interface State Electron Capture Cross-Section in MOS Systems by DLTS Technique
(a) Vol. **80**, No. 2, p. K209–K212 (1983).
See also *13.3*, p. *171*; *14.3.2*, p. *235*.

22.1.3 Group VI Elements

Review Article

L. N. Aleksandrov and F. L. Edelman
Shock Crystallization of Films
(a) Vol. **76**, No. 2, p. 409–427 (1983).
See also 1.3, p. 21; *1.5*, p. *35*; *3*, p. *57*; 22.1.1, p. 418; 22.1.2, p. 423; 22.6, p. 490.

Original Papers and Short Notes

H. Atmani, P. Michon, and C. Vautier
Determination of the Crystallization Parameters of Amorphous Selenium Thin Films by Optical Microscopy
(a) Vol. **75**, No. 1, p. K5–K8 (1983).
See also *1.5*, p. *35*.

G. P. Srivastava, K. N. Tripathi, and N. K. Sehgal
Effect of Laser Irradiation on the Electrical Properties of Polycrystalline Thin Films of Tellurium
(a) Vol. **76**, No. 2, p. K137–K140 (1983).
See also *11*, p. *127*; *14.3.1*, p. *231*.

G. Juška and K. Arlauskas
Features of Hot Carriers in Amorphous Selenium
(a) Vol. **77**, No. 1, p. 387–391 (1983).
See also *2*, p. *48*; *14.3.4*, p. *245*.

A. K. Sharma
Thickness Dependence of the Thermoelectric Power of Tellurium Films
(a) Vol. **77**, No. 1, p. K81–K85 (1983).
See also *15*, p. *261*.

A. A. El-Sharkawy, M. S. Zaghloul, and A. M. Sanad
Study of Thermal Properties, Electrical Conductivity, and the Crystallization Process of the Glassy Alloy $SeGe_{20}$
(a) Vol. **78**, No. 2, p. 457–462 (1983).
See also *3*, p. *59*; 8, p. 89; 14.3, p. 222; 22.1.1, p. 421.

K. H. Goetz and W. Richter
Interaction of Free Hole Intraband Excitations with the A_1-Phonon in Highly Doped (p-Type) Tellurium
(b) Vol. **119**, No. 2, p. 631–638 (1983).
See also *6*, p. *79*; *13.1*, p. *163*; 20.1, p. 327.

D. Hasselkamp and A. Scharmann
Ion-Induced Electron Emission from Carbon
(a) Vol. **79**, No. 2, p. K197–K200 (1983).
See also *17*, p. *270*.

I. E. Bolotov and V. L. Prilepo

On the Vacancy Mechanism of the Occurrence of Compressive Stresses during the Crystallization of Amorphous Films

(a) Vol. **80**, No. 1, p. K67 – K70 (1983).

See also *1.5*, p. *38*; 3, p. 59; 10.2, p. 120; 12.2, p. 144; 22.4.1, p. 465.

P. Koštial, L. Malík, and L. Both

New Aspect of Ultrasonic Detection of Some Significant Structural Parameters of Amorphous Se–Te Alloys

(a) Vol. **80**, No. 2, p. K167 – K170 (1983).

See also*2*, p. *56* ; *7*, p. *84*.

22.2 III-V-Compounds

Original Papers and Short Notes

HOANG NGOC CAM, NGUYEN VAN HIEU, and HA VINH TAN
On the Theory of the Nonlinear Acousto-Optical Effect in Semiconductors
(b) Vol. **116**, No. 1, p. 25–29 (1983).
See also *7*, p. *82*; 13.5.1, p. 195; *20*, p. *307*; 22.4, p. 460.

U. RENSCH and G. EICHHORN
Investigations on the Structure of MOCVD AlN Layers on Silicon
(a) Vol. **77**, No. 1, p. 195–199 (1983).
See also *1.5*, p. *37*; *3*, p. *58*.

E. P. POKATILOV and S. I. BERIL
Electron–Phonon Interaction in Periodic Two-Layer Structures
(b) Vol. **118**, No. 2, p. 567–573 (1983).
See also *6*, p. *76*; 13.3, p. 169.

T. ITO
A Pseudopotential Approach to the Lattice Parameters for GaAs Containing Impurity Elements
(b) Vol. **119**, No. 1, p. 307–314 (1983).
See also *10.2*, p. *117*; 12.2, p. 143; 22.2.1, p. 446.

T. SOMA and H.-MATSUO KAGAYA
Binding Force Including Ionic Interactions and Bulk Properties for Tetrahedral Compounds
(b) Vol. **119**, No. 1, p. K41–K45 (1983).
See also *4*, p. *63*; 12.2, p. 143; 22.4, p. 461.

J. WALACHOVÁ and D. NOHAVICA
Determination of p–n Junctions in InP/GaInAsP Laser Heterostructures Using a Profilometer
(a) Vol. **79**, No. 1, p. K69–K72 (1983).
See also *14.3.3*, p. *241*.

T. SOMA and H.-MATSUO KAGAYA
Local Heine-Abarenkov Model Potential for III–V and II–VI Covalent Compounds
(b) Vol. **119**, No. 1, p. 547–553 (1983).
See also *4*, p. *63*; 13.1, p. 163; 22.4, p. 461.

A. C. SHARMA, N. M. RAVINDRA, S. AULUCK, and V. K. SRIVASTAVA
Temperature-Dependent Effective Masses in III–V Compound Semiconductors
(b) Vol. **120**, No. 2, p. 715–721 (1983).
See also *13.1*, p. *165*.

22.2.1 Arsenides

Original Papers and Short Notes

A. I. Bobrysheva, S. I. Beryl, V. T. Zyukov, S. A. Moskalenko, and E. P. Pokatylov
The One- and Two-Photon Absorption Due to Surface State Excitons
(b) Vol. **115**, No. 1, p. 153–160 (1983).
See also *13.3*, p. *167*; *13.5.1*, p. *194*.

Yen-chu Wang and Zheng-an Qiu
Velocity Overshoot of Electrons in GaAs with Space Charge and Non-Uniform Field
(a) Vol. **75**, No. 1, p. 67–72 (1983).
See also *14.3.3*, p. *236*.

O. G. Alaverdova, M. Ya. Fuks, L. S. Khazan, L. P. Koval, L. A. Matveeva, I. F. Mikhailov, N. N. Soldatenko, and Yu. A. Tkhorik
Coherency of the Interphase Boundary and Elastic Strain in the Epitaxial System Ge/GaAs
(a) Vol. **75**, No. 2, p. 367–371 (1983).
See also *1.5*, p. *35*; *10.2*, p. *109*; 12.2, p. 141; 22.1.1, p. 419.

H. Gotoh, K. Sasamoto, S. Kuroda, and M. Kimata
Molecular Beam Epitaxy of AlSb on GaAs and GaSb on AlSb Films
(a) Vol. **75**, No. 2, p. 641–645 (1983).
See also *1.5*, p. *36*; *3*, p. *57*; 14.3.1, p. 230; 22.2.3, p. 452.

G. V. Berenshtein, A. M. Dyachenko, and I. E. Maronchuk
The Degradation of Semiconductor Mechanical Properties Induced by Microinclusions
(a) Vol. **75**, No. 2, p. K129–K132 (1983).
See also *10.2*, p. *110*; *12.2*, p. *141*.

S. Sarkar
Second Harmonic Suppression in Delayed Domain Mode of GaAs Samples
(a) Vol. **75**, No. 2, p. K161–K163 (1983).
See also *14.3.4*, p. *244*.

Yu. B. Bolkhovityanov
The Lattice-Pulling Effect Induced by the Film Stresses. Conditions of Existence and Discovery
(a) Vol. **76**, No. 1, p. 85–93 (1983).
See also *1.5*, p. *36*; *3*, p. *57*; 22.2.2, p. 448.

R. S. Bhattacharya, A. K. Rai, S. C. Ling, and P. P. Pronko
Epitaxial Regrowth of Si Implanted (100) and (211) GaAs
(a) Vol. **76**, No. 1, p. 131–136 (1983).
See also *1.4*, p. *30*; *11*, p. *125*.

S. Nojima
Isolation Characteristics in Selectively O^+ and Cr^+ Implanted GaAs
(a) Vol. **76**, No. 1, p. K9–K12 (1983).
See also *11*, p. *125*; *14.3.3*, p. *238*.

J. Bonnafé, M. de Murcia, M. Castagné, and J. P. Fillard
Infrared Photomodulation of the Band Edge Transmission in Semi-Insulating GaAs
(a) Vol. **76**, No. 2, p. K147 – K150 (1983).
See also *20.1*, p. *317*.

D. Flamm and E.-H. Weber
Adsorption of Oxygen on GaAs (001) Surfaces by UPS
(a) Vol. **76**, No. 2, p. K163 – K166 (1983).
See also 13.3, p. 168; *17*, p. *269*.

I. Thurzo, K. Gmucová, J. Červenák, and T. Lalinský
GaAs–Anodic Oxide Interface Examination by Deep-Level Transient Current Spectroscopy
(a) Vol. **77**, No. 1, p. 323 – 330 (1983).
See also 13.3, p. 168; *14.3.2*, p. *234*.

J. Betko and K. Měřínský
Geometrical Magnetoresistance Effect in Semiconductors with Mixed Conductivity. Determination of the Electron Hall Mobility in Semi-Insulating GaAs
(a) Vol. **77**, No. 1, p. 331 – 337 (1983).
See also *14.3*, p. *219*.

V. N. Brudnyi, A. I. Potapov, Yu. V. Rud, and M. Serginov
Electrical Properties of Electron-Irradiated GaAs and $ZnSiAs_2$ Solid Solutions
(a) Vol. **77**, No. 1, p. K33 – K36 (1983).
See also *11*, p. *128*; 13.4, p. 179; *14.3*, p. *219*; 22.8, p. 505.

Th. Harbich and G. Mahler
Bistability of the Electron–Hole Plasma Density in a Strong Electromagnetic Field
(b) Vol. **117**, No. 2, p. 635 – 644 (1983).
See also *13.1*, p. *159*; *13.5*, p. *193*.

K. P. Ghatak and A. N. Chakravarti
Carrier Statistics in Superlattices of Small-Gap Semiconductors in the Presence of a Quantizing Magnetic Field
(b) Vol. **117**, No. 2, p. 707 – 715 (1983).
See also *13.1*, p. *159*; 22.2.3, p. 452.

P. Kleinert
Cluster–Bethe–Lattice Treatment of the Silicon Pair Defect in Gallium Arsenide
(b) Vol. **117**, No. 2, p. K91 – K94 (1983).
See also *6*, p. *74*; *20.1*, p. *320*.

O. Breitenstein and T. Wosinski
Scanning-DLTS Investigation of the EL 2 Level in Plastically Deformed GaAs
(a) Vol. **77**, No. 2, p. K107 – K110 (1983).
See also *10.2*, p. *113*; *13.4*, p. *181*; 14.3, p. 221.

R. V. KONAKOVA, YU. A. TKHORIK, I. L. ZAITSEVSKII, P. KORDOS, M. MORVIČ, and J. CERVENAK
The Influence of Stress on the Fine Structure of Schottky Diode *I–U* Characteristics in the Breakdown Region
(a) Vol. **77**, No. 2, p. K125–K128 (1983).
See also *12.2*, p. *142*; *14.3.4*, p. *246*; 22.2.4, p. 454.

K. N. KANG and S. CRISTOLOVEANU
On the Role of Exact Boundary Conditions in Phototransport Phenomena: Case of SI–GaAs
(a) Vol. **77**, No. 2, p. K155–K159 (1983).
See also *14.3.2*, p. *234*; *16*, p. *266*.

R. CHAOUI and B. TUCK
Very Short Diffusions of Silver in the III–V Semiconductors
(a) Vol. **77**, No. 2, p. K189–K192 (1983).
See also *9*, p. *94*; 22.2.2, p. 450.

TH. JUNG and E. NEBAUER
Interdiffusion Profiles of AuGe/n-GaAs Ohmic Contacts Studied by AES
(a) Vol. **77**, No. 2, p. K203–K206 (1983).
See also *9*, p. *94*; 14.3.4, p. 246.

V. YA. PRINZ and S. N. RECHKUNOV
Influence of a Strong Electric Field on the Carrier Capture by Nonradiative Deep-Level Centers in GaAs
(b) Vol. **118**, No. 1, p. 159–166 (1983).
See also *13.4*, p. *181*; *14.3.4*, p. *246*.

T. SOMA and H.-MATSUO KAGAYA
The Metallic and Ionic Contributions to Lattice Vibrations of III–V Covalent Crystals
(b) Vol. **118**, No. 1, p. 245–254 (1983).
See also *4*, p. *61*; 6, p. 75; 22.2.3, p. 452.

F. BECHSTEDT, R. ENDERLEIN, and D. REICHARDT
Electronic Relaxation Effects in Core Level Spectra of Surfaces and Interfaces
(b) Vol. **118**, No. 1, p. 327–336 (1983).
See also *13.3*, p. *169*; 14.4, p. 252; 21.6, p. 398; 22.1.2, p. 432; 22.6, p. 494.

P. MÜLLER
Contactless Determination of the Conductivity in Semi-Insulators
(a) Vol. **78**, No. 1, p. 41–51 (1983).
See also *14.3.3*, p. *239*; *14.3.4*, p. *246*; 22.5.1, p. 476.

R. S. BHATTACHARYA, A. K. RAI, and P. P. PRONKO
Oxygen Implantation in GaAs: Damage Studies
(a) Vol. **78**, No. 1, p. 283–286 (1983).
See also *1.4*, p. *32*; *11* p. *129*.

A. K. SHARMA and P. JAYARAMA REDDY
Thermoelectric Power in InAs Films Obtained by Flash Evaporation
(a) Vol. **78**, No. 1, p. K27–K29 (1983).
See also *15*, p. *262*.

O. K. Gorodnichenko, V. A. Presnov, and N. D. Vasilenko
Deformation of Epitaxial Structures on the Basis of GaAs in the Process of Their Production
(a) Vol. **78**, No. 2, p. K171 – K173 (1983).
See also *1.5*, p. *38*; *10.2*, p. *117*.

T. Ito
A Pseudopotential Approach to the Lattice Parameters for GaAs Containing Impurity Elements
(b) Vol. **119**, No. 1, p. 307 – 314 (1983).
See also *10.2*, p. *117*; 12.2, p. 143; 22.2, p. 442.

P. Kleinert
Cluster Bethe Lattice Treatment for the Localized Vibrational Mode of Carbon in GaAs
(b) Vol. **119**, No. 1, p. K37 – K40 (1983).
See also *6*, p. *77*.

O. K. Gorodnichenko, E. A. Glushkov, V. F. Kovalenko, and N. D. Vasilenko
Strengthening of GaAs by Impurity Atom–Vacancy Complexes
(a) Vol. **79**, No. 1, p. 331 – 335 (1983).
See also *10.2*, p. *118*; 20.3, p. 344.

J. Grňo, Š. Bederka, and M. Veselý
Pulse Radiation Annealing of Capless GaAs
(a) Vol. **79**, No. 1, p. K41 – K44 (1983).
See also *11*, p. *132*; *14.3*, p. *224*.

N. S. Orlova
Variation of Phonon Dispersion Curves with Temperature in Indium Arsenide Measured by X-Ray Thermal Diffuse Scattering
(b) Vol. **119**, No. 2, p. 541 – 546 (1983).
See also *6*, p. *79*; 20, p. 310.

E. E. Velmre and B. P. Freidin
Numerical Modeling of the Electrothermal Transient Process in Diode Structures Based on Direct-Gap Semiconductors
(a) Vol. **79**, No. 2, p. K131 – K134 (1983).
See also 8, p. 90; *14.3.3*, p. *241*.

D. V. Korbutyak and V. G. Litovchenko
Expansion of the Electron–Hole Liquid Plasma at GaAs Surface
(b) Vol. **120**, No. 1, p. 87 – 97 (1983).
See also *13.3*, p. *170*; 13.5.1, p. 198; *20.3*, p. *346*.

L. Wendler and E. Jäger
Phonon-Polaritons in Bilayer Systems. Surface Phonon-Polaritons
(b) Vol. **120**, No. 1, p. 235 – 247 (1983).
See also *6*, p. *80*; 13.5, p. 193; *20.1*, p. *328*; 22.2.2, p. 451; 22.5.3, p. 487.

M. Mondal and K. P. Ghatak
Effect of Size Quantization on the Electronic Heat Capacity in Ultrathin Films of Semiconductors
(b) Vol. **120**, No. 1, p. K63 – K68 (1983).
See also *8*, p. *91*; *13.3*, p. *170*.

U. Pietsch and K. Unger
The Influence of Free Carriers on the Equilibrium Lattice Parameter of Semiconductor Materials
(a) Vol. **80**, No. 1, p. 165 – 172 (1983).
See also *1.1*, p. *12*; 4, p. 64; 22.1.1, p. 422; 22.1.2, p. 438; 22.2.2, p. 451; 22.2.4, p. 455.

P. Bogdanski, R. Carin, O. Gardais, and J. Vizot
Characterisation of MIS Structures (Al–Si_3N_4–GaAs) by Different Capacitive Methods
(a) Vol. **80**, No. 1, p. 185 – 193 (1983).
See also *13.3*, p. *171*; *14.3.2*, p. *234*.

T. Jung, L. Roth, and H. Wittrich
Auger Analysis of Plasma-Grown GaAs Oxide Films
(a) Vol. **80**, No. 1, p. K11 – K14 (1983).
See also *11*, p. *134*; 22.6, p. 497.

I. Mojzes, R. Veresegyházy, V. Obodnikov, Yu. Vaulin, and S. Vasilev
Mapping the Redistributed Components of the Au–Ge/GaAs System after Scanning Laser Annealing
(a) Vol. **80**, No. 1, p. K33 – K36 (1983).
See also *14.3.2* p. *235*; 14.3.4, p. 248.

V. S. Arakelyan and A. A. Avakyán
Diffusion of Interstitial Admixture in Solids Due to Standing Ultrasound Waves
(a) Vol. **80**, No. 1, p. K71 – K74 (1983).
See also *7*, p. *84*; *9*, p. *96*; 22.1.2, p. 438.

M. Bhattacharya, S. Mukhopadhyay, and T. K. Mitra
Study of a Bound Surface Optical Polaron by a Non-Perturbative Method
(b) Vol. **120**, No. 2, p. 555 – 564 (1983).
See also *13.3*, p. *171*; *13.5.3*, p. *202*; 22.4.1, p. 465; 22.5.2, p. 485.

O. Z. Alekperov, V. G. Golubev, V. I. Ivanov-Omskii, and A. Sh. Mekhtiev
Plasma Shift of Cyclotron Resonance Line in n-GaAs
(b) Vol. **120**, No. 2, p. K179 – K181 (1983).
See also *13*, p. *154*; *14.3*, p. *227*; 18, p. 276.

22.2.2 Phosphides

Original Papers and Short Notes

V. Schulz and L. Merten
Dispersion Curves of Stimulated Raman Effect in Piezoelectric Cubic Crystals
(b) Vol. **115**, No. 1, p. 225–233 (1983).
See also *6*, p. *69*; 13.5, p. 192; *20*, p. *307*.

C. Engler and W. Lorenz
Erratum to: Calculation of Local Densities of States, Local Charges, and Adsorption Energy on Large Semiconductor Substrate with Single Adsorbate
(b) Vol. **115**, No. 1, p. 325 (1983).
See also 4, p. 60; *13.1*, p. *155*; 13.3, p. 167.

K. Žďánský, L. Kratěna, M. Matyáš, Jr., and J. Zelinka
Absolute Energy of the Nitrogen-Related Electron Trap in Gallium Phosphide
(a) Vol. **75**, No. 1, p. 335–339 (1983).
See also *13.4*, p. *173*; *14.3.3*, p. *236*.

S. A. Azimov, D. A. Aronov, V. M. Rubinov, and M. Tuichiev
Quantum Yield of Current Carriers Excited by Electron Beams and Electron Penetration Depth in Thin GaP Films
(a) Vol. **75**, No. 1, p. K63–K68 (1983).
See also *11*, p. *123*; *14.3.1*, p. *229*.

A. Mazel, B. Legros-De Mauduit, and F. Reynaud
Electron Microscopic Investigation of the Contrast on Dislocation Loops Seen Edge-on in InP (B3 Structure)
(a) Vol. **75**, No. 2, p. K121–K123 (1983).
See also *10.2*, p. *109*.

O. Boehnke, E. Nowak, P. Paufler, and H. Neumann
On a Slip-Induced Anisotropy of the Electrical Resistance of GaP
(a) Vol. **75**, No. 2, p. K137–K139 (1983).
See also *10.2*, p. *110*; *14.3*, p. *217*.

Yu. B. Bolkhovityanov
The Lattice-Pulling Effect Induced by the Film Stresses. Conditions of Existence and Discovery
(a) Vol. **76**, No. 1, p. 85–93 (1983).
See also *1.5*, p. *36*; *3*, p. *57*; 22.2.1, p. 443.

G. Eftekhari, D. de Cogan, and B. Tuck
Electrical Conduction through Anodic Oxides on InP
(a) Vol. **76**, No. 1, p. 331–336 (1983).
See also *14.3.1*, p. *230*; *14.3.2*, p. *233*.

H. Weinert
Decay Behaviour of High-Excitation Luminescence in GaP:N (I): Model for the Decay of Two Coexisting Phases
(b) Vol. **116**, No. 2, p. 525–532 (1983).
See also *13.5*, p. *192*; *20.3*, p. *337*.

B. Pödör, J. Pfeifer, L. Csontos, N. Nádor, and F. Deák
Electrical Properties of Donors in Gallium Phosphide
(a) Vol. **76**, No. 2, p. 695–704 (1983).
See also *13.4*, p. *177*; *14.3*, p. *218*.

A. D'Amico, G. Petrocco, G. Stochino, A. Salsano, and G. Luzzi
Grain Boundaries as Possible Origin of Flicker Noise in GaP Polycrystalline Material
(a) Vol. **76**, No. 2, p. K125–K129 (1983).
See also *1.4*, p. *31*; *14.3.2*, p. *233*.

W. Lorenz, C. Engler, A. Nechwatal, and G. Adler
Quantumchemical Studies of Chemisorption at Anion or Cation Sites of the GaP (110) Surface. Coverage Dependence of Adsorption Energy and Equilibrium in Cases of Positive and Negative Partial Charge Injection
(b) Vol. **117**, No. 1, p. 203–212 (1983).
See also *13.3*, p. *168*.

H. Weinert, R. Bindemann, and K. Kreher
Decay Behaviour of the High-Excitation Luminescence in GaP:N (II): Experimental Results on the Densities of the Coexisting Phases and the Relevant Recombination Processes
(b) Vol. **117**, No. 1, p. 247–253 (1983).
See also *13.5*, p. *192*; *20.3*, p. *338*.

W. Siegel, E. Ziegler, and G. Kühnel
Impurity Parameters of Undoped and Zn-Doped p-Type GaP with Two Acceptor Levels Determined by Hall Measurements
(a) Vol. **77**, No. 1, p. 167–173 (1983).
See also *13.4*, p. *178*; *14.3*, p. *219*.

A. S. Popov, D. B. Kushev, and Sh. Chadraabal
A Model for Photoconductivity of Fe-Doped GaP
(a) Vol. **77**, No. 1, p. K53–K57 (1983).
See also *13.4*, p. *179*; *16*, p. *265*.

J. Jimenez, M. A. Gonzalez, V. Carbayo, and J. Bonnafe
On the Thermal Evolution of the Photoconductivity in Bulk Fe-Doped Semi-Insulating InP
(a) Vol. **77**, No. 1, p. K69–K73 (1983).
See also *13.4*, p. *179*; *16*, p. *266*.

R. Schwabe, I. Streit, F. Thuselt, and K. Unger
Experimental Investigations of Electron–Hole Plasma in Tellurium-Doped GaP
(b) Vol. **117**, No. 2, p. 625–634 (1983).
See also *13.5*, p. *192*; *20.3*, p. *339*.

SH. CHADRAABAL, A. S. POPOV, and D. B. KUSHEV
Photoconductivity in n-Type InP:Fe
(a) Vol. **77**, No. 2, p. 709–714 (1983).
See also *13.4*, p. *181*; *16*, p. *266*.

B. RHEINLÄNDER
Diffusion-Limited Carrier Lifetime in GaP:N LED
(a) Vol. **77**, No. 2, p. K111–K114 (1983).
See also *14.3.3*, p. *239*; *20.3*, p. *340*.

O. YU. BORKOVSKAYA, N. L. DMITRUK, R. V. KONAKOVA, V. I. FINEBERG, and V. I. SHAKHOVTSOV
The Radiation Effects on the Metal–GaP Interface
(a) Vol. **77**, No. 2, p. K129–K133 (1983).
See also *11*, p. *129*; *14.3.4*, p. *246*.

R. CHAOUI and B. TUCK
Very Short Diffusions of Silver in the III–V Semiconductors
(a) Vol. **77**, No. 2, p. K189–K192 (1983).
See also *9*, p. *94*; 22.2.1, p. 445.

M. RÖSLER, F. THUSELT, and R. ZIMMERMANN
Theory of Electron–Hole Liquid in Doped Semiconductors. Application to GaP
(b) Vol. **118**, No. 1, p. 303–317 (1983).
See also *13.5*, p. *193*.

S. A. AZIMOV, D. A. ARONOV, V. M. RUBINOV, and M. TUICHIEV
The Influence of Heating on the Characteristics of the Electron-Excited Conduction in Gallium Phosphide Films
(a) Vol. **78**, No. 1, p. K53–K58 (1983).
See also *11*, p. *130*; *14.3.1*, p. *231*.

E. YU. BRAILOVSKII, G. N. ERITSYAN, and N. E. GRIGORYAN
Optical Absorption in InP Crystals with Point Radiation Defects
(a) Vol. **78**, No. 2, p. K113–K115 (1983).
See also *11*, p. *131*; *20.1*, p. *325*.

O. GOEDE and D. HENNIG
Isovalent-Impurity Clusters in Koster-Slater Approximation
(b) Vol. **119**, No. 1, p. 261–268 (1983).
See also *13.1*, p. *162*; *13.4*, p. *185*; 22.4.1, p. 464.

G. IRMER, V. V. TOPOROV, B. H. BAIRAMOV, and J. MONECKE
Determination of the Charge Carrier Concentration and Mobility in n-GaP by Raman Spectroscopy
(b) Vol. **119**, No. 2, p. 595–603 (1983).
See also 6, p. 79; 13.5.2, p. 200; *14.3*, p. *224*; *20.1*, p. *327*.

L. Z. MIRETS, G. P. PEKA, S. I. RADAUTSAN, and E. V. RUSSU
Photoelectric Investigation of Recombination Centre Parameters in InP(Fe)
(a) Vol. **79**, No. 2, p. 593–597 (1983).
See also 13.4, p. 187; *16*, p. *267*.

B. Pödör
Zone Edge Phonons in Gallium Phosphide
(b) Vol. **120**, No. 1, p. 207–213 (1983).
See also *6*, p. *80*.

L. Wendler and E. Jäger
Phonon-Polaritons in Bilayer Systems. Surface Phonon-Polaritons
(b) Vol. **120**, No. 1, p. 235–247 (1983).
See also *6*, p. *80*; 13.5, p. 193; *20.1*, p. *328*; 22.2.1, p. 446; 22.5.3, p. 487.

A. N. Georgobiani, A. V. Mikulyonok, I. G. Stoyanova, and J. M. Tiginyanu
Non-Equilibrium Carrier Radiative Recombination in Indium Phosphide Single Crystals
(a) Vol. **80**, No. 1, p. 109–118 (1983).
See also *13.4*, p. *189*; *20.3*, p. *346*.

G. Kühnel, W. Siegel, and E. Ziegler
Electrical and Photoelectrical Properties of Semi-Insulating GaP
(a) Vol. **80**, No. 1, p. 159–164 (1983).
See also *13.4*, p. *189*; *14.3*, p. *226*; 16, p. 267.

U. Pietsch and K. Unger
The Influence of Free Carriers on the Equilibrium Lattice Parameter of Semiconductor Materials
(a) Vol. **80**, No. 1, p. 165–172 (1983).
See also *1.1*, p. *12*; 4, p. 64; 22.1.1, p. 422; 22.1.2, p. 438; 22.2.1, p. 447; 22.2.4, p. 455.

K. Löhnert and E. Kubalek
On the Degradation of Electroluminescence Efficiency in Gallium Phosphide Green Light Emitting Diodes
(a) Vol. **80**, No. 1, p. 173–183 (1983).
See also *14.3.3*, p. *242*; *20.3*, p. *346*.

M. Kusaka, M. Hirai, and S. Okazaki
Change in Capacitance of InP MIS Diodes Caused by Piezoelectric Effect
(a) Vol. **80**, No. 1, p. K25–K28 (1983).
See also *14.3.2*, p. *235*; *14.3.3*, p. *242*.

P. Deus, U. Voland, and H. A. Schneider
Thermal Expansion of GaP within 20 to 300 K
(a) Vol. **80**, No. 1, p. K29–K32 (1983).
See also *8*, p. *91*.

B. Rheinländer, R. Pickenhain, H. Haefner, and G. Oelgart
Degradation of LPE–GaP:N LED
(a) Vol. **80**, No. 2, p. K203–K207 (1983).
See also *14.3.4*, p. *249*; *20.3*, p. *348*.

22.2.3 Antimonides

Original Papers and Short Notes

L. Kończewicz and W. Trzeciakowski
Highly Localized Donor States in InSb and Their Pressure Dependence
(b) Vol. **115**, No. 2, p. 359–367 (1983).
See also 12.2, p. 141; *13.4*, p. *174*; 14.3, p. 216.

H. Gotoh, K. Sasamoto, S. Kuroda, and M. Kimata
Molecular Beam Epitaxy of AlSb on GaAs and GaSb on AlSb Films
(a) Vol. **75**, No. 2, p. 641–645 (1983).
See also *1.5*, p. *36*; *3*, p. *57*; 14.3.1, p. 230; 22.2.1, p. 443.

A. Chovet, S. Cristoloveanu, Y. Daher, and V. K. Malyutenko
Useful Relations for Highly Sensitive Sensors Based on Carrier Concentration
(a) Vol. **76**, No. 1, p. K43–K48 (1983).
See also *14.3.3*, p. *238*; 22.1.1, p. 419.

R. Meisels and F. Kuchar
Far-Infrared Photoconductivity Spectroscopy of Shallow Acceptors in InSb
(b) Vol. **116**, No. 2, p. 557–560 (1983).
See also *13.4*, p. *177*; *16*, p. *265*.

A. M. Berezhkovskii and A. A. Ovchinnikov
Temperature Dependence of the Wannier-Stark Level Width in a Semiconductor in a Strong Electric Field
(b) Vol. **117**, No. 1, p. 289–299 (1983).
See also *6*, p. *73*; *13*, p. *148*; 14.3.4, p. 244.

K. D. Chaudhuri, A. Malik, S. Kukreti, and P. C. Mathur
Impurity Band Conduction in Single Crystals of p-Type GaSb
(b) Vol. **117**, No. 1, p. 407–416 (1983).
See also *13.4*, p. *178*; *14.3*, p. *219*.

K. P. Ghatak and A. N. Chakravarti
Carrier Statistics in Superlattices of Small-Gap Semiconductors in the Presence of a Quantizing Magnetic Field
(b) Vol. **117**, No. 2, p. 707–715 (1983).
See also *13.1*, p. *159*; 22.2.1, p. 444.

E. W. Kreutz
Size Effects on the Magnetoresistance of p-InSb
(a) Vol. **77**, No. 2, p. 583–594 (1983).
See also *14.3.2*, p. *234*.

T. Soma and H.-Matsuo Kagaya
The Metallic and Ionic Contributions to Lattice Vibrations of III–V Covalent Crystals
(b) Vol. **118**, No. 1, p. 245–254 (1983).
See also *4*, p. *61*; 6, p. 75; 22.2.1, p. 445.

L. M. Kanskaya, S. I. Kokhanovskii, R. P. Seysyan, and Al. L. Efros
Direct Determination of Indium Antimonide Energy Band Parameters from Diamagnetic Exciton Spectra
(b) Vol. **118**, No. 1, p. 447–452 (1983).
See also *13.1*, p. *160*; *13.5.1*, p. *197*; 20.1, p. 322.

E. W. Kreutz
Magnetoresistance Thickness Dependence of n-InSb
(a) Vol. **78**, No. 1, p. K71–K75 (1983).
See also *14.3*, p. *222*.

O. Ziep and M. Mocker
New Arguments to the Problem of Screening in Auger Recombination in Semiconductors
(b) Vol. **119**, No. 1, p. 299–306 (1983).
See also *13.1*, p. *163*; 22.8, p. 508.

S. Ghosh and S. Khan
Parametric Conversion of an Electron Plasma Wave into an Electromagnetic Wave in Longitudinally Magnetized n-Type Piezoelectric Semiconductor Plasmas
(b) Vol. **119**, No. 1, p. 363–373 (1983).
See also *13.5.2*, p. *200*; 18, p. 275.

A. K. Sharma and P. Jayarama Reddy
Temperature and Thickness Dependence of Electrical Conductivity of Flash-Evaporated InSb Films
(a) Vol. **79**, No. 1, p. 183–188 (1983).
See also *14.3.1*, p. *232*.

B. R. Sethi, P. K. Goyal, O. P. Sharma, and P. C. Mathur
Negative Magnetoresistance in Heavily Doped and Compensated n-Type InSb
(b) Vol. **119**, No. 2, p. 721–726 (1983).
See also *14.3*, p. *225*.

22.2.4 Mixed Crystals

Original Papers and Short Notes

G. Oelgart and H. Scholz
Determination of Electron Depth–Dose Function for Kilovolt Electrons on GaAsP
(a) Vol. **75**, No. 2, p. 547–553 (1983).
See also *11*, p. *124*; *14.3.3*, p. *237*.

D. Tjapkin and V. Milanović
The Influence of Parameter Nonuniformity on the Two-Dimensional Einstein Relation in Periodic and Non-Periodic Semiconductor Structures
(b) Vol. **116**, No. 2, p. 653–658 (1983).
See also *13.1*, p. *157*; *14.3.2*, p. *233*.

A. I. Basyk, V. F. Kovalenko, and G. P. Peka
Transfer Mechanisms of Nonequilibrium Charge Carriers in Graded Band-Gap Semiconductors
(a) Vol. **76**, No. 2, p. 429–435 (1983).
See also *13.1*, p. *157*; *20.3*, p. *337*.

W. Kowalsky, H.-H. Wehmann, F. Fiedler, and A. Schlachetzki
Optical Absorption and Refractive Index near the Bandgap for InGaAsP
(a) Vol. **77**, No. 1, p. K75–K80 (1983).
See also *20.1*, p. *319*.

W. Drabik and J. Toruń
On the Analytical Methods of Line Defect and Beam Direction Determinations
(a) Vol. **77**, No. 2, p. 513–519 (1983).
See also *1.4*, p. *32*.

B. Rheinländer, G. Oelgart, R. Reulke, H. Haefner, and R. Stegmann
Degradation of Yellow Light-Emitting $GaAs_{0.1}P_{0.9}$:N Diodes (I): Implanted Diodes
(a) Vol. **77**, No. 2, p. 527–534 (1983).
See also *14.3.3*, p. *239*; 14.3.4, p. 245; *20.3*, p. *340*.

J. C. Lou, M. S. Lin, J. I. Chyi, M. D. Lin, and S. Sche
Electrical and Optical Characteristics of $GaAs_{0.6}P_{0.4}$ LEDs Fabricated by Zn Semi-Closed Diffusion Method
(a) Vol. **77**, No. 2, p. 741–747 (1983).
See also *14.3.3*, p. *239*; 14.3.4, p. 245; *20.3*, p. *340*.

R. V. Konakova, Yu. A. Tkhorik, I. L. Zaitsevskii, P. Kordos, M. Morvič, and J. Cervenak
The Influence of Stress on the Fine Structure of Schottky Diode *I–U* Characteristics in the Breakdown Region
(a) Vol. **77**, No. 2, p. K125–K128 (1983).
See also *12.2*, p. *142*; *14.3.4*, p. *246*; 22.2.1, p. 445

P. Kleinert
Vibrational Behaviour of Mixed Diatomic Crystals $AB_{1-c}C_c$ within the Bethe Lattice
(b) Vol. **118**, No. 1, p. 283–291 (1983).
See also *6*, p. *75*; *20.1*, p. *322*; 22.4.4, p. 471.

G. Oelgart, B. Rheinländer, R. Reulke, H. Haefner, and R. Stegmann
Degradation of Yellow Light-Emitting $GaAs_{0.1}P_{0.9}$:N Diodes (II): Diffused Diodes
(a) Vol. **78**, No. 1, p. 169–176 (1983).
See also *14.3.3*, p. *240*; 14.3.4, p. 246; *20.3*, p. *342*.

Yu. B. Bolkhovityanov, B. A. Bobylev, V. I. Korzhov, V. I. Yudaev, and E. H. Hairi
Growth Peculiarities of Graded Band Gap GaAsSb/GaAs Heterostructures and Effective Light Modulation in the 1.06 nm Wavelength Region
(a) Vol. **78**, No. 1, p. 343–347 (1983).
See also *3*, p. *58*; 20.1, p. 323.

N. R. Nurtdinov, R. Stegmann, and M. Griepentrog
The Influence of the Variation of the Chemical Composition and of the Nitrogen Doping Level on the Energy Spectra of the Bound Excitons in $GaAs_{1-x}P_x$:N
(b) Vol. **118**, No. 2, p. 613–621 (1983).
See also 10.2, p. 115; *13.5.1*, p. *197*; *20.3*, p. *342*.

R. Pickenhain, B. Rheinländer, and S. Brehme
P_{Ga} Antisite Defect in Zn-Implanted $GaAs_{0.1}P_{0.9}$ LED
(a) Vol. **79**, No. 1, p. K61–K63 (1983).
See also *13.4*, p. *186*; *20.3*, p. *344*.

I. D. Henning and H. Thomas
Process-Induced Defect States in GaAsP Light Emitting Diodes
(a) Vol. **79**, No. 2, p. 567–574 (1983).
See also *13.4*, p. *187*; 14.3.3, p. 241; 20.3, p. 345.

U. Pietsch and K. Unger
The Influence of Free Carriers on the Equilibrium Lattice Parameter of Semiconductor Materials
(a) Vol. **80**, No. 1, p. 165–172 (1983).
See also *1.1*, p. *12*; 4, p. 64; 22.1.1, p. 422; 22.1.2, p. 438; 22.2.1, p. 447; 22.2.2, p. 451.

22.3 IV-IV, II-V, and III-VI Compounds

Original Papers and Short Notes

M. SINGH, J. LEOTIN, and P. R. WALLACE
Theory of the Diffusivity–Mobility Ratio in Cadmium Arsenide
(b) Vol. **115**, No. 1, p. 105 – 113 (1983).
See also *14.3*, p. *215*; 18, p. 273.

H. SOBOTTA, H. NEUMANN, N. N. SYRBU, and V. RIEDE
Infrared Lattice Vibration Spectra of Monoclinic ZnP_2
(b) Vol. **115**, No. 1, p. K55 – K58 (1983).
See also *6*, p. *70*; *20.1*, p. *313*.

D. V. KRISHNA SASTRY and P. JAYARAMA REDDY
Hall Effect Measurements in High-Temperature Deposited and Heat Treated Films of InTe and InSe
(a) Vol. **75**, No. 2, p. K183 – K185 (1983).
See also *14.3.1*, p. *230*.

A. N. CHAKRAVARTI, K. P. GHATAK, K. K. GHOSH, and H. M. MUKHERJEE
Effect of Varying the Orientation of a Quantizing Magnetic Field on the Effective Electron Mass in Degenerate Tetragonal Semiconductors
(b) Vol. **116**, No. 1, p. 17 – 24 (1983).
See also *13.1*, p. *156*; 18, p. 273.

V. G. FEDOTOV, A. G. BYCHKOV, D. N. KARLIKOV, L. M. GORYNYA, A. U. SHELEG, E. M. SMOLYARENKO, and Z. K. ORLIK
Temperature Studies of Optical Properties of Cadmium Diphosphide Crystals
(a) Vol. **76**, No. 1, p. 121 – 126 (1983).
See also *20.1*, p. *316*.

A. N. GEORGOBIANI, M. V. GLUSHKOV, E. S. LOGOZINSKAYA, ZH. A. PUKHLII, I. M. TIGINYANU, and I. A. SHCHERBAKOV
Radiative Recombination in γ-La_2S_3 Single Crystals
(a) Vol. **76**, No. 1, p. 311 – 317 (1983).
See also *13.4*, p. *176*; *20.3*, p. *337*.

D. V. KRISHNA SASTRY and P. JAYARAMA REDDY
Influence of Nitrogen Ion Implantation on the Electrical Transport Properties of InTe and InSe Thin Films
(a) Vol. **76**, No. 1, p. 345 – 350 (1983).
See also 11, p. 125; *14.3.1*, p. *230*.

K. TAKARABE and K. WAKAMURA
Polarized Raman Spectra of InS Single Crystal
(b) Vol. **116**, No. 2, p. K79 – K83 (1983).
See also *6*, p. *73*; *20.1*, p. *317*.

Yu. P. Gnatenko, Z. D. Kovalyuk, P. A. Skubenko, and Yu. I. Zhirko
Emission of Free and Bound Excitons in Layered GaSe Crystals
(b) Vol. **117**, No. 1, p. 283–287 (1983).
See also *13.5.1*, p. *196*; *20.3*, p. *338*.

B. Vengalis and K. Valatska
Phase Segregation in Tellurium-Saturated GeTe
(b) Vol. **117**, No. 2, p. 471–476 (1983).
See also *1.2*, p. *16*; 8, p. 88; 13.4, p. 179; 14.3, p. 220; 20.1, p. 319.

G. Babonas, N. S. Korets, and S. Marcinkevičius
Dispersion of Optical Constants in CdP_2
(b) Vol. **117**, No. 2, p. 477–481 (1983).
See also *4*, p. *61*; *20.1*, p. *319*.

A. K. Sarychev and A. P. Vinogradov
Effective Medium Theory for the Magnetoconductivity Tensor of Disordered Materials
(b) Vol. **117**, No. 2, p. K113–K118 (1983).
See also *14.3*, p. *220*.

B. G. Tagiev, G. M. Niftiev, and S. A. Abushov
Intrashell Yb^{3+} Transitions in Gallium Sulphide Single Crystals
(b) Vol. **118**, No. 1, p. K13–K16 (1983).
See also *13.4*, p. *182*; *20.3*, p. *341*.

C. De Blasi, C. Manfredotti, G. Micocci, and A. Tepore
Thermoluminescence of n-GaS
(a) Vol. **78**, No. 1, p. K31–K34 (1983).
See also 13.4, p. 183; *20.3*, p. *342*.

A. N. Chakravarti, K. P. Ghatak, K. K. Ghosh, and H. M. Mukherjee
Effect of Crystal-Field Splitting on the Einstein Relation in Tetragonal Semiconductors
(b) Vol. **118**, No. 2, p. 843–852 (1983).
See also 13.1, p. 161; *14.3*, p. *222*.

B. G. Tagiev, G. M. Niftiev, and S. A. Abushov
Recombination and Trapping Levels in GaS:Mn Single Crystals
(a) Vol. **78**, No. 2, p. K183–K187 (1983).
See also *13.4*, p. *184*; *16*, p. *266*.

B. G. Tagiev, G. M. Niftiev, G. K. Aslanov, I. M. Askerov, and F. N. Sultanov
Effect of Ytterbium on the Electrophysical Properties of Ga_2Se_3 Single Crystals
(a) Vol. **78**, No. 2, p. K189–K193 (1983).
See also *13.4*, p. *185*; *14.3*, p. *223*.

M. Gatulle, M. Fischer, and A. Chevy
Elastic Constants of the Layered Compounds GaS, GaSe, InSe, and Their Pressure Dependence (I): Experimental Part
(b) Vol. **119**, No. 1, p. 327–336 (1983).
See also *12.2*, p. *143*.

V. A. Korshunov, M. A. Aldzhanov, K. K. Mamedov, and T. G. Kerimov
Phonon Densities of States of GaS and GaSe Layer Crystals
(b) Vol. **119**, No. 1, p. K47 – K50 (1983).
See also *8*, p. *78*.

B. T. Kolomiets, T. N. Mamontova, A. V. Chernyshov, G. Z. Vinogradova, and N. V. Demokritova
The Effect of Low Oxygen Concentrations on the Photoluminescence of Vitreous Selenium
(a) Vol. **79**, No. 1, p. K89 – K92 (1983).
See also 2, p. 52; *10.2*, p. *118*; *20.3*, p. *345*.

G. Micocci, A. Rizzo, A. Tepore, and F. Zuanni
Isothermal Currents in InSe, GaSe, and GaS Single Crystals
(a) Vol. **80**, No. 1, p. 263 – 267 (1983).
See also *13.4*, p. *189*; *14.3*, p. *226*.

L. Červinka, V. Cháb, and S. Koc
XPS Investigation of the Amorphous $TlSe_x$ System
(a) Vol. **80**, No. 2, p. 685 – 689 (1983).
See also *2*, p. *56*; *4*, p. *64*; 17, p. 271.

M. Nowak
The Photomagnetoelectric Effect and Photoconductivity for Non-Normal Incidence of Radiation
(a) Vol. **80**, No. 2, p. 691 – 702 (1983).
See also *14.3.1*, p. *232*; *16*, p. *268*.

22.3.1 Mixed Crystals

Original Papers and Short Notes

S. G. Abdullaeva, N. T. Mamedov, and G. S. Orudzhev
Band Structure of $TlGaSe_2$
(b) Vol. **119**, No. 1, p. 41–48 (1983).
See also *13.1*, p. *162*.

B. G. Tagiev, G. M. Niftiev, S. A. Abushov, and A. O. Khalilov
Optical Properties of $InSe_{1-x}S_x$ Single Crystals in the Region of the Fundamental Absorption Edge
(b) Vol. **119**, No. 2, p. K143–K148 (1983).
See also 13.5.1, p. 198; 14.3.4, p. 247; *20.1*, p. *328*; *20.3*, p. *345*.

N. M. Gasanly, A. F. Goncharov, N. N. Melnik, and A. S. Ragimov
Optical Phonons in $GaS_{1-x}Se_x$ Layer Mixed Crystals
(b) Vol. **120**, No. 1, p. 137–147 (1983).
See also *6*, p. *79*; *20.1*, p. *328*.

22.4 II–VI Compounds

Original Papers and Short Notes

K.-H. Bäther, W. Brückner, W. Moldenhauer, and H.-P. Brückner
Microscopic Degradation Model for ZnO Varistors
(a) Vol. **75**, No. 2, p. 465–472 (1983).
See also *1.4*, p. *29*; *14.3.3*, p. *237*.

V. B. Bikbaev, S. Č. Karpinskas, and J. J. Vaitkus
Electrical Properties of Mo–Cadmium Chalcogenide Schottky Barrier Diodes
(a) Vol. **75**, No. 2, p. 583–590 (1983).
See also *14.3.2*, p. *233*; 14.3.3, p. 237.

Hoang Ngoc Cam, Nguyen Van Hieu, and Ha Vinh Tan
On the Theory of the Nonlinear Acousto-Optical Effect in Semiconductors
(b) Vol. **116**, No. 1, p. 25–29 (1983).
See also *7*, p. *82*; 13.5.1, p. 195; *20*, p. *307*; 22.2, p. 442.

M. Scrocco
Satellite Structure of ZnO and Zn(II) Halides Studied by X-Ray Photoelectron Spectroscopy
(b) Vol. **116**, No. 1, p. 137–144 (1983).
See also *13.1*, p. *157*; 17, p. 269; 22.5, p. 473.

T. L. Tansley, D. F. Neely, and C. P. Foley
Optical Dispersion in Zinc Oxide
(a) Vol. **77**, No. 2, p. 491–496 (1983).
See also 13.1, p. 159; *20.1*, p. *320*.

J. M. Hvam, G. Blattner, M. Reuscher, and C. Klingshirn
The Biexciton Levels and Nonlinear Optical Transitions in ZnO
(b) Vol. **118**, No. 1, p. 179–189 (1983).
See also *13.5.1*, p. *197*; *20.3*, p. *341*.

G. Dräger, R. Wippermann, O. Brümmer, and A. Šimůnek
Polarized ZnK- and OK-Emission Bands of ZnO
(b) Vol. **118**, No. 2, p. K113–K116 (1983).
See also *13.1*, p. *161*; *20*, p. *309*.

Vu Xuan Quang, Nguyen Quang Liem, Nguyen Cong Thanh, Thuong Van Chuong, and Le Thi Le Thanh
Photoluminescence of Rare-Earth Centers in ZnO Phosphors
(a) Vol. **78**, No. 2, p. K161–K165 (1983).
See also *20.3*, p. *343*.

B. H. Bairamov, A. Heinrich, G. Irmer, V. V. Toporov, and E. Ziegler
Raman Study of the Phonon Halfwidths and the Phonon–Plasmon Coupling in ZnO
(b) Vol. **119**, No. 1, p. 227–234 (1983).
See also *6*, p. *77*; 13.5.3, p. 202; *20.1*, p. *325*.

J. Puls, F. Henneberger, and J. Voigt
Excited States of Bound Excitons in Wurtzite-Type Semiconductors
(b) Vol. **119**, No. 1, p. 291 – 298 (1983).
See also *13.5.1*, p. *198*; *20.3*, p. *343*.

T. Soma and H.-Matsuo Kagaya
Binding Force Including Ionic Interactions and Bulk Properties for Tetrahedral Compounds
(b) Vol. **119**, No. 1, p. K41 – K45 (1983).
See also *4*, p. *63*; 12.2, p. 143; 22.2, p. 442.

T. Soma and H.-Matsuo Kagaya
Local Heine-Abarenkov Model Potential for III–V and II–VI Covalent Compounds
(b) Vol. **119**, No. 2, p. 547 – 553 (1983).
See also *4*, p. *63*; 13.1, p. 163; 22.2, p. 442.

D. Suisky
Multiphonon Structure of Optical Transitions to Impurity Resonance States. The Influence of Anharmonicity
(b) Vol. **120**, No. 2, p. 645 – 654 (1983).
See also 6, p. 81; *13.4*, p. *190*; *20.1*, p. *330*.

22.4.1 Sulphides

Original Papers and Short Notes

M. P. Lisitsa, S. F. Terekhova, and N. A. Onishchenko
Two-Beam Interferometry in the Exciton Spectroscopy of CdS
(b) Vol. **115**, No. 1, p. 187 – 194 (1983).
See also *13.5.1*, p. *194*; *20.1*, p. *312*.

H. Yoshida and S. Shionoya
Luminescence of High Density Electron–Hole Plasma in CdS and CdSe in a Wide Temperature Range
(b) Vol. **115**, No. 1, p. 203 – 208 (1983).
See also 13.5, p. 192; *20.3*, p. *334*; 22.4.2, p. 466.

H. Rossmann, F. Henneberger, and J. Voigt
Memory Effect in the Excitonic Transmission of CdS
(b) Vol. **115**, No. 1, p. K63 – K67 (1983).
See also *13.5.1*, p. *194*.

B. R. Sethi, P. K. Goyal, O. P. Sharma, and P. C. Mathur
High-Field DC Conductivity in n-Type CdS Single Crystals Annealed in Molten Indium
(a) Vol. **75**, No. 1, p. 83 – 89 (1983).
See also 13.4, p. 173; *14.3*, p. *215*.

P. L. Kukk and A. J. Erm
Luminescent Centre Formation Processes in Activated CdS Phosphors
(a) Vol. **75**, No. 1, p. 91 – 100 (1983).
See also *10.2*, p. *108*; 13.4, p. 173; *20.3*, p. *334*.

R. Mach, G. O. Müller, W. Gericke, and J. von Kalben
The Efficiency of Electron Impact Excited Luminescence in ZnS Thin Film Devices
(a) Vol. **75**, No. 1, p. 187 – 194 (1983).
See also 14.3.1, p. 229; *14.3.3*, p. *236*; *20.3*, p. *335*; 22.4.2, p. 466.

D. Nesheva and E. Vateva
Some Peculiarities of High Temperature Sensitization in CdS
(a) Vol. **75**, No. 1, p. K45 – K48 (1983).
See also *13.4*, p. *174*; *16*, p. *264*.

M. S. Brodin, N. A. Davydova, and I. Yu. Shablii
Relation of Recombination Radiation Spectra to the Concentration of Structure Defects in CdS Single Crystals
(b) Vol. **115**, No. 2, p. 641 – 647 (1983).
See also *13.5.1*, p. *195*; 20.3, p. 336.

P. L. Kukk and A. J. Erm
Isochronal Annealing of Quenched CdS Phosphors
(a) Vol. **75**, No. 2, p. 393 – 399 (1983).
See also *10.2*, p. *109*; 13.4, p. 175; *20.3*, p. *336*.

A. HAUFE, K. HENNEBERGER, J. RÖSELER, and H.-J. WÜNSCHE
Parameters of Bound Multiexciton Complexes in Direct Polar Semiconductors
(b) Vol. **116**, No. 1, p. 191–203 (1983).
See also *13.5.1*, p. *195*; 20.3, p. 336; 22.4.2, p. 466.

R. BAUMERT, I. BROSER, J. GUTOWSKI, and A. HOFFMANN
Forbidden Luminescence and Resonance Raman Scattering of Bound Exciton States in CdS
(b) Vol. **116**, No. 1, p. 261–267 (1983).
See also *13.5.1*, p. *195*; *20.3*, p. *337*.

G. F. DOBRZHANSKII, O. V. FIALKOVSKAYA, L. I. KHIRUNENKO, and S. P. SMIRNOV
Optical Absorption of CdS:Nd Crystals
(b) Vol. **116**, No. 1, p. K31–K34 (1983).
See also *20.1*, p. *315*.

J. M. NORAS
Fluctuations in Light Transmission of Zinc Sulphide
(a) Vol. **76**, No. 1, p. K53–K56 (1983).
See also *20.1*, p. *316*.

S. PAKEVA and R. DAFINOVA
Electroluminophors on the Basis of ZnS:Sn
(a) Vol. **76**, No. 1, p. K77–K80 (1983).
See also *20.3*, p. *337*.

W. BUSSE, H.-E. GUMLICH, R. O. TÖRNQVIST, and V.-P. TANNINEN
Zero-Phonon Lines in Electroluminescence and Photoluminescence of ZnS:Mn Thin Films Grown by Atomic Layer Epitaxy
(a) Vol. **76**, No. 2, p. 553–558 (1983).
See also *20.3*, p. *338*.

H. ZIMMERMANN, R. BOYN, and N. NAGEL
Photoinduced Absorption Spectra of ZnS:Fe Crystals
(b) Vol. **117**, No. 1, p. 229–239 (1983).
See also 13.4, p. 178; *20.1*, p. *318*.

H. ZIMMERMANN and R. BOYN
On the Configuration Coordinate Diagram of ZnS:Fe
(b) Vol. **117**, No. 2, p. 611–618 (1983).
See also 6, p. 74; *13.4*, p. *179*; 20.1, p. 320; 20.3, p. 339.

V. P. GRIBKOVSKII, A. A. GLADYSHCHUK, V. V. ZUBRITSKII, V. V. PARASHCHUK, and G. P. YABLONSKII
Streamer Discharges in Semiconductors
(a) Vol. **77**, No. 2, p. 765–774 (1983).
See also *14.3.4*, p. *245*; 20.3, p. 340.

G. O. Müller and R. Mach
On the Mechanism of Electron Impact Excited Luminescence Devices
(a) Vol. **77**, No. 2, p. K179 – K183 (1983).
See also *14.3.4*, p. *246*; *20.3*, p. *341*.

S. Asano, N. Yamashita, and Y. Ogawa
Luminescence and Vibrational Structure of the Luminophors CaS:Cr^{3+} and CaSe:Cr^{3+}
(b) Vol. **118**, No. 1, p. 89 – 100 (1983).
See also *13.4*, p. *181*; *20.3*, p. *341*; 22.4.2, p. 467.

D. Verity, D. Shaw, F. J. Bryant, and C. G. Scott
DLTS Investigation of Electron Traps in As-Grown and Cd-Fired CdS
(a) Vol. **78**, No. 1, p. 267 – 275 (1983).
See also *13.4*, p. *183*; *14.3.4*, p. *247*.

M. S. Brodin, N. I. Vitrikhovskii, A. A. Kypen, S. G. Shevel, and N. I. Yanushevskii
Spatial and Spectral Characteristics and a New Model of Laser Generation for CdS-Type Single Crystals under One-Photon Excitation
(a) Vol. **78**, No. 1, p. 349 – 363 (1983).
See also *20.2*, p. *332*; 22.4.4, p. 471.

G. D. Shibarshina, S. A. Moskalenko, and P. I. Khadzhi
The Polariton Soliton Formation Due to the Effect of the Dipole Momentum Saturation
(b) Vol. **119**, No. 1, p. 153 – 158 (1983).
See also *13.5*, p. *193*; 13.5.1, p. 197; *20.1*, p. *325*.

L. Stauffer, B. Stébé, and G. Munschy
Optical Absorption by an Exciton Bound to an Ionized Donor Impurity in Anisotropic Semiconductors
(b) Vol. **119**, No. 1, p. 193 – 198 (1983).
See also *13.4*, p. *185*; *13.5.1*, p. *197*; 20.1, p. 325.

O. Goede and D. Hennig
Isovalent-Impurity Clusters in Koster-Slater Approximation
(b) Vol. **119**, No. 1, p. 261 – 268 (1983).
See also *13.1*, p. *162*; *13.4*, p. *185*; 22.2.2, p. 450.

J. Flohrer and M. Porsch
Effect of Valence Band Splitting on the Biexciton Binding Energy in Wurtzite-Type Semiconductors
(b) Vol. **119**, No. 2, p. 639 – 649 (1983).
See also *13.1*, p. *163*; *13.5.1*, p. *198*.

V. X. Quang, T. Kim Anh, N. Q. Liem, K. Maass, and B. Selle
A New Luminescence Band Related to Radiation Defects in Ion-Implanted Zinc Sulphide
(a) Vol. **79**, No. 2, p. K181 – K185 (1983).
See also *11*, p. *133*; *20.3*, p. *345*.

G. O. Müller, S. L. Karpenko, R. P. Seisyan, and M. A. Yakobson
Influence of Doping on Excitonic Luminescence Spectra in CdS
(b) Vol. **120**, No. 1, p. K93 – K95 (1983).
See also 13.5.1, p. 198; *20.3*, p. *346*.

I. E. Bolotov and V. L. Prilepo
On the Vacancy Mechanism of the Occurrence of Compressive Stresses during the Crystallization of Amorphous Films
(a) Vol. **80**, No. 1, p. K67 – K70 (1983).
See also *1.5*, p. *38*; 3, p. 59; 10.2, p. 120; 12.2, p. 144; 22.1.3, p. 441.

M. Bhattacharya, S. Mukhopadhyay, and T. K. Mitra
Study of a Bound Surface Optical Polaron by a Non-Perturbative Method
(b) Vol. **120**, No. 2, p. 555 – 564 (1983).
See also *13.3*, p. *171*; *13.5.3*, p. *202*; 22.2.1, p. 447; 22.5.2, p. 485.

A. Ferreira da Silva, R. Micnas, I. C. da Cunha Lima, and R. Kishore
The Impurity Conduction of n-Type CdS: A Two Band Approach
(b) Vol. **120**, No. 2, p. K101 – K106 (1983).
See also *13.4*, p. *190*; *14.3*, p. *227*.

Dang Dinh Thong and O. Goede
Optical Study of Highly Mn-Doped ZnS Crystals
(b) Vol. **120**, No. 2, p. K145 – K150 (1983).
See also *20.3*, p. *347*.

H. Zimmermann und R. Boyn
Excitation Behaviour of Blue ($^1G_4 \rightarrow {}^3H_6$) and Infrared ($^3F_4 \rightarrow {}^3H_6$) Photoluminescence of ZnS:Tm, Ag Crystals
(b) Vol. **120**, No. 2, p. K189 – K193 (1983).
See also *13.4*, p. *191*; *20.3*, p. *347*.

22.4.2 Selenides

Original Papers and Short Notes

H. Yoshida and S. Shionoya
Luminescence of High Density Electron–Hole Plasma in CdS and CdSe in a Wide Temperature Range
(b) Vol. **115**, No. 1, p. 203–208 (1983).
See also 13.5, p. 192; *20.3*, p. *334*; 22.4.1, p. 462.

V. V. Aristov, A. V. Zaretskii, Yu. A. Osipyan, V. F. Petrenko, G. K. Strukova, and I. I. Khodos
Dislocations in Deformed ZnSe Single Crystals as Studied by Weak-Beam Electron Microscopy
(a) Vol. **75**, No. 1, p. 101–106 (1983).
See also *10.2*, p. *109*.

R. Mach, G. O. Müller, W. Gericke, and J. von Kalben
The Efficiency of Electron Impact Excited Luminescence in ZnS Thin Film Devices
(a) Vol. **75**, No. 1, p. 187–194 (1983).
See also 14.3.1, p. 229; *14.3.3*, p. *236*; *20.3*, p. *335*; 22.4.1, p. 462.

W. Westera
Electro-Acoustic Effects in CdSe
(a) Vol. **75**, No. 2, p. 405–408 (1983).
See also *7*, p. *82*; *14.3.4*, p. *243*.

R. Mach, J. van Kalben, W. Gericke, G. O. Müller, and G. U. Reinsperger
ZnSe:Mn in AC Thin Film Electroluminescence Devices
(a) Vol. **75**, No. 2, p. 489–494 (1983).
See also 14.3.1, p. 230; *14.3.3*, p. *237*; *20.3*, p. *336*.

A. Haufe, K. Henneberger, J. Röseler, and H.-J. Wünsche
Parameters of Bound Multiexciton Complexes in Direct Polar Semiconductors
(b) Vol. **116**, No. 1, p. 191–203 (1983).
See also *13.5.1*, p. *195*; 20.3, p. 336; 22.4.1, p. 463.

Y. Watanabe, J. Takeuchi, and M. Oka
Microwave Transport Properties around the Structural Phase Transition of MnSe
(a) Vol. **76**, No. 1, p. K13–K15 (1983).
See also *1.2*, p. *15*; *14.3*, p. *218*.

S. M. Wasim, B. Fernández, and R. Aldana
Thermal Conductivity of HgSe
(a) Vol. **76**, No. 2, p. 743–746 (1983).
See also *8*, p. *88*.

I. E. Türe, F. Poulin, A. W. Brinkman, and J. Woods
Electron Traps and Deep Levels in Cadmium Selenide
(a) Vol. **77**, No. 2, p. 535–544 (1983).
See also *13.4*, p. *180*; *14.3.3*, p. *239*.

S. Asano, N. Yamashita, and Y. Ogawa
Luminescence and Vibrational Structure of the Luminophors CaS:Cr^{3+} and CaSe:Cr^{3+}
(b) Vol. **118**, No. 1, p. 89–100 (1983).
See also *13.4*, p. *181*; *20.3*, p. *341*; 22.4.1, p. 464.

T. Amand, A. Cornet, M. Pugnet, J. Collet, M. Brousseau, and B. S. Razbirin
Picosecond Study of the Exciton Gas Cooling
(b) Vol. **118**, No. 1, p. 113–119 (1983).
See also *13.5.1*, p. *197*; 20.3, p. 341; 22.5, p. 474.

S. W. Biernacki
Splitting of the 3d Levels of Iron Group Impurities in Crystals with Zincblende Structure
(b) Vol. **118**, No. 2, p. 525–533 (1983).
See also *13.4*, p. *183*; *20.1*, p. *323*.

M. S. Brodin, V. M. Bandura, and M. G. Matsko
Direct Luminescence Evidence for the Degenerate Electron-Hole Plasma Formation in $A^{II}B^{VI}$ Semiconductors
(b) Vol. **119**, No. 1, p. 83–87 (1983).
See also *13*, p. *151*; 13.5.1, p. 197; *20.3*, p. *343*; 22.4.3, p. 468.

22.4.3 Tellurides

Original Papers and Short Notes

D. Verity, J. Nahum, and F. J. Bryant
Donor-Related Edge Luminescence in Ion-Implanted ZnTe
(b) Vol. **115**, No. 2, p. 499–508 (1983).
See also *13.4*, p. *174*; *20.3*, p. *335*.

J. Toušková, D. Kindl, J. Toušek, and E. Klier
Some Electrical Properties of $Au_xTe/CdTe$ Heterojunctions on CdTe Thick Films
(a) Vol. **76**, No. 1, p. 365–374 (1983).
See also *14.3.1*, p. *230*; *14.3.3*, p. *238*.

M. Nagabhooshanam and V. Hari Babu
Conductivity and Hall Coefficient Studies in Plastically Deformed CdTe Single Crystals
(a) Vol. **76**, No. 1, p. K89–K92 (1983).
See also *10.2*, p. *111*; *14.3*, p. *218*.

M. Ya. Valakh, A. P. Litvinchuk, and G. G. Tarasov
Effect of the Pumping Intensity on the Resonant Secondary Emission Spectra of Optically Aligned Hot Excitons in ZnTe Crystals
(b) Vol. **116**, No. 2, p. K61–K65 (1983).
Erratum
(b) Vol. **118**, No. 2, p. K167 (1983).
See also *6*, p. *73*; 13.5.1, p. 196; *20.3*, p. *337*.

Z. Dziuba
Impurity Band Conduction in HgTe (I): Experimental Data
(b) Vol. **117**, No. 2, p. 531–540 (1983).
See also *13.4*, p. *179*; *14.3*, p. *220*.

T. Huang and A. L. Ruoff
Pressure-Induced Phase Transitions of HgTe
(a) Vol. **77**, No. 2, p. K193–K197 (1983).
See also *1.2*, p. *17*; *12.2*, p. *143*.

Z. Dziuba
Impurity Band Conduction in HgTe (II): Theoretical Analysis
(b) Vol. **118**, No. 1, p. 319–326 (1983).
See also *13.4*, p. *182*; *14.3*, p. *221*.

M. S. Brodin, V. M. Bandura, and M. G. Matsko
Direct Luminescence Evidence for the Degenerate Electron–Hole Plasma Formation in $A^{II}B^{VI}$ Semiconductors
(b) Vol. **119**, No. 1, p. 83–87 (1983).
See also *13*, p. *151*; 13.5.1, p. 197; *20.3*, p. *343*; 22.4.2, p. 467.

V. I. Ivanov-Omskii, A. Sh. Mekhtiev, S. A. Rustambekova, and E. N. Ukraintsev
Plasma Reflection and Hole Optical Effective Mass in p-HgTe
(b) Vol. **119**, No. 1, p. 159–169 (1983).
See also *13.1*, p. *162*; *20.1*, p. *325*.

Z. Dziuba and R. Kowalczyk
The Analysis of the Conductivity Tensor Components in HgTe-Type Semiconductors
(b) Vol. **119**, No. 1, p. K11–K15 (1983).
See also *14.3*, p. *223*.

R. Baltramiejūnas, E. Kuokštis, and A. Žukauskas
Self-Heating of Dense Exciton Gas in CdTe Single Crystals
(b) Vol. **119**, No. 2, p. 453–457 (1983).
See also *13.5.1*, p. *198*; 20.3, p. 345.

H. Müller, P. Haasen, and W. Schröter
Thermopower of Plastically Deformed Cadmium Telluride
(a) Vol. **80**, No. 1, p. 223–230 (1983).
See also *10.2* p. *120*; 14.3.4, p. 248; *15*, p. *263*.

22.4.4 Mixed Crystals

Original Papers and Short Notes

E. Gurauskas, J. Kavaliauskas, G. Krivaitė, and A. Šileika
Infrared Absorption Spectra of $Cd_{0.2}Hg_{0.8}Te$ Crystals
(b) Vol. **115**, No. 1, p. 195 – 202 (1983).
See also *6*, p. *69*; *20.1*, p. *312*.

A. V. Malyavkin
Optically Detected Magnetic Resonance in $Zn_{1-x}Mn_xTe$
(b) Vol. **115**, No. 2, p. 353 – 358 (1983).
See also 13.5.1, p. 194; *19*, p. *299*; *20.3*, p. *335*.

A. P. Litvinchuk and N. I. Vitrikhovskii
The Variation of the Energy Gap with Composition in the Quaternary Alloy System $ZnTe_{1-2x}S_xSe_x$
(b) Vol. **115**, No. 2, p. K151 – K154 (1983).
See also *13.1*, p. *156*; 13.5.1, p. 195; *20.3*, p. *336*.

R. A. Abreu, J. Stankiewicz, and W. Giriat
Temperature Dependence of the Absorption Edge in $Cd_{1-y}Mn_ySe$
(a) Vol. **75**, No. 2, p. K153 – K156 (1983).
See also *20.1*, p. *315*.

V. Agekyan, R. Bindemann, R. Schwabe, Yu. Stepanov, and I. Streit
Luminescence of Excitons Bound by Disorder-Potential Fluctuations in $CdS_{1-x}Se_x$: Spatial Diffusion, Excitation Level Dependence, and Kinetics
(b) Vol. **116**, No. 1, p. K43 – K46 (1983).
See also *13.5.1*, p. *196*; *20.3*, p. *337*.

V. I. Ivanov-Omskii, V. K. Ogorodnikov, and V. D. Rozumnyi
Surface Defects of Closed-System-Grown $Cd_xHg_{1-x}Te_x$ Epitaxial Layers
(a) Vol. **76**, No. 1, p. 71 – 74 (1983).
See also *1.5*, p. *36*; 20.1, p. 316.

Nguyen The Khoi, J. Ginter, and A. Twardowski
Magnetooptical Study of the Fundamental Absorption Edge in $Cd_{1-x}Mn_xTe$
(b) Vol. **117**, No. 1, p. 67 – 74 (1983).
See also 13.5.1, p. 196; *18*, p. *274*; *20*, p. *308*.

P. Wiśniewski and M. Nawrocki
Composition Dependence of the Energy Gap in $Cd_{1-x}Mn_xSe$
(b) Vol. **117**, No. 1, p. K43 – K45 (1983).
See also *13.5.1*, p. *196*; 20.1, p. 318.

Mukesh Jain, H. K. Sehgal, and A. V. R. Warrier
The Variation of Optical Energy Gaps with Composition in $Pb_{1-x}Hg_xS$ Alloy Films
(a) Vol. **77**, No. 1, p. 67 – 70 (1983).
See also *20.1*, p. *319*.

J. Zhang and J. C. Thuillier
The Effect of Annealing Temperature on the Electron Concentration of $Hg_{0.3}Cd_{0.7}Te$
(a) Vol. **77**, No. 2, p. 649 – 653 (1983).
See also *14.3*, p. *221*.

T. Taguchi
Crystal Growth and Neutral-Acceptor Bound-Exciton Emission of ZnCdTe by THM with Te Solvent
(a) Vol. **77**, No. 2, p. K115 – K119 (1983).
See also *3*, p. *58*; 13.4, p. 181; 13.5.1, p. 196; *20.3*, p. *341*.

P. Kleinert
Vibrational Behaviour of Mixed Diatomic Crystals $AB_{1-c}C_c$ within the Bethe Lattice
(b) Vol. **118**, No. 1, p. 283 – 291 (1983).
See also *6*, p. *75*; *20.1*, p. *322*; 22.2.4, p. 455.

K. E. Mironov, V. K. Ogorodnikov, V. D. Rozumnyi, and V. I. Ivanov-Omskii
Closed System LPE Growth of $Cd_xHg_{1-x}Te$
(a) Vol. **78**, No. 1, p. 125 – 131 (1983).
See also *1.5*, p. *37*; *3*, p. *58*.

M. S. Brodin, N. I. Vitrikhovskii, A. A. Kypen, S. G. Shevel, and N. I. Yanushevskii
Spatial and Spectral Characteristics and a New Model of Laser Generation for CdS-Type Single Crystals under One-Photon Excitation
(a) Vol. **78**, No. 1, p. 349 – 363 (1983).
See also *20.2*, p. *332*; 22.4.1, p. 464.

P. Chandrasekharam, D. Raja Reddy, and B. K. Reddy
Optical Properties of $(ZnSe)_x(CdTe)_{1-x}$ Single Crystals
(b) Vol. **119**, No. 1, p. K5 – K9 (1983).
See also *13.1*, p. *163*; *20.1*, p. *326*.

M. T. Sebastian and P. Krishna
Mechanism of Solid State Transformations in Single Crystals of $Zn_xCd_{1-x}S$
(a) Vol. **79**, No. 1, p. 271 – 280 (1983).
See also *1.2*, p. *18*; 10.2, p. 118.

P. Chandrasekharam, D. Raja Reddy, and B. K. Reddy
Electrical Properties of Vapour Phase Grown $(ZnSe)_x(CdTe)_{1-x}$ Single Crystals
(a) Vol. **79**, No. 1, p. K103 – K107 (1983).
See also *2*, p. *52*; *14.3*, p. *224*.

A. J. Szadkowski and A. Lubomirska-Wittlin
Imperfection Photoconductivity of $Cd_{1-x}Mn_xTe$
(a) Vol. **79**, No. 2, p. K127 – K129 (1983).
See also *13.4*, p. *188*; *16*, p. *267*.

J. SZATKOWSKI and K. SIERAŃSKI
Theory of Free-Carrier Magnetoabsorption in Mixed Zincblende Narrow-Gap Semiconductors
(b) Vol. **120**, No. 1, p. 263–272 (1983).
See also 13.1, p. 164; *20.1*, p. *329*.

J. P. VITTON, J. P. DUFOUR, R. MACHET, and J. C. THUILLIER
Magnetoconduction of n-Inversion Layers on $Hg_{1-x}Cd_xTe$
(b) Vol. **120**, No. 1, p. K53–K57 (1983).
See also *14.3*, p. *226*.

A. ZARĘBA and A. J. NADOLNY
Photoconductivity Originated from Dissociation of Excitons in $Cd_{1-x}Mn_xTe$
(b) Vol. **120**, No. 2, p. K97–K100 (1983).
See also *13.5.1*, p. *199*; *16*, p. *268*.

V. KOUBELE, M. ZVÁRA, and P. HLÍDEK
Compositional Variations of the Nonresonant Interband Faraday Rotation in Cd_xHg_{1-x}-Te Single Crystals
(b) Vol. **120**, No. 2, p. K173–K177 (1983).
See also 13, p. 153; *18*, p. *276*; *20.1*, p. *330*.

J. H. BASSON and H. BOOYENS
The Introduction of Misfit Dislocations in HgCdTe Epitaxial Layers
(a) Vol. **80**, No. 2, p. 663–668 (1983).
See also *1.5*, p. *38*; *10.2*, p. *121*.

W. UCHIDA
Photoluminescence in Self-Activated $Zn_xCd_{1-x}S$ Powders
(a) Vol. **80**, No. 2, p. K199–K202 (1983).
See also *20.3*, p. *348*.

22.5 Halides

Review Articles

K. N. Shrivastava
Theory of Spin–Lattice Relaxation
(b) Vol. **117**, No. 2, p. 437–458 (1983).
See also *6*, p. *69*; *13*, p. *146*; 20.1, p. 312; 22.8.1, p. 512.

B. Pałósz
Reasons for Polytypism of Crystals of the Type MX_2 (II): Classification of Faults and Structural Series of Polytypes; Conditions of Polytypic Growth of CdI_2, PbI_2, $CdBr_2$, SnS_2, $SnSe_2$ and $Ti_{1.2}S_2$
(a) Vol. **80**, No. 1, p. 11–41 (1983).
See also *1*, p. *9*; *1.1*, p. *11*; 10.2, p. 108; 22, p. 403.

Original Papers and Short Notes

J. Shanker and R. Sundaraj
Dielectric and Anharmonic Behaviour of Silver Halides, Thallous Halides, Alkaline Earth Fluorides, and Lead Fluoride Crystals
(b) Vol. **115**, No. 1, p. 67–74 (1983).
See also 4, p. 60; 13, p. 146; *14.4*, p. *250*; 22.5.1, p. 476; 22.5.3, p. 486.

W. Korczak, M. Paradowski, and M. Subotowicz
EPR Linewidth Studies of Gd^{3+}-Doped $La_xCe_{1-x}F_3$ Single Crystals
(b) Vol. **115**, No. 1, p. 89–94 (1983).
See also 13.4, p. 172; *19*, p. *299*.

B. Subramanyam and Y. V. G. S. Murti
Elastic Constant Correlations in Ionic Crystals of Rocksalt Structure
(b) Vol. **115**, No. 1, p. K29–K31 (1983).
See also *12.2*, p. *141*; 22.6, p. 490.

V. B. Sharan
Absorption, Fluorescence, and Excitation Spectra of Thallium-Activated Rubidium Bromide
(b) Vol. **116**, No. 1, p. 41–47 (1983).
See also 13.4, p. 175; *20.1*, p. *315*; *20.3*, p. *336*.

M. Scrocco
Satellite Structure of ZnO and Zn(II) Halides Studied by X-Ray Photoelectron Spectroscopy
(b) Vol. **116**, No. 1, p. 137–144 (1983).
See also *13.1*, p. *157*; 17, p. 269; 22.4, p. 460.

Ch. Opitz and F. Dübler
Modelling and Illustration of a Penetration Mechanism for Hydrogen into Metal
(a) Vol. **76**, No. 2, p. 505–512 (1983).
See also 4, p. 61; *9*, p. *94*; 13, p. 148.

K. SOMAIAH and V. HARI BABU
Ionic Conductivity of BaFCl
(b) Vol. **117**, No. 1, p. 75–79 (1983).
See also *10.2*, p. *112*; *14.4*, p. *251*.

K. M. HÄUSSLER, J. BRANDMÜLLER, L. MERTEN, H. FINSTERHÖLZL, and A. LEHMEYER
Observation of Dispersion on Magnetic Polaritons
(b) Vol. **117**, No. 1, p. 225–228 (1983).
See also *13.5.4*, p. *203*; 18, p. 274; *20.1*, p. *318*.

M. HEMING, G. LEHMANN, K. RECKER, and F. WALLRAFEN
EPR and Luminescence of Mn^{2+} in Distorted Tetrahedral Symmetry in $CaZnF_4$ Crystals
(b) Vol. **117**, No. 1, p. 271–276 (1983).
See also 3, p. 58; *19*, p. *301*; *20.3*, p. *338*.

J. BRANDMÜLLER, A. LEHMEYER, K. M. HÄUSSLER, and L. MERTEN
Electromagnetic Wave Propagation in the Uniaxial Antiferromagnetic Crystals FeF_2, CoF_2, and MnF_2
(b) Vol. **117**, No. 1, p. 323–328 (1983).
See also *18.4*, p. *296*; *20.1*, p. *318*.

Y. ITOH and C. HAMAGUCHI
Resonant Brillouin Scattering in HgI_2 Investigated by Acoustic Domain Injection Method
(b) Vol. **117**, No. 2, p. 725–732 (1983).
See also 7, p. *83*; *20.1*, p. *320*.

M. S. BRODIN, I. V. BLONSKII, and B. M. NITSOVICH
Anisotropic Exciton Scattering by Bending Waves in Layer Crystals
(b) Vol. **117**, No. 2, p. K99–K103 (1983).
See also *13.5.1*, p. *196*; 20.1, p. 320.

P. PIPINYS, A. KIVERIS, and A. RIMEIKA
Temperature Dependence of Electroluminescence in CdF_2 Thin Films Doped with Rare Earth
(a) Vol. **77**, No. 2, p. 471–476 (1983).
See also *20.3*, p. *340*.

T. AMAND, A. CORNET, M. PUGNET, J. COLLET, M. BROUSSEAU, and B. S. RAZBIRIN
Picosecond Study of the Exciton Gas Cooling
(b) Vol. **118**, No. 1, p. 113–119 (1983).
See also *13.5.1*, p. *197*; 20.3, p. 341; 22.4.2, p. 467.

A. V. MALAKHOVSKII and G. G. VASILEV
Covalency as Source of Intensity for d–d Transitions in Octahedral Complexes
(b) Vol. **118**, No. 1, p. 337–344 (1983).
See also 4, p. 61; *20.1*, p. *322*.

B. RAMBABU, C. RAMASASTRY, and B. V. R. CHOWDARI
Radiation Damage Studies of Co^{2+} and Ni^{2+} Doped NH_4Cl Crystals
(b) Vol. **118**, No. 1, p. 381–387 (1983).
See also *10.2*, p. *114*; 19, p. 302; *20.1*, p. *322*.

B. RAMBABU, C. RAMASASTRY, and B. V. R. CHOWDARI
Cobalt Centres in Ammonium Chloride: Optical Absorption Studies
(b) Vol. **118**, No. 2, p. 541–549 (1983).
See also *10.2*, p. *115*; 13.4, p. 183; *20.1*, p. *323*.

A. LEHMEYER, L. MERTEN, and K. M. HÄUSSLER
On the Reflectivity for Oblique Incidence of the Uniaxial Antiferromagnetic Crystals FeF_2, CoF_2, and MnF_2
(b) Vol. **118**, No. 2, p. 675–679 (1983).
See also *18.4*, p. *296*; *20.1*, p. *324*.

A. V. MALAKHOVSKII, T. P. MOROZOVA, and V. I. YUZVAK
Magneto-Optical Properties of $RbMnF_3$ in the Paramagnetic State
(b) Vol. **119**, No. 1, p. 411–417 (1983).
See also 13.1, p. 163; *18*, p. *275*; *20.1*, p. *326*.

T. MOHAMMED-BRAHIM, A. FRIANT, and J. MELLET
Characterization of Deep Energy Levels in Mercuric Iodide
(a) Vol. **79**, No. 1, p. 71–82 (1983).
See also *13.4*, p. *186*; 16, p. 266.

K. SOMAIAH and V. HARI BABU
Thermoluminescence of Thallium-Doped BaFCl
(a) Vol. **79**, No. 1, p. 237–242 (1983).
See also 13.4, p. 186; 20.1, p. 326; *20.3*, p. *344*.

J. W. BRIGHTWELL, L. S. MILLER, A. MUNDAY, and B. RAY
The Silver Iodide–Thallium Iodide Pseudo-Binary System
(a) Vol. **79**, No. 1, p. 293–300 (1983).
See also *1.2*, p. *19*; 14.4, p. 253; 22.5.1, p. 476.

R. B. YADAV
Brillouin Scattering in Single Crystals of Carbon Tetraiodide
(a) Vol. **79**, No. 2, p. 345–350 (1983).
See also *12.2*, p. *144*.

M. N. GHATIKAR
Chemical Transferability of Phase Shifts in Extended X-Ray Absorption Fine Structure
(b) Vol. **120**, No. 1, p. 445–450 (1983).
See also *20*, p. *311*; 21.1, p. 376; 21.4, p. 393; 22.6, p. 497; 22.7, p. 499.

22.5.1 Silver Halides

Original Papers and Short Notes

J. SHANKER and R. SUNDARAJ
Dielectric and Anharmonic Behaviour of Silver Halides, Thallous Halides, Alkaline Earth Fluorides, and Lead Fluoride Crystals
(b) Vol. **115**, No. 1, p. 67–74 (1983).
See also 4, p. 60; 13, p. 146; *14.4*, p. *250*; 22.5, p. 473; 22.5.3, p. 486.

A. L. KARTUZHANSKII, V. M. ULANOV, and A. F. YURCHENKO
Latent-Image and Fog Formation in Silver Halide Microcrystals by the Action of a Sequence of Light and Electric Pulses
(a) Vol. 75, No. 2, p. 669–673 (1983).
See also *14.3.4*, p. *244*; 16, p. 265.

A. N. ERMOSHKIN, R. A. EVARESTOV, S. A. KUCHINSKII, and V. K. ZAKHAROV
The Quasimolecular Approach to the Electronic Structure Calculations for Silver and Copper Halides
(b) Vol. **118**, No. 1, p. 191–203 (1983).
See also *13.1*, p. *159*; 13.4, p. 182.

P. MÜLLER
Contactless Determination of the Conductivity in Semi-Insulators
(a) Vol. 78, No. 1, p. 41–51 (1983).
See also *14.3.3*, p. *239*; *14.3.4*, p. *246*; 22.2.1, p. 445.

B. A. DEVLIN and J. CORISH
Impurity–Vacancy Dipoles in Cadmium-Doped Silver Bromide Crystals
(a) Vol. 78, No. 1, p. 53–60 (1983).
See also *10.2*, p. *114*.

F. CALLENS and W. MAENHOUT-VAN DER VORST
The Influence of Triazaindolisine on the Microwave Photoconductivity of AgBr Emulsion Grains
(a) Vol. 78, No. 1, p. K47–K52 (1983).
See also *16*, p. *266*.

J. W. BRIGHTWELL, L. S. MILLER, A. MUNDAY, and B. RAY
The Silver Iodide–Thallium Iodide Pseudo-Binary System
(a) Vol. 79, No. 1, p. 293–300 (1983).
See also *1.2*, p. *19*; 14.4, p. 253; 22.5, p. 475.

22.5.2 Alkali Halides

Original Papers and Short Notes

E. A. Radzhabov
A Model of Titanium Luminescence Centers in Lithium and Sodium Fluorides
(b) Vol. **115**, No. 1, p. 83–88 (1983).
See also *13.4*, p. *172*; 20.1, p. 312; *20.3*, p. *334*.

J. Prakash
Effect of Ultraviolet Irradiation on Se^{2-}–Anion Vacancy Dipoles in Potassium Iodide
(b) Vol. **115**, No. 1, p. 131–139 (1983).
See also *10.2*, p. *108*; *20.1*, p. *312*.

P. Guerin and A. Laforgue
Theoretical Studies on the Selfdiffusion in Ionic Crystals
(b) Vol. **115**, No. 1, p. 235–244 (1983).
See also *9*, p. *93*.

E. A. Radzhabov
Luminescence Centers in LiF:Li_2O
(b) Vol. **115**, No. 1, p. K25–K28 (1983).
See also *13.4*, p. *173*; *20.3*, p. *334*.

G. Hoebeeck, J. Provoost, and G. Jacobs
The F' Centre in RbBr
(b) Vol. **115**, No. 1, p. K33–K35 (1983).
See also *10.2*, p. *108*; *20.1*, p. *313*.

K. Takiyama, T. Fujita, H. Nojima, and M. Nishi
Vacuum Ultraviolet Absorption in γ-Irradiated LiCl
(b) Vol. **115**, No. 1, p. K59–K61 (1983).
See also 10.2, p. 108; *20.1*, p. *313*.

A. R. Lakshmanan, Bhuwan Chandra, and R. C. Bhatt
Thermoluminescence of Z Centres in LiF (TLD-100) at Elevated Irradiation Temperatures
(a) Vol. **75**, No. 1, p. 263–272 (1983).
See also *10.2*, p. *109*; 11, p. 123; *20.3*, p. *335*.

H. Koizumi and T. Suzuki
Low-Temperature Plasticity of CsBr and Inertial Effect of Dislocation Motion
(a) Vol. **75**, No. 1, p. 301–309 (1983).
See also *10.2*, p. *109*; 12.2, p. 141.

J. Shanker and K. Singh
Analysis of the Equation of State of Lithium and Sodium Halides
(b) Vol. **115**, No. 2, p. 381–387 (1983).
See also *4*, p. *60*.

YU. A. EKMANIS and I. S. RADCHENKO
Optical Properties of Inhomogeneous Colloidal Centres in Alkali Halides
(b) Vol. **115**, No. 2, p. 389–397 (1983).
See also *20.1*, p. *314*.

M. ELANGO, J. PRUULMANN, and A. P. ZHURAKOVSKII
Recombination Luminescence and Energy Transfer in Ionic Crystals at XUV Excitation by Synchrotron Radiation
(b) Vol. **115**, No. 2, p. 399–407 (1983).
See also *20.3*, p. *335*; 22.6, p. 491.

Z. BŁASZCZAK, M. LUDWICZAK, and F. KACZMAREK
Bleaching of F_2 Centers in LiF:F_2 Color Center Lasers
(a) Vol. **75**, No. 2, p. K195–K198 (1983).
See also *10.2*, p. *110*; *20.1*, p. *315*; 20.2, p. 332.

D. BOSE, A. GHOSH, and A. N. BASU
Perturbation-Theoretical Model Calculation of the Lattice Mechanics of Low Polarisability Ionic Crystals
(b) Vol. **116**, No. 1, p. 313–322 (1983).
See also *6*, p. *72*; 12.2, p. 142.

E. L. ANDRONIKASHVILI and M. V. GALUSTASHVILI
Influence of the Orientation of Anisotropic Impurity–Vacancy Complexes upon Dislocation Charging in NaCl Crystals
(b) Vol. **116**, No. 1, p. 335–338 (1983).
See also *10.2*, p. *110*.

N. TAKEUCHI, K. INABE, and S. NAKAMURA
Paramagnetic Species in NaCl:Ca, Cu Crystals Irradiated at RT
(b) Vol. **116**, No. 1, p. K5–K8 (1983).
See also *11*, p. *124*; *19*, p. *301*.

C. K. ONG and J. ADETUNJI
The Optical Absorption of F and F_H Centres in Cesium Halides
(b) Vol. **116**, No. 1, p. K19–K21 (1983).
See also *10.2*, p. *110*; *20.1*, p. *315*.

R. B. PODE, D. G. WAKDE, and B. T. DESHMUKH
Effect of Z_1 Centres on the Isothermal Stability of Colour Centres in Microcrystalline Powder of KCl
(b) Vol. **116**, No. 1, p. K39–K42 (1983).
See also *10.2*, p. *110*; *20.1*, p. *316*.

V. B. DUDNIKOVA and V. S. URUSOV
Solvus of KCl:Ca^{2+} Solid Solution by Flotation Measurements of Density of Single Crystals
(a) Vol. **76**, No. 1, p. 217–221 (1983).
See also *1.2*, p. *15*; *10.2*, p. *111*.

N. L. Sizova, V. D. Natsik, I. I. Gorina, and V. R. Regel
The Investigation of Heat Release in Slip Bands in Alkali Halide Crystals
(a) Vol. **76**, No. 2, p. 447 – 451 (1983).
See also *10.2*, p. *111*.

K. N. Reddy, Md. Iqbal Ahmed, and V. H. Babu
On Production of Z_1 Centres in NaCl:Sm^{2+} Crystals
(a) Vol. **76**, No. 2, p. K167 – K170 (1983).
See also *10.2*, p. *112*; 20.1, p. 317; 20.3, p. 338.

J. Prakash
Photo-Chemical Reaction and Photo-Induced Polarization Studies in Sulphur-Doped Potassium Bromide
(b) Vol. **117**, No. 1, p. 161 – 169 (1983).
See also *10.2*, p. *112*; *20.1*, p. *317*.

S. B. S. Sastry and S. Nagarajan
Thermoluminescence and Optical Absorption Studies on Rubidium Bromide Doped with Barium
(b) Vol. **117**, No. 1, p. 171 – 175 (1983).
See also 10.2, p. 112; *20.1*, p. *318*; *20.3*, p. *338*.

A. Subrahmanyam
Effect of DC Field and X-Irradiation on the Optical Absorption and Thermoluminescence of NaCl Single Crystals
(a) Vol. **77**, No. 1, p. 45 – 50 (1983).
See also *20.1*, p. *319*; *20.3*, p. *338*.

K. N. Subramanian
Deformation Structure in LiF Single Crystals
(a) Vol. **77**, No. 1, p. 361 – 367 (1983).
Erratum
(a) Vol. **78**, No. 2, p. 760 (1983).
See also *10.2*, p. *112*.

S. V. Moharil
A Phenomenological Model for Supralinearity and Sensitization in LiF-TLD 100
(a) Vol. **77**, No. 1, p. 413 – 420 (1983).
See also 13.4, p. 179; *20.3*, p. *339*.

M. Casalboni, R. Francini, U. M. Grassano, and R. Pizzoferrato
Nonlinear Excitation of New Emission Bands in KCl Crystals Containing Color Centers
(b) Vol. **117**, No. 2, p. 493 – 498 (1983).
See also *10.2*, p. *113*; *20.3*, p. *339*.

P. A. Frugoli and C. A. Pimentel
Point Defect Aggregates in γ-Irradiated LiF Single Crystals
(b) Vol. **117**, No. 2, p. 549 – 556 (1983).
See also *10.2*, p. *113*.

S. P. Kuila

Analysis of the 215 °C Glow Peak in Li-Doped KCl Single Crystals
(a) Vol. 77, No. 2, p. 477–482 (1983).
See also *13.4*, p. *180*; *20.3*, p. *340*.

J. Wollbrandt, U. Brückner, and E. Linke

Investigations of Mechanically Induced Excited States on Cleavage Planes of Ionic Crystals (I): Method and Apparatus for Charge and Electron Emission Measurements
(a) Vol. 77, No. 2, p. 545–552 (1983).
See also *13.3*, p. *169*; *14.4*, p. *251*; 17, p. 270.

K. N. Reddy, Md. Iqbal Ahmed, U. V. Subba Rao, and V. Hari Babu

Thermoluminescence and Optical Absorption Studies of Z_1-Centres in NaCl Crystals Doped with Terbium
(a) Vol. 77, No. 2, p. 643–648 (1983).
See also 10.2, p. 113; *20.1*, p. *321*; *20.3*, p. *340*.

G. A. Andreev and V. A. Klimov

Preferential Precipitation on Extended Edge Dislocations in NaCl Crystals
(a) Vol. 77, No. 2, p. 793–798 (1983).
See also *10.2*, p. *113*.

I. Sildos, O. Sild, and I. Dolindo

Manifestation of Spin–Rotation Coupling in Intersystem Crossing in a Rotating Impurity Molecule
(b) Vol. **118**, No. 1, p. 25–29 (1983).
See also *13.4*, p. *181*; 20.1, p. 321.

U. Pilatus and F. Fischer

Thermally Stimulated Depolarization of AsO_2^- Centres in Alkali Halide Crystals
(b) Vol. **118**, No. 1, p. 217–221 (1983).
See also *10.2*, p. *113*; *14.4*, p. *252*.

J. Prakash and Rahul

A Method for the Evaluation of Dielectric Relaxation Parameters in Electronic Excited State
(b) Vol. **118**, No. 1, p. 229–237 (1983).
See also 10.2, p. 113; *14.4*, p. *252*.

K. C. Gupta, R. M. Agrawal, and R. P. S. Rathore

Incomplete Gamma Functions and Lattice Sums for B.C.C. Structures
(b) Vol. **118**, No. 1, p. 399–404 (1983).
See also *4*, p. *62*; *6*, p. *75*.

F. Callens, P. Matthys, and E. Boesman

ESR of a New Sulphur Centre in KCl
(b) Vol. **118**, No. 1, p. K35–K37 (1983).
See also *10.2*, p. *114*; *19*, p. *302*.

R. Capelletti, O. Carranza Torres, J. García Solé, and M. Manfredi
On the 430 nm Emission Band in the NaCl:Pb^{++} System
(a) Vol. 78, No. 1, p. 35–39 (1983).
See also *10.2*, p. *114*; 13.4, p. 182; *20.3*, p. *342*.

J. Poźniak and G. Berg
Theoretical Analysis of the Dipole Polarization and Depolarization in Me^{++}-Doped NaCl-Type Alkali Halide Crystals (II): Next-Nearest Neighbour Approach
(a) Vol. 78, No. 1, p. 69–76 (1983).
See also *10.2*, p. *114*; 14.4, p. 252.

G. Berg, A. Pippel, and F. Fröhlich
Influence of the Variation of the Association Energy on the Equilibrium Concentration of Dipole Aggregates in Ionic Crystals
(a) Vol. 78, No. 1, p. 77–83 (1983).
See also *10.2*, p. *114*.

J. Garciá Solé, C. Zaldo, and F. Agulló-López
Optical Spectra of Pb^{2+} to Monitor the Precipitation Behavior of NaCl:Cd
(a) Vol. 78, No. 1, p. 85–91 (1983).
See also *10.2*, p. *114*; *20.1*, p. *323*; 20.3, p. 342.

J. Wollbrandt, U. Brückner, and E. Linke
Investigations of Mechanically Induced Excited States on Cleavage Planes of Ionic Crystals (II): Investigations of the Mechanisms of Electron Emission from Cleaved Alkali Halide Crystals and of the Influence of Surface Charges
(a) Vol. 78, No. 1, p. 163–168 (1983).
See also *13.4*, p. *183*; 14.4, p. 252; 17, p. 270.

B. S. Acharya and A. K. Panda
Optical Absorption Spectra of Mercury Centres in KCl and NaCl
(a) Vol. 78, No. 1, p. K81–K85 (1983).
See also *10.2*, p. *115*; *20.1*, p. *323*.

N. Takeuchi
On the General-Order Kinetics in Thermoluminescence
(a) Vol. 78, No. 1, p. K93–K96 (1983).
See also *20.3*, p. *342*.

A. Kumar, A. K. Sinha, and M. A. Ansari
Thermal Conductivity of LiF Containing Dislocations at Low Temperatures
(b) Vol. 118, No. 2, p. 483–491 (1983).
See also 6, p. 76; *8*, p. *88*.

K. Regiński
Phonon Polaritons in the Presence of a Magnetic Field
(b) Vol. 118, No. 2, p. 587–593 (1983).
See also *6*, p. *76*; *13.5*, p. *193*; 18, p. 275; 20, p. 309.

DEBENDRANATH SAHOO
Heterodyne Correlation of Scattered Light from a Sr^{++}-Doped NaCl Crystal. A Feasibility Study
(b) Vol. **118**, No. 2, p. 623–630 (1983).
See also *10.2*, p. *115*; *20*, p. *309*.

S. ZAZUBOVICH, A. USAROV, and ZH. EGEMBERDIEV
Photothermal Creation of $Cu^{\circ}v_a^+$ Centres and Mobility of Anion Vacancies in X-Irradiated KCl:Cu Crystals
(b) Vol. **118**, No. 2, p. 789–798 (1983).
See also *10.2*, p. *116*; 20.1, p. 324; *20.3*, p. *342*.

H. OPYRCHAŁ, K. D. NIERZEWSKI, and H. DRULIS
Effect of γ-Irradiation on EPR Spectra of Eu^{2+} Doped KCl and NaCl Crystals
(b) Vol. **118**, No. 2, p. K125–K127 (1983).
See also 10.2, p. 116; *11*, p. *130*; *19*, p. *303*.

D. N. POPOV, N. M. NANCHEVA, B. N. ZLATOV, and G. P. GEORGIEV
Positron Angular Correlation in Uncoloured and Neutron Irradiated LiF and LiF:OH
(b) Vol. **118**, No. 2, p. K129–K132 (1983).
See also *10.2*, p. *116*; *11*, p. *130*.

A. SUBRAHMANYAM
Optical Absorption and Thermoluminescence of KCl Single Crystals Coloured under DC Fields
(a) Vol. 78, No. 2, p. 665–670 (1983).
See also *20.1*, p. *324*; *20.3*, p. *343*.

J. MANAM, R. K. GARTIA, and V. V. RATNAM
Thermoluminescence of Z_3 Centres in X-Irradiated Ca-Doped KCl Crystals
(a) Vol. 78, No. 2, p. 685–688 (1983).
See also *10.2*, p. *116*; *20.3*, p. *343*.

V. ZIRAPS
Ionic and Electronic Processes during Frenkel Defect Annealing in KBr Crystals above LHeT
(b) Vol. **119**, No. 1, p. 49–59 (1983).
See also *10.2*, p. *117*; 11, p. 131; 13.4, p. 185; 20.1, p. 325.

K.-P. OBOTH and K. SCHMITT
Effective Dynamical Coupling Constants of the A-Band of Tl^+ Centers in NaCl, KCl, KBr, and KI
(b) Vol. **119**, No. 1, p. 215–219 (1983).
See also *13.4*, p. *185*; *20.1*, p. *325*.

S. BENCI, F. FERMI, and G. LENZI
Dimer Center Luminescence in Low-Doped KI:Tl^+ Crystals
(b) Vol. **119**, No. 1, p. 337–342 (1983).
See also 13.4, p. 185; *20.3*, p. *344*.

N. Dass and B. M. Mustafa
A Study of the Anderson-Grüneisen Parameter in Sodium Chloride
(b) Vol. **119**, No. 1, p. K27–K30 (1983).
See also *8*, p. *90*; 12.2, p. 143.

B. P. Chandra, M. Elyas, and B. Majumdar
Pressure and Temperature Dependence of the Mechanoluminescence of γ-Irradiated Alkali Halide Crystals
(a) Vol. **79**, No. 1, p. 95–102 (1983).
See also 12.2, p. 143; *20.3*, p. *344*.

J. Manam, R. K. Gartia, and V. V. Ratnam
Effect of Post-Irradiation Deformation on Thermoluminescence Spectra of KCl:Ca and NaCl:Ca Crystals
(a) Vol. **79**, No. 1, p. 167–171 (1983).
See also *10.2*, p. *118*; *20.3*, p. *344*.

L. B. Harris and S. L. Bewlay
Diffusion of Water in NaCl Bicrystals
(a) Vol. **79**, No. 1, p. 243–250 (1983).
See also *9*, p. *95*.

V. I. Barbashov and Yu. B. Tkachenko
Dislocation Mobility in Hydrostatically Compressed KBr Crystals
(a) Vol. **79**, No. 1, p. K17–K19 (1983).
See also *10.2*, p. *118*; *12.2*, p. *144*.

S. P. Kathuria and C. M. Sunta
Fractional Order of Kinetics and Its Variation with Trap Occupancy in LiF TLD-100
(a) Vol. **79**, No. 1, p. K77–K80 (1983).
See also *13.4*, p. *186*; *20.3*, p. *345*.

S. V. Moharil
Reply to "Fractional Order of Kinetics and Its Variation with Trap Occupancy in LiF TLD-100" by S. P. Kathuria and C. M. Sunta
(a) Vol. **79**, No. 1, p. K81–K84 (1983).
See also *13.4*, p. *187*; *20.3*, p. *345*.

V. G. Plekhanov, V. V. Shepelev, and A. U. Grinfelds
Spectroscopic Manifestation of Self-Trapping Excitons in Ionic Insulators
(b) Vol. **119**, No. 2, p. 493–502 (1983).
See also 6, p. 78; *13.5.1*, p. *198*; 20.3, p. 345.

Z. G. Akhvlediani, I. G. Akhvlediani, and T. L. Kalabegishvili
Effect of Combined Mechanical Load and Irradiation on Hydrogen Stabilization in LiF:OH
(b) Vol. **119**, No. 2, p. 503–506 (1983).
See also *10.2*, p. *119*; 19, p. 304; 20.1, p. 327.

A. A. Kiselev
Circular Phonons and Optical Magnetization of Crystals
(b) Vol. **119**, No. 2, p. 525–530 (1983).
See also *6*, p. *78*; *18*, p. *275*; 20, p. 310.

N. A. Tyapunina and Yu. A. Ivashkin
Excess Concentration of Point Defects in Alkali Halide Crystals Exposed to Ultrasonic Waves
(a) Vol. **79**, No. 2, p. 351–359 (1983).
See also 7, p. *83*; *10.2*, p. *119*.

G. Ya. Akimov and I. Yu. Prokhorov
Fracture of Hydrostatically Compressed NaCl Single Crystals
(a) Vol. **79**, No. 2, p. 423–431 (1983).
See also *10.2*, p. *119*; *12.2*, p. *144*.

S. V. Moharil
Track Interaction Model for Supralinearity and Sensitization in LiF-TLD 100
(a) Vol. **79**, No. 2, p. 599–605 (1983).
See also 10.2, p. 119; 13.4, p. 187; *20.3*, p. *345*.

L. Bosi, P. Gagliardelli, and M. Nimis
More on the F-Center Decay in Additively Colored NaF:LiF
(b) Vol. **120**, No. 1, p. 41–48 (1983).
See also *10.2*, p. *119*; *20.3*, p. *346*.

V. Hizhnyakov, G. Liidya, V. Nagirnyi, T. Soovik, and S. Zazubovich
Effect of Magnetic Field on the Decay Kinetics and Polarization of the A_T Emission of KCl:Ga at 0.38 K
(b) Vol. **120**, No. 1, p. 105–115 (1983).
See also *13.4*, p. *188*; *20.3*, p. *346*.

V. Nagirnyi, T. Soovik, S. Zazubovich, and N. Janson
Polarized A_T Emission of In^+ Centres in KI:In
(b) Vol. **120**, No. 1, p. 117–122 (1983).
See also *13.4*, p. *188*; *20.3*, p. *346*.

S. Bhan
Erratum to: Thermoluminescence of NaF:Ca^{2+} Single Crystals
(b) Vol. **120**, No. 1, p. 451 (1983).
See also *10.2*, p. *119*; *20.3*, p. *346*.

V. B. Dudnikova, V. S. Urusov, and L. N. Bannykh
Solvus of KCl:Ba^{2+} Solid Solution from Flotation Measurements of Single Crystal Density
(a) Vol. **80**, No. 1, p. 399–402 (1983).
See also *1.3*, p. *27*; *10.2*, p. *120*.

L. Bragina, V. Hizhnyakov, G. Liidya, V. Nagirnyi, T. Soovik, S. Zazubovich, and N. Janson
Effect of Magnetic Field on the Decay Kinetics and Polarization of the A_T Emission of KCl:Tl and KI:In Crystals at 0.4 to 0.7 K
(b) Vol. **120**, No. 2, p. 463–470 (1983).
See also *13.4*, p. *190*; *20.3*, p. *347*.

M. Bhattacharya, S. Mukhopadhyay, and T. K. Mitra
Study of a Bound Surface Optical Polaron by a Non-Perturbative Method
(b) Vol. **120**, No. 2, p. 555–564 (1983).
See also *13.3*, p. *171*; *13.5.3*, p. *202*; 22.2.1, p. 447; 22.4.1, p. 465.

K. Budde and H. Richtering
Incorporation and Diffusion of Cu(I)-Ions in CsI by NMR Measurements
(b) Vol. **120**, No. 2, p. 595–599 (1983).
See also *9*, p. *96*; 10.2, p. 120; *19*, p. *305*.

A. B. Gerasimov, M. M. Merkin, and A. A. Tsertsvadze
Calculation of the Exoelectron Energy Spectrum for NaCl
(b) Vol. **120**, No. 2, p. 711–714 (1983).
See also *17*, p. *271*.

C. Pedrini, H. Chermette, A. B. Goldberg, D. S. McClure, and B. Moine
Optical Absorption and Electronic Structure of NaCl:Ag^+. A Multiple Scattering Xα Study of the $(AgCl_6)^{5-}$ Cluster
(b) Vol. **120**, No. 2, p. 753–766 (1983).
See also *13.4*, p. *190*; *20.1*, p. *330*.

P. K. Chakraborty and K. V. Rao
F-Band Absorption and Thermoluminescence of KBr Single Crystals X-Ray Irradiated under High AC Fields
(b) Vol. **120**, No. 2, p. K107–K110 (1983).
See also *10.2*, p. *120*; 11, p. 134; *20.3*, p. *347*.

P. D. Alekseev, G. I. Baranov, E. P. Kurakina, and K. A. Maltsev
Formation of Hydrogen Bonding in Doped OH-Group Alkali Halide Crystals by the Action of γ-Radiation
(b) Vol. **120**, No. 2, p. K119–K121 (1983).
See also 10.2, p. 121; *11*, p. *134*; *20.1*, p. *330*.

K. Asami and M. Ishiguro
Pressure Effects on the Absorption and Emission of F_A Centers in KCl
(b) Vol. **120**, No. 2, p. K205–K209 (1983).
See also *10.2*, p. *121*; *13.4*, p. *191*; 20.1, p. 330; 20.3, p. 347.

G. A. Andreev
Annealing of Deformation-Induced Vacancy Clusters in NaCl Crystals
(a) Vol. **80**, No. 2, p. K129–K132 (1983).
See also *10.2*, p. *121*

22.5.3 Alkali-Earth Halides

Original Papers and Short Notes

J. Shanker and R. Sundaraj
Dielectric and Anharmonic Behaviour of Silver Halides, Thallous Halides, Alkaline Earth Fluorides, and Lead Fluoride Crystals
(b) Vol. **115**, No. 1, p. 67–74 (1983).
See also 4, p. 60; 13, p. 146; *14.4*, p. *250*; 22.5, p. 473; 22.5.1, p. 476.

R. K. Singh and N. K. Pandey
Study of Thermophysical and Anharmonic Properties of Fluorite Compounds
(b) Vol. **115**, No. 2, p. 555–562 (1983).
See also *4*, p. *60*; 6, p. 70; 12.2, p. 141.

N. Guskos, J. Kuriata, I. H. Salikhov, N. I. Silkin, and S. Yagadin
EPR Study of Eu^{2+} in PbF_2 Single Crystals
(b) Vol. **115**, No. 2, p. K133–K135 (1983).
See also *19*, p. *300*.

J. Kuriata and N. Guskos
Superhyperfine Structure in the EPR of $PbF_2:Eu^{2+}$ Single Crystals
(b) Vol. **115**, No. 2, p. K137–K140 (1983).
See also *19*, p. *300*.

W. Pastusiak
A Microscopic Interpretation of the Spin–Lattice Coupling Coefficients for S-State $4f^7$ Ions
(b) Vol. **116**, No. 1, p. 63–70 (1983).
See also *13.4*, p. *176*; *19*, p. *300*.

N. C. Amaral, B. Maffeo, and D. Guenzburger
Self-Consistent Embedded-Cluster Calculations of the Electronic Structure of Alkaline Earth Fluorides in the Hartree-Fock-Slater Approximation
(b) Vol. **117**, No. 1, p. 141–153 (1983).
See also *13.1*, p. *158*.

R. K. Chaudhary, A. V. Krishna Rao, and K. V. Rao
Dielectric Properties of Sm or Gd Doped CaF_2 Single Crystals
(a) Vol. 77, No. 2, p. K161–K165 (1983).
See also *14.4*, p. *252*.

G. Jeandel, A. Warin, and G. Morlot
Analysis of the Infrared Absorption Spectrum of $BaF_2:Y^{3+}$
(b) Vol. **118**, No. 1, p. 167–172 (1983).
See also *6*, p. *75*; *20.1*, p. *321*.

H. J. Fuchs, W. Karthe, and S. Börner
On the Kinetics of a Paramagnetic Oxygen Centre in CaF_2
(b) Vol. **118**, No. 1, p. 211–215 (1983).
See also 9, p. 94; *10.2*, p. *113*; 19, p. 302.

L. Wendler and E. Jäger
Phonon-Polaritons in Bilayer Systems. Surface Phonon-Polaritons
(b) Vol. **120**, No. 1, p. 235–247 (1983).
See also *6*, p. *80*; 13.5, p. 193; *20.1*, p. *328*; 22.2.1, p. 446; 22.2.2, p. 451.

L. E. Kholodenkov and A. G. Makhanek
Laser-Induced Reversible Deformation of Europium Stark Levels in Fluorite
(b) Vol. **120**, No. 1, p. K5–K8 (1983).
See also 13.4, p. 189; *20.1*, p. *329*; *20.3*, p. *346*.

J. Kuriata, J. M. Baker, T. Rewaj, and N. Guskos
Temperature Dependence of the "Intrinsic" $\bar{b}_2$ Parameter for $CdF_2:Eu^{2+}$
(b) Vol. **120**, No. 2, p. K135–K139 (1983).
See also *19*, p. *305*.

W. Pastusiak
The Relativistic Field Lattice Contributions to the Spin–Lattice Coupling Coefficients of the s-State $4f^7$ Ions
(b) Vol. **120**, No. 2, p. K151–K155 (1983).
See also *19*, p. *305*.

22.5.4 Mixed Crystals

Original Papers and Short Notes

T. Warmiński, M. Maier, and B. A. Orłowski
Energy Levels in $Cd_{1-x}Mn_xF_2$ Crystals Obtained from the Auger Spectra
(b) Vol. **115**, No. 1, p. K7 – K10 (1983).
See also *13*, p. *146*; *17*, p. *269*.

S. Ketolainen
On Mollwo-Ivey Relations of Mixed Alkali Halides
(b) Vol. **115**, No. 1, p. K41 – K44 (1983).
See also *10.2*, p. *108*; *20.1*, p. *313*.

J. W. Brightwell, C. N. Buckley, L. S. Miller, and B. Ray
Electrical Transport and Phase Stability in Silver Iodide–Cadmium Iodide
(a) Vol. **76**, No. 1, p. 391 – 398 (1983).
See also *1.2*, p. *15*; *14.4*, p. *250*.

H. B. Lal and O. P. Srivastava
Electrical Transport Studies in Copper Mercuric Iodide
(a) Vol. 77, No. 1, p. 405 – 412 (1983).
See also 10.2, p. 112; *14.4*, p. *251*; *15*, p. *261*.

A. E. Nikiforov, S. Yu. Shashkin, M. L. Levitan, and T. H. Agamalyan
Cooperative Jahn-Teller Orbital Ordering in $KCuF_3$ and K_2CuF_4 Crystals
(b) Vol. **118**, No. 1, p. 419 – 426 (1983).
See also *13.4*, p. *182*.

R. U. Bauer, J. R. Niklas, and J.-M. Spaeth
Two Types of F Centres in Alkaline Earth Fluoro-Halides (I): Superhyperfine Interactions and Electronic Structure
(b) Vol. **118**, No. 2, p. 557 – 566 (1983).
See also *13.4*, p. *184*; *19*, p. *302*.

N. Agrawal, R. P. Goyal, B. R. K. Gupta, and R. K. Singh
Evaluation of Cohesive Energies for NaCl–NaBr Mixed Crystals
(b) Vol. **118**, No. 2, p. K141 – K146 (1983).
See also *4*, p. *62*.

G. D. Jain and J. Shanker
Use of Harrison's Potential in Mixed Alkali Halide Crystals
(b) Vol. **119**, No. 1, p. 79 – 82 (1983).
See also *4*, p. *63*; 12.2, p. 143.

J. R. Niklas, R. U. Bauer, and J.-M. Spaeth
Two Types of F Centres in Alkaline Earth Fluoro-Halides (II): Double ENDOR Experiments
(b) Vol. **119**, No. 1, p. 171 – 178 (1983).
See also *13.4*, p. *185*; *19*, p. *303*.

J. Shanker and G. D. Jain
Analysis of the Dielectric Properties of Mixed Alkali Halide Crystals
(b) Vol. **119**, No. 1, p. 185–191 (1983).
See also *14.4*, p. *253*; 20, p. 310.

22.6 Simple Oxides (CdO, ZnO See 22.4)

Review Articles

R. E. Hummel
Differential Reflectometry and Its Application to the Study of Alloys, Ordering, Corrosion, and Surface Properties
(a) Vol. **76**, No. 1, p. 11–44 (1983).
See also *1.3*, p. *21*; 10.1, p. 99; 13.1, p. 155; 13.3, p. 167; *20.1*, p. *312*; 21, p. 349; 21.1, p. 363.

L. N. Aleksandrov and F. L. Edelman
Shock Crystallization of Films
(a) Vol. **76**, No. 2, p. 409–427 (1983).
See also 1.3, p. 21; *1.5*, p. *35*; *3*, p. *57*; 22.1.1, p. 418; 22.1.2, p. 423; 22.1.3, p. 440.

Original Papers and Short Notes

G. A. Bordovskii, N. L. Gordeev, A. N. Ermoshkin, V. A. Izvozchikov, and R. A. Evarestov
Self-Consistent Energy Band Structure of Tetragonal Lead Oxide
(b) Vol. **115**, No. 1, p. K15–K19 (1983).
See also *13.1*, p. *156*.

B. Subramanyam and Y. V. G. S. Murti
Elastic Constant Correlations in Ionic Crystals of Rocksalt Structure
(b) Vol. **115**, No. 1, p. K29–K31 (1983).
See also *12.2*, p. *141*; 22.5, p. 473.

H.-J. Fitting and J. Boyde
Monte-Carlo Calculation of Electron Attenuation in SiO_2
(a) Vol. **75**, No. 1, p. 137–142 (1983).
See also *6*, p. *70*; *14.3.4*, p. *243*.

G. D. Khattak
Specific Heat of Uranium Dioxide (UO_2) between 0.3 and 50 K
(a) Vol. **75**, No. 1, p. 317–321 (1983).
See also *6*, p. *70*; *8*, p. *85*; 18.4, p. 295.

A. V. Salker, K. Seshan, and H. V. Keer
Phase Transition Behaviour of VO_2
(a) Vol. **75**, No. 1, p. K37–K40 (1983).
See also *1.2*, p. *14*; 8, p. 85.

C. N. Duckworth, A. W. Brinkman, and J. Woods
Deep Level Transient Spectroscopy in TiO_2:Nb
(a) Vol. **75**, No. 1, p. K99–K102 (1983).
See also *13.4*, p. *174*; 14.3.3, p. 236.

M. Elango, J. Pruulmann, and A. P. Zhurakovskii
Recombination Luminescence and Energy Transfer in Ionic Crystals at XUV Excitation by Synchrotron Radiation
(b) Vol. **115**, No. 2, p. 399–407 (1983).
See also *20.3*, p. *335*; 22.5.2, p. 478.

H. Köster, Jr., I. N. Iassievich, and K. Hübner
Thermally Stimulated Excitation of Electrons from Deep Traps of Insulators in Electric Fields
(b) Vol. **115**, No. 2, p. 409–414 (1983).
See also *6*, p. *70*; *13.4*, p. *174*; 14.3.4, p. 243.

V. S. Grunin, I. B. Patrina, and Z. N. Zonn
Impurities and Electron Localization in V_2O_5 Crystals
(b) Vol. **115**, No. 2, p. 545–553 (1983).
See also *13.4*, p. *174*; *19*, p. *300*.

M. L. Kalra
Role of the Electron Scattering Processes in the Hubbard Model in the Insulator-to-Metal Transition
(b) Vol. **115**, No. 2, p. 575–585 (1983).
See also *14*, p. *204*.

M. D. Smolin, A. F. Andreeva, and V. A. Ogorodnik
Electroconduction and Thermoelectricity of Terbium Oxide Films
(b) Vol. **115**, No. 2, p. 587–592 (1983).
See also 13.5.3, p. 201; *14.3.1*, p. *229*; *15*, p. *261*.

V. Ya. Antonchenko, A. S. Davydov, and A. V. Zolotariuk
Solitons and Proton Motion in Ice-Like Structures
(b) Vol. **115**, No. 2, p. 631–640 (1983).
See also *4*, p. *60*; *6*, p. *71*; 13.5.4, p. 203.

H. Gruber and E. Krautz
Magnetoresistance and Conductivity in the Binary System Titanium–Oxygen (II): Semiconductive Titanium Oxides
(a) Vol. 75, No. 2, p. 511–518 (1983).
See also 8, p. 86; *14.1*, p. *206*; *14.3*, p. *216*; 15, p. 261; 18, p. 273.

V. D. Rodriguez, V. M. Orera, Y. Chen, and R. Alcala
Thermoluminescence of Low-Temperature X-Irradiated MgO and MgO:Li Single Crystals
(a) Vol. 75, No. 2, p. 577–582 (1983).
See also *20.3*, p. *336*.

P. Paroli, L. Ryabova, V. Salun, and A. Tucciarone
Optical Filters Based on Pyrolytic Films of Metal Oxides
(a) Vol. 75, No. 2, p. K209–K210 (1983).
See also *20.1*, p. *315*.

W. GUNSSER and K. ROHWER
*Determination of the Correlation between the Crystal Field Axis System and the Crystallographic Axes in Chromium-Doped β-Ga_2O_3 by EPR
(b) Vol. **116**, No. 1, p. 275–278 (1983).
See also *19*, p. *300*.

I. T. GODMANIS, A. N. TRUKHIN, and K. HÜBNER
Exciton–Phonon Interaction in Crystalline and Vitreous SiO_2
(b) Vol. **116**, No. 1, p. 279–287 (1983).
See also *6*, p. *72*; *13.5.1*, p. *195*; 20.1, p. 315; 22.1.1, p. 419.

E. V. ANUFRIEV and V. A. GURTOV
Accumulation and Discharging of the Hole Charge in Oxide Layers of MIS Structures
(a) Vol. **76**, No. 1, p. 107–111 (1983).
See also 11, p. 125; *13.3*, p. *168*; *14.3.3*, p. *237*.

L. KOMITOV and A. G. PETROV
Optical Method for Determination of Anchoring Energy of Tilted Nematic Layers
(a) Vol. **76**, No. 1, p. 137–143 (1983).
See also *1.6*, p. *39*; *2*, p. *45*.

S. B. NEWCOMB, W. M. STOBBS, and J. A. LITTLE
The Transmission Electron Microscopy of 1 to 2 nm Rhodium Particles Dispersed on γ-Al_2O_3
(a) Vol. **76**, No. 1, p. 191–196 (1983).
See also *1.4*, p. *30*.

H. FISCHER, G. GÖTZ, and H. KARGE
Radiation Damage in Ion-Implanted Quartz Crystals (I): Nuclear and Electronic Energy Deposition
(a) Vol. **76**, No. 1, p. 249–256 (1983).
See also *11*, p. *125*.

U. MESSERSCHMIDT, Y. NISHINO, T. IMURA, and H. SAKA
X-Ray Topographic In-Situ Observation of Slip Band Propagation in MgO Single Crystals
(a) Vol. **76**, No. 1, p. 277–284 (1983).
Erratum
(a) Vol. **80**, No. 1, p. 411 (1983).
See also 1.4, p. 30; *10.2*, p. *111*.

D. SCHULZE, J. FINSTER, E. HENSEL, W. SKORUPA, and U. KREISSIG
XPS Studies of SiO_2 Surface Layers Formed by Oxygen Ion Implantation into Silicon
(a) Vol. **76**, No. 1, p. K21–K24 (1983).
See also *1.6*, p. *39*; *17*, p. *269*.

V. G. BARYSHEVSKII, S. A. KUTEN, and V. I. RAPOPORT
Quadrupole Splitting of Hydrogen-Like Atom Levels in Crystals
(b) Vol. **116**, No. 2, p. 489–500 (1983).
Erratum
(b) Vol. **119**, No. 2, p. 755 (1983).
See also *13.4*, p. *177*; *19*, p. *301*.

H. FISCHER, G. GÖTZ, and H. KARGE
Radiation Damage in Ion-Implanted Quartz Crystals (II): Annealing Behaviour
(a) Vol. 76, No. 2, p. 493–498 (1983).
See also *11*, p. *126*.

S. A. Y. AL-ISMAIL and C. A. HOGARTH
Temperature Dependence of Voltage Controlled Negative Resistance in an Electroformed Cu–SiO_x–Cu Structure
(a) Vol. 76, No. 2, p. 559–563 (1983).
See also *14.4*, p. *250*.

C. SANCHEZ, R. MORINEAU, and J. LIVAGE
Electrical Conductivity of Amorphous V_2O_5
(a) Vol. 76, No. 2, p. 661–666 (1983).
See also *2*, p. *47*; *14.3*, p. *218*.

B. BANERJEE and S. LAHIRY
Superparamagnetism in γ-Mn_2O_3–α-Fe_2O_3–α-Mn_2O_3 System
(a) Vol. 76, No. 2, p. 683–694 (1983).
See also *18.3*, p. *291*; *18.4*, p. *295*; 22.8.2, p. 520.

H. BEEZ, D. FASOLD, and H. KARGE
Refractive Index Changes of Ion-Implanted Quartz
(a) Vol. 76, No. 2, p. K171–K174 (1983).
See also *11*, p. *127*; *20.1*, p. *317*.

A. F. ANDREEVA and V. A. OGORODNIK
Optical Absorption of Terbium Oxide
(b) Vol. 117, No. 1, p. K57–K59 (1983).
See also *20.1*, p. *318*.

A. HÜBLER and D. MARTON
Energy Dependence of the Si–SiO_2 Interface State Cross-Section Measured by the TSC Method
(a) Vol. 77, No. 1, p. K17–K21 (1983).
See also *13.3*, p. *168*; 22.1.2, p. 430.

A. REMÓN, J. PIQUERAS, J. LLOPIS, and C. BALLESTEROS
Cathodoluminescence and Optical Absorption in 20 kV Electron Irradiated MgO Single Crystals
(a) Vol. 77, No. 1, p. K29–K32 (1983).
See also 11, p. 128; *20.1*, p. *319*; *20.3*, p. *339*.

K. BIEDRZYCKI, J. CHRZANOWSKI, and B. SUJAK
Some Electrical and Optical Effects Observed in Polycrystalline Ice at 200 K
(a) Vol. 77, No. 1, p. K95–K98 (1983).
See also *1.2*, p. *16*; 14, p. 204; 20, p. 308.

A. LEHMANN, L. SCHUMANN, and K. HÜBNER
Optical Phonons in Amorphous Silicon Oxides (I): Calculation of the Density of States and Interpretation of LO–TO Splittings of Amorphous SiO_2
(b) Vol. **117**, No. 2, p. 689–698 (1983).
See also 2, p. 48; *6*, p. *74*; *20.1*, p. *320*.

V. M. ORERA, V. D. RODRIGUEZ, R. ALCALÁ, and Y. CHEN
Thermoluminescence of X-Irradiated CaO and CaO:Li Single Crystals
(a) Vol. **77**, No. 2, p. 625–631 (1983).
See also *20.3*, p. *340*.

B. BRIDGE, N. D. PATEL, and D. N. WATERS
On the Elastic Constants and Structure of the Pure Inorganic Oxide Glasses
(a) Vol. **77**, No. 2, p. 655–668 (1983).
See also *2*, p. *49*; *12.2*, p. *142*.

L. NANAI, I. HEVESI, L. MICHAILOVITS, I. APOSTOL, M. DINESCU, I. N. MIHAILESCU, and A. V. SIDORIN
Damage of V_2O_5 Single Crystals by Powerful Pulsed CO_2 Laser Irradiation
(a) Vol. **77**, No. 2, p. 679–684 (1983).
See also *11*, p. *129*.

K. HÜBNER, A. V. SHENDRIK, and A. M. PRAULINSH
Decomposition of SiO_x Films Due to Internal Stress
(b) Vol. **118**, No. 1, p. 239–244 (1983).
See also *1.5*, p. *37*; *4*, p. *61*; 19, p. 302; 20.1, p. 321.

H. KÖSTER, JR. and K. HÜBNER
Statistics and Transport Behaviour of Electrons in SiO_2
(b) Vol. **118**, No. 1, p. 293–301 (1983).
See also *6*, p. *75*; *13*, p. *150*; 14.4, p. 252.

F. BECHSTEDT, R. ENDERLEIN, and D. REICHARDT
Electronic Relaxation Effects in Core Level Spectra of Surfaces and Interfaces
(b) Vol. **118**, No. 1, p. 327–336 (1983).
See also *13.3*, p. *169*; 14.4, p. 252; 21.6, p. 398; 22.1.2, p. 432; 22.2.1, p. 445.

B. R. K. GUPTA and R. P. GOYAL
A Theoretical Analysis of Interionic Potentials in Rutile-Type Crystals
(b) Vol. **118**, No. 1, p. K7–K12 (1983).
See also *4*, p. *62*; *12.2*, p. *143*.

HJ. BERNHARDT
Vacuum-UV Absorption and Thermoluminescence of X-Rayed Quartz Crystals
(a) Vol. **78**, No. 1, p. 61–67 (1983).
See also 10.2, p. 114; *20.1*, p. *323*; *20.3*, p. *342*.

P. H. BEZIRGANYAN and V. G. ASLANYAN
About a Feature of X-Ray Diffraction in Imperfect Crystals
(a) Vol. **78**, No. 1, p. 157–161 (1983).
See also *1.1*, p. *12*; 22.1.2, p. 432.

C. Boulesteix and B. Yangui
Star Patterns and Dissociation of Interfaces without Spontaneous Strain Compatibility between Orientation Domains in Ferroelastic Crystals. Case of a Monoclinic Rare Earth Sesquioxide
(a) Vol. 78, No. 1, p. 303 – 308 (1983).
See also *1.4*, p. *32*.

W. Gehlhoff and K. H. Segsa
Determination of Stresses in Si–SiO_2 and Mechanically Affected Si by EPR
(b) Vol. **118**, No. 2, p. 703 – 712 (1983).
See also *12.2*, p. *143*; *19*, p. *303*; 22.1.2, p. 433.

F. López-Aguilar and J. Costa-Quintana
Renormalized Atom Model (RAM) Applied to the Heavy Rare-Earth Sesquioxides in Cubic Crystal Structure
(b) Vol. **118**, No. 2, p. 779 – 787 (1983).
See also *13.1*, p. *161*.

K. Hübner, H. Köster, Jr., B. Derlich, and W. Ecke
DLTS Investigations of Si–SiO_2 Interface States of Electron-Beam Irradiated MOS Structures
(b) Vol. **118**, No. 2, p. K133 – K136 (1983).
See also *11*, p. *130*; *13.3*, p. *169*; 14.3.2, p. 234; 22.1.2, p 433.

H. Ezbakhe and A. Donnadieu
Surface Roughness and Selectivity of Molybdenum Dioxide Thin Films
(a) Vol. 78, No. 2, p. 607 – 614 (1983).
See also *1.5*, p. *38*; *20.1*, p. *324*.

J. S. Kumar, G. Narayana, M. Chandra Shekar, U. V. Subba Rao, and V. Hari Babu
Dielectric and Electrical Properties of SnO_2, Sb_2O_3, and Their Mixed Films
(a) Vol. 78, No. 2, p. 647 – 653 (1983).
See also *14.3.1*, p. *231*; 14.4, p. 253; 22.6.1, p. 498.

J. Llopis, C. Ballesteros, J. Piqueras, A. Remon, and R. González
Thermally Induced Changes in the Cathodoluminescence Image of Deformed MgO
(a) Vol. 78, No. 2, p. 679 – 684 (1983).
See also *20.3*, p. *343*.

G. Kameswara Rao
$a\langle 100\rangle$ Dislocation Loops in MgO. Unusual Configurations
(a) Vol. 78, No. 2, p. 697 – 703 (1983).
See also *10.2*, p. *116*.

K. Kawabata
Intense Exoelectron Emission from Fine Particles of Al_2O_3 and Its Dependence on the Conductivity of the Substrate Material
(a) Vol. 78, No. 2, p. K127 – K130 (1983).
See also 11, p. 131; 13.3, p. 170; *17*, p. *270*.

D. F. Lynch and A. E. Smith
A Scattering Matrix Calculation of Very Low Energy Electron Reflection Intensities for Barium and Magnesium Oxides
(b) Vol. **119**, No. 1, p. 355–361 (1983).
See also *1.6*, p. *41*.

F. Campabadal, V. Milián, and X. Aymerich-Humet
Trap-Assisted Tunneling in MIS and Schottky Structures
(a) Vol. **79**, No. 1, p. 223–236 (1983).
See also 13.4, p. 186; *14.3.3*, p. *240*; *14.3.4*, p. *247*.

A. Wolkenberg
A Mechanism for the Effect of Doping on the Silicon Native Oxide Thickness
(a) Vol. **79**, No. 1, p. 313–322 (1983).
See also *1.6*, p. *41*; 10.2, p. 118; 22.1.2, p. 435.

G. S. Popkirov
An Electrochemical Evidence of the O_2 Photodesorption from n-TiO_2 Surface
(a) Vol. **79**, No. 1, p. 323–326 (1983).
See also *1.6*, p. *41*; *14.3.3*, p. *240*.

A. Boeuf, R. Caciuffo, J. M. Fournier, L. Manes, J. Rebizant, F. Rustichelli, J. C. Spirlet, and A. Wright
Study of the 25 K NpO_2 Phase Transition by a Neutron Oscillation Camera
(a) Vol. **79**, No. 1, p. K1–K3 (1983).
See also *1.2*, p. *19*.

R. Olivella, E. Molins, and J. Tejada
Mössbauer Second-Order Doppler Shift in MgO:^{57}Co
(b) Vol. **119**, No. 2, p. K165–K168 (1983).
See also *5*, p. *68*.

W. Gräfe
A Method for the Determination of a Static Fatigue Limit of Glasses
(a) Vol. **79**, No. 2, p. 403–410 (1983).
See also *2*, p. *53*; *10.2*, p. *119*.

E. Gmelin, N. Lenge, and H. Kronmüller
Vacancy Effect on the Transition Temperature of Magnetite
(a) Vol. **79**, No. 2, p. 465–475 (1983).
See also *8*, p. *90*; 18.3, p. 293.

H. Bahadur and R. Parshad
Some New Results on Irradiation Characteristics of Synthetic Quartz Crystals and Their Application to Radiation Hardening
(a) Vol. **79**, No. 2, p. 639–644 (1983).
See also *11*, p. *133*; *12.2*, p. *144*.

MIN-SHENG LEI and ZHI-CHENG WANG
Space Distribution of the Chemical Composition of the Thermo-Grown SiO_2–SiO_x–Si System
(b) Vol. **120**, No. 1, p. 189 – 195 (1983).
See also *1.3*, p. *26*; 1.6, p. 42; *4*, p. *64*.

M. N. GHATIKAR
Chemical Transferability of Phase Shifts in Extended X-Ray Absorption Fine Structure
(b) Vol. **120**, No. 1, p. 445 – 450 (1983).
See also *20*, p. *311*; 21.1, p. 376; 21.4, p. 393; 22.5, p. 475; 22.7, p. 499.

J. C. BERNÈDE
Influence of the Anode Metal on the Existence of a Negative Differential Resistance
(a) Vol. **80**, No. 1, p. 85 – 93 (1983).
See also *14.3.4*, p. *248*.

T. JUNG, L. ROTH, and H. WITTRICH
Auger Analysis of Plasma-Grown GaAs Oxide Films
(a) Vol. **80**, No. 1, p. K11 – K14 (1983).
See also *11*, p. *134*; 22.2.1, p. 447.

H. JAIN
Haven Ratio for Sodium in Synthetic α-Quartz
(a) Vol. **80**, No. 1, p. K15 – K20 (1983).
See also *9*, p. *96*; 10.2, p. 120; 14.4, p. 254.

J. L. JANSONS, P. A. KULIS, Z. A. RACHKO, M. J. SPRINGIS, I. A. TALE, and J. A. VALBIS
Luminescence of Ga-Doped α-Al_2O_3 Crystals
(b) Vol. **120**, No. 2, p. 511 – 518 (1983).
See also *20.3*, p. *347*.

A. KUMAR and P. PADHI
Phonon Conductivity of Ice Single Crystals
(b) Vol. **120**, No. 2, p. 679 – 683 (1983).
See also *6*, p. *81*; *14*, p. *205*.

B. SUSŁA, R. CZAJKA, and J. RAUŁUSZKIEWICZ
Electronic Excitation of Ni^{2+} and Co^{2+} by Tunneling Electrons
(b) Vol. **120**, No. 2, p. K129 – K134 (1983).
See also *13.4*, p. *190*; *14.3.3*, p. *242*; 19, p. 305; 20.1, p. 330.

22.6.1 Mixed Crystals

Original Papers and Short Notes

C. S. SUNANDANA and R. SINGH
An Infrared Study of the Glass System $CaO \cdot B_2O_3 \cdot Al_2O_3 \cdot x\ Fe_2O_3$ ($0 \leqq x \leqq 1$)
(a) Vol. 75, No. 1, p. K91–K94 (1983).
See also *2*, p. *44*; *20.1*, p. *314*.

K. SAKATA
Study of Electronic Structure of Reduced TiO_2 and $V_xTi_{1-x}O_2$ Crystals by ESCA and Optical Absorption
(b) Vol. **116**, No. 1, p. 145–153 (1983).
See also *13.4*, p. *176*; 17, p. 269; *20.1*, p. *315*.

S. NEOV, I. GERASSIMOVA, and B. SYDZHIMOV
Neutron Diffraction Study of 2 $TeO_2 \cdot V_2O_5$ Glass
(a) Vol. 76, No. 1, p. 297–301 (1983).
See also *2*, p. *46*; 4, p. 61.

J. CLAVERIE, G. CAMPET, D. CONTE, G. LE FLEM, and P. HAGENMULLER
Influence of Covalent Bonding on the Photoelectrochemical Properties of Some Perovskite-Type Related Compounds
(a) Vol. 77, No. 2, p. 603–609 (1983).
See also *4*, p. *61*; 13.4, p. 180; *16*, p. *266*.

A. J. COHEN and G. G. JANEZIC
Relationships among Trapped Hole and Trapped Electron Centers in Oxidized Soda–Silica Glasses of High Purity
(a) Vol. 77, No. 2, p. 619–624 (1983).
See also 2, p. 48; *13.4*, p. *180*; *20.1*, p. *321*.

J. SZCZYRBOWSKI, A. DIETRICH, and H. HOFFMANN
Optical and Electrical Properties of RF-Sputtered Indium-Tin Oxide Films
(a) Vol. 78, No. 1, p. 243–252 (1983).
See also *14.3.1*, p. *231*; *20.1*, p. *323*.

J. PEBLER
Paramagnetic Hyperfine Structure of Iron (57) in $V_{1-x}Fe_xO_2$
(a) Vol. 78, No. 2, p. 589–594 (1983).
See also *5*, p. *67*; *18.1*, p. *279*.

J. S. KUMAR, G. NARAYANA, M. CHANDRA SHEKAR, U. V. SUBBA RAO, and V. HARI BABU
Dielectric and Electrical Properties of SnO_2, Sb_2O_3, and Their Mixed Films
(a) Vol. 78, No. 2, p. 647–653 (1983).
See also *14.3.1*, p. *231*; 14.4, p. 253; 22.6, p. 495.

22.7 Semiconducting Intermetallic Compounds

Original Papers and Short Notes

L. G. Khvostantsev and V. A. Sidorov
Thermoelectric Properties of GeTe at High Hydrostatic Pressure up to 8.5 GPa and Its Valence-Band Structure
(b) Vol. **116**, No. 1, p. 83–89 (1983).
See also 13.1, p. 157; 14.3, p. 217; *15*, p. *261*.

Y. Iwama and Y. Takeno
Formation Process of MnBi Thin Films by Williams' Method
(a) Vol. **76**, No. 1, p. 75–83 (1983).
See also *3*, p. *57*; *18.2.1*, p. *290*.

M. K. Agarwal, P. D. Patel, and O. Vijayan
Electrical Studies on (Mo/W)Se_2 Single Crystals (I): Electrical Resistivity
(a) Vol. **78**, No. 1, p. 133–136 (1983).
See also *14.3*, p. *221*.

M. K. Agarwal, P. D. Patel, and O. Vijayan
Electrical Studies on (Mo/W)Se_2 Single Crystals (II): Temperature Dependence of Hall Effect
(a) Vol. **78**, No. 2, p. 469–473 (1983).
See also *14.3*, p. *222*.

J. Horák, P. Lošťák, and A. Montaner
Point Defects in Ga-Doped Bi_2Se_3 Single Crystals
(b) Vol. **119**, No. 1, p. K17–K21 (1983).
See also *10.2*, p. *117*; 14.3, p. 223; 20.1, p. 326.

M. K. Agarwal, P. D. Patel, and O. Vijayan
Electrical Studies on (Mo/W)Se_2 Single Crystals (III): Anisotropy in Resistivity
(a) Vol. **79**, No. 1, p. 103–108 (1983).
See also *14.3*, p. *224*.

M. N. Ghatikar
Chemical Transferability of Phase Shifts in Extended X-Ray Absorption Fine Structure
(b) Vol. **120**, No. 1, p. 445–450 (1983).
See also *20*, p. *311*; 21.1, p. 376; 21.4, p. 393; 22.5, p. 475; 22.6, p. 497.

A. Maciel, S. G. Fries, J. A. H. da Jornada, R. P. Livi, and E. R. Fraga
Study of the Nuclear Quadrupole Interaction in Tl_5Te_3
(a) Vol. **80**, No. 1, p. 245–248 (1983).
See also *1.3*, p. *27*; *13.4*, p. *189*.

A. Priemuth and H. Süssmann
On the Relation between the Carrier Concentration and the Effective Ion Charge of Interstitial Cu at the Electromigration in V_2VI_3 Compounds
(a) Vol. **80**, No. 1, p. K7–K9 (1983).
See also *9*, p. *95*; *14.3*, p. *227*.

22.7.1 Mixed Crystals

Original Papers and Short Notes

P. Lošťák, J. Horák, and L. Koudelka
Some Physical Properties and Point Defects in $Bi_{1-x}In_xTe_3$ Mixed Crystals
(a) Vol. **76**, No. 1, p. K71–K75 (1983).
See also *1.3*, p. *22*; 14.3, p. 218; *20.1*, p. *316*.

A. Dragūnas, K. Makariūnas, and M. Balčiūniene
Correlations of the ^{119}Sn Mössbauer Isomer Shift with Interatomic Distances for Complex Tetrahedral Semiconductors
(a) Vol. **77**, No. 2, p. 463–466 (1983).
See also *5*, p. *66*.

R. Požėra, A. Sutkus, and R. Tolutis
The Influence of Hydrostatic Pressure on the Effective Concentration and Cyclotron Mobility of Electrons in Semiconducting $Bi_{1-x}Sb_x$ Alloys
(b) Vol. **120**, No. 1, p. 165–171 (1983).
See also *13.1*, p. *164*; 14.3, p. 226.

H. Süssmann and W. Heiliger
Seebeck Coefficient and Electrical Conductivity in p-$(Bi_{1-x}Sb_x)_2Te_3$ at Room Temperature
(a) Vol. **80**, No. 2, p. 535–539 (1983).
See also *14.3*, p. *228*; *15*, p. *263*.

22.8 Tertiary and Higher Compounds

Original Papers and Short Notes

O. Ziep
Estimation of Second-Order Auger Recombination in Lead Chalcogenides
(b) Vol. **115**, No. 1, p. 161 – 170 (1983).
See also *6*, p. *69*; *13.1*, p. *155*; 13.5.2, p. 200.

C. W. M. Timmermans, S. O. Cholakh, R. L. van der Woude, and G. Blasse
Some Optical and Electrical Measurements on $Cs_3Bi_2Br_9$ Single Crystals
(b) Vol. **115**, No. 1, p. 267 – 271 (1983).
See also 13.1, p. 155; *16*, p. *264*; *20.1*, p. *312*.

V. Kalesinskas, J. Grigas, R. Jankevičius, and A. Audzionis
Soft Mode in the Microwave Dielectric Spectra of the SbSI–BiSI System
(b) Vol. **115**, No. 1, p. K11 – K13 (1983).
See also 6, p. 70; *14.4.1*, p. *256*.

K. N. Shrivastava
Nuclear Relaxation in Antiferromagnets at Low Temperatures
(b) Vol. **115**, No. 1, p. K37 – K40 (1983).
See also *19*, p. *299*.

G. M. Niftiev, B. G. Tagiev, and G. K. Aslanov
Injection and Thermodepolarization Currents in $YbGa_2Se_4$ Single Crystals
(a) Vol. **75**, No. 1, p. 59 – 65 (1983).
See also 13.4, p. 173; *14.3.4*, p. *243*.

A. A. Kaminskii, N. R. Agamalyan, L. P. Kozeeva, V. F. Nesterenko, and A. A. Pavlyuk
New Data on Stimulated Emission of Nd^{3+} Ions in Disordered Crystals with Scheelite Structure
(a) Vol. **75**, No. 1, p. K1 – K4 (1983).
See also 13.4, p. 173; *20.3*, p. *335*.

G. Blasse
Energy Transfer from Ce^{3+} to Eu^{3+} in $(Y, Cd)F_3$
(a) Vol. **75**, No. 1, p. K41 – K43 (1983).
See also *13.4*, p. *173*; *20.3*, p. *335*.

S. M. Wasim
On the Electron Effective Mass in n–$CuInTe_2$
(a) Vol. **75**, No. 1, p. K69 – K72 (1983).
See also *13.1*, p. *156*; *14.3*, p. *215*.

V. N. Brudnyi, A. I. Potapov, and Yu. V. Rud
Electrical Properties of H^+-Irradiated p–$ZnGeAs_2$
(a) Vol. **75**, No. 1, p. K73 – K76 (1983).
See also *11*, p. *123*; *14.3*, p. *215*.

M. Mocker and O. Ziep
Intrinsic Recombination in Dependence on Doping Concentration and Excitation Level. Application to Lead Chalcogenides
(b) Vol. **115**, No. 2, p. 415–425 (1983).
See also *13.1*, p. *156*; *14.3*, p. *216*; 20.3, p. 335.

H. Rauh, R. Geick, N. Lehner, J. Bouillot, and G. Eckold
Phonon Dispersion in the Perovskite-Type Layer Material K_2ZnF_4
(b) Vol. **115**, No. 2, p. 463–470 (1983).
See also *6*, p. *70*.

I. P. Aleksandrova, S. Grande, Yu. N. Moskvich, A. I. Krieger, and V. A. Koptsik
Second-Order Quadrupole Shift in the Incommensurate Structures of Rb_2ZnBr_4 and Rb_2ZnCl_4
(b) Vol. **115**, No. 2, p. 603–613 (1983).
See also *1.2*, p. *14*; 14.4.1, p. 256; *19*, p. *300*.

M. A. Aldzhanov, A. M. Aliev, R. K. Veliev, K. K. Mamedov, M. I. Mekhtiev, and V. Ya. Steinshreiber
Phonon Density of States and Heat Capacity of $CdIn_2S_4$
(b) Vol. **115**, No. 2, p. K75–K78 (1983).
See also *6*, p. *71*; *8*, p. *86*.

O. Kersten, A. Rost, and G. Schmidt
Dielectric Dispersion of Relaxor Ferroelectrics (SBN 75 and PLZT 8/65/35)
(a) Vol. 75, No. 2, p. 495–500 (1983).
See also *14.4.1*, p. *256*.

W. Thulke, R. Frahm, R. Haensel, and P. Rabe
X-Ray Study of Niobium Disulfide Intercalated with Copper
(a) Vol. 75, No. 2, p. 501–509 (1983).
See also *1.3*, p. *21*; *4*, p. *60*; 20, p. 307.

A. N. Anagnostopoulos
Influence of Composition Faults on the AC Conductivity of $ZnIn_2S_4$ (III)
(a) Vol. 75, No. 2, p. 595–599 (1983).
See also *14.3*, p. *217*.

V. M. Garmash, A. A. Kaminskii, M. I. Polyakov, S. E. Sarkisov, and A. A. Filimonov
Luminescence and Stimulated Emission of Nd^{3+} Ions in $LaMgAl_{11}O_{19}$ Crystals in the $^4F_{3/2} \rightarrow {}^4I_{11/2}$ and $^4F_{3/2} \rightarrow {}^4I_{13/2}$ Transitions
(a) Vol. 75, No. 2, p. K111–K116 (1983).
See also 13.4, p. 175; *20.2*, p. *332*; *20.3*, p. *336*.

M. I. Mitkowa and W. Fallmann
Some Properties of Silver Photodiffused Chalcogenide Glasses
(a) Vol. 75, No. 2, p. K145–K148 (1983).
See also *2*, p. *45*; *14.3.3*, p. *237*; 20.1, p. 315.

G. A. Petrakovskii, K. A. Sablina, and V. P. Ikonnikov
Amorphization of $RbFeS_2$
(a) Vol. **75**, No. 2, p. K165 – K167 (1983).
See also *2*, p. *45*; *5*, p. *65*.

H. Neumann, R. D. Tomlinson, N. Avgerinos, and E. Nowak
Electrical Properties of $CuInSe_2$ Single Crystals Grown by the Vertical Bridgman Technique
(a) Vol. **75**, No. 2, p. K199 – K203 (1983).
See also 13.4, p. 175; *14.3*, p. *217*.

M. Mocker and M. Beiler
Theory of Intrinsic Recombination at Zero Temperature in Small Gap Semiconductors. Estimations for PbSSe
(b) Vol. **116**, No. 1, p. 205 – 215 (1983).
See also *13.4*, p. *176*; *14.3*, p. *217*; 20.3, p. 336.

Yu. N. Moskvich, A. M. Polyakov, S. P. Prozorov, V. E. Zobov, and A. V. Ponomarenko
The Spin–Lattice Relaxation of the Moving Two-Spin Group in Multipulse Spin Locking
(b) Vol. **116**, No. 1, p. K35 – K38 (1983).
See also *19*, p. *301*.

M. M. Hafiz, A. A. Ammar, A. I. Al-Adl, and Aboutaleb Mohamed
Effect of Heat Treatment on the Conduction and Structure of Ge–Se–Te Amorphous Alloys
(a) Vol. **76**, No. 1, p. 319 – 325 (1983).
See also 1.2, p. 15; *2*, p. *46*; *14.3*, p. *217*.

D. V. Gitsu, I. N. Grincheshen, and N. S. Popovich
The Electric and Photoelectric Properties of $TlSbSe_2$ Crystals in the Region of Impurity Conduction
(a) Vol. **76**, No. 1, p. K5 – K7 (1983).
See also 13.4, p. 176; *14.3*, p. *217*; *16*, p. *265*.

D. Dădârlat, Rodica M. Cândea, and M. Bârlea
Majority Carrier Concentration and Quasi-Fermi Level for $Pb_{1-x}Sn_xTe$ DH Diode Lasers
(a) Vol. **76**, No. 1, p. K61 – K65 (1983).
See also *14.3.3*, p. *238*; *20.2*, p. *332*.

K. H. Herrmann and K.-P. Möllmann
Characteristic Steps in the Relaxation Curves of Field-Quenched Persistent Photoconductivity of $Pb_{0.78}Sn_{0.22}Te$:In
(a) Vol. **76**, No. 1, p. K67 – K70 (1983).
See also *16*, p. *265*.

N. M. Gasanly, A. F. Goncharov, N. N. Melnik, A. S. Ragimov, and V. I. Tagirov
Optical Phonons and Structure of $TlGaS_2$, $TlGaSe_2$, and $TlInS_2$ Layer Single Crystals
(b) Vol. **116**, No. 2, p. 427 – 443 (1983).
See also *6*, p. *72*; *20.1*, p. *316*.

L. M. Avkhutskii, R. L. Davidovich, L. A. Zemnukhova, P. S. Gordienko, V. Urbonavičius, and J. Grigas
Peculiarities of Phase Transitions and Physical Properties in $(NH_4)_2SbF_5$
(b) Vol. **116**, No. 1, p. 483–488 (1983).
See also 8, p. 87; *14.4.1*, p. *256*; 19, p. 301.

K. Kambas
Preparation and Optical Investigation of the $CdInGaS_4$ Compound
(a) Vol. 76, No. 2, p. 735–741 (1983).
See also 1.3, p. 23; *6*, p. *73*; *20.1*, p. *317*.

F. Glasse
The Influence of Thermal Measurement Conditions on the Results of Dielectric Measurements in PLZT Ceramic with Diffused Phase Transition
(a) Vol. **76**, No. 2, p. K121–K123 (1983).
See also *14.4*, p. *251*.

I. S. Edelman, G. Dustmuradov, and V. P. Kononov
Polarized Light Absorption of the Magnetic Semiconductor $CdCr_2Se_4$
(b) Vol. 117, No. 1, p. 351–354 (1983).
See also 13.1, p. 158; *20.1*, p. *318*.

D. Gerritzen and D. Siebert
Second Order Shift of ^{71}Ga NMR in the Two Different Sites of a $CdGa_2S_4$ Single Crystal
(b) Vol. 117, No. 1, p. K5–K7 (1983).
See also *1*, p. *9*; *19*, p. *301*.

Ramani, M. P. Sathyavathiamma, N. G. Puttaswamy, an R. M. Mallya
Mössbauer Effect Study of the Thermal Decomposition in Hydrogen of Homogeneously Mixed Ferrous Nickel Oxalates with Different Iron to Nickel Ratios
(a) Vol. 77, No. 1, p. 87–96 (1983).
See also *5*, p. *66*.

A. A. Kaminskii, V. A. Timofeeva, N. R. Agamalyan, A. B. Bykov, S. E. Sarkisov, and Yu. M. Gerasimov
Luminescence Properties of Nd^{3+} Ions in Disordered $Na_5YSi_4O_{12}$ Crystals
(a) Vol. 77, No. 1, p. 107–114 (1983).
See also *20.1*, p. *319*; *20.3*, p. *339*.

C. Lacour, A. Dubon, and P. M. Grangeon
Random Aspects of the Formation of Compounds by Reactive Sintering
(a) Vol. 77, No. 1, p. 309–322 (1983).
See also *1.4*, p. *31*.

V. N. Brudnyi, A. I. Potapov, Yu. V. Rud, and M. Serginov
Electrical Properties of Electron-Irradiated GaAs and $ZnSiAs_2$ Solid Solutions
(a) Vol. 77, No. 1, p. K33–K36 (1983).
See also *11*, p. *128*; 13.4, p. 179; *14.3*, p. *219*; 22.2.1, p. 444.

R. Borromei, E. Cavalli, and G. Ingletto
Electronic Spectrum of the Co(II) Ion in the Columbite $CoNb_2O_6$ Host Lattice at 4.2 K
(b) Vol. **117**, No. 2, p. 733–742 (1983).
See also *13.4*, p. *180*; *20.1*, p. *320*.

K. R. Allakhverdiev, T. G. Mamedov, E. Yu. Salaev, and I. K. Efendieva
The Fundamental Absorption Spectra of $TlInSe_2$ Crystals under Pressure
(b) Vol. **117**, No. 2, p. K109–K111 (1983).
See also *13.1*, p. *159*; *20.1*, p. *320*.

C. Rincón, L. González, and G. Sánchez Pérez
Temperature Dependence of the Energy Gap in $CuInSe_2$
(b) Vol. **117**, No. 2, p. K123–K126 (1983).
See also *13.1*, p. *159*; *20.1*, p. *320*.

A. N. Anagnostopoulos, C. Manolikas, and D. Papadopoulos
Comparison of the Structure and the Electric Properties of $ZnIn_2S_4$(III)- and $CdInGaS_4$-Layered Crystals
(a) Vol. 77, No. 2, p. 595–601 (1983).
See also *1*, p. *10*; *14.3*, p. *221*.

B. Pistoulet, D. Coquillat, J. C. Tedenac, G. Brun, and M. Maurin
Hole Effective Mass and Impurity Levels in Undoped AgTlTe. Optimum Thermoelectric Figure of Merit
(a) Vol. 77, No. 2, p. 669–677 (1983).
See also *13.1*, p. *159*; *13.4*, p. *181*; 14.3, p. 221; 15, p. 262.

A. A. Bahgat and M. K. Fayek
Erratum to: Absolute Evaluation of Ferrous and Ferric Concentration in Ca Amphibole
(a) Vol. 77, No. 2, p. 809 (1983).
See also *5*, p. *67*.

R. Hrabański
EPR of Mn^{2+} in $[Mg(H_2O)_6]SnCl_6$ Single Crystals
(a) Vol. 77, No. 2, p. K143–K147 (1983).
See also *19*, p. *302*.

C. Jäger and M. Müller
Crystallization in a Fluorozirconate Glass: Determination of the ^{19}F Chemical Shift in β-$BaZrF_6$
(a) Vol. 77, No. 2, p. K167–K170 (1983).
See also *2*, p. *49*; *19*, p. *302*.

S. Deonarine
The Nature of Stage I–II Structural Phase Transition in the Layer Compound $RbVF_4$
(a) Vol. 77, No. 2, p. K199–K201 (1983).
See also *1.2*, p. *17*.

C. W. M. TIMMERMANS and G. BLASSE
The Luminescence of $K_4Bi_2Br_{10} \cdot 4\ H_2O$ and Rb_3BiBr_6 Single Crystals
(b) Vol. **118**, No. 1, p. 353–361 (1983).
See also *20.3*, p. *341*.

P. M. TOMCHUK, V. A. SHENDEROVSKII, P. N. GORLEY, and N. P. GAVALESHKO
Hall Coefficient and Magnetoresistance of Semiconductors with Anisotropic Energy Dispersion
(b) Vol. **118**, No. 1, p. 433–439 (1983).
See also 13.1, p. 160; 13.4, p. 182; *14.3*, p. *221*.

C. RINCÓN and J. GONZÁLEZ
Residual Optical Absorption below the Band Gap in $CuInSe_2$
(b) Vol. **118**, No. 1, p. K21–K24 (1983).
See also *13.1*, p. *160*; *20.1*, p. *322*.

A. MANOOGIAN and B. W. CHAN
Electron Spin Resonance and Electron Nuclear Double Resonance of $^{53}Cr^{3+}$ in RbIn Alum
(b) Vol. **118**, No. 1, p. K31–K34 (1983).
See also *19*, p. *302*.

A. P. DESHPANDE, V. B. SAPRE, and C. MANDE
A Study of the Chemical Shifts in X-Ray Absorption Spectra and the Sublattice Distortion in Some Ternary Chalcopyrite Compounds of the Type $CuMS_2$
(b) Vol. **118**, No. 1, p. K39–K42 (1983).
See also *4*, p. *62*; *20*, p. *309*.

H. NEUMANN, R. D. TOMLINSON, W. KISSINGER, and N. AVGERINOS
Polarization-Dependent Infrared Reflectivity Spectra of $CuInSe_2$
(b) Vol. **118**, No. 1, p. K51–K56 (1983).
See also *6*, p. *76*; *20.1*, p. *322*.

A. A. KAMINSKII, B. P. SOBOLEV, and T. V. UVAROVA
Stimulated Emission of Ho^{3+} in $BaTm_2F_8$ and Tm^{3+} Ions in $BaYb_2F_8:Er^{3+}$ Crystals
(a) Vol. 78, No. 1, p. K13–K17 (1983).
See also 13.4, p. 183; *20.2*, p. *332*; *20.3*, p. *342*.

G. I. MAKOVETSKII
Magnetic Properties of the $Cr_{1-x}Fe_xTe_8$ ($x = 0$ to 3) System
(a) Vol. 78, No. 1, p. K39–K41 (1983).
See also *18*, p. *275*.

O. B. TAGIEV, G. M. NIFTIEV, and I. M. ASKEROV
The Frenkel Thermal-Field Emission in $EuGa_2S_4$ Single Crystals
(a) Vol. 78, No. 1, p. K43–K46 (1983).
See also *13.4*, p. *183*; *14.3.4*, p. *247*.

I. V. Bodnar and N. S. Orlova
X-Ray Study of Thermal Expansion in $CuAlS_2$, $CuGaS_2$, and $CuInS_2$ Compounds over the Temperature Range from 90 to 650 K
(a) Vol. **78**, No. 1, p. K59–K63 (1983).
See also *8*, p. *88*.

C. Razzetti, P. P. Lottici, L. Zanotti, and M. Curti
A Raman Study of Compositional Disorder in Defect Chalcopyrites
(b) Vol. **118**, No. 2, p. 743–749 (1983).
See also *6*, p. *76*; *20.1*, p. *324*.

R. Tovar Barradas, J. González, and C. Rincón
Impurity States near the Fundamental Absorption Edge in p-$CuInS_2$
(b) Vol. **118**, No. 2, p. K103–K106 (1983).
See also 13.1, p. 161; 13.4, p. 184; *20.1*, p. *324*.

S. M. Wasim, G. Marcano, and G. Sánchez Pérez
Electrical Properties of $CuGaTe_2$
(a) Vol. **78**, No. 2, p. 423–430 (1983).
See also *13.4*, p. *184*; *14.3*, p. *222*.

W. Thulke, R. Haensel, and P. Rabe
Anisotropic Absorption of Unpolarized X-Radiation in Copper-Intercalated Niobium Disulfide
(a) Vol. **78**, No. 2, p. 539–545 (1983).
See also *1.3*, p. *25*; *4*, p. *63*; 20, p. 310.

M. Krupski
Rigid Sphere Model Applied to the Phase Transitions of Complex Compounds of the Type K_2PtCl_6
(a) Vol. **78**, No. 2, p. 751–758 (1983).
See also *1.2*, p. *18*; 6, p. 77.

O. Ziep and M. Mocker
New Arguments to the Problem of Screening in Auger Recombination in Semiconductors
(b) Vol. **119**, No. 1, p. 299–306 (1983).
See also *13.1*, p. *163*; 22.2.3, p. 453.

P. Muralt, E. Voit, and R. Kind
Cl-NQR Study of the Low Temperature Phase of Rb_2ZnCl_4
(b) Vol. **119**, No. 1, p. K65–K67 (1983).
See also *1.2*, p. *18*; *19*, p. *304*.

D. K. Chakrabarty, A. Bandyopadhyay, S. B. Patil, and S. N. Shringi
Magnetic and Mössbauer Studies of Some Mixed Cobaltites of the Lanthanides and Alkali Metals
(a) Vol. **79**, No. 1, p. 213–222 (1983).
See also *5*, p. *67*; *18.2*, p. *286*.

M. Hidaka, Y. Okamoto, and Y. Zikumaru
Structural Phase Transition of $CsPbCl_3$ below Room Temperature
(a) Vol. **79**, No. 1, p. 263–269 (1983).
See also *1.2*, p. *18*; 6, p. 78.

A. N. Anagnostopoulos and A. Karoutis
Cation Mixing and the AC Conductivity of $CdInGaS_4$
(a) Vol. **79**, No. 1, p. K29–K33 (1983).
See also *14.3*, p. *224*.

G. M. Guseinov, G. M. Niftiev, and M. G. Shakhtakhtinskii
Effect of Illumination on Current–Voltage Characteristics of $GdIn_3S_6$ Single Crystals
(a) Vol. **79**, No. 1, p. K53–K55 (1983).
See also *14.3.4*, p. *247*.

S. M. Wasim and G. Sánchez Porras
On the Hole Mobility of $CuGaSe_2$
(b) Vol. **79**, No. 1, p. K65–K68 (1983).
See also *14.3*, p. *224*.

S. K. Decheva-Plachkova
On the Fundamental Energy Gap in the System $(GeTe)_{1-x}(AgSbTe_2)_x$
(a) Vol. **119**, No. 2, p. K97–K100 (1983).
See also 1.2, p. 19; *13.1*, p. *164*.

P. Deus and H. A. Schneider
Thermal Expansion of the Diamond-Like Semiconducting Compound $ZnSiP_2$ within the Temperature Range 20 to 300 K
(a) Vol. **79**, No. 2, p. 411–415 (1983).
See also *8*, p. *90*.

A. N. Anagnostopoulos and Ch. Lioutas
Voltage-Dependent Conductivity along the *c*-Axis of $ZnIn_2S_4$(III)
(a) Vol. **79**, No. 2, p. 513–516 (1983).
See also *14.3*, p. *225*.

R. Z. Sadykhov, L. M. Valiev, D. A. Guseinov, and A. O. Ismailov
Magnetic Properties of the $Fe_xZn_{1-x}Cr_2S_4$ $(0 < x < 1)$ System
(a) Vol. **79**, No. 2, p. K147–K150 (1983).
See also *18.2*, p. *287*; *18.4*, p. *297*.

P. Deus, H. Neumann, G. Kühn, and B. Hinze
Low-Temperature Thermal Expansion in $CuInSe_2$
(a) Vol. **80**, No. 1, p. 205–209 (1983).
See also *8*, p. *91*.

D. Lapraz and A. Baumer
Thermoluminescent Properties of Synthetic and Natural Fluorapatite, $Ca_5(PO_4)_3F$
(a) Vol. **80**, No. 1, p. 353–366 (1983).
See also *10.2*, p. *120*; *20.3*, p. *347*.

G. M. GUSEINOV, G. M. NIFTIEV, and M. G. SHAKHTAKHTINSKII
Effect of External Factors on the Thermally Stimulated Conductivity of $GdIn_3S_6$ Single Crystals
(a) Vol. **80**, No. 1, p. K21 – K24 (1983).
See also *13.4*, p. *189*; 14.3, p. 227.

A. N. GEORGOBIANI, A. N. GRUZINTSEV, S. I. RADAUTSAN, and I. M. TIGINYANU
The Contribution of Iodine Impurity to the Formation of Centers in $CdGa_2S_4$ Single Crystals
(a) Vol. **80**, No. 1, p. K45 – K48 (1983).
See also *13.4*, p. *190*; 16, p. 268; 20.1, p. 329.

YA. M. OLIKH
Characteristics of Sound Waves Propagating in the Symmetry Plane of Pyrargyrite and Proustite Crystals
(a) Vol. **80**, No. 1, p. K81 – K85 (1983).
See also *7*, p. *84*.

S. K. PLACHKOVA
Thermoelectric Power in the System $(GeTe)_{1-x}(AgSbTe_2)_x$
(a) Vol. **80**, No. 1, p. K97 – K100 (1983).
See also 8, p. 91; *14.3*, p. *227*; *15*, p. *263*.

G. V. LOSEVA and G. M. ABRAMOVA
Anomalies of Electric and Magnetic Properties of Ternary Cobalt Sulfides
(a) Vol. **80**, No. 1, p. K109 – K111 (1983).
See also *14.3*, p. *227*; *18*, p. *276*.

I. SAVATINOVA and E. ANACHKOVA
Correlations between OH(OD) Stretching Vibrations and Structural Characteristics of Hydrogen Bonds in $K_4Fe(CN)_6 \cdot 3\,H_2O$ Crystal
(b) Vol. **120**, No. 2, p. 539 – 545 (1983).
See also 4, p. 64; *6*, p. *80*; *20.1*, p. *329*.

J. M. WESSELINOWA
Dynamic Behaviour of the s–d Model
(b) Vol. **120**, No. 2, p. 585 – 593 (1983).
See also *13.5.4*, p. *203*; *18.2*, p. *288*.

O. UEMURA, Y. WADA, T. TAKAHARA, and T. SATOW
The Semiconduction in the Molten Tl–$CuTlTe_2$ System
(b) Vol. **120**, No. 2, p. 685 – 693 (1983).
See also *2*, p. *55*; *14.3*, p. *227*; 15, p. 263; 18, p. 276.

C. MANOKILAS and A. N. ANAGNOSTOPOULOS
Electrical Effects Associated with the Ordering Process in $CdInGaS_4$ Crystals (I): Electron Microscopy and Electron Diffraction Study
(a) Vol. **80**, No. 2, p. 503 – 511 (1983).
See also *1.3*, p. *27*; 14.3, p. 228.

L. A. Zemnukhova, R. L. Davidovich, P. S. Gordienko, J. Grigas, A. N. Kovrianov, S. I. Kuznetsov, T. A. Kaidalova, and V. Urbonavičius
Investigation of Phase Transitions in K_2SbF_5
(a) Vol. **80**, No. 2, p. 553 – 558 (1983).
See also *1.2*, p. *20*; 8, p. 92; *14.4.1*, p. *259*; 19, p. 306.

F. Vasiliu, P. Gr. Lucuta, and F. Constantinescu
Structure–Sintering Temperature Relationship in PZT Solid Solutions at the Morphotropic Phase Boundary
(a) Vol. **80**, No. 2, p. 637 – 642 (1983).
See also *1.3*, p. *28*; *14.4.1*, p. *260*.

J. Kansy, T. J. Panek, and M. Szuszkiewicz
Influence of Non-Homogeneity of Sample Thickness on the Determination of Absolute Mössbauer Fraction
(a) Vol. **80**, No. 2, p. 643 – 655 (1983).
See also *5*, p. *68*.

K. G. Nikiforov, V. P. Buzhor, S. I. Radautsan, and V. E. Tezlevan
Switching Effect Based on N-Shaped Negative Differential Resistance in n-$CdCr_2Se_4$ Single Crystals
(a) Vol. **80**, No. 2, p. K149 – K151 (1983).
See also *14.3.4*, p. *249*.

22.8.1 Salts of Oxy-Acids (Carbonates, Phosphates, Silicates Including Mica etc.)

Review Article

K. N. Shrivastava
Theory of Spin–Lattice Relaxation
(b) Vol. **117**, No. 2, p. 437–458 (1983).
See also *6*, p. *69*; *13*, p. *146*; 20.1, p. 312; 22.5, p. 473.

Original Papers and Short Notes

Fat Duen Ho
Approximate Wave Functions for the F^+-Centers in $LiNbO_3$ and $LiTaO_3$
(b) Vol. **115**, No. 1, p. 299–303 (1983).
See also *10.2*, p. *108*; 13.4, p. 172; *20.1*, p. *313*.

Y. Y. Abolinsh, S. V. Karpov, and A. A. Shultin
Raman Studies of Ammonium Ion Motion in NH_4NO_3–V Crystals
(b) Vol. **115**, No. 1, p. K1–K5 (1983).
See also *1.2*, p. *14*; *6*, p. *70*; 20.1, p. 313.

P. Hari, K. Purandar, J. Lakshmana Rao, and S. V. J. Lakshman
Optical Absorption Spectrum of Co^{2+} in Lithium Ammonium Sulphate Single Crystals
(b) Vol. **115**, No. 1, p. K45–K50 (1983).
See also *13.4*, p. *173*; *20.1*, p. *313*.

H. Beige, A. Kühnel, and B. Lorenz
Anomalous Behaviour of Nonlinear Dielectric, Elastic, and Electromechanical Coefficients at Ferroelectric Phase Transitions. Rochelle Salt
(a) Vol. **75**, No. 1, p. 129–136 (1983).
See also 12.2, p. 141; *14.4.1*, p. *256*.

R. Stumpe, D. Wagner, and D. Bäuerle
Influence of Bulk and Interface Properties on the Electric Transport in ABO_3 Perovskites
(a) Vol. **75**, No. 1, p. 143–154 (1983).
See also *14.3*, p. *215*.

W. Kucharczyk and P. Górski
Measurement of Nonlinear Electrooptic Effects in KDP Crystals
(a) Vol. **75**, No. 1, p. K87–K90 (1983).
See also *14.4.1*, p. *256*; *20.1*, p. *313*.

V. K. Jain, V. P. Seth, and S. K. Yadav
EPR of VO^{2+} in $ZnSeO_4 \cdot 6H_2O$ Single Crystals
(b) Vol. **115**, No. 2, p. K113–K115 (1983).
See also *19*, p. *300*.

A. Sternberg, M. Zvara, and P. Hlídek
Influence of Permanent Magnetic Field on Dielectric Properties of Transparent Ferroelectric PLZT Ceramics
(a) Vol. **75**, No. 2, p. K179 – K182 (1983).
See also *14.4.1*, p. *256*; 18, p. 273.

M. Ishigame and S. Yamashita
Raman Scattering in K_2SO_4 at High Temperatures
(b) Vol. **116**, No. 1, p. 49 – 56 (1983).
See also 1.2, p. 14; *6*, p. *71*; *20.1*, p. *315*.

J. M. Wesselinowa
Spin–Phonon Interactions in the Ising Model with a Transverse Field
(b) Vol. **116**, No. 1, p. 71 – 82 (1983).
See also *6*, p. *71*; *14.4.1*, p. *256*.

S. Grosswig, J. Härtwig, U. Schellenberger, and W.-D. Zimmer
Theoretical and Experimental Study of Growth Sector Boundaries in Deuterated Potassium Dihydrogen Phosphate Single Crystals
(a) Vol. **76**, No. 1, p. 241 – 247 (1983).
See also *1.5*, p. *36*; *3*, p. *57*; 10.2, p. 111.

N. Kamegashira and Y. Miyazaki
Jahn-Teller Transition in $NdMnO_3$
(a) Vol. **76**, No. 1, p. K39 – K42 (1983).
See also *1.2*, p. *15*; *14.3*, p. *218*.

S. Pačesová, B. Březina, and L. Jastrabík
Optical Properties of K_2SeO_4 in the Vicinity of Phase Transitions
(b) Vol. **116**, No. 2, p. 645 – 652 (1983).
See also *6*, p. *73*; 14.4.1, p. 257; *20.1*, p. *316*.

Ph. Colomban and D. Vivien
EPR Study of Ordering in Stoichiometric β-Aluminate
(a) Vol. **76**, No. 2, p. 565 – 574 (1983).
See also *19*, p. *301*.

F. Ekpo Eno, R. Grasser, and A. Scharmann
Radiation Effects in $CaWO_4$ Crystals
(a) Vol. **76**, No. 2, p. K113 – K116 (1983).
See also *11*, p. *126*; *20.3*, p. *338*.

G. Borchardt, J. von Cieminski, and G. Schmidt
Switching Properties of SBN Single Crystals
(a) Vol. **76**, No. 2, p. K141 – K145 (1983).
See also *14.4.1*, p. *257*.

Z. Czapla, O. Czupiński, L. Sobczyk, J. Mróz, and H. Pykacz
Dielectric Properties and Ferroelectric Phase Transitions in Mixed $Rb_x(NH_4)_{1-x}HSeO_4$ Crystals
(a) Vol. **76**, No. 2, p. K155 – K157 (1983).
See also *1.2*, p. *16*; *14.4.1*, p. *257*.

A. L. Tronconi, K. Skeff Neto, and M. A. Amato
Piezoelectric Detection of the Photoacoustic Signal
(a) Vol. **76**, No. 2, p. K209–K211 (1983).
See also *7*, p. *82*.

S. Jetschke, H. Karge, and K. Hehl
Anisotropic Effects in N^+-Implanted $LiNbO_3$
(a) Vol. **77**, No. 1, p. 207–214 (1983).
See also *11*, p. *128*; 20.1, p. 319.

L. A. Shebanov, V. Ya. Fritsberg, and A. R. Gaevskis
Crystallographic Properties and Superconductivity of Solid Solutions of the $BaBi_xPb_{1-x}O_3$ System
(a) Vol. **77**, No. 1, p. 369–373 (1983).
See also *1.2*, p. *16*; *14.2*, p. *213*.

S. I. Golovkova, A. M. Gurvich, A. I. Kravchenko, V. V. Mikhailin, and A. N. Vasilev
Enhancing and Quenching of Intrinsic Luminescence and Characteristic Features of Calcium Tungstate Phosphorescence in the Presence of Lanthanides
(a) Vol. **77**, No. 1, p. 375–380 (1983).
See also *13.4*, p. *179*; *20.3*, p. *339*.

V. M. Mukhortov, Yu. I. Golovko, V. A. Aleshin, E. V. Sviridov, Vl. M. Mukhortov, V. P. Dudkevich, and E. G. Fesenko
Ferroelectric Phase Transition in Heteroepitaxial Films of (Ba, Sr) TiO_3
(a) Vol. **77**, No. 1, p. K37–K40 (1983).
See also 1.5, p. 37; *14.4.1*, p. *257*.

P. V. Reddy and T. S. Rao
A Study of the Semiconducting Properties of Mixed Mn–Mg Ferrites
(a) Vol. **77**, No. 1, p. K63–K67 (1983).
See also *14.3*, p. *219*; *15*, p. *261*.

S. V. J. Lakshman and A. Sundar Jacob
Absorption Spectrum of VO^{2+} in $LiKSO_4$ Single Crystals
(b) Vol. **117**, No. 2, p. K119–K122 (1983).
See also *13.4*, p. *180*; 19, p. 301; *20.1*, p. *320*.

R. N. Mukherjee, D. N. S. Srivastwa, and S. C. Sen
Ultraviolet Absorption of Tl^+ Ion in K_2O–KBr–B_2O_3 and Na_2O–NaBr–B_2O_3 Glasses
(a) Vol. **77**, No. 2, p. 799–802 (1983).
See also 2, p. 49; *20.1*, p. *321*.

A. A. Kaminskii, A. G. Petrosyan, K. L. Ovanesyan, and M. I. Chertanov
Stimulated Eimission of Pr^{3+} Ions in $YAlO_3$ Crystals
(a) Vol. **77**, No. 2, p. K173–K178 (1983).
See also 20.1, p. 321; *20.2*, p. *332*; 20.3, p. 341.

A. P. Fedotov and V. F. Shabanov
Deuteron Dynamics in Crystals of Sodium Trideuterium Selenite
(b) Vol. **118**, No. 1, p. 345–351 (1983).
See also *6*, p. *75*; 20.1, p. 322.

M. El Sherif, F. Brehat, and B. Wyncke
A Simple Best Fit Method for the Far Infrared Reflectivity Spectra by Using the Four-Parameter Semi-Quantum Model: Application on the Sodium Nitrite Spectrum
(b) Vol. **118**, No. 1, p. K47–K50 (1983).
See also *6*, p. *76*; *20.1*, p. *322*.

V. V. Aristov, L. S. Kokhanchik, K.-P. Meyer, and H. Blumtritt
Scanning Electron Microscopic Investigation of Peculiarities of the $BaTiO_3$ Ferroelectric Domain Contrast
(a) Vol. **78**, No. 1, p. 229–236 (1983).
See also *14.4.1*, p. *257*.

V. M. Mukhortov, Yu. I. Golovko, V. A. Aleshin, E. V. Sviridov, Vl. M. Mukhortov, V. P. Dudkevich, and E. G. Fesenko
$(Ba, Sr)TiO_3$ Heteroepitaxial Films
(a) Vol. **78**, No. 1, p. 253–257 (1983).
See also 1.5, p. 37; *14.4.1*, p. *258*.

T. Kala
Contribution to the Study of Tetragonal and Rhombohedral Phase Coexistence in the $PbZrO_3$–$PbTiO_3$ System
(a) Vol. **78**, No. 1, p. 277–282 (1983).
See also *1.2*, p. *17*; *14.4.1*, p. *258*.

L. Malicskó and R. Scholz
Microtwins in $LiNbO_3$ Crystals Observed by High-Voltage and High-Resolution Electron Microscopy
(a) Vol. **78**, No. 1, p. K1–K4 (1983).
See also *1.4*, p. *32*; 10.2, p. 115.

M. Saakes and G. Blasse
The Luminescence of Bi^{3+} and Mn^{2+} in Lanthanum Magnesium Borate ($LaMgB_5O_{10}$)
(a) Vol. **78**, No. 1, p. K97–K99 (1983).
See also *20.3*, p. *342*.

H. Hülsing and A. Kasten
Magnetic Properties of Dysprosium Oxisulphate ($Dy_2O_2SO_4$)
(b) Vol. **118**, No. 2, p. 725–741 (1983).
See also *18.3*, p. *293*; *18.4*, p. *296*.

H. Mashiyama
Modified Phase Modulation-Only Approximation of Incommensurate Phases in K_2SeO_4-Type Crystals
(b) Vol. **118**, No. 2, p. 757–767 (1983).
See also *1.2*, p. *18*; 8, p. 89; 14.4, p. 252.

A. L. Aleksandrovskii and A. I. Nagaev
Electrooptic Diffraction in Polydomain Barium–Sodium Niobate Crystals
(a) Vol. **78**, No. 2, p. 431 – 438 (1983).
See also *14.4.1*, p. *258*; *20*, p. *310*.

P. Kurek and W. Bogusz
Electrical Conductivity of $Li_{16-2x}Zn_x(GeO_4)_4$
(a) Vol. **78**, No. 2, p. K117 – K119 (1983).
See also *13.4*, p. *184*.

I. Henning, E. Hegenbarth, and B. Brezina
Specific Heat of a K_2SeO_4 Single Crystal in Temperature Range 0.6 to 7 K
(a) Vol. **78**, No. 2, p. K167 – K170 (1983).
See also *8*, p. *89*; *14.4.1*, p. *258*.

K. S. Dubey
Effect of Empty Spaces on the Lattice Thermal Conductivity of Non-Crystalline Material at Low Temperatures. Application to Borosilicate Glass
(a) Vol. **78**, No. 2, p. K195 – K200 (1983).
See also 6, p. 77; *8*, p. *89*.

P. H. Müller, A. Kasten, and M. Schienle
Relaxation in the Ferromagnetic Phase of Terbium Arsenate ($TbAsO_4$)
(b) Vol. **119**, No. 1, p. 239 – 249 (1983).
See also *18.2*, p. *286*.

D. M. Daraselia and D. L. Japaridze
Determination of the Structure of the Co^{2+} Paramagnetic Centers in α-$LiIO_3$ by the RFDS Method
(b) Vol. **119**, No. 1, p. K57 – K58 (1983).
See also *13.4*, p. *186*; *19*, p. *303*.

C. S. Sunandana
Low Temperature ESR Studies on $KHSO_4$
(b) Vol. **119**, No. 1, p. K59 – K64 (1983).
See also *19*, p. *304*.

G. Koschek and E. Kubalek
Micron-Scaled Spectral-Resolved Cathodoluminescence of Grains in Bariumtitanate Ceramics
(a) Vol. **79**, No. 1, p. 131 – 139 (1983).
See also *20.3*, p. *344*.

A. A. Demidenko, A. P. Zdebskii, V. N. Piskovoi, and A. U. Savchuk
Determination of Nonlinear Elastic Coefficients by Resonant Methods
(a) Vol. **79**, No. 1, p. K73 – K76 (1983).
See also *12.2*, p. *144*; *14.4.1*, p. *258*.

R. Styrkowiec
The Hampered, Sideways Motion of a Single Domain Wall in Rochelle Salt Crystals
(a) Vol. **79**, No. 1, p. K99 – K102 (1983).
See also *14.4.1*, p. *258*.

M. Schienle, A. Kasten, and P. H. Müller
Magnetic Relaxation in Dysprosium Phosphate ($DyPO_4$)
(b) Vol. **119**, No. 2, p. 611 – 620 (1983).
See also *18.4*, p. *297*.

V. K. Jain, V. P. Seth, R. K. Malhotra, and S. K. Yadav
Electron Paramagnetic Resonance of Mn^{2+} and VO^{2+} in $Ba_2Zn(HCOO)_6 \cdot 4\,H_2O$ Single Crystals
(b) Vol. **119**, No. 2, p. 659 – 666 (1983).
See also *13.4*, p. *187*; *19*, p. *304*.

P. Bhattacharya and A. R. Chetal
Chemical Shift and EXAFS Studies of Some Copper Complexes
(b) Vol. **119**, No. 2, p. 735 – 740 (1983).
See also *13.1*, p. *163*; 20, p. 310.

O. S. Torosyan
Hyperfine Structure of the EPR Spectra of Mn^{4+} Ions in a Strong Trigonal Crystal Field
(b) Vol. **119**, No. 2, p. K101 – K104 (1983).
See also *19*, p. *304*.

I. Gregora, V. Vorlíček, and R. Uecker
Determination of the Fresnel Indicatrix of the Monoclinic $NaVO_3$ Crystal in the Visible and Near Infrared Spectral Range
(a) Vol. **79**, No. 2, p. 497 – 501 (1983).
See also *20.1*, p. *328*.

A. D. Reddy, S. G. Sathyanarayan, and G. S. Sastry
Thermally Stimulated Depolarisation Current Studies on Some Hydrogen-Bonded Crystals
(a) Vol. **79**, No. 2, p. 575 – 583 (1983).
See also *14.4*, p. *253*.

V. S. Gorelik, B. S. Umarov, and M. Umarov
On the Connection between Isofrequency Temperature Dependences of Inelastic Light Scattering and Dielectric Anomalies in Lithium Niobate Crystals
(b) Vol. **120**, No. 1, p. 131 – 136 (1983).
See also *6*, p. *79*; 14.4.1, p. 259; *20.1*, p. *328*.

J. Berdowski
The Influence of Isotope Effects on Anomalous Changes of Dielectric and Electric Properties of DADA Crystals in the Range of the Phase Transition
(a) Vol. **80**, No. 1, p. 65 – 72 (1983).
See also *1.2*, p. *19*; *14.4.1*, p. *259*.

A. ARESTI, F. MELONI, G. G. PEGNA, and M. SERRA
Microwave Radiometric Study of Ferro–Paraelectric Phase Transitions in KNO_3
(a) Vol. **80**, No. 1, p. 119–126 (1983).
See also *1.2*, p. *19*; *14.4.1*, p. *259*.

A. A. KAMINSKI, B. V. MILL, G. G. KHODZHABAGYAN, A. F. KONSTANTINOVA, A. I. OKOROCHKOV, and I. M. SILVESTROVA
Investigation of Trigonal $(La_{1-x}Nd_x)_3Ga_5SiO_{14}$ Crystals (I): Growth and Optical Properties
(a) Vol. **80**, No. 1, p. 387–398 (1983).
See also *3*, p. *59*; *20.1*, p. *329*.

G. I. ROSENMAN, V. I. PECHORSKII, YU. L. CHEPELEV, E. I. BOIKOVA, and L. E. ISSAKOVA
Exoemission of Pyroelectrics
(b) Vol. **120**, No. 2, p. 667–670 (1983).
See also *14.4.1*, p. *259*; *17*, p. *271*.

H. J. A. KOOPMANS, M. M. A. PERIK, B. NIEUWENHUIJSE, and P. J. GELLINGS
The Interpretation of Broadened Gd^{3+} EPR Spectra of Polycrystalline Ceramics
(b) Vol. **120**, No. 2, p. 745–751 (1983).
See also *19*, p. *305*.

E. BĂRBULESCU, D. BARB, A. BĂRBULESCU, and D. GEANA
Computer Calculated F–P Diagram for the $BaTiO_3$–$PbTiO_3$ System
(a) Vol. **80**, No. 2, p. 431–437 (1983).
See also *1.2*, p. *20*; *14.4.1*, p. *259*.

S. K. MISRA, J. BANDET, G. BACQUET, and T. E. MCENALLY
Electron Spin Resonance Studies of Defect Centers Containing Boron Atoms in Natural Danburite Crystals
(a) Vol. **80**, No. 2, p. 581–588 (1983).
See also 13.4, p. 191; *19*, p. *306*.

A. A. KAMINSKII, I. M. SILVESTROVA, S. E. SARKISOV, and G. A. DENISENKO
Investigation of Trigonal $(La_{1-x}Nd_x)_3Ga_5SiO_{14}$ Crystals (II): Spectral Laser and Electromechanical Properties
(a) Vol. **80**, No. 2, p. 607–620 (1983).
See also 12.2, p. 144; *14.4*, p. *254*; 20.1, p. 331; *20.2*, p. *333*; 20.3, p. 347.

J. TANAKA, K. TAKAHASHI, K. YUKINO, and S. HORIUCHI
Electrical Conduction of $(La_{0.8}Ca_{0.2})MnO_3$ with Homogeneous Ionic Distribution
(a) Vol. **80**, No. 2, p. 621–630 (1983).
See also *1.3*, p. *27*; *14.3*, p. *228*.

K. KISHAN RAO and D. B. SIRDESHMUKH
Effect of X-Ray Irradiation on the Hardness of $NaBrO_3$ Crystals
(a) Vol. **80**, No. 2, p. K117–K119 (1983).
See also *11*, p. *135*; *12.2*, p. *144*.

22.8.2 Spinels, Garnets, Ferrites

Original Papers and Short Notes

K. A. Krezhov, M. T. Mikhov, A. M. Kadomtseva, I. B. Krinetskii, and M. M. Lukina
Magnetic and Neutron Diffraction Study of the Phase Transition in $TmFe_{1-x}Mn_xO_3$
(b) Vol. **115**, No. 1, p. 47–51 (1983).
See also *18*, p. *273*.

M. Ya. Flyagin and A. N. Men
Calculation of Magnetic Hyperfine Interactions in the Rare-Earth Orthoferrites with Model Covalency Parameter
(b) Vol. **115**, No. 1, p. 277–284 (1983).
See also *5*, p. *65*; *18.4*, p. *295*.

H. Kawakami, S. Eguchi, M. Ikeda, S. Matsumoto, and T. Niimi
Negative Differential Resistances in Ferromagnetic Semiconductor $CdCr_2Se_4$ Single Crystals
(a) Vol. **75**, No. 2, p. 433–439 (1983).
See also *14.3.4*, p. *244*.

A. D. Franklin and H. J. de Bruin
The Fourier Analysis of Impedance Spectra for Electroded Solid Electrolytes
(a) Vol. **75**, No. 2, p. 647–656 (1983).
See also *14.3.3*, p. *237*.

L. M. Dedukh, V. S. Gornakov, and V. I. Nikitenko
One-Direction Motion of Bloch Lines during Their Nonlinear Oscillation
(a) Vol. **75**, No. 2, p. K117–K119 (1983).
See also *18.3*, p. *291*.

V. K. Vlasko-Vlasov and L. S. Uspenskaya
Dynamic Changes of the Domain Structure Period in Magnetic Dielectric
(a) Vol. **75**, No. 2, p. K125–K127 (1983).
See also *18.3*, p. *291*.

V. P. Barkhatov, V. F. Balakirev, Yu. V. Golikov, and E. G. Kostitsin
X-Ray Diffraction Investigation of Cation Distribution in High-Temperature Cubic Spinels of Mg–Mn–O System
(a) Vol. **76**, No. 1, p. 57–63 (1983).
See also *1.2*, p. *15*; *1.3*, p. *22*.

G. P. Kramar, Ya. I. Panova, and V. V. Passynkov
The Influence of Microstructure on the Dielectric Spectra of Ferrites
(a) Vol. **76**, No. 1, p. 95–100 (1983).
See also *14.4*, p. *250*.

C. Pasnicu, D. Condurache, and E. Luca
Influence of Substitution and Addition of Calcium on Magnetic Properties of $Ni_{0.245}Zn_{0.755}Fe_2O_4$ Ferrite
(a) Vol. **76**, No. 1, p. 145–150 (1983).
See also *18.3*, p. *291*.

M. Halbwachs, P. Mazot, and J. Woirgard
Anelastic Relaxation Phenomena in Plasma-Sprayed $(Al_2O_3)_3MgO$ Spinel with Reference to Diffusion Process
(a) Vol. **76**, No. 1, p. 157–164 (1983).
See also 9, p. 93; *10.2*, p. *110*; *12.2*, p. *142*.

Y. Kawai and T. Ogawa
ΔE Effect of a Mn Ferrite Single Crystal at Temperatures between 150 and 300 K
(a) Vol. **76**, No. 1, p. 375–381 (1983).
See also 7, p. 82; *12.2*, p. *142*; *18.3*, p. *291*.

V. G. Kleparskii and I. Pintér
Direct Observation of Domain Nucleation in Garnet Films
(a) Vol. **76**, No. 1, p. K1–K4 (1983).
See also *18.3*, p. *291*.

D. Adlienė, L. Pranevičius, and G. Zubauskas
Acoustic Emission Initiated by Ion Implantation in Bubble Garnet Films
(a) Vol. **76**, No. 1, p. K35–K37 (1983).
See also *7*, p. *82*; *11*, p. *125*; 18.3, p. 291.

A. F. Rakov
Neutron Induced Colour and Luminescence Centers in Yttrium–Aluminium Garnet Single Crystals
(a) Vol. **76**, No. 1, p. K57–K60 (1983).
See also *10.2*, p. *111*; *11*, p. *126*; 20.1, p. 316; 20.3, p. 337.

B. Gillot and F. Jemmali
Dependence of Electrical Properties in Iron–Cobalt, Iron–Zinc Ferrites near Stoichiometry on Firing Temperature and Atmosphere
(a) Vol. **76**, No. 2, p. 601–608 (1983).
See also *14.3*, p. *218*; *15*, p. *261*.

B. Banerjee and S. Lahiry
Superparamagnetism in γ-Mn_2O_3–α-Fe_2O_3–α-Mn_2O_3 System
(a) Vol. **76**, No. 2, p. 683–694 (1983).
See also *18.3*, p. *291*; *18.4*, p. *295*; 22.6, p. 493.

J. Baszyński
Measurements of Magnetic Parameters of YBiFe Garnet Films by Ferrimagnetic Resonance and Ferrimagnetic Antiresonance
(a) Vol. **76**, No. 2, p. K189–K191 (1983).
See also *18.2.1*, p. *290*; *18.3*, p. *292*.

U. Heinecke, C. Cruz, E. Wieser, and C. Bernhardt
Structural Transformations and Connected Changes of Magnetic Properties by Mechanical Treatment of $SrFe_{12}O_{19}$
(a) Vol. **77**, No. 1, p. 225–232 (1983).
See also *1.2*, p. *16*; 5, p. 66; *18.3*, p. *292*.

B. Gillot and F. Jemmali
Electrical Conduction Mechanism in Al-, Cr-, and Zn-Substituted Magnetites over the Temperature Range from 77 to 1000 K
(a) Vol. **77**, No. 1, p. 339–346 (1983).
See also *14.3*, p. *219*.

A. Niklas and W. Jeleński
X-Ray Luminescence of YAG:Nd^{3+}
(a) Vol. **77**, No. 1, p. 393–398 (1983).
See also *13.4*, p. *179*; *20.3*, p. *339*.

G. Vértesy, I. Tomas, Z. Vértesy, and M. Balaskó
Temperature Hysteresis of Demagnetized Domain Structures
(a) Vol. **77**, No. 1, p. K87–K90 (1983).
See also *18.2*, p. *284*.

S. Ligenza, M. Łukasiak, Z. Kucharski, and J. Suwalski
Study of the $3d^5$ Shell Deformation of Fe^{3+} Ions in $ZnFe_2O_4$ by High-Field Mössbauer Spectroscopy
(b) Vol. **117**, No. 2, p. 465–470 (1983).
See also *5*, p. *66*; 18.3, p. 292.

G. P. Kramar and Ya. I. Panova
Resonance and Relaxation of Domain Walls in Polycrystalline Ferrites
(a) Vol. **77**, No. 2, p. 483–489 (1983).
See also *18.3*, p. *292*.

O. P. Smirnov and V. P. Plakhty
On the Role of Intrasublattice Interactions in YIG with Diamagnetically Substituted a-Sublattice
(a) Vol. **77**, No. 2, p. 505–511 (1983).
See also *18.3*, p. *292*.

J. L. Dormann, A. Tomas, and M. Nogues
Cation Ordering in $LiFe_5O_8$ Studied by Mössbauer Spectroscopy and X-Ray Crystallography
(a) Vol. **77**, No. 2, p. 611–618 (1983).
See also *1.3*, p. *24*; *5*, p. *66*.

Y. Kawai and T. Ogawa
Domain Wall Motions Due to External Stresses in a Mn–Zn Ferrite Single Crystal
(a) Vol. **77**, No. 2, p. K103–K106 (1983).
See also *12.2*, p. *142*; *18.3*, p. *292*.

L. Pranevičius, A. Sakalas, and G. Zubauskas
Ion Beam Induced Mixing in Bubble Garnet Films
(a) Vol. **77**, No. 2, p. K171 – K172 (1983).
See also *11*, p. *129*; *18.3*, p. *292*.

A. G. Arias and F. Calderón
Effect of Some Additives in Nickel–Zinc Ferrite Losses
(a) Vol. **77**, No. 2, p. K185 – K188 (1983).
See also *18.3*, p. *293*.

H. Luchian
Magnetic Symmetry in Several Cubic Compounds
(b) Vol. **118**, No. 1, p. 31 – 39 (1983).
See also *13*, p. *149*; *18*, p. *274*.

N. L. Pakhomova, L. M. Kasimenko, and V. Christoph
Ferromagnetic Intra-Sublattice Exchange in Multicomponent Ferrites
(a) Vol. **78**, No. 1, p. 111 – 114 (1983).
See also *18.3*, p. *293*.

M. Wurlitzer
Richter Type Magnetic After-Effect in YIG at Helium Temperature
(a) Vol. **78**, No. 1, p. K87 – K91 (1983).
See also 10.2, p. 115; *18.3*, p. *293*.

G. Dehe and U. Behn
Contribution of the Direct B–B Interaction to the Supertransferred Hyperfine Field at Diamagnetic Antimony in Oxidic Spinels Determined by ^{121}Sb Mössbauer Spectroscopy
(a) Vol. **78**, No. 2, p. 485 – 488 (1983).
See also *5*, p. *67*; *18.3*, p. *293*.

A. Torres, J. Jimenez, V. Carbayo and J. A. de Saja
Thermally Stimulated Depolarization Currents in Cerium Sulfate Hydrates
(a) Vol. **78**, No. 2, p. 671 – 677 (1983).
See also *14.4*, p. *253*.

A. A. Kaminskii, B. V. Mill, and A. V. Butashin
Growth and Stimulated Emission Spectroscopy of $Ca_3Ga_2Ge_3O_{12}$–Nd^{3+} Garnet Crystals
(a) Vol. **78**, No. 2, p. 723 – 732 (1983).
See also 20.1, p. 325; *20.2*, p. *333*; *20.3*, p. *343*.

L. Půst
Influence of Magnetostriction on FMR in Thin Garnet (111) Films
(a) Vol. **78**, No. 2, p. 741 – 749 (1983).
See also *18.2.1*, p. *290*; 19, p. 303.

P. Novotný
Bubble Film Characterization by Differential Susceptibility Measurements
(a) Vol. **78**, No. 2, p. K101 – K104 (1983).
See also *18.3*, p. *293*.

P. Venugopal Reddy, R. Satyanarayana, and T. Seshagiri Rao
Electrical Conduction in Nickel Ferrite
(a) Vol. **78**, No. 2, p. K109–K111 (1983).
See also *14.3*, p. *223*; *15*, p. *262*.

V. P. Miroshkin and Ya. I. Panova
Temperature Dependence and Frequency Dispersion Mechanisms of the Magnetic Permeability of Monocrystalline Mn–Zn Ferrite
(a) Vol. **78**, No. 2, p. K179–K182 (1983).
See also *18.4*, p. *297*.

G. Dehe and J. Suwalski
^{121}Sb and ^{57}Fe Mössbauer Effect in Sb Substituted Lithium Ferrite in the Presence of a Magnetic Field
(b) Vol. **119**, No. 2, p. K155–K158 (1983).
See also *5*, p. *68*; *18.3*, p. *293*.

Yu. G. Chukalkin, V. R. Shtirts, and B. N. Goshchitskii
Amorphization of Neutron-Irradiated Yttrium Iron Garnet
(a) Vol. **79**, No. 2, p. 361–366 (1983).
See also 2, p. 53; *11*, p. *132*; 18.3, p. 293.

V. K. Vlasko-Vlasov and L. S. Uspenskaya
Direct Experimental Observation of a Localized Wave of Magnetization in a Ferrimagnetic
(b) Vol. **120**, No. 1, p. K1–K4 (1983).
See also *18.3*, p. *294*.

A. E. Nosenko, A. I. Otko, V. V. Kravchishyn, and L. V. Kostyk
Growth Macrodefects in $Ca_3Ga_2Ge_3O_{12}$ Single Crystals
(a) Vol. **80**, No. 1, p. 373–386 (1983).
See also 1.4, p. 34; *10.2*, p. *120*.

M. Wurlitzer
The Influence of Irradiation on Richter-Type Magnetic After-Effects in YIG:Si at Helium Temperature
(a) Vol. **80**, No. 2, p. 457–461 (1983).
See also *18.3*, p. *294*.

S. R. Murthy
ΔE Effect in Co–Zn Ferrites
(a) Vol. **80**, No. 2, p. K161–K165 (1983).
See also *12.2*, p. *145*; 18.3, p. 294.

Sh. A. Vakhidov and A. F. Rakov
Radiation Stimulated Recombination Processes in $Y_3Al_5O_{12}$ Crystals
(a) Vol. **80**, No. 2, p. K175–K178 (1983).
See also *11*, p. *135*; 13.4, p. 191; 13.5.1, p. 199; *20.3*, p. *348*.

M. Pardavi-Horváth and M. Osvay
Thermoluminescent Properties of Gadolinium Gallium Garnet Crystals Containing Ca^{2+} Impurity
(a) Vol. **80**, No. 2, p. K183–K185 (1983).
See also *20.3*, p. *348*.

P. Novák, M. Maryško, S. Krupička, J. Kub, J. Simšova, J. Čermák, and M. Nevřίva
Noncubic Anisotropy in YIG:Mn^{3+} Thin Films
(a) Vol. **80**, No. 2, p. K213–K215 (1983).
See also *13.4*, p. *191*; *18*, p. *277*.

22.9 Organic Semiconductors

Review Article

V. KH. BRIKENSHTEIN, V. A. BENDERSKII, and P. G. FILIPPOV
Fluorescence of Molecular Crystals at High Levels of Optical Pumping
(b) Vol. **117**, No. 1, p. 9–39 (1983).
See also 6, p. 69; *13.5.1*, p. *194*; 20.2, p. 332; *20.3*, p. *334*.

Original Papers and Short Notes

T. TAKASE, E. MORIKAWA, and M. KOTANI
Triplet Exciton Trapping in p-Terphenyl Crystals Studied by Delayed Fluorescence and Phosphorescence
(b) Vol. **115**, No. 1, p. 293–298 (1983).
See also *13.5.1*, p. *194*; *20.3*, p. *334*.

W. REHWALD, A. VONLANTHEN, and W. MEYER
Single Crystal 2,4-Hexadiynylene-Bis (p-Toluenesulfonate): Elastic Properties and Phase Transitions in the Monomer and Polymer
(a) Vol. **75**, No. 1, p. 219–226 (1983).
See also *1.2*, p. *14*; 7, p. 82; *12.2*, p. *141*.

W. MEDYCKI
Mean Charge Depth in Non-Metallized Electrets
(a) Vol. **75**, No. 1, p. K21–K24 (1983).
See also *14.4*, p. *250*.

A. N. NIKITINA, I. S. OSADKO, I. A. VASILIEVA, and L. A. YANOVSKAYA
Vibronic and Electron–Phonon Coupling in Spectra of Substituted Polyenes in Rigid Matrices
(b) Vol. **115**, No. 2, p. 493–498 (1983).
See also *6*, p. *70*; *13.4*, p. *174*; 20.1, p. 314; 20.3, p. 335.

E. YA. GLUSHKO and N. A. EFREMOV
Diffusion of Frenkel Excitons in Anisotropic Crystals
(b) Vol. **115**, No. 2, p. 529–538 (1983).
See also *13.5.1*, p. *195*.

S. V. MARISOVA, E. N. MYASNIKOV, and A. N. LIPOVCHENKO
Phonon-Assisted Decay of Excitons in Anthracene Crystals
(b) Vol. **115**, No. 2, p. 649–654 (1983).
See also *13.5.1*, p. *195*; *20.1*, p. *314*.

A. P. TYUTNEV, G. S. MINGALEEV, A. V. VANNIKOV, A. F. DUNAEV, V. S. SAENKO, and V. S. LIKHOVIDOV
Kinetics of Radiation-Induced Conductivity in Some Polymers
(a) Vol. **75**, No. 2, p. 385–392 (1983).
See also *11*, p. *123*; *14.3*, p. *216*.

G. J. Ashwell, J. G. Allen, E. P. Goodings, D. A. Kennedy, and I. W. Nowell
Crystal Structure of (DEPA) $(TCNQ)_4$ and Stoichiometric Variation of Some Related TCNQ Salts
(a) Vol. **75**, No. 2, p. 663–667 (1983).
See also *1.1*, p. *11*; 14.3, p. 217.

S. K. Khanna
On the Elastic Anisotropy of Ferroelectric Triglycine Selenate
(b) Vol. **116**, No. 1, p. 173–178 (1983).
See also *1.2*, p. *14*; 12.2, p. 142; *14.4.1*, p. *256*.

N. Crisa
Anomalous Transient-Time Dispersion in Amorphous Solids: Oxadiazole
(b) Vol. **116**, No. 1, p. 269–274 (1983).
See also *14.3.1*, p. *230*.

E. A. Silinsh, V. A. Kolesnikov, I. J. Muzikante, and D. R. Balode
Erratum to: On Charge Carrier Photogeneration Mechanisms in Organic Molecular Crystals
(b) Vol. **116**, No. 1, p. 347 (1983).
See also 14.3.1, p. 230; *16*, p. *265*.

B. S. Tošić, G. S. Davidović-Ristovski, and Lj. M. Ristovski
Excitons in a Finite Polymer Chain
(b) Vol. **116**, No. 1, p. K1–K4 (1983).
See also *13.5.1*, p. *195*.

A. Rajewska, B. Pura, J. Przedmojski, and R. Dąbrowski
X-Ray Study of the Smectic A–Nematic Phase Transition in 4-n-Pentylphenyl-4-(Trans-4′-n-Pentylcyclohexyl) Benzoate
(a) Vol. **76**, No. 1, p. 127–130 (1983).
See also *1.2*, p. *15*; 2, p. 45.

H. Sodolski
Depolarization Current in a Dielectric with Deep Trapped Charge Layers
(a) Vol. **76**, No. 1, p. 303–309 (1983).
See also *14.4*, p. *250*.

H. von Berlepsch and L. Brehmer
Invalidity of Scaling Property of Surface Potential Kinetics in HDPE
(a) Vol. **76**, No. 1, p. K27–K29 (1983).
See also *14.3.1*, p. *230*; *14.3.4*, p. *244*.

W. Stark and R. Danz
Investigations of the Thermally Stimulated Change of the Surface Potential of FEP Electrets
(a) Vol. **76**, No. 1, p. K101–K103 (1983).
See also *14.4*, p. *250*.

K. Handrich
Phonon Lifetime and Thermal Conductivity in Amorphous Solids
(b) Vol. **116**, No. 2, p. 533–538 (1983).
See also 2, p. 46; *6*, p. *72*; *8*, p. *87*.

V. I. Sugakov and Yu. D. Shtepa
Spin–Lattice Relaxation Theory of Excited Triplet-State Impurities in Molecular Crystals
(b) Vol. **116**, No. 2, p. 633–638 (1983).
See also *13.4*, p. *177*.

B. Jha, B. Bhattacharjee, S. Paul, and R. Paul
X-Ray Diffraction Studies on the Mesophases of n–p-Cyanobenzylidene-p-Octyloxy Aniline in Magnetic Field
(a) Vol. **76**, No. 2, p. 461–466 (1983).
See also *2*, p. *47*.

T. G. Abdel-Malik, A. A. Aly, A. M. Abdeen, and H. M. El-Labany
Electric and Photoelectric Investigations of β-Nickel Phthalocyanine Thin Films
(a) Vol. **76**, No. 2, p. 651–659 (1983).
See also *13.4*, p. *177*; *14.3.1*, p. *231*; 16, p. 265.

M. A. Semary, M. F. Mostafa, and M. M. Abdel-Kader
Magnetic Susceptibility of the Canted Spin Compounds $(CH_2)_n(NH_3)_2FeCl_4$; $n = 4, 5$, and 7
(a) Vol. **76**, No. 2, p. K131–K136 (1983).
See also *18.4*, p. *296*.

T. G. Abdel-Malik
Effect of Oxygen on the Electrical Properties of β-Zinc Phthalocyanine Single Crystals
(a) Vol. **76**, No. 2, p. K159–K162 (1983).
See also *13.4*, p. *177*; *14.3*, p. *218*.

V. N. Myasnikov, S. V. Marisova, and A. N. Lipovchenko
Optical Properties of Polaritons in Thin Crystal Films
(b) Vol. **117**, No. 1, p. 109–116 (1983).
See also *13.5*, p. *192*; 20, p. 308; 20.3, p. 338.

B. J. Schmid and P. Reineker
On the Theory of Singlet and Triplet Charge-Transfer Excitons
(b) Vol. **117**, No. 1, p. 117–126 (1983).
See also *13.5.1*, p. *196*.

R. P. Bhardwaj, J. K. Quamara, K. K. Nagpaul, and B. L. Sharma
The Non-Isothermal Depolarization Current Studies in Kapton-H Thermoelectrets
(a) Vol. **77**, No. 1, p. 347–354 (1983).
Erratum
(a) Vol. **78**, No. 2, p. 759 (1983).
See also 14.3.4, p. 245; *14.4*, p. *251*.

A. U. Shareef, K. K. Saraf, and A. P. Srivastava
Spontaneous Electric Current from Polyvinyl Alcohol Induced by the First Heating Run
(a) Vol. **77**, No. 1, p. 381–386 (1983).
See also *14.3.3*, p. *238*.

A. K. Batra, S. C. Mathur, and A. Mansingh
Dielectric Dispersion in Triglycine Sulphate Single Crystals
(a) Vol. **77**, No. 1, p. 399–404 (1983).
See also *14.4.1*, p. *257*.

V. N. Semkin, R. M. Vlasova, and G. Yu. Yashin
Absence of 3.39 μm Photoconductivity in the Quasi-One-Dimensional Organic Semiconductor $Cs_2(TCNQ)_3$
(a) Vol. **77**, No. 1, p. 421–425 (1983).
See also *16*, p. *265*.

B. P. Bespalov and A. M. Vainrub
Thermal Conductivity of Pressed Pellets of TCNQ Complexes. Manifestation of Amorphous Structure
(a) Vol. **77**, No. 1, p. K13–K15 (1983).
See also *8*, p. *88*.

V. Železný, L. Jastrabík, and L. Dobiášová
The Temperature Dependence of Emission Spectra in α-9,10-Dichloroanthracene
(b) Vol. **117**, No. 2, p. K95–K97 (1983).
See also *13.5.1*, p. *196*; *20.3*, p. *340*.

P. K. C. Pillai, G. K. Narula, A. K. Tripathi, and R. G. Mendiratta
A Study of Absorption Currents in Polycarbonate: Polyporpylene Blend
(a) Vol. **77**, No. 2, p. 693–698 (1983).
See also *14.4*, p. *251*.

H. Sodolski
DC Conductivity and Electric Relaxation Currents in a Polyester Polymer
(a) Vol. **77**, No. 2, p. 749–758 (1983).
See also *14.4*, p. *251*.

N. M. Plakida, A. V. Belushkin, I. Natkaniec, and T. Wasiutyński
The Internal Soft-Mode Phase Transition in Solid Biphenyl
(b) Vol. **118**, No. 1, p. 129–133 (1983).
See also *6*, p. *75*.

K. Kamarás, R. Swietlik, and A. Graja
Polarized IR Reflectance Spectra of Pure and Neutron Irradiated $Qn(TCNQ)_2$ Mosaics
(b) Vol. **118**, No. 1, p. K43–K46 (1983).
See also 11, p. 129; *20.1*, p. *322*.

B. R. Bulka
On the Electronic Structure of Interacting Solitons with Impurities in Polyacetylene
(b) Vol. **118**, No. 1, p. K57–K59 (1983).
See also *13.4*, p. *182*.

A. R. ARAKELYAN, R. G. GABRIELYAN, A. R. MKRTCHYAN, H. A. ASLANYAN, and KH. V. KOTANDYAN
On a Certain Inverse Problem in γ-Acoustic Spectroscopy
(b) Vol. **118**, No. 1, p. K69–K74 (1983).
See also *5*, p. *67*; *7*, p. *83*.

A. I. PROKHVATILOV and A. P. ISAKINA
Lattice Parameters, Thermal Expansion Coefficients, and Vacancy Density in Solid CD_4
(a) Vol. **78**, No. 1, p. 147–155 (1983).
See also 1.1, p. 11; 8, p. 88; *10.2*, p. *114*; 12.2, p. 143.

J. TYCZKOWSKI and M. KRYSZEWSKI
Poole-Frenkel Centres in Dielectrics. Kind and Space Density in Plasma-Polymerized Organosilicon Films
(a) Vol. **78**, No. 1, p. 259–265 (1983).
See also *13.4*, p. *183*; 14.3.1, p. 231; *14.3.4*, p. *246*.

CH. SEIDEL and V. N. PRIGODIN
The Impurity Effect on the Phase Transitions in Quasi-One-Dimensional Conductors.
(b) Vol. **118**, No. 2, p. 713–724 (1983).
See also *1.2*, p. *18*; *10.2*, p. *115*; 14.2, p. 213.

J. KALINOWSKI and R. SIGNERSKI
Exciton-Enhanced Double Injection Currents in Tetracene Crystals
(b) Vol. **118**, No. 2, p. K147–K150 (1983).
See also *13.5.1*, p. *197*; *14.3.4*, p. *247*; 20.3, p. 343.

A. P. TYUTNEV, V. S. SAENKO, A. V. VANNIKOV, and G. S. MINGALEEV
Radiation-Induced Conductivity in Polyethylene
(a) Vol. **78**, No. 2, p. 689–696 (1983).
See also 11, p. 130; *14.3*, p. *222*.

M. SALEH
Drift Mobility Measurements on Single Crystals of Orthorhombic Benzophenone Using the Electron Bombardment Technique
(a) Vol. **78**, No. 2, p. K137–K140 (1983).
See also 13.5.3, p. 201; *14.3*, p. *223*.

R. JAKUBAS and L. SOBCZYK
Dielectric Relaxation Study of the Incommensurate Phase of $[(CH_3)_4N]_2 \cdot CuBr_4$ near the Transition Temperature
(a) Vol. **78**, No. 2, p. K147–K150 (1983).
See also *14.4.1*, p. *258*.

K. G. CHAKRABORTY and J. F. BARET
Spin-One Lattice Gas Model for Isotherm-Reversal in Spin-Labelled Lipid Monolayers
(b) Vol. **119**, No. 1, p. 61–71 (1983).
See also *2*, p. *51*; *13*, p. *150*.

A. S. T. Pires and F. C. Sá Barreto
Solitons in Ferroelectric Liquid Crystal
(b) Vol. **119**, No. 1, p. 121–130 (1983).
See also *2*, p. *51*; *14.4.1*, p. *258*.

P. Bhattacharya and A. R. Chetal
Study of Cu K Edge in Some Copper (II) Octahedral Complexes
(b) Vol. **119**, No. 1, p. 179–183 (1983).
See also *13.1*, p. *162*; *20*, p. *310*.

S. Sastry, K. V. Lingam, and M. J. Rao
EPR of Cu(II) Doped in Trisethylenediamine Cadmium Chloride
(b) Vol. **119**, No. 1, p. K51–K55 (1983).
See also *19*, p. *303*.

S. Różański
Determination of Pitch and Chiral Smectic C DOBAMBC
(a) Vol. **79**, No. 1, p. 309–312 (1983).
See also 2, p. 51; *14.4.1*, p. *258*.

G. J. Ashwell and K. Osborne
Electrical and Structural Characterisation of (DBzBP) $(TCNQ)_4$
(a) Vol. **79**, No. 1, p. K35–K39 (1983).
See also *1.1*, p. *12*; *14.3*, p. *224*.

J. Muto and Y. Tsunekawa
Photovoltaic Behaviour of Rhodamine 6G Schottky-Type Cells
(a) Vol. **79**, No. 1, p. K109–K112 (1983).
See also *14.3.3*, p. *241*; *16*, p. *267*.

B. Lorenz
Thermodynamic Properties of Uniform and Dimerized Hubbard Chains
(b) Vol. **119**, No. 2, p. 555–563 (1983).
See also *8*, p. *90*; 18, p. 275.

I. Muševič, B. Žekš, R. Blinc, Th. Rasing, and P. Wyder
Dielectric Study of the Modulated Smectic C*–Uniform Smectic C Transition in a Magnetic Field
(b) Vol. **119**, No. 2, p. 727–733 (1983).
See also 2, p. 52; *14.4.1*, p. *259*.

G. J. Ashwell, J. G. Allen, G. H. Cross, and I. W. Nowell
Electrical, Magnetic, and Structural Characterisation of the 1:2 and 1:3 TCNQ Salts of 1,2-Bis(4-Pyridinium)Ethane
(a) Vol. **79**, No. 2, p. 455–463 (1983).
See also *1.1*, p. *12*; *14.3*, p. *225*; 15, p. 262; 18.1, p. 279.

B. Gross, H. von Seggern, and D. A. Berkley
Long Term Behaviour of Radiation-Induced Currents in Fluorinated Ethylene Propylene Copolymer
(a) Vol. **79**, No. 2, p. 607–615 (1983).
See also *11*, p. *133*; *14.4*, p. *253*.

L. Komitov and A. G. Petrov
Quasihomeotropic, Reversely Pretilted Nematic Liquid Crystal Layers. Electrooptical Behaviour
(a) Vol. **79**, No. 2, p. 623–629 (1983).
See also *2*, p. *53*; 14.4, p. 253; 20.1, p. 328.

Yu. A. Vidadi, A. V. Ragimov, B. A. Mamedov, S. A. Guseinov, A. A. Medzhidov, and B. I. Liogonkii
On the Contribution of Naphthoxyl-Type Paramagnetic Centres to the Electric Conductiviti of Oligonaphthols
(a) Vol. **79**, No. 2, p. 645–650 (1983).
See also 13.4, p. 188; *14.3*, p. *225*.

A. P. Tyutnev, G. S. Mingaleev, V. S. Saenko, and A. V. Vannikov
Radiation Induced Conductivity of Polyethyleneterephthalate and Polystyrene
(a) Vol. **79**, No. 2, p. 651–659 (1983).
See also *11*, p. *133*; *14.3*, p. *225*.

G. Pretzsch
Charge State of PTFE Thermoelectrets
(a) Vol. **79**, No. 2, p. K139–K142 (1983).
See also *14.4*, p. *253*.

P. Fedorko and F. Masin
DC Conductivity and Optical Absorption Spectra of Polyacetylene Doped with HSO_3F
(a) Vol. **79**, No. 2, p. K155–K158 (1983).
See also *14.4*, p. *254*; *20.1*, p. *328*.

P. Petelenz
Optical Detrapping in Molecular Crystals (I): Vibrational Structure of Detrapping Transitions
(b) Vol. **120**, No. 1, p. 31–40 (1983).
See also *13.4*, p. *188*; 20.1, p. 328.

M. T. Dove, I. U. Heilmann, J. K. Kjems, J. Kurittu, and G. S. Pawley
A Neutron Scattering Study of Phonons in Per-Deuterated S-Triazine
(b) Vol. **120**, No. 1, p. 173–181 (1983).
See also *1.2*, p. *19*; *6*, p. *79*; 12.2, p. 144.

A. Algeria, J. M. Barandiarán, and J. Colmenero
Dielectric Relaxation at the Glass Transition as a Free Volume Process. A Single Relaxation Time Approach
(b) Vol. **120**, No. 1, p. 349–360 (1983).
See also *2*, p. *54*; 14.4, p. 254.

M. N. Sapozhnikov and V. I. Alekseev
Study of Optical Relaxation and Inhomogeneous Distribution Function of Impurity Organic Molecules by Site-Selective Laser Spectroscopy
(b) Vol. **120**, No. 1, p. 435–444 (1983).
See also *6*, p. *80*; 13.4, p. 189; 20.1, p. 329; *20.3*, p. *346*.

G. J. ASHWELL, S. JANSEN, M. T. JONES, and C. H. L. KENNARD
(Diethyldiphenylphosphonium)$^+$ (7,7,8,8-tetracyano-p-quinodimethanide)$_2^-$, a Novel Quasi-One-Dimensional Organic Conductor
(a) Vol. **80**, No. 1, p. 127–134 (1983).
See also *1.1*, p. *12*; 14.3, p. 226; 18.1, p. 279.

R. P. BHARDWAJ, J. K. QUAMARA, K. K. NAGPAUL, and B. L. SHARMA
Thermally Stimulated Spontaneous Currents from Kapton Polyimide–Metal Contacts
(a) Vol. **80**, No. 1, p. 325–332 (1983).
See also *14.3.4*, p. *248*; *14.4*, p. *254*.

A. N. BOTVICH, V. G. PODOPRIGORA, V. F. SHABANOV, N. P. SHESTAKOV, and A. N. VTYURIN
Optical Properties of Molecular Crystals. Phenomenological Approach
(b) Vol. **120**, No. 2, p. 491–501 (1983).
See also *6*, p. *80*; *20.1*, p. *329*.

A. BRAU, J. P. FARGES, A. FILHOL, and H. GRASSI
Order–Disorder Phase Transition in TEA$^+$(TCNQ)$_2^-$
(b) Vol. **120**, No. 2, p. 547–554 (1983).
See also *1*, p. *10*; 8, p. 92; *14.3*, p. *227*.

R. B. LAL, M. D. AGGARWAL, R. L. KROES, and W. R. WILCOX
A New Technique of Solution Crystal Growth
(a) Vol. **80**, No. 2, p. 547–552 (1983).
See also *3*, p. *59*; 14.4.1, p. 259.

R. P. BHARDWAJ, J. K. QUAMARA, K. K. NAGPAUL, and B. L. SHARMA
Field-Induced Thermally Stimulated Currents in Kapton-H Polyimide Films
(a) Vol. **80**, No. 2, p. 559–566 (1983).
See also *14.3.4*, p. *248*; *14.4*, p. *254*.

A. T. OZA
Electrical Conduction in Some Inclusion Compounds of Iodine
(a) Vol. **80**, No. 2, p. 573–579 (1983).
See also *13.4*, p. *191*; *14.3*, p. *228*.

J. HAŁADYJ, J. MICHALSKI, and A. SZYMAŃSKI
On the Anomalous Field Dependence of Drift Mobility in Thin Organic Films
(a) Vol. **80**, No. 2, p. K137–K139 (1983).
See also *14.3.1*, p. *232*; *14.3.4*, p. *249*.

J. K. JESZKA, J. ULAŃSKI, and M. KRYSZEWSKI
TTT · TCNQ CT Complex Crystallized in Polymer Matrix
(a) Vol. **80**, No. 2, p. K141–K143 (1983).
See also *3*, p. *59*; *14.3*, p. *228*.

P. SIMOVA and M. PETROV
On the Electrical Conductivity in the Nematic Phase of 4-n-Alkyloxybenzoic Acids HOBA, OOBA, and NOBA
(a) Vol. **80**, No. 2, p. K153–K156 (1983).
See also *14.3*, p. *228*.

23. Solidified Gases

Original Paper

R. K. Singh, D. K. Neb, and S. P. Sanyal
Effect of Three-Body Forces on Thermophysical and Anharmonic Properties of Rare Gas Solid Mixtures
(b) Vol. **116**, No. 1, p. 289–297 (1983).
See also *4*, p. *60*; 6, p. 72; 8, p. 86; 12.2, p. 142.

Author Index

19. Magnetic Resonance

20. Optical Properties
    - 20.1 Spectra. Optical Constants (X-Ray Spectra See 20)
    - 20.2 Lasers
    - 20.3 Luminescence (X-Ray Spectra See 20)

21. Metals. Alloys
    - 21.1 Transition Metals (Group Sc to Zn) and Their Alloys
        - 21.1.1 Fe and Fe Alloys
    - 21.2 Alkali-Metals
    - 21.3 Alkali-Earth Metals
    - 21.4 Rare-Earth Metals
    - 21.5 Actinides
    - 21.6 Noble Metals
    - 21.7 Semi-Metals (Semiconducting Alloys See 22.7)

22. Semiconductors and Ionic Crystals
    - 22.1 Elements
        - 22.1.1 Germanium
        - 22.1.2 Silicon
        - 22.1.3 Group VI Elements
    - 22.2 III–V Compounds
        - 22.2.1 Arsenides
        - 22.2.2 Phosphides
        - 22.2.3 Antimonides
        - 22.2.4 Mixed Crystals
    - 22.3 IV–IV, II–V, and III–VI Compounds
        - 22.3.1 Mixed Crystals
    - 22.4 II–VI Compounds
        - 22.4.1 Sulphides
        - 22.4.2 Selenides
        - 22.4.3 Tellurides
        - 22.4.4 Mixed Crystals
    - 22.5 Halides
        - 22.5.1 Silver Halides
        - 22.5.2 Alkali Halides
        - 22.5.3 Alkali-Earth Halides
        - 22.5.4 Mixed Crystals
    - 22.6 Simple Oxides (CdO, ZnO See 22.4)
        - 22.6.1 Mixed Crystals
    - 22.7 Semiconducting Intermetallic Compounds
        - 22.7.1 Mixed Crystals
    - 22.8 Tertiary and Higher Compounds
        - 22.8.1 Salts of Oxy-Acids (Carbonates, Phosphates, Silicates Including Mica etc.)
        - 22.8.2 Spinels, Garnets, Ferrites
    - 22.9 Organic Semiconductors

23. Solidified Gases

For comments on the use of the classification scheme see p. 6.

32913/32914

Manuscripts and letters for physica status solidi (b) — basic research and physica status solidi (a) — applied research

should be sent to

DDR-1086 Berlin, Leipziger Str. 3–4, P. B. 1233

or

Internationaler Buchversand GmbH, 1000 Berlin 19, Eschenallee 21.

The object of "physica status solidi (b)" and "physica status solidi (a)" is to publish rapidly important and topical results concerning fundamental research in the field of solid state physics and aspects of application and experimental techniques of fundamental solid state research, respectively. The issues of "physica status solidi (b)" and "physica status solidi (a)" will be published on the 1st and the 16th day of each month, respectively. Either series will contain the following categories:

1. Review articles written by arrangement with the editors. They will be remunerated.
2. Original papers selected from those which have not been submitted for publication elsewhere, after being reviewed. (The latest date for acceptance of contributions is the 8th and the 23rd day of each month for "physica status solidi (b)" and "physica status solidi (a)", respectively, 50 days before the delivery of the journal.)
3. Short notes published in the off-set part of the journal. They are, if necessary after being reviewed, selected by the editorial staff. (The latest date for acceptance of manuscripts for this part is the 12th and the 27th day of each month for "physica status solidi (b)" and "physica status solidi (a)", respectively, 16 days before the delivery of the journal.)
4. Pre-printed titles of papers to be published in the next issue of both series.

Editorial Information

Manuscripts should be submitted preferentially in English. Besides, papers in German, French, and Russian are admitted. The final decision on the incorporation of the papers in the series (b) or (a) is taken by the editorial staff after consultation of the referee. Authors are asked to observe the following rules:

1. Manuscripts have to be written as concisely as possible. As most, 20 and 4 pages (28 lines per page; abstracts, tables, references, figures, and captions of figures included, where 3 figures are counted as about one page) are admitted for original papers and short notes, respectively. All manuscripts should be type-written on one side only. Two copies are required. Badly arranged manuscripts are returned.
2. The author's name, initials, and institution should be given.
3. One abstract of about ten lines in English and one in German, French, or Russian are required for original papers.
4. Because of lack of space conclusions (if any) must not have the character of a summary (i. e. substantially a repetition of the contents of the abstract).
5. Formulae should not be written too small and not with pencil. Separate lines for formulae are desirable. Indistinct writing will lead to a return of the manuscript.
6. Drawings should be enclosed on separate sheets of paper, not exceeding the maximum and minimum sizes of 20×30 and 10×15 cm^2, respectively. Photographs should be submitted, on white paper, only for review articles and original papers. Duplicates are required.
7. Captions for the figures should be submitted on separate sheets of paper, and special instructions with regard to the size of illustrations should be given on these.
8. References should be numbered in order of appearance and should be given, on a separate sheet, in the form used in physica status solidi.
9. SI-units should be used.
10. Each paper should contain a proposal for the subject classification according to the classification scheme appearing on the second and third cover of each issue.

The authors will be sent galley proofs of review articles and original papers. Proof reading of short notes and pre-printed titles will be done by the editorial staff.

For each original paper, and short note, a total of 75 reprints is available free of charge. The number of reprints free of charge per review article is 25. Requests for additional reprints (against payment) should be given together with the submitted manuscript. **The deadline for orders of additional reprints is the 15th and the 30th day of each month** for "physica status solidi (b)" and "physica status solidi (a)", respectively, two weeks before the delivery of the journal.

Kolibri 360 GmbH
Ferdinand-Jühlke-Straße 7
99095 Erfurt, Deutschland
produktsicherheit@kolibri360.de